AF838602

Robotics 2010: Current and Future Challenges

Edited by

Houssem Abdellatif

Robotics 2010: Current and future challenges

Edited by Houssem Abdellatif

Republished by InTech

Janeza Trdine 9, 51000 Rijeka, Croatia

Copyright © 2010 InTech

Abstracting and non-profit use of the material is permitted with credit to the source. Statements and opinions expressed in the chapters are these of the individual contributors and not necessarily those of the editors or publisher. No responsibility is accepted for the accuracy of information contained in the published articles. Publisher assumes no responsibility liability for any damage or injury to persons or property arising out of the use of any materials, instructions, methods or ideas contained inside. After this work has been published by the Intech, authors have the right to republish it, in whole or part, in any publication of which they are an author or editor, and the make other personal use of the work.

Technical Editor Maja Jakobovic

Cover Designer Dino Smrekar

Additional hard copies can be obtained from orders@intechopen.com

Robotics 2010: Current and future challenges, Edited by Houssem Abdellatif
p. cm.
ISBN 978-953-7619-78-7

We are IntechOpen, the world's leading publisher of Open Access books

Built by scientists, for scientists

4,200+
Open access books available

116,000+
International authors and editors

125M+
Downloads

151
Countries delivered to

Our authors are among the
Top 1%
most cited scientists

12.2%
Contributors from top 500 universities

CLARIVATE ANALYTICS BOOK CITATION INDEX INDEXED

WEB OF SCIENCE™

Selection of our books indexed in the Book Citation Index in Web of Science™ Core Collection (BKCI)

Interested in publishing with us?
Contact book.department@intechopen.com

Numbers displayed above are based on latest data collected.
For more information visit www.intechopen.com

For EU product safety concerns:
IN TECH d.o.o., Prolaz Marije Krucifikse Kozulić 3, 51000 Rijeka, Croatia,
info@intechopen.com or visit our website at intechopen.com.

Meet the editor

Dr. Houssem Abdellatif has joined TÜV SÜD in December 2016, as global head of autonomous driving and ADAS. The topics of his work are the effectiveness assessment of automated driving systems and connected vehicles. It includes the functional safety and cyber security. His team is globally acting, in Germany, Europe and Asia pacific. Dr. Abdellatif is a system engineer and holds phd in intelligent and control systems from the university of Hannover, Germany. Prior to TÜV Süd he has occupied different responsibilities with different automotive supplier and IT companies

Preface

Without doubt, robotics has made an incredible progress over the last decades. The vision of developing, designing and creating technical systems that help humans to achieve hard and complex tasks, intelligently has led to an incredible variety of solutions. There are barely technical fields that could exhibit more interdisciplinary interconnections like robotics. This fact is generated by the highly complex challenges imposed by robotic systems, especially the requirement on intelligent and autonomous operation.

When attending an international conference on robotics or visiting a robotics and automation fair, wide range of robotic systems and applications are presented: manipulators, walking humanoids, autonomous helicopters, exoskeletons and so on. These systems, appearing completely different and not having much in common at first glance do in fact share the same vision of a universal robot, helping and also outperforming humans. All different autonomous and robotic systems known to date have the same history and are solely different occurrences of a common aim. This book wants to underline the aspect of such kindredship. It presents current and future challenges that may seem of a different nature and tries to motivate their tight technical and historical relationships.

The progress and variety achieved in robotics are actually the results of a classic process chain that is necessary for the development of reliable and safe technical products. One begins with a rudimentary task or profile of the desired system and keeps the lowest possible complexity by holding a high number of constraints on the system. Once the technical challenge is mastered, one moves to the next development step by enhancing the system features or eliminating some technical constraints. Of course new challenges arise that may lead to different solutions, new concepts or to the reconsideration of previous assumptions and postulates. A simple example can be considered: the development of wheeled mobile robots as a necessary and logical consequence to make classic industrial manipulators movable. Traditionally, industrial manipulators have been supposed to be fixed in a safe area such a production cell. They have to execute a monotonous task repeatedly in a well predictable, structured and human-free area. It is a logical step forward wanting to overcome the constraint of fixing the robot. A movable manipulator would promise more flexibility in the production, so why not mounting it on an autonomous mobile platform? At that moment, development and research efforts have started for the new technology of wheeled mobile platforms, yielding a new variety of systems. New challenges arise, i.e. sensors are then necessary to collect more precise information on the environment. Algorithms need to be developed, that solve navigation and localization issues, etc. First generations of mobile platforms would keep old constraints, such as human-free and structured industrial-fields. Next generations would develop e.g. more sophisticated locomotion techniques to operate in different areas, and why not in deep sea or in the air. It is an evolutionary process that occurs, yielding milestones, generations of robotic systems, new categories and species of robots, but also to dead-end branches. This is quite normal and makes the development and research in robotics even more exciting and animated. It is a continuous progress, moving from one challenge to the other and mastering difficulties of different impacts and complexities.

To my opinion, the idea of making robotic manipulators mobile was one of the fundamental impulses in the development of modern robotic systems. Of course, pioneer works in mobile robotics have been always carried out within the last century. But the necessity of mobility prevailed at a time, as the classic industrial robots have reached a certain degree of maturity. Principle challenges have been mastered and the technique of industrial robotics was standardized and rolled out over a wide field of automation applications. In the eighties industrial robots became a leading technique and without doubt one of the most favourite references and signs of the prosperity of the western civilisation. Once the industrial challenges have been overcome, the horizon of robotics started to grow rapidly. The focus in academic research has moved from a mostly mechanical and a control point of view to a more interdisciplinary sight. In a short time, mobile and service robotics attracted the attention of information and computer scientists and many others that may not meet the profile of robotics expert at that time. Since then the challenges on robotics research grew rapidly and so the proposed solutions. With mobile robots, the problem of localization became a central issue. It was necessary to develop more sophisticated sensors to achieve an appropriate and reliable capturing of the environment, e.g. range finders, laser scanner, cameras or video cameras. The methods of image processing and pattern recognition found their way in robotics. Additionally, appropriate algorithms were needed to manage data of different kinds and at different rates and to extract position information for localization. Data fusion and simultaneous localization and mapping (SLAM) built the new challenges for new generations of robot species. Analogously to the industrial manipulators, mobile robots started in the industrial field and in a well structured environment. Further and further, the robot systems have been carried out into a human-centered environment. Unlike the beginning, robots are then supposed to interact with humans. Once again, new challenges took over, like secure and safe interaction. Such challenges could not be solved like forty years ago by encapsulating robots from human. In contrast, the physical interaction became one of the most key-challenges in modern robotics. Here, old research issues on contact modelling or compliance control have experienced a new revival, although the question is now quite different. Additionally, higher safety standards and more restrictive requirements affect the development of new systems. They accentuate the necessity of new thoughts and solutions for software-architectures and -components that allow more intelligent but also more fault tolerant and safer robots.

This book tries to give a sight into the evolutionary process that takes place in robotics. It provides articles covering a wide range of this exciting area. The progress of technical challenges and concepts may illuminate the relationship between developments that seem to be completely different at first sight. The book begins by classic manipulation that still provides high potentials of development, especially in application fields, like high-load telemanipulation (Dubus et al.) or space manipulators (Abiko and Hirzinger). It is shown in (Aghili) and (Abdellatif et al.) that there are still theoretical and practical control challenges needing to be addressed. A very interesting study in (Kobayashi and Hosoe) shows the typical interdisciplinary solutions in robotics, that addresses contact mechanical problems by using soft computation techniques.

The next group of papers treats the enhancement of manipulation skills of conventional manipulators. The proposed examples are: ball-throwing control (senoo et al.), tactile shape classification (Nakamoto et al.), neural control (Yang et al.) and rope knotting (Yamakawa et al.). The obtained methodologies provide valuable knowledge that can be carried out to the field of surgical robotics, humanoids, entertainment robotics and similar human-centered applications.

One emerging area consists of anthropomorphic and medical assistance robotics. Manipulator-like systems are developed to assist or to help humans to achieve physical tasks more easily. This field exploits knowledge and research from classical mechanics and haptics and develops it toward applications in medical rehabilitation. The old science fiction vision of powered human exoskeletons is realised by real wearable robots for military, medical or industrial purposes. (Garrec) provides details on design of novel screw and cable actuators for such application. Two examples of assistive robotic systems are presented in (Yves et al.) and (Baklouti et al.).

The trend of making robot mobile persists also for assistive robots like it is shown by (Nihei et al.). Nowadays highly interesting works focus on new locomotion techniques to make mobile robots more flexible and more reliable to access difficult terrains or areas. Often, nature and biology serve as important sources of inspiration, as the examples shown in (Kinugasa et al.), (Heidari et al.) and (Omori et al.) that discuss respectively the locomotion of flexible mono-tread tracks, crawling robots and earthworm crawling technique. Examples of unmanned aerial vehicles (Koyuncu and Inalhan) and autonomous underwater vehicle (Kim and Ura) are subsequently given.

As mentioned above, one of the most demanding challenges for human-near robots is providing them with a certain level of intelligence that ensures a secure interaction. A key role is played by the computer software and by the respective architecture. Providing appropriate frameworks remains a central issue for the next years. Two examples are provided in this book (Brice et al.) and (Nishio et al.). Sensor networking and data processing are considered to be primordial for the design of intelligent robots. The recognition of environmental situation as well as the quality of the related decision making, depend highly on the sensor integration and data handling. This may be the case for interaction issues with humans (Chikaraishi et al.), localization problems (Brscic and Hashimoto) or navigation and locomotion (Leppanen). The entire previous evolution of research from the old days of mechanics and automatic control until the progress made by computers and pattern recognition have led to outstanding new progress of intelligent robots (Ng-Thow-Hinget al.) and (Hashimoto et al.). The robotics remains an exciting scientific and engineering field. The commnity looks optimistically ahead and also looks forward for the future challenges and new development.

Munich, January 2010

Houssem Abdellatif

Contents

1

From oil to pure water hydraulics, making cleaner and safer force feedback high payload telemanipulators

Gregory Dubus, Olivier David and Yvan Measson
CEA LIST – Interactive Robotics Unit
France

1. Introduction

One redundant characteristic of dismantling operations of nuclear facilities is the lack of exhaustive and accurate data relating to the actual state of the facilities. Most of the time the harsh working conditions (heat, dust, radiological contamination...) are rated far too severe for human workers to carry out the work. As a consequence robots are set to take over from human staff. It is necessary to use flexible, powerful and remotely-operated manipulator arms that are fitted with specially-designed processes and tools for cutting, handling and cleaning-up.

For similar reasons the maintenance of fusion reactors is another kind of application which will be carried out with help of robotic means. The International Thermonuclear Experimental Reactor (ITER) is an experimental fusion reactor based on the Russian "tokamak" concept and is the next generation of fusion machines. It will benefit of the research results on the actual existing fusion reactors to experiment long lasting pulses at high energy level. Owing to plasma interactions, some in-vessel components are expected to erode to such an extent that they will require replacement several times during the lifetime of the machine. Among these components the divertor is one of the most challenging. At the same time it has to exhaust the impurities of the plasma and to work as an actively cooled thermal shield for the lower part of the torus. But fusion reactions between deuterium and tritium isotopes produce high-energy neutron fluxes that irradiate the structure of the torus and forbid direct human access inside the reactor. As a consequence the maintenance of the in-vessel components requires the use of Remote Handling (RH) technology.

Hydraulic technology provides compact and powerful manipulators compared to electrical actuating technology. For that reason they become interesting solutions to complete maintenance and dismantling heavy duty tasks (Gravez, 2002). But decommissioning contaminated areas and operating in a fusion reactor both require a cleanliness level that oil hydraulics cannot ensure: any drop of oil inside the controlled zone must be avoided. Therefore pure water hydraulics proposes a good alternative to oil. Indeed demineralised water self evaporates in case of leakage and cannot become radioactive after radiations exposure. That's why developments are today focusing on that direction and the

development of a water hydraulic manipulator has become a key issue of both French decommissioning program and ITER maintenance program (Siuko, 2003); (Mattila, 2006).
Although basic hydraulic elements like pumps, on-off valves, filters running with pure water are already available on the market, actuators are not so many and generally limited to linear jacks. Fine control of the joint is achieved with help of servovalves. Today's off the shelf products are only adaptations from standard oil servovalves and are not specifically designed for water use. Operational experience for these products shows short lifetime expectancy and could not last a complete operating time.
Starting from the standard oil hydraulic Maestro arm, a six-degrees-of-freedom hydraulic manipulator manufactured by Cybernetix and used in decommissioning and offshore activities, CEA LIST redesigned for water applications the elbow vane actuator of the arm. Endurance tests of the Maestro vane actuator powered with demineralised water were started for identification of long term issues. Moreover, servovalves are essential components of the joint's control loop. CEA LIST evaluated the feasibility to accommodate the existing design of the Maestro oil servovalve to a prototype running with water. This prototype is a pressure-control valve. To a current input this servovalve supplies a very accurate pressure difference output instead of a flow rate in the case of flow control servovalves that are generally used in that kind of applications. The advantage is the improvement of the performances and stability of the force control loop. In addition, architecture of hydraulic manipulators with force feedback capabilities available on the market is today based on a serial arrangement of rotational joints (generally six). The replacement of one rotational degree of freedom by a linear joint, or the addition of a linear joint within the joint arrangement, could significantly improve the working range of such systems which are considered at the present time as a limiting factor for many specific RH tasks. Designing a hydraulic manipulator with a prismatic joint could therefore lead to a heavy duty multi-purpose manipulator with extended reach capabilities and alternative access to space constrained area. As a consequence, in parallel of the above-mentioned works, a new linear joint concept has been designed and proposed by CEA LIST.
This chapter first presents the complete Maestro system and then gives an overview of the development activities currently carried out to adapt its hydraulic manipulator so that it works with water instead of oil. In parts 3 and 4 both static and dynamic performances are given for the modified vane actuator and the new servovalve respectively. About the joint we also describe the results of the endurance test campaign that has been carried out. Then a design update is proposed to adapt the present design to water operating constraints with a minimum of changes. Basis of a numerical model of the servovalve is proposed in order to identify its driving parameters and validate the projected evolutions of its design. Part 5 concerns the new linear joint concept. We describe the mock-up manufactured on the proposed joint concept and the first trials with this new driving axis.

2. Overview of the Maestro system

2.1 Overall system description

The Maestro telerobotic system belongs to the class of servomanipulators, which appeared in the early 80's with the progress on computer assisted teleoperation. Compared to traditional through-the-wall workstations equipped with mechanical master-slave systems, these systems provide innovative features and improved capabilities including:

- operation from a remote control room located in an unrestricted access (cold) area
- use of different arm morphologies and technologies for the master and the slave
- work in cartesian coordinates
- compensation of the handled tools' weight
- adjustable force and speed ratios in the force feedback loop
- automatic robot modes (tool picking, return to rest position...)
- virtual mechanisms to assist operator in tricky tasks
- virtual reality to improve operator viewing
- real-time collision avoidance to protect both environment and manipulators

But if power of electric motors is enough to supply good force feedback capabilities to the operator in master arm stations, operations in the hot zone sometimes require the capability to supply high forces that standard electric motors are unable to provide in a limited space.
Starting from a hydraulic manipulator developed for offshore applications, CEA LIST developed the remote handling system Maestro (Modular Arm and Efficient System for TeleRObotics) (Dubus, 2008) for heavy duty nuclear operations (see Fig. 1).
The Maestro telerobotic system is composed of:

- a master station including:
  - a Haption Virtuose 6D master-arm
  - a master-arm controller
  - the 3D graphical supervision interface MagritteWorks, based on Solidworks
  - video display monitors
- a slave station including:
  - a 6 DOF hydraulic manipulator
  - a rad-hardened embedded slave-arm controller
  - an embedded hydraulic power pack
  - a remotely controlled PTZ video camera with tool tracking capabilities

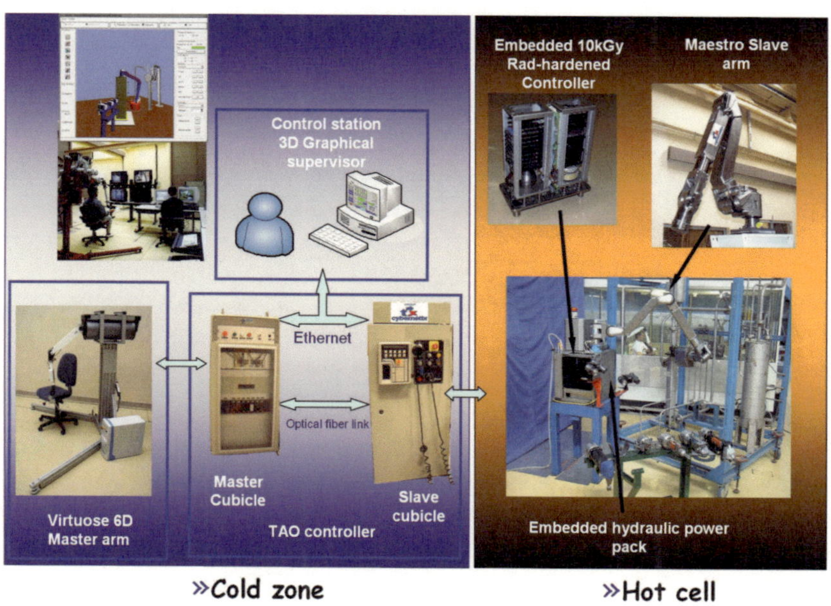

Fig. 1. Description of the Maestro system

2.2 Design of the slave manipulator

Built in titanium, the Maestro slave-arm is a 6-DOF, 2.4m-long hydraulic manipulator (see Fig. 2). Its payload capacity is up to 100 kg for 90 kg own weight. The actuator technology is based on rotary hydraulic joints. The fluid, traditionally oil, is supplied through the arm at a 210 bars pressure and a 15 L/min flow rate. The monitoring of the pressure difference between the two chambers of each joint makes it possible to drive the arm in a traditional force reflective master-slave configuration.

The system specifications were defined according to the requirements of decommissioning activities in existing nuclear facilities and maintenance scenarios of the fusion reactor ITER. Although rad-resistance of the joint itself is higher, a qualification campaign in an irradiation facility already proved resistance of the joint and its rad-hardened embedded electronic-controller to a cumulated dose of 10.65 kGy under a mean dose rate of 74 Gy/h. Special attention was paid to satisfy easy decontamination requirements, preferring smooth surfaces and avoiding any contamination traps in the design.

Qualification of the complete system for RH operations in nuclear facilities ran through a validation process including long term reliability testing. Endurance tests were carried out with different payloads during 1000 hrs. This operating time should be close to ITER needs between two shutdowns. The trajectory was defined according to position records during a representative teleoperation task including tool picking, task completion with tool, and tool removal.

Fig. 2. The Maestro manipulator

2.3 Servovalves

Servovalves are, in servo controlled hydraulic systems, the equivalent of amplifiers for electrical servomotors. Each joint is equipped with a servovalve, which controls the in and out fluid flows through the joint chambers. Servovalves generally used in that kind of robotic applications are flow control servovalves, which supply a flow rate to a current input. This category of valve is interesting in position control loops, but it needs additional sensor information when used in force control loops.

A good alternative to flow control servovalves in force control modes is the use of pressure control servovalves. In that scheme, the controlled parameter is directly linked to the force and this has a direct impact on the control loop stability. Indeed to a current input this servovalve supplies a very accurate pressure difference output instead of a flow rate in the case of flow control servovalves. From a control point of view the scheme is highly simplified as the inner loop previously needed to compute the flow according to the measured pressure is no longer needed. Therefore, improvement of force control performance (better stability and duration of the loop highly decreased) and tuning time (less parameters to adjust) is achieved.

Moreover this technical choice is also interesting from a security point of view. Indeed using these components allows removal of all pressure sensors and therefore reduces the probability of failure of the system. In the case of an electrical failure of the pressure servovalve, no pressure will be sent to the joints and the arm will fall down slowly with a minimum impact on its environment thanks to mechanical safety valves. With a flow control scheme, a pressure sensor failure would make the control system unstable, trying to compensate the "virtual" lack of pressure. The result would be a full speed movement of the concerned joint until the reception of an emergency signal, which could be harmful for the arm itself and its surroundings.

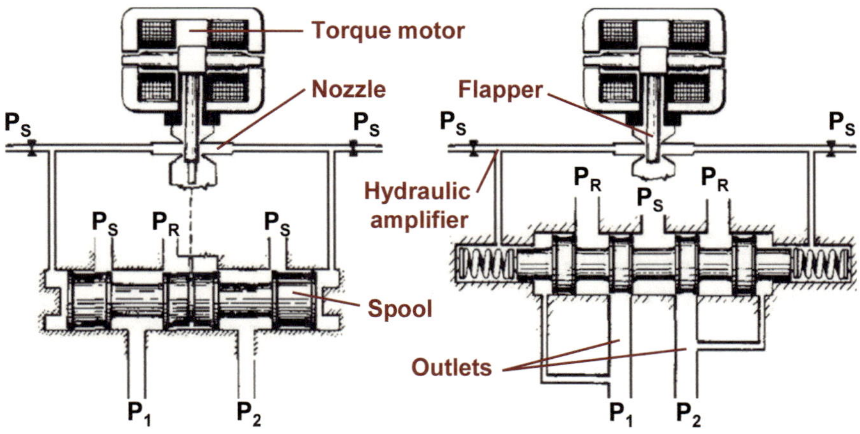

Fig. 3. Principles of flow servovalves (a) and pressure servovalves (b)

The main difference between flow and pressure servovalves is the pressure feedback exerted on the spool. The two principles are shown in Fig. 3. As for a flow servovalve, the first stage of a pressure servovalve is composed of a torque motor in which the input current creates magnetic forces on both ends of the armature. The assembly {armature + flapper} rotates around a flexure tube support which moves the flapper between the two nozzles. It builds up a differential pressure proportional to the torque induced by the input current. This pressure moves the spool and opens one control port to supply pressure P_S and the other control port to return pressure P_R. The particularity of the pressure servovalve is that building-up the differential pressure (P_2-P_1) creates a feedback force on the spool, which moves backward to balance forces giving proportionality.

Prototypes of oil pressure servovalves with space and performance requirements needed by a Maestro manipulator were developed by CEA LIST. Their operating pressure was 210 bars and was obtained for a 10 mA current. The maximum linearity error was close to 10 bars and the threshold was about 3 bars, which was also the value of the hysteresis error. Their maximal flow rate (outlet to the atmosphere) was close to 11.5 L/min and the leak rate was less than 0.5 L/min. The bandwidth (167 Hz) was far beyond our requirements (20 Hz).
The integration of a complete set of pressure servovalves in the arm proved the feasibility of the concept. Achieved force control performance was better than observed with flow servovalves and it allowed a reduction of the total control loop period by a factor of two.

2.4 Force feedback

Accurate remote handling operations rely on good force feedback capabilities of the remote handling tools. Indirect vision of the operating scene introduces difficulties during maintenance tasks that can be successfully overcome with this extra sense of touch. Force feedback is provided to the operator by means of a hybrid force-position control scheme. As shown in previous works (Bidard, 2004), high quality force control can only be achieved with a good real-time compensation of all the manipulator mechanical joints imperfections, the arm inertia and the gravity (own weight, payload, tool...).

3. Redesign of the vane actuator

3.1 Specification and test rig description

The elbow joint of the Maestro manipulator is a 1300N.m. compact vane actuator with a 270° stroke, designed to withstand high radiations environment and to minimize duration of decontamination procedures.
Traditionally used with oil, the joint was analysed to adapt its design to water. Driving requirements during this adaptability study were:

- To use corrosion resistant materials
- To reduce clearances (direct impact on internal leaks due to water's low viscosity)
- To prevent contact between water and components with poor corrosion resistance
- To adapt seal materials and properties to water

The characterization of the joint was made on the test rig of Fig. 4 (Dubus, 2007).
It was composed of a Danfoss Nessie power pack, resins tanks to demineralise water directly coming from the tap, a Maestro elbow joint, a Moog flow control servovalve (type 30-417), an Arthus pancake resolver and four pressure sensors (Entran EPXT) measuring the supplied pressure, the pressure in the back-loop and the two output pressures at the outlets of the servovalve. To assess its performance different torques could be applied to the joint by means of an adjustable payload attached at its tip.
In addition, particular attention was paid to control properties and quality of the water used during the trials. Water was filtered and demineralised in a secondary circuit. The most efficient filter was a 1μm filter and conductivity was kept between 0.1μS/cm and 1μS/cm. This upper value was only obtained occasionally, when resins were saturated and needed to be replaced.

Fig. 4. Water hydraulic test bench

3.2 Characterization and performance of the hybrid force-position control

To implement a force control on the joint and assess its dynamical performance, a parametric model has been identified. As explained in paragraph 2.4, the main interest of this stage was the modelling and the identification of the friction and gravity torques, which are compensated in the force loop. A classical torque model was proposed as follows:

$$T_0 = J.\ddot{\theta} + C_v.\dot{\theta} + \text{sign}(\dot{\theta}).C_s + offset + M_x.\sin(\theta) + M_y.\cos(\theta) \quad (1)$$

In this expression J is the arm inertia, θ, $\dot{\theta}$ and $\ddot{\theta}$ are the angular position and its derivatives, C_v and C_s are respectively the viscous and dry friction coefficients, M_x and M_y represent the load among x and y axes. Being given the actuation torque, the position, the velocity and the acceleration during a position controlled sequence, the parameters were estimated thanks to a least square method. More complex models of the friction were tested, considering the joint efficiency and the effects of backdrivability as a function of the payload. But this approach had no significant impact on the identification of the main parameters. It is interesting to notice that both viscous and dry friction coefficients are 30% lower when using water instead of oil (see Table 1).

The final control scheme of the joint took into account the following compensation models: friction, gravity and rated flow (converted into torque units).

Table 2 and Table 3 present the performance for both oil and water. Obviously internal leakage is far higher in the water device. Nevertheless it seems to have a positive damping impact on the force loop dynamic performance.
Regarding the position control loop, a good tuning gives an overshoot close to 3% and the time response for a 2 rad step is close to 1 s. This value is due to the speed limitation assessed in Table 2. It corresponds to the maximum flow rate supplied by the servovalve. But compared to the 0.6 rad/s mean speed for rotary joints during standard teleoperation tasks, this performance is in agreement with the requirements.

	Oil device	Water device
Cv (N.m.s/rad)	93.0	60.1
Cs (N.m)	28.6	17.3

Table 1. Comparison between frictions of oil and water devices

	Oil device	Water device
Maximal torque (N.m)	1280	1250
Mean value of internal leak rate[a] (L/min)	0.3	1.1
Speed saturation[b] (rad/s)	2.4	2.4

[a] For the system {servovalve + joint}.
[b] Corresponds to the maximum flow rate supplied by the servovalve.

Table 2. Comparison of the static performance for oil and water hydraulic joints

	Oil device	Water device
Overshoot (%)	82	48
Time response (ms)	175	6

Table 3. Force loop performance for a 160N.m step, for oil and water hydraulic joints

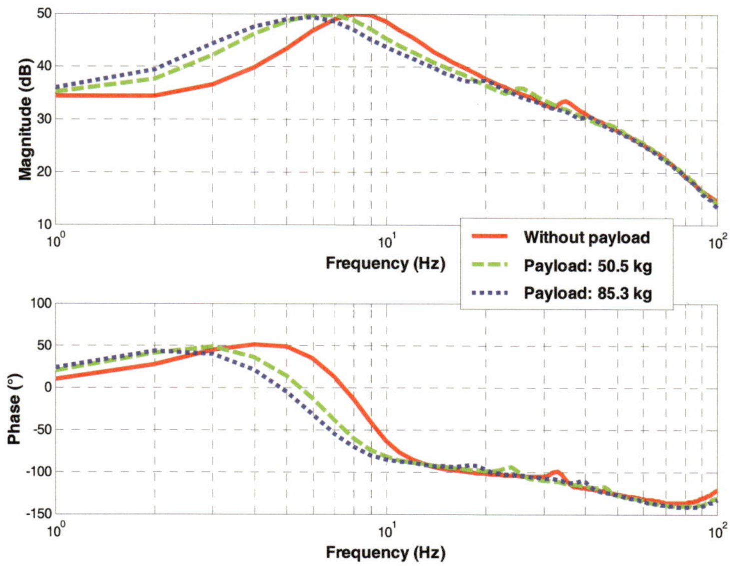

Fig. 5. Comparison of transfer functions according to payload

The torque dynamic response of the system to different payloads is given in Fig. 5. There is a reduction of the bandwidth when the payload increases, which means that it is necessary to adjust the control loop with the most critical configuration.

To evaluate the position resolution of the joint, tests were carried out at very low speed (see Fig. 6). Although the resolver resolution is very high, the position resolution of the joint is close to 0.65 mrad which is equivalent to 0.80 mm at the end-effector of the manipulator. This is due to the residual dry friction and stick slip effect that lowers the whole performance of the joint.

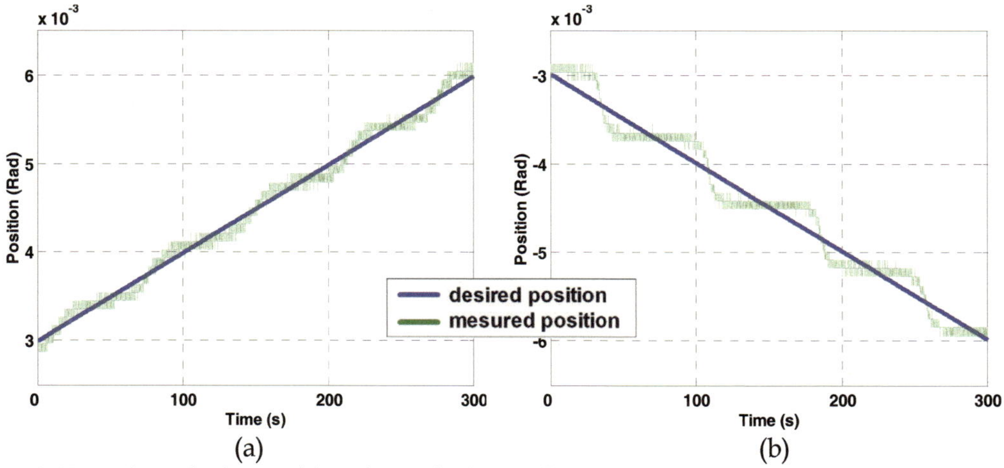

Fig. 6. Very slow clockwise (a) and anticlockwise (b) movements

Reversibility tests provided a good representation of the force control loop quality when all compensation models were active (see Fig. 7). The torque peaks observed during these trials occurred during high speed transient and they were rapidly corrected by the control scheme. Performance achieved with water was equivalent or even better than with oil.

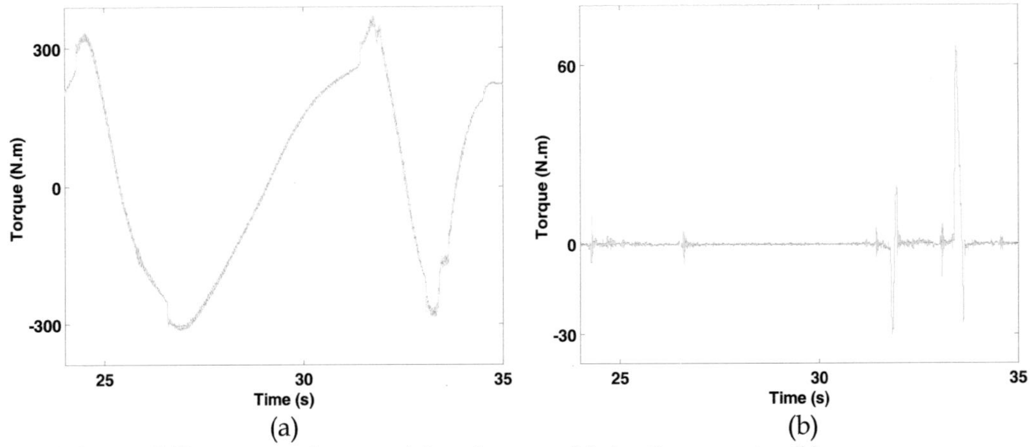

Fig. 7. Reversibility test: real torque (a) and torque felt by the operator (b)

3.3 Endurance tests

As for the complete oil hydraulics arm, qualification of the joint for RH operations had to run through a validation process including long term reliability testing. 1000 hours of operation are the usual specification for the oil version of the Maestro manipulator between two stops for maintenance. This value should be close to ITER needs between two shutdowns.

The endurance tests that we performed consisted of the repetition of a single trajectory with different payload in order to simulate different manipulator configurations: with or without tool, performing a task with tool. For safety reasons, a security chain containing two limit switches and an optical watchdog were added to the test rig. Presence detection of the bar in front of the watchdog (see Fig. 8) every two minutes was necessary to avoid emergency stop.

Fig. 8. Actuator in the 50daN equivalent payload configuration during endurance tests

The reference trajectory (see Fig. 9 (a)) was chosen to be representative of the movement of the Maestro elbow joint during a standard RH task such as using a shear or a circular saw. Its duration was 65 s, with mean and max speed values respectively equal to 0.21 rad/s and 0.75 rad/s. The tools' presence was simulated with adjustments of the payload. Three payloads equally distributed with time were used, each of them generating a maximal torque of 260 N.m, 545 N.m and 833 N.m respectively simulating complete manipulator configurations without tool, with a 25 kg payload, and with a 50 kg payload. Every 70 hrs the load configuration was changed.

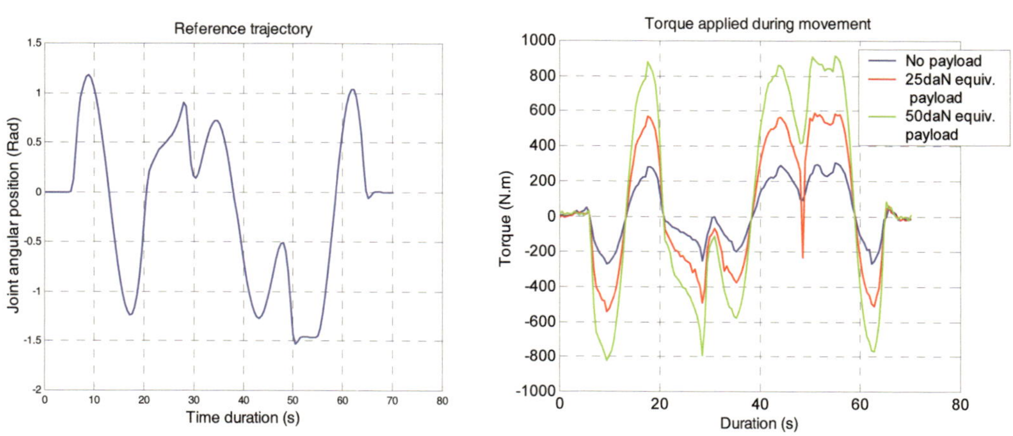

(a) (b)

Fig. 9. Reference trajectory (a) and different torque configurations (b) during endurance test

In order to detect any loss of performances, records of the current sent to the servovalves were made regularly during the trials. It was expected to detect any wear of the actuator by an increase of this current. Indeed wear of the actuator would rapidly increase the internal leak rate, thus increasing the water flow demand to the servovalve and the current as well.

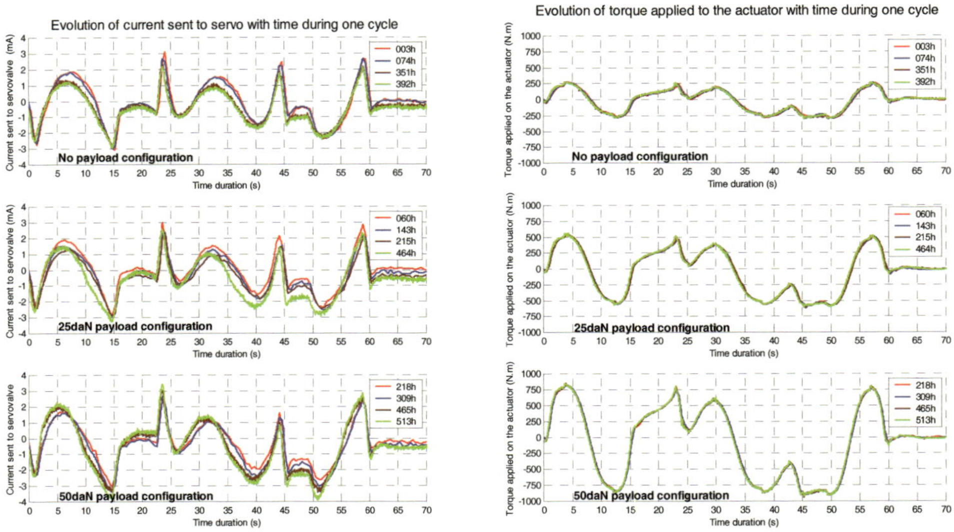

(a) (b)

Fig. 10. Current sent to servovalve (a) and measured torque (b) at different time steps

Fig. 10 presents for some cycles (random selection) the evolution of the current and the torque for all three loading configurations. Although the trajectory was always the same for the three configurations, the differences between the three configurations can be explained by the absence of inertia compensation in our control scheme.

Before every change of load a mechanical identification of the joint (dry friction, viscous friction...) was performed using the same technique as in paragraph 3.2. Variations of these mechanical parameters can usually be related to the degradation of the actuator. A modification of the viscous friction may mean an increase of the internal leakage whereas an augmentation of the dry friction may indicate a mechanical degradation of actuator.

Tests were stopped prematurely after 533h due to a consequential power pack failure. That was the second power pack failure as the distribution plate in the high pressure pump first broke around 50 hours of tests. Up to that level the MOOG servovalve was still behaving properly with no external indications of forthcoming failure.

No significant degradation of the rotary joint from both performance and mechanical points of view were noticed. Indeed the variation of both the servovalve current and the applied torque were negligible after 500 hrs. After disassembly the visual examination of the actuator didn't show any wear of the internal parts of the system. The cylinder, the vane and the flanges, which represent the main elements of the actuator, were presenting clean surfaces without corrosion spots. Seals also seemed to behave properly and were still in good physical condition. Bearings only showed that a few drops of water leaked through the drain arrangement and mixed itself with the grease but it had no consequences on the material state.

3.4 Design update

Concerning hydraulic manipulators, thorough control of the manufacturing quality and minimisation of the clearances are the main elements having an impact on the leaks. The endurance tests we performed showed that the present design of the Maestro joint was able to run at least 500 hrs without any observable degradation of its performance. To ensure the reliability of the arm up to 1000 hrs of operations, minor design updates may be considered.
New design arrangements are currently studied to minimize the impact of the change of fluid in the Maestro manipulator. These modifications have to be compatible with the existing overall design, so that the modified joint remains in a size envelope comparable to the oil version.
The main trouble that has been identified for the development of a water hydraulic arm concerns the viscosity of water. Indeed lower the viscosity, higher the potential for leaks. And external leaks would obviously result in bearing seizure. In the present Maestro design tapered roller bearings are used to withstand both the internal loads due to the fluid pressure and the external loads depending on the task in progress. These bearings are made of standard bearing steel and therefore are corrosion sensitive. Even if no corrosion has been noticed on the bearings during our tests, using such bearings without any modification of the present design could potentially affect the lifetime of the actuator.
Two technical solutions are therefore being studied to overcome this trouble:

- Integration of water compatible bearings (ceramic materials or stainless steel)
- Modification of the tightness arrangement to protect the present bearings

Results of these studies in progress should be published in future publications.

4. Development of a water hydraulic servovalve

4.1 Specification

Small size off the shelf servovalves specially developed for water hydraulic applications are unavailable on the market at the present time. The only existing products are adaptations of oil components without long term guarantee on performance and lifetime. Starting from previous results (Measson, 2003), CEA LIST launched the development of a pressure control servovalve dedicated to water hydraulic applications that fits the space constraints of a Maestro manipulator.
To meet the performance of a Maestro arm, requirements were set as follow:

- Pressure gain: 210 bars for 10 mA
- Resolution: 2 bars
- Flow rate on open ports: mini 6 L/min
- Internal leak rate: close to 1 L/min
- Bandwidth > 20 Hz

4.2 Characterization of two prototypes

As a first step, CEA LIST evaluated the feasibility to accommodate the existing design of the oil version of the servovalve to a prototype running with water. Two prototypes were manufactured. Tests were carried out on the mock-up shown in Fig. 11. This test rig was composed of a drilled block supporting the servovalve and 4 pressure sensors. Servovalve performance is traditionally measured on closed apertures. But due to fluid compressibility, the fluid volume in both chambers acts as a spring + damper unit and affects the performance of the servovalve. That's why it was possible to connect dead volumes simulating the actuator chambers on the outlets of the servovalve.

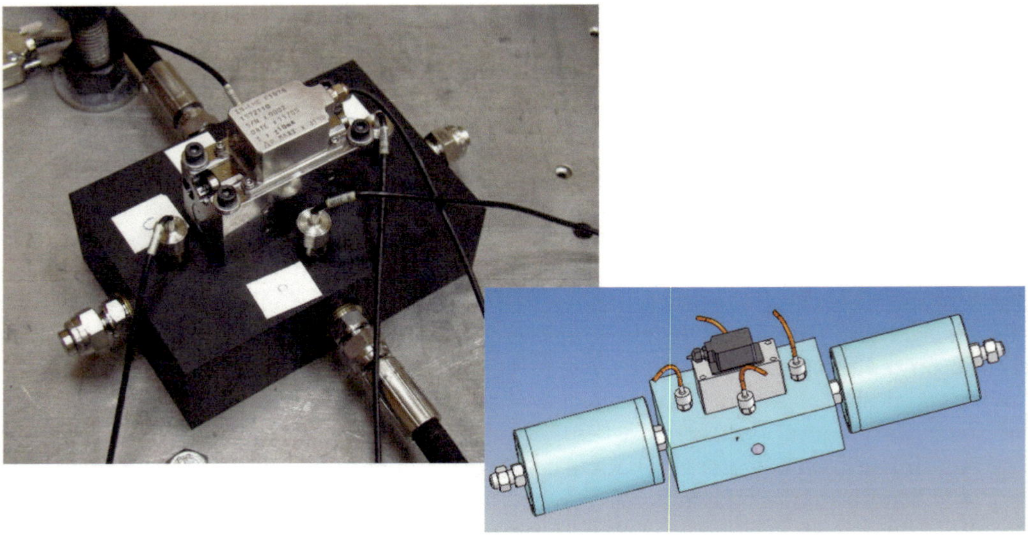

Fig. 11. Test mock-up of the water hydraulic pressure servovalve

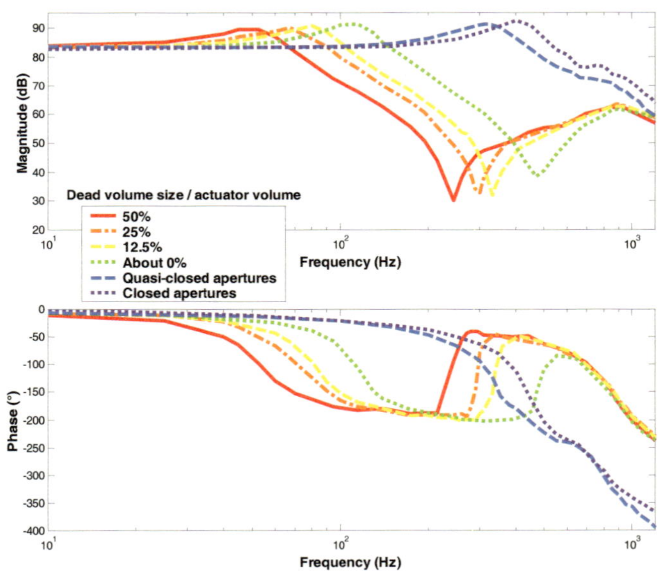

Fig. 12. Bode diagram of the servovalve for different connected volumes

The Bode diagram of the servovalve (see Fig. 12) shows a significant reduction of the valve bandwidth (ie its dynamic performance) as the dead volumes connected to the outlets of the valve increase. But it never passes below the 20 Hz requirement. Leak rate (1.2 L/min) and flow rate (22 L/min) of the valve are close or better than the specifications but a reduction of the gain was observed. Indeed the prototype only managed to provide a 150 bars pressure difference between the 2 outlets instead of the expected 210 bars. This loss of performance was presumed to be due to an underestimated internal leakage.

4.3 Dynamic model of the pressure servovalve

To validate the above assumption, numerical models were built to identify all driving parameters of the servovalve. Starting from previous works on hydraulic systems (Merrit, 1967); (Guillon, 1992), the servovalve was divided in four subsystems. Let's establish the equations that describe the dynamic behaviour of each of these subsystems.

- Torque motor

The pilot stage consists in a torque motor and its dynamics mainly depends on the behavior of the armature-flapper assembly. The free body diagram of the armature-flapper is shown in Fig. 13 (a). We assume that the assembly moves around the pivot point *O*. The armature linear displacement is then deduced by the relation $x_f = L_n.\theta_f$. The armature-flapper is subjected to the magnetic force F_g, the pressure force at the nozzles F_n, a damping moment and a moment due to the pivot stiffness. Usually, the magnetic force F_g is found by analyzing the magnetic circuit created by the armature, the magnetic plate and the pole pieces of the torque motor. At our level, we assume that this force linearly depends on the input current and the armature displacement. At the nozzle, we can write the static pressure force as being $F_n = A_n.(P'' - P')$, where P' and P'' are the pressures that drive the spool and A_n the nozzle cross section.

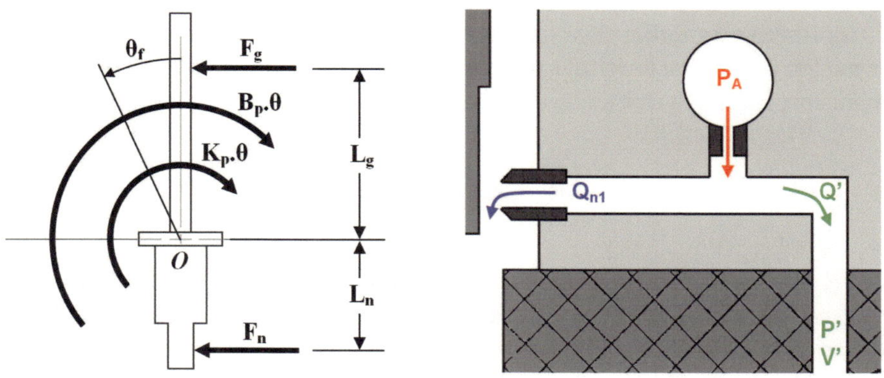

(a) (b)

Fig. 13. Free body diagram of the flapper (a) and schema of the hydraulic amplifier (b)

Summing the different moments around the pivot (Urata, 1998); (Kim, 2000), we obtain:

$$J_p\ddot{\theta}_f + B_p\dot{\theta}_f + K_p\theta_f = A_n\left(P''-P'\right)L_n + \left(K_{gf}x_f + K_{gf}i\right)L_g \qquad (2)$$

According to the relation between x_f and θ, (2) gives:

$$M_f\ddot{x}_f + B_f\dot{x}_f + K_f x_f = A_n\left(P''-P'\right) + \left(K_{gf}\frac{L_g}{L_n}x_f + K_{gi}\frac{L_g}{L_n}i\right) \qquad (3)$$

and leads to:

$$M_f \ddot{x}_f + B_f \dot{x}_f + K'_f x_f = A_n \left(P'' - P'\right) + K_g i \tag{4}$$

As a conclusion, for the armature-flapper assembly, we get the linear relation:

$$\ddot{x}_f = f(x_f, \dot{x}_f, P', P'', i) \tag{5}$$

- Hydraulic amplifier

Let's consider the pilot differential pressure $\Delta P_1 = P' - P''$. Pressures P' and P'' are determined by the basic hydraulic compressibility equations and the flow balance in the hydraulic amplifier (see Fig. 13 (b)). For the right part of the amplifier we get:

$$\dot{P}' = \beta \frac{Q' - \dot{V}'}{V'} \tag{6}$$

where Q' is the hydraulic flow towards the right spool chamber and V' is the volume of the chamber between the spool and the right flapper face. Its volume is therefore given by $V' = V'_0 - A_S . x_S$.

Moreover, the flow Q' into the chamber includes the flow from the supply orifice, the flow past the nozzle and the leakage past the spool. Combining the three contributions (Anderson, 2002), we get:

$$Q' = \underbrace{C_{do} A_o \sqrt{\frac{2}{\rho}\left(P_A - P'\right)}}_{\text{from supply orifice}} \underbrace{- C_{df} \pi . D_m \left(x_{fo} - x_f\right) \sqrt{\frac{2}{\rho}\left(P' - P_T\right)}}_{\text{past the nozzle}} \underbrace{- \frac{\pi . D_b C_r^3 P'}{12 \mu \left(L_{lo} - x_S\right)}}_{\text{past the spool}} \tag{7}$$

C_{do} and C_{df} are both discharge coefficient respectively for the supply orifice and the nozzle orifice. In this relation, the leakage is modeled to be a laminar flow in an annulus between an annular shaft and a concentric cylinder which initial length is L_{lo}, as it is done in (Guillon, 1992). C_r represents the radial clearance. μ is the dynamic viscosity of the fluid. In the same way, for the left part of the amplifier we get the anti-symmetric expression of Q''.

As a conclusion, for the hydraulic amplifier, we get the two nonlinear relations:

$$\dot{P}' = f(x_S, \dot{x}_S, P', x_f) \tag{8}$$

$$\dot{P}'' = f(x_S, \dot{x}_S, P'', x_f) \tag{9}$$

- Spool

The spool is subjected to the pilot differential pressure ΔP_1, a feedback force due to the load differential pressure ΔP_L, a force due to the centring springs, viscous friction and flow forces (Li, 2002). Equating these forces on the spool gives:

$$M \ddot{x}_S = \underbrace{\left(P' - P''\right) A_1}_{\Delta P_1 \text{ force}} + \underbrace{\left(P_1 - P_2\right) A_2}_{\Delta P_L \text{ feedback}} \underbrace{- 2K_S x_S}_{\substack{\text{Spring} \\ \text{feedback}}} - \underbrace{F_V}_{\substack{\text{Viscous} \\ \text{friction}}} - \underbrace{F_Q}_{\substack{\text{Flow} \\ \text{forces}}} \tag{10}$$

Assuming that the spool is perfectly centred in the bore, the viscous damping force is:

$$F_V = \mu A \frac{\Delta v}{\Delta y} = \mu \underbrace{\left(\pi d_S L_S\right)}_{\text{contact area}} \frac{\dot{x}_S}{C_r} \tag{11}$$

As previously, μ and C_r are respectively the dynamic viscosity of the fluid and the radial clearance. The contact area is a cylinder of diameter d_s and length L_s. The flow forces F_Q are sometimes called Bernoulli Forces and are due to the dynamics of the fluid in the spool chambers. These forces can be split in two kinds: steady-state flow forces and transient flow forces. Steady-state flow forces are due to the angle of the average stream line when the fluid is going in or out the spool chamber. Transient flow forces are the reactive forces associated with the acceleration of the fluid in the spool chamber. According to (Merrit, 1967), these flow forces on the spool are given by:

$$F_Q = \begin{cases} \left.\begin{array}{l} C_{dj}\, C_V\, w\left(P_A - P_1\right)\cos\alpha\, x_S \;+\; L_p\, C_{dj}\, w\sqrt{2\rho\left(P_A - P_1\right)}\,\dot{x}_S \\ +\; C_{dj}\, C_V\, w\left(P_2 - P_T\right)\cos\alpha\, x_S \;-\; L_p\, C_{dj}\, w\sqrt{2\rho\left(P_2 - P_T\right)}\,\dot{x}_S \end{array}\right\} \text{if } x_s \geq 0 \\ \\ \left.\begin{array}{l} C_{dj}\, C_V\, w\left(P_1 - P_T\right)\cos\alpha\, x_S \;+\; L_p\, C_{dj}\, w\sqrt{2\rho\left(P_1 - P_T\right)}\,\dot{x}_S \\ +\; C_{dj}\, C_V\, w\left(P_A - P_2\right)\cos\alpha\, x_S \;-\; L_p\, C_{dj}\, w\sqrt{2\rho\left(P_A - P_2\right)}\,\dot{x}_S \end{array}\right\} \text{if } x_s < 0 \end{cases} \tag{12}$$

In this expression the terms of the left column correspond to the steady-state forces whereas the terms of the right column are those of the transient forces. C_{dj} is the jet discharge coefficient and C_v corresponds to a velocity coefficient. Typical values for these parameters are $C_{dj} = 0.61$ and $C_v = 2$. L_p is called the damping length and represents the length of the fluid column that undergoes the acceleration. For our servo-valve, this parameter depends on the spool displacement. The angle a, which corresponds to the average stream line angle, is theoretically a non-linear function of x_S/C_r and varies from 21° to 69° (Merrit, 1967). A first approximation consists in fixing this parameter at its maximum value. At last, in the steady-state forces, x_S can be replaced by $\text{sign}(x_S).(x_S^2+C_r^2)^{1/2}$ in order to take into account the effect of clearance.

As a conclusion, for the spool equilibrium, we get the single nonlinear relation:

$$\ddot{x}_S = f(x_S, \dot{x}_S, P', P'', P_1, P_2) \tag{13}$$

- Controlled ports

As the spool moves in its bore, the fluid is either sucked into or out of the valve. These fluid movements have a non negligible impact on the behaviour of the servovalve. According to the flows defined in Fig. 14 (a) the flow balance in the boost stage can be written:

$$Q_1 + Q_{1F} = Q_{1S} - Q_{1R} \tag{14}$$

$$Q_2 + Q_{2F} = Q_{2R} - Q_{2F} \tag{15}$$

Q_{1F} and Q_{2F} are determined by the basic hydraulic compressibility equation:

$$Q_{1F} = \frac{dV_{1F}}{dt} + \frac{V_{1F}}{\beta}\frac{dP_1}{dt} \tag{16}$$

$$Q_{2F} = \frac{dV_{2F}}{dt} + \frac{V_{2F}}{\beta}\frac{dP_2}{dt} \tag{17}$$

To clarify the expressions of Q_{1S}, Q_{1R}, Q_{2S} and Q_{2R}, we make the choice to combine leakage and orifice flows in a single continuous relation (Eryilmaz, 2000). Therefore we get:

$$Q_{1S} = K_{1S} \sqrt{P_S - P_1} \begin{cases} (x_O + x_S) & (x_s \geq 0) \\ x_O^{\,2} (x_O - k_{1S}\, x_S)^{-1} & (x_s < 0) \end{cases} \tag{18}$$

$$Q_{1R} = K_{1R} \sqrt{P_1 - P_R} \begin{cases} x_O^{\,2} (x_O + k_{1R}\, x_S)^{-1} & (x_s \geq 0) \\ (x_O - x_S) & (x_s < 0) \end{cases} \tag{19}$$

$$Q_{2S} = K_{2S} \sqrt{P_S - P_2} \begin{cases} x_O^{\,2} (x_O + k_{2S}\, x_S)^{-1} & (x_s \geq 0) \\ (x_O - x_S) & (x_s < 0) \end{cases} \tag{20}$$

$$Q_{2R} = K_{2R} \sqrt{P_2 - P_R} \begin{cases} (x_O + x_S) & (x_s \geq 0) \\ x_O^{\,2} (x_O - k_{2R}\, x_S)^{-1} & (x_s < 0) \end{cases} \tag{21}$$

At last, theses considerations about the flow rates in the spool chambers point out an important aspect of the servovalve behaviour: the servovalve dynamics highly depends on the actuator dynamics.

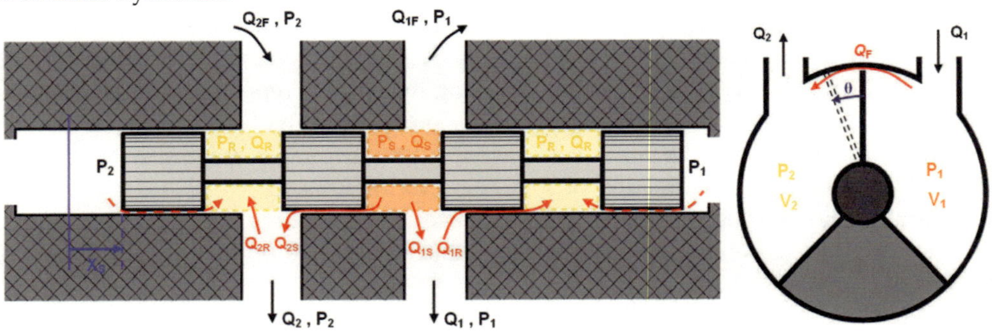

(a) (b)

Fig. 14. Free body diagrams of the spool (a) and of the rotary actuator (b)

From the hydraulic compressibility equation, we get:

$$Q_1 - Q_F = \frac{dV_1}{dt} + \frac{V_1}{\beta}\frac{dP_1}{dt} \tag{22}$$

$$-Q_2 + Q_F = \frac{dV_2}{dt} + \frac{V_2}{\beta}\frac{dP_2}{dt} \tag{23}$$

In these expressions, Q_F corresponds to the leakage from one actuator chamber to the other (see Fig. 14 (b)). Because this leakage occurs through a constant rectangular area, a simple expression for it is:

$$Q_F = C_{df}\, A_f\, sign(P_1 - P_2)\sqrt{\frac{2}{\rho}|P_1 - P_2|} \tag{24}$$

As the volumes of the two actuator chambers depend on θ the dynamics of the servovalve load differential pressure ΔP_L do as well. We can use (1) to describe the actuator dynamics.

As a conclusion, this study on the flow rates in the spool chambers leads to the eight following nonlinear relations:

$$Q_i = f(P_i, x_S, Q_{iF})_{i=(1,2)} \tag{25}$$

$$Q_{iF} = f(x_S, \dot{x}_S, P_i)_{i=(1,2)} \tag{26}$$

$$Q_i = f(\theta, \dot{\theta}, P_i, Q_F)_{i=(1,2)} \tag{27}$$

$$\ddot{\theta} = f(\theta, \dot{\theta}, P_1, P_2) \tag{28}$$

$$Q_F = f(P_1, P_2) \tag{29}$$

From the balance equations of all four subsystems, non linear systems of equations were assembled in a block diagram (see Fig. 15). Numerical solving methods were then applied to study the influence of each design parameter of the valve.

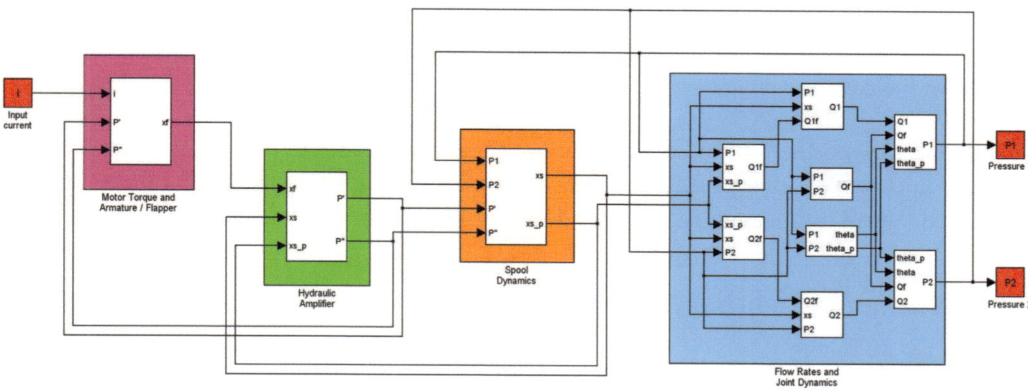

Fig. 15. Physical model block diagram of system {servovalve + joint}

This model made it possible to highlight three main dynamics corresponding to the pilot stage, the spool and the fluid compression at the outlets. We propose to express the dynamics of the entire servovalve as a simplified linear model, given by the following transfer function:

$$H(s) = \frac{K\left(\frac{s^2}{f_1^2} + 2m_1\frac{s}{f_1} + 1\right)}{\left(\frac{s^2}{f_2^2} + 2m_2\frac{s}{f_2} + 1\right)\left(\frac{s^2}{f_3^2} + 2m_3\frac{s}{f_3} + 1\right)} \tag{30}$$

To validate this model, parameters of the transfer function were evaluated thanks to the physical parameters of the block diagram system. Then the frequency response was compared with experimental results under similar conditions (see Fig. 16).

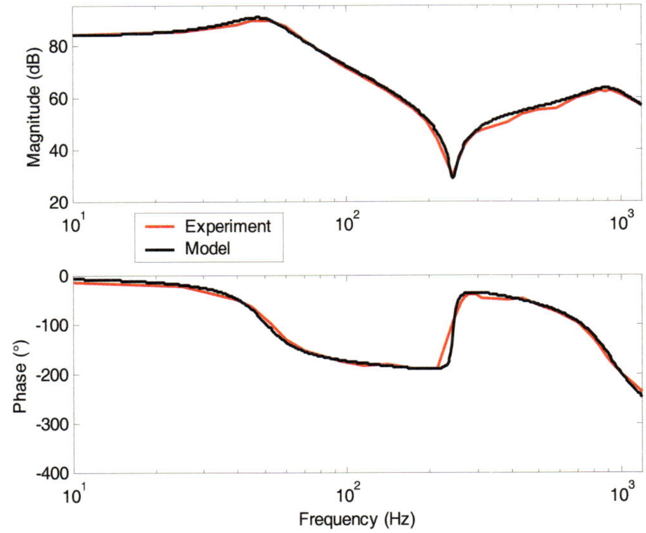

Fig. 16. Frequency response of the servovalve connected to dead volumes (75% of the volume of the actuator chamber)

This model proved that an underestimation of the leakage from the outlets toward the pilot pressure area would effectively limit the pilot forces on the spool. Solutions to reduce this effect were found but needed redesign and machining of a new prototype.

5. Development of a linear hydraulic joint

5.1 Motivation

Advanced robots architectures rely on parallel or serial arrangements of articulations. Parallel structures are used where operations require mechanics with high stiffness, transfer of high loads and high positioning accuracy. But with this kind of architecture, the workspace of the machine is limited. This is the main reason why tasks requiring high dexterity prefer serial kinematics even if wiring of the complete machine becomes a challenging task. Manipulators used for RH applications need to address a large variety of tasks and that's why dexterity is one key element for this kind of equipment.

The most common serial architecture is composed of six rotational joints in series. Orientation of all axes relatively to each other and segments' lengths are different according to the model and manufacturer. For higher dexterity the axis of three last rotations composing what is called the wrist need to be secant.

High payload Master Slave Manipulators (MSM, Fig. 17 (a)) use another alternative where the third axis is composed of a prismatic joint. The MT200 La Calhène (see Fig. 17 (b)), the CRL Model 8 or the A100 Wälischmiller are all manipulators equipped with a telescopic joint offering a 1.5 to 4m reach and a 20daN payload capacity. But limitations exist due to the cable (or metal tape) mechanisms used for both movement and force transmission. The stiffness of cables is too low to avoid extreme wrist deflection (up to 60° at full load) in some configurations during manipulation.

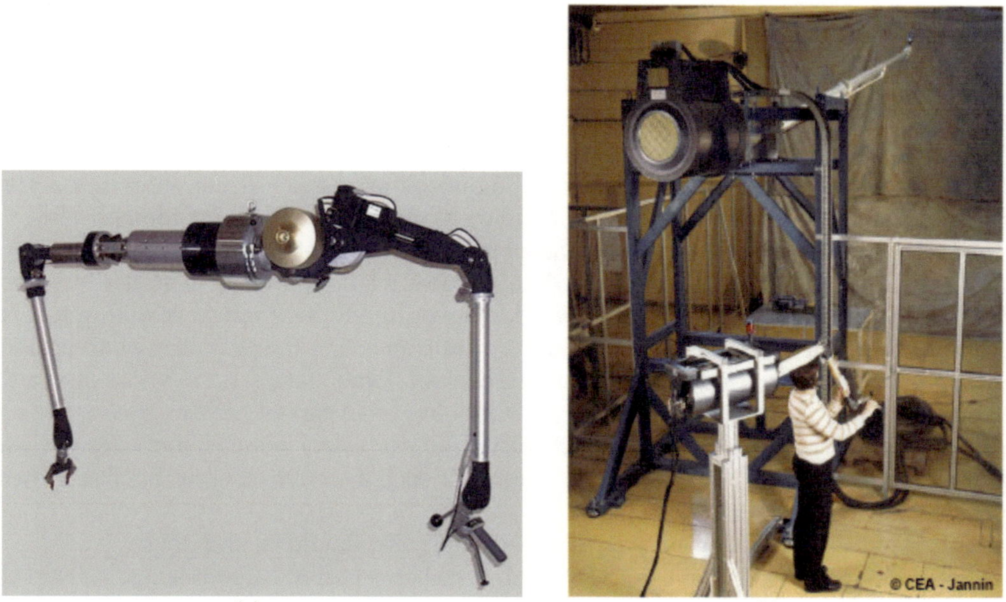

(a) (b)

Fig. 17. Examples of high payload master slave manipulator: a mechanical MSM (a) and the MT200 TAO telerobotic system (b)

The main advantage of this kinematics is to increase the workspace of the manipulator and sometimes to provide an easier access to the operating area. On the other hand, sensitivity of the force feedback is usually lower. But to keep on operating with a high dexterity level, it is commonly accepted that the position of the prismatic joint has to be upstream the three axis of the wrist.

A hydraulic manipulator having a prismatic joint has also been identified as one of the requirements for the maintenance of ITER's divertor cassettes and hot cells. To be relevant for man-in-the-loop tasks, the considered system should have the following requirements:

- Including a prismatic joint with a 1 meter stroke minimum
- High speed performance: max speed = 0.8 m/s, 0.5 m/s speed being a common value for standard displacements

Dealing with prismatic joints in the middle of a kinematics often generates a lot of difficulties for all equipment and axes that are located downstream this linear joint. Whether it is for cables (for cable powered articulations) or wires (for electric motors or measurement systems), overcoming the problem of the length adjustment is a major challenge that is not so critical for rotational joint. For hydraulic systems, the problem is the same. Supplying with fluid all downstream axes through a linear joint is far from trivial.

Two options can be followed:

- All pilot valves are located upstream the linear joint and take into account the movements of the linear joint in addition to the fluid demand of their respective axes. The control of each actuator is then linked to the control of the linear axis. Moreover the design of a proper hydraulic line for each axis is necessary. Considering that the linear joint is placed in the third position, it would mean two pipes for each three remaining axes plus four additional pipes for the gripper and the tool changer: i.e. a total of ten pipes passing through the linear joint.

- Each pilot valve is located close to its actuator. Each actuator is independent, but a "pressure bus" with supply and return canals must run through the manipulator.

A usual solution with electric wires is to coil them in such a manner that the entire stroke of the prismatic joint is compensated by a coiling-uncoiling motion of the wire. In the same way cables in MSM systems are relying on pulley blocks arrangements for compensating the extra length adjustment during movements of the prismatic joint.

Such solutions cannot be adapted to hydraulic hoses which cannot be bent easily around small radius, especially when powered on. Moreover due to their radial flexibility it is not recommended to use hoses in the hydraulic circuit connecting the chambers of an actuator and the pre-actuator (proportional valve, servovalve). Flow and pressure variations in the chambers create second order terms in the control loop that cannot be overcome easily and decrease global stability of the system. Additionally, for safety reasons, tools designed for nuclear applications usually avoid any component such as wires or hoses outside moving bodies.

At the end of the day an efficient prismatic hydraulic joint should be such that:

- Its design avoids the use of any hydraulic hoses and must be oriented toward the use of telescopic pipes arrangement such as hydraulic jacks.
- Hydraulic circuits for supplying downstream elements should be made of two pipes providing the pressure supply at 210bars from the power pack and the return loop towards the tank.

5.2 Proposed concept

The proposed concept for the linear joint follows the principle of Fig. 18. In this arrangement the linear joint is composed of three linear jacks:

- Two passive jacks (considered as a hydraulic link between the different sections of the manipulator):
    - one jack for the hydraulic power supply
    - one jack for the return loop towards the tank
- 1 jack controlling motion and power inside the axis

The two passive jacks play the role of an extendable hydraulic circuit. They replace the rotating seal arrangement found in rotary axis. Within the joint assembly, a servovalve is connected to the pressure supply and tank return loop of the two passive jacks and controls the in and out movement of the third jack. This jack controls both force and position of the whole joint.

The main difficulty is to design a passive jack acting only as a "hydraulic link" between the two parts of the manipulator and with a minimal impact on the load supplied to the system. The difficulty is mainly due to the volume variation in each passive jack between the retracted and extended position due to the presence of the shaft on one side of the piston.

The impact of this variation for the return loop is low due to air compressibility. The air contained in the tank can adjust its volume without problem to deal with flow variations. For the high pressure fluid, the problem is more difficult during the transition from the extended configuration to the retracted configuration. During that movement a reduction of the fluid volume is necessary. Fluid can not return back within the pump of the power pack and dealing with this extra volume requires the design of an extra system which is presented as the passive jack in the Fig. 18.

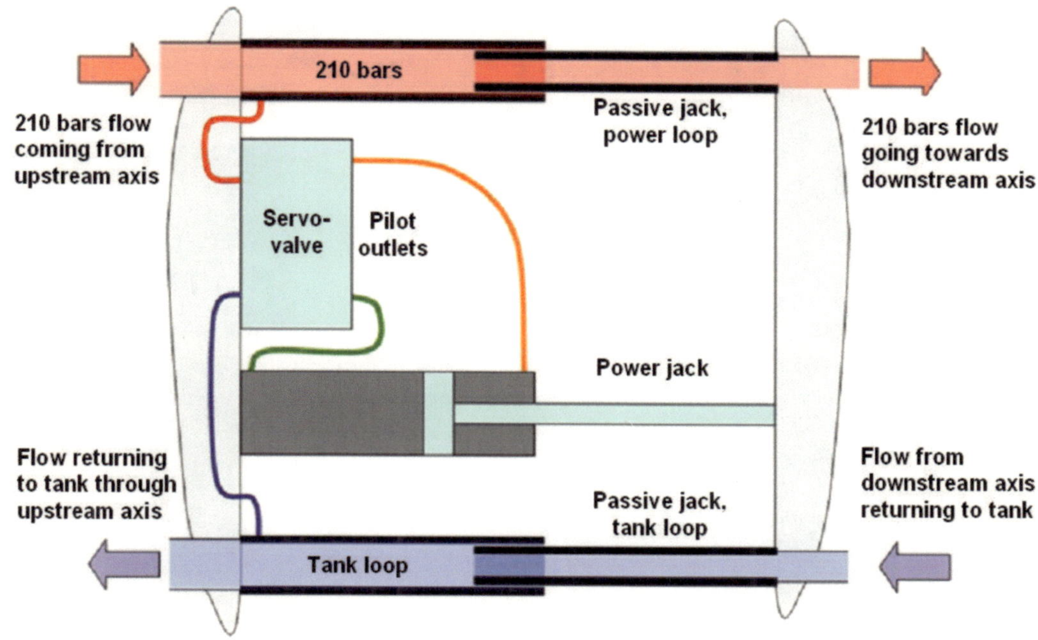

Fig. 18. Model architecture for linear joint

5.3 Preliminary tests

The test rig was designed to be modular and adaptable. The goal of these preliminary tests was to test the functions of the components and not to save space and build a fully integrated system.

Sensors and pre-actuators used for the test rig are:

- Pressures sensors : ENTRAN model EPXM-N22-350b
- Cable potentiometer: µ-epsilon WDS-1500-P60-SR-U
- Servovalve: MOOG model 4633116000

Although no force control loop has been implemented for this characterization of the joint, pressure sensors can provide information on the axial force delivered by the primary jack.

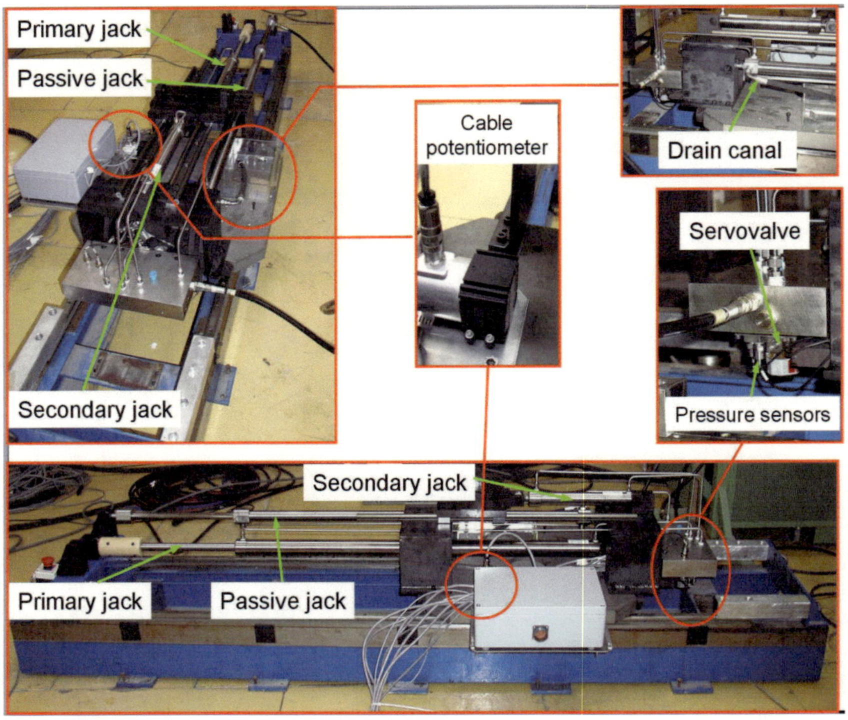

Fig. 19. Test rig

In the present design, the passive jack is one of the main components of the actuator. Due to its design and location within the system's kinematics it will act a damping system. It is therefore interesting to test the performances of the system with and without this component to characterize its influence on the whole behaviour.

Response of the actuator to a step signal is shown Fig. 20 (a). As in section 3.2 concerning the rotational link, the speed saturation is a consequence of the servovalve limited flow rate.

These results are interesting because even at the highest speed the presence of the passive jack do not seem to seriously affect the performance of the system.

Fig. 20 (b) presents the force within the primary jack when operated with and without passive jack. The reconstruction of the force was made according to the pressure values within the chambers. The results are in agreement with the expectations: both dry and viscous frictions are higher with the passive jack.

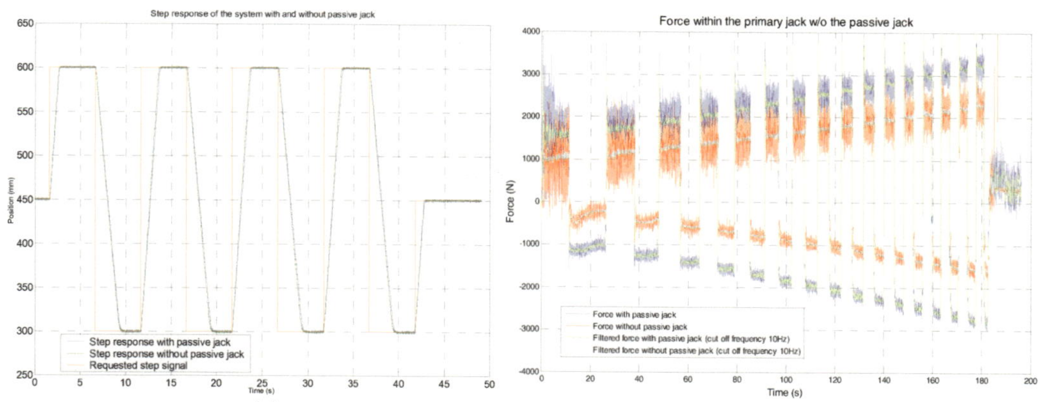

(a) (b)

Fig. 20. Position step response (a) and force measured in the primary jack (b) with and without passive jack

Asymmetry of the signal is due to the offset of the servovalve that has not been compensated yet. Moreover the oscillations noted on this force signal are due to the poor quality of the position measurement, which lead to a low quality control loop and speed oscillations. These oscillations should be reduced by a better speed measurement but creating an internal leak within the passive jack could also be another good option. Internal leaks are acting as pressure dampers and would therefore naturally filter the force within the primary jack.

As previously an identification of the system parameters has been performed to assess the force feedback capabilities of the proposed system. The test bench was configured to be used with and without the passive jack. The following table gives the values of all parameters in both configurations.

As it could be expected viscous and dry friction are higher when the passive jack is mounted on the bench. Due to its design itself (long guiding length, two concentric pipes sliding one into each other) it is not surprising to see that most of the dry friction comes from the passive jack. Viscous friction of the passive jack itself is not that high.

Parameter	Test with passive jack	Test without passive jack
Viscous friction N/(m/s)	24600	20600
Dry friction (N)	738	214
Offset (N)	305	-378

Table 4. Mechanical parameters issued from identification process

6. Conclusions

In this chapter we have tried to give the reader an overview of the studies currently carried out at CEA LIST to make hydraulic manipulators work with demineralised water instead of oil as a power fluid.

We showed that force and position performances of a Maestro elbow joint running with water are globally similar or better than the performances of the one running with oil. Minor design updates may be executed even if endurance tests proved that the joint is reliable up

to 500 hrs of operation without any observable degradation of its performance. Therefore it seems clear that the Maestro actuator becomes a very good candidate for the design of a complete water hydraulic manipulator.
Beside, pressure control water servovalve prototypes were tested with closed apertures and connected to dead volumes for qualification and characterization with water. Although most of the requirements are better or close to the expected values, the maximal pressure difference between the two ports is lower than expected. A physical model was proposed in order to identify which parameters could be responsible for this effect. Taking into account that these tests are the first ones on the first prototype generation, these results are encouraging and should foster developments in the area of water hydraulic servovalves.
At last the issue of integrating linear joint in serial architecture of hydraulic actuators has been considered. Assessment of the performances required during standard operations showed that creating a "pressure bus" within the manipulator to allow each servovalve to obtain its required fluid flow was the best answer to the problem. An innovative design was proposed. Preliminary tests on a functional mock-up have been presented and discussed.
In the next step, qualification of the water hydraulic joint equipped with a pressure servovalve instead of the flow control pre-actuator will be made. This modification would provide significant improvement of the force control loop in terms of accuracy, stability and tuning procedures. Concerning the linear actuator performance of the position measurement needs improvements to overcome limitations in the tuning of the control loop and provide a speed signal compatible with force control requirements. It is proposed to investigate the possibility to introduce data fusion procedures between two distinct sensors to reach the requested quality level.
Then the integration of all these technologies to build an extended 6DOF water hydraulic manipulator should be conceivable around 2010-2011.

7. Acknowledgements

This work, supported by the European Communities under the contract of association between EURATOM and CEA, was carried out within the framework of the European Fusion Development Agreement (EFDA). The views and opinions expressed herein do not necessary reflect those of the European Commission.

8. References

Anderson, R.T. & Perry, L.Y. (2002). Mathematical Modeling of a Two Spool Flow Control Servovalve Using a Pressure Control Pilot, *Journal of Dynamic Systems, Measurement, and Control*, Vol. 124, Issue 3, September 2002, pp. 420-427.

Bidard, C.; Libersa, C.; Arhur, D.; Measson, Y.; Friconneau, J.-P. & Palmer, J. (2004). Dynamic identification of the hydraulic MAESTRO manipulator - Relevance for monitoring, *Proceedings of SOFT23*, Venice, 2004.

Dubus, G.; David, O.; Nozais, F.; Measson, Y.; Friconneau, J.-P. & Palmer, J. (2007). Assessment of a water hydraulics joint for remote handling operations in the divertor region, *Proceedings of ISFNT8*, Heidelberg, 2007.

Dubus, G.; David, O.; Nozais, F.; Measson, Y. & Friconneau, J.-P. (2008). Development of a water hydraulics remote handling system for ITER maintenance, *Proceedings of the IARP/EURON Workshop on Robotics for Risky Interventions and Environmental Surveillance*, Benicàssim, 2008.

Eryilmaz, B. & Wilson, B.H. (2000). Combining leakage and orifice flows in a Hydraulic Servovalve Model, *ASME*, 2000.

Gravez, P.; Leroux, C.; Irving, M.; Galbiati, L.; Raneda, A.; Siuko, M.; Maisonnier, D. & Palmer, J. (2002). Model-based remote handling with the MAESTRO hydraulic manipulator, *Proceedings of SOFT22*, Helsinki, 2002.

Guillon, M. (1992). Commande et asservissement hydrauliques et électrohydrauliques, Lavoisier, Paris, 1992.

Kim, D.H. & Tsao, T.-C. (2000). A Linearized Electrohydraulic Servovalve Model for Valve Dynamics Sensitivity Analysis and Control System Design, *Journal of Dynamic Systems, Measurement, and Control*, Vol. 122, Issue 1, March 2000, pp. 179-187.

Li, P.Y. (2002). Dynamic Redesign of a Flow Control Servo-valve using a Pressure Control Pilot, *ASME Journal of Dynamic Systems, Measurement and Control*, Vol. 124, No. 3, Sept 2002.

Mattila, J.; Siuko, M.; Vilenius, M.; Muhammad, A.; Linna, O.; Sainio, A.; Mäkelä, A.; Poutanen, J. & Saarinen, H. (2006). Development of water hydraulic manipulator, *Fusion Yearbook, Association Euratom-Tekes, Annual Report 2005*, VTT Publications.

Measson, Y.; David, O.; Louveau, F. & Friconneau, J.P. (2003). Technology and control for hydraulic manipulators, *Fusion Engineering and Design*, vol.69, September 2003.

Merrit, H. E. (1967). Hydraulic Control Systems, Wiley, New York, 1967.

Siuko, M.; Pitkäaho, M.; Raneda, A.; Poutanen, J.; Tammisto, J.; Palmer, J. & Vilenius, M. (2003). Water hydraulic actuators for ITER maintenance devices, *Fusion Engineering and Design*, Vol.69, September 2003.

Urata, E. & Yamashina C. (1998). Influence of flow force on the flapper of a water hydraulic servovalve, *JSME international journal. Series B, fluids and thermal engineering*, vol. 41, no2, 1998, pp. 278-285.

2

Operational Space Dynamics of a Space Robot and Computational Efficient Algorithm

Satoko Abiko and Gerd Hirzinger
Institute of Robotics and Mechatronics,German Aerospace Center (DLR)
82334, Weßling, Germany

1. Introduction

On-orbit servicing space robot is one of the challenging applications in space robotic field. Main task of the on-orbit space robot involves the tracking, the grasping and the positioning of a target. The dynamics in operational space is useful to achieve such tasks in Cartesian space. The operational space dynamics is a formulation of the dynamics of a complex branching redundant mechanism in task or operational points. Khatib proposed the formulation of a serial robot manipulator system on ground in (Khatib, 1987). Russakow *et. al.* modified it for a branching manipulator system in (Russakow et al., 1995). Chang and Khatib introduced efficient algorithms for this formulation, especially for operational space inertia matrix in (Chang & Khatib, 1999; 2000).

The operational space dynamics of the space robot is more complex than that of the ground-based manipulator system since the base-satellite is inertially free. However, by virtue of no fixed-base, the space robot is invertible in its modeling and arbitrary operational points to control can be chosen in a computational efficient manner. By making use of this unique characteristic, we firstly propose an algorithm of the dynamics of a single operational point in the space robot system. Then, by using the concept of the articulated-body algorithm(Featherstone, 1987), we propose a recursive computation of the dynamics of multi-operational points in the space robot. The numerical simulations are carried out using a two-arm space robot shown in Fig. 1.

This chapter is organized as follows. Section 2 describes basic dynamic equations of free-flying and free-floating space robots. Section 3 derives the operational space formulation of both types of space robots. Section 4 briefly introduces spatial notation to represent complex robot kinematics and dynamics, which is used for the derivation of the proposed algorithms. Section 5 describes recursive algorithms of the generalized Jacobian matrix(Xu & Kanade, 1993), that is a Jacobian matrix including dynamical coupling between the base body and the robot arm. Section 6 proposes computational efficient algorithms of the operational space dynamics. Section 7 shows the simulation example of the proposed algorithms. Section 8 summarizes the conclusions.

2. Basic Equations

This section presents basic dynamic equations of the space robot. The main symbols used in this section are defined in table 1.

Fig. 1. Chaser-robot and target scenario

2.1 Linear and Angular Momentum Equations

The motion of the space robot is generally governed by the principle of the conservation of momentum. When the spatial velocity of the base body, $\dot{\boldsymbol{x}}_b = (\boldsymbol{v}_b^T, \boldsymbol{\omega}_b^T)^T \in R^{6\times1}$, and the motion rate of the joints, $\dot{\boldsymbol{\phi}} \in R^{n\times1}$, are considered as the generalized coordinates, total linear and angular momentum, $\mathcal{M}_0 \in R^{6\times1}$, are expressed as follows:

$$\mathcal{M}_0 = \boldsymbol{H}_b\dot{\boldsymbol{x}}_b + \boldsymbol{H}_{bm}\dot{\boldsymbol{\phi}} \,. \tag{1}$$

Note that $\mathcal{M}_0$ represents the total momentum around the center of mass of the base body. In the absence of external forces, the total momentum is conserved. From eq. (1), the motion of the base body is expressed by $\dot{\boldsymbol{\phi}}$ and $\mathcal{M}_0$ as:

$$\dot{\boldsymbol{x}}_b = \boldsymbol{J}_b^*\dot{\boldsymbol{\phi}} + \boldsymbol{H}_b^{-1}\mathcal{M}_0 \in R^{6\times1}, \tag{2}$$

where

$$\boldsymbol{J}_b^* = -\boldsymbol{H}_b^{-1}\boldsymbol{H}_{bm} \in R^{6\times n} \tag{3}$$

represents the generalized Jacobian matrix of the base body (Yokokohji et al., 1993). By introducing the kinematic mapping of the i-th operational point, $\dot{\boldsymbol{x}}_{e_i} = \boldsymbol{J}_{b_i}\dot{\boldsymbol{x}}_b + \boldsymbol{J}_{m_i}\dot{\boldsymbol{\phi}}$, eq. (1) provides the velocity of the operational point as follows:

$$\dot{\boldsymbol{x}}_{e_i} = \boldsymbol{J}_{m_i}^*\dot{\boldsymbol{\phi}} + \boldsymbol{J}_{b_i}\boldsymbol{H}_b^{-1}\mathcal{M}_0 \in R^{6\times1}, \tag{4}$$

where

$$\boldsymbol{J}_{m_i}^* = \boldsymbol{J}_{m_i} - \boldsymbol{J}_{b_i}\boldsymbol{H}_b^{-1}\boldsymbol{H}_{bm} \in R^{6\times n} \tag{5}$$

is called the generalized Jacobian matrix of the operational point (Umetani & Yoshida, 1989). The above generalized Jacobian matrix, (5), is for the case that a single point is selected as an operational point. This matrix is simply extended to the case of the multi-operational points

n		:	number of joints
p		:	number of operational points
$\dot{\boldsymbol{x}}_b$	$\in R^{6\times 1}$	:	linear and angular velocity of the base.
$\dot{\boldsymbol{\phi}}$	$\in R^{n\times 1}$	:	motion rate of the arms.
$\dot{\boldsymbol{x}}_e = \begin{bmatrix} \dot{\boldsymbol{x}}_{e_1} \\ \vdots \\ \dot{\boldsymbol{x}}_{e_p} \end{bmatrix}$	$\in R^{6p\times 1}$	:	linear and angular velocity of the operational points ($i = 1 \cdots p$).
$\boldsymbol{H}_b$	$\in R^{6\times 6}$	:	inertia matrix of the base.
$\boldsymbol{H}_m$	$\in R^{n\times n}$	:	inertia matrix of the arms.
$\boldsymbol{H}_{bm}$	$\in R^{6\times n}$	:	coupling inertia matrix between the base and the arms.
$\boldsymbol{c}_b$	$\in R^{6\times 1}$	:	non-linear velocity dependent term of the base.
$\boldsymbol{c}_m$	$\in R^{n\times 1}$	:	non-linear velocity dependent term of the arms.
$\mathcal{F}_b$	$\in R^{6\times 1}$	:	force and moment exerted on the base.
$\mathcal{F}_e$	$\in R^{6p\times 1}$	:	force and moment exerted on the operational points.
$\boldsymbol{\tau}$	$\in R^{n\times 1}$	:	torque on joints.
$\boldsymbol{J}_{b_i}$	$\in R^{6\times 6}$	:	Jacobian matrix of the base in terms of the i-th operational point.
$\boldsymbol{J}_{m_i}$	$\in R^{6\times n}$	:	Jacobian matrix of the arms in terms of the i-th operational point.
$\boldsymbol{J}_b^*$	$\in R^{6\times 6}$	:	Generalized Jacobian matrix of the base body.
$\boldsymbol{J}_{m_i}^*$	$\in R^{6\times n}$	:	Generalized Jacobian matrix of the arms in terms of the i-th operational point.

Table 1. Main Notation

by augmenting the Jacobian matrix of each operational point. In section 5, we derive recursive calculations of the matrices, (3) and (5).

2.2 Equations of Motion

The general dynamic equation of the space robot is described by the following expression (Xu & Kanade, 1993):

$$\begin{bmatrix} \boldsymbol{H}_b & \boldsymbol{H}_{bm} \\ \boldsymbol{H}_{bm}^T & \boldsymbol{H}_m \end{bmatrix} \begin{bmatrix} \ddot{\boldsymbol{x}}_b \\ \ddot{\boldsymbol{\phi}} \end{bmatrix} + \begin{bmatrix} \boldsymbol{c}_b \\ \boldsymbol{c}_m \end{bmatrix} = \begin{bmatrix} \mathcal{F}_b \\ \boldsymbol{\tau} \end{bmatrix} + \begin{bmatrix} \boldsymbol{J}_b^T \\ \boldsymbol{J}_m^T \end{bmatrix} \mathcal{F}_e. \tag{6}$$

where $\dot{\boldsymbol{x}}_b = (\boldsymbol{v}_b^T, \boldsymbol{\omega}_b^T)^T \in R^{6\times 1}$, and the motion rate of the joints, $\dot{\boldsymbol{\phi}} \in R^{n\times 1}$ are considered as the generalized coordinates. When $\mathcal{F}_b$ is actively generated (*e.g.* jet thrusters or reaction wheels etc.), the system is called a *free-flying* robot. If no active actuators are applied on the base, the system is termed a *free-floating* robot. The integral of the upper part of eq. (6) describes the total linear and angular momentum around the center of mass of the base body and corresponds to the equation (1).

2.3 Dynamics of a Free-Floating Space Robot

The dynamic equation of the free-floating space robot can be furthermore reduced a form expressed with only joint acceleration, $\ddot{\boldsymbol{\phi}}$, by eliminating the base body acceleration, $\ddot{\boldsymbol{x}}_b$, from eq. (6):

$$\boldsymbol{H}_m^* \ddot{\boldsymbol{\phi}} + \boldsymbol{c}_m^* = \boldsymbol{\tau} + \boldsymbol{J}_b^{*T} \mathcal{F}_b + \boldsymbol{J}_m^{*T} \mathcal{F}_e \tag{7}$$

where $\boldsymbol{H}_m^* = \boldsymbol{H}_m - \boldsymbol{H}_{bm}^T \boldsymbol{H}_b^{-1} \boldsymbol{H}_{bm} \in R^{n \times n}$ and $\boldsymbol{c}_m^* = \boldsymbol{c}_m - \boldsymbol{H}_{bm}^T \boldsymbol{H}_b^{-1} \boldsymbol{c}_b \in R^{n \times 1}$ represent generalized inertia matrix and generalized non-linear velocity dependent term, respectively.

3. Operational Space Formulation

The operational space dynamics is useful to control the system in the operational space, which represents the dynamics projected from the joint space to the operational space. The two types of space robot dynamics are described in the following subsections. One is for the free-flying space robot and the other is for the free-floating space robot. This section derives the equations of motion for the space robots consisting of n-links with p operational points.

3.1 Free-Flying Space Robot

The operational space dynamics of the free-flying space robot is described in the following form:

$$\boldsymbol{\Gamma}_e \ddot{\boldsymbol{x}}_e + \boldsymbol{\mu}_e = \mathcal{F}_e^{in} + \mathcal{F}_e \,, \tag{8}$$

where

$$\begin{bmatrix} \mathcal{F}_b \\ \boldsymbol{\tau} \end{bmatrix} = \boldsymbol{J}_e^T \mathcal{F}_e^{in} \,.$$

$\mathcal{F}_e \in R^{6p \times 1}$ consists of the 6×1 external force of each of p operational points. $\boldsymbol{J}_e \in R^{6p \times (6+(n-1))}$ consists of Jacobian matrix of each operational point.

$$\mathcal{F}_e = \begin{bmatrix} \mathcal{F}_{e_1} \\ \vdots \\ \mathcal{F}_{e_p} \end{bmatrix} \text{ and } \boldsymbol{J}_e = \begin{bmatrix} \boldsymbol{J}_{b_1}, & \boldsymbol{J}_{m_1} \\ \vdots & \vdots \\ \boldsymbol{J}_{b_p}, & \boldsymbol{J}_{m_p} \end{bmatrix}.$$

The operational space inertia matrix of the free-flying space robot, $\boldsymbol{\Gamma}_e$, is an $6p \times 6p$ symmetric positive definite matrix. Its inverse matrix can be expressed as :

$$\boldsymbol{\Gamma}_e^{-1} = \boldsymbol{J}_e \boldsymbol{H}^{-1} \boldsymbol{J}_e^T \,,\; \boldsymbol{H} = \begin{bmatrix} \boldsymbol{H}_b & \boldsymbol{H}_{bm} \\ \boldsymbol{H}_{bm}^T & \boldsymbol{H}_m \end{bmatrix}. \tag{9}$$

The operational space centrifugal and Coriolis forces, $\boldsymbol{\mu}_e$, is expressed as :

$$\boldsymbol{\mu}_e = \boldsymbol{J}_e^{T+} \begin{bmatrix} \boldsymbol{c}_b \\ \boldsymbol{c}_m \end{bmatrix} - \boldsymbol{\Gamma}_e \frac{d}{dt} \boldsymbol{J}_e \begin{bmatrix} \dot{\boldsymbol{x}}_b \\ \dot{\boldsymbol{\phi}} \end{bmatrix}, \tag{10}$$

where $\boldsymbol{J}_e^+$ is the dynamically consistent generalized inverse of the Jacobian matrix $\boldsymbol{J}_e$ for the free-flying space robot to minimize the instantaneous kinetic energy of the space robot:

$$\boldsymbol{J}_e^+ = \boldsymbol{H}^{-1} \boldsymbol{J}_e^T \boldsymbol{\Gamma}_e. \tag{11}$$

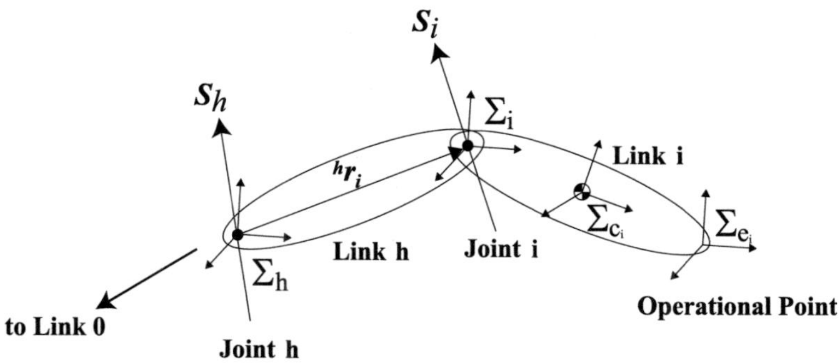

Fig. 2. Notation Representation

3.2 Free-Floating Space Robot

In the free-floating space robot, no active forces exist on the base (*e.g.* $\mathcal{F}_b = \mathbf{0}$). Then, the system can be described as the reduced form in the joint space by using eq. (7). Its operational space dynamics can be derived from eqs. (5) and (7):

$$\boldsymbol{\Gamma}_e \ddot{\boldsymbol{x}}_e + \boldsymbol{\Gamma}_e \boldsymbol{\mu} = \boldsymbol{\Gamma}_e \boldsymbol{\Lambda}^{-1} \boldsymbol{J}_m^{*T+} \boldsymbol{\tau} + \mathcal{F}_e , \tag{12}$$

where

$$\boldsymbol{\Gamma}_e^{-1} = \boldsymbol{\Lambda}^{-1} + \boldsymbol{\Lambda}_b^{-1} \in R^{6p \times 6p} ,$$

$$\boldsymbol{\Lambda}^{-1} = \boldsymbol{J}_m^* \boldsymbol{H}_m^{*-1} \boldsymbol{J}_m^{*T} , \ \boldsymbol{\Lambda}_b^{-1} = \boldsymbol{J}_b \boldsymbol{H}_b^{-1} \boldsymbol{J}_b^T .$$

The matrix, $(\boldsymbol{\Lambda}^{-1} + \boldsymbol{\Lambda}_b^{-1})$, corresponds to the inertia matrix described in eq. (9). The vector, $\boldsymbol{\mu}$, expresses the bias acceleration vector resulting from the Coriolis and centrifugal forces as:

$$\boldsymbol{\mu} = \boldsymbol{\Lambda}^{-1} \boldsymbol{J}_m^{*+} \boldsymbol{c}_m^* - \dot{\boldsymbol{J}}_m^* \dot{\boldsymbol{\phi}} - \frac{d}{dt}(\boldsymbol{J}_b \boldsymbol{H}_b^{-1}) \mathcal{M}_0 \in R^{6p \times 1}.$$

where

$$\boldsymbol{J}_m^{*+} = \boldsymbol{H}_m^{*-1} \boldsymbol{J}_m^{*T} \boldsymbol{\Lambda} \tag{13}$$

represents the dynamically consistent generalized inverse of the Jacobian matrix $\boldsymbol{J}_m^*$ for the free-floating space robot. Compared with eq. (8), the relationship $\boldsymbol{\Gamma}_e \boldsymbol{\mu} = \boldsymbol{\mu}_e$ is obtained. Note that each dynamic equation described in this section is expressed in the inertial frame. Section 6 describes the efficient algorithms for the operational space dynamics represented in this section.

4. Spatial Notation

The *Spatial Notation* is well-known and intuitive notation in modeling kinematics and dynamics of articulated robot systems, introduced by Featherstone (Featherstone (1987); Chang & Khatib (1999)). This section concisely reviews the basic spatial notation. The main symbols used in the spatial notation are defined in Table 2. The symbols are expressed in the frame fixed at each link. (See. Fig. 2).

$\mathbf{v}_i$	$\in R^{6\times1}$	:	spatial velocity of link i.
$\mathbf{a}_i$	$\in R^{6\times1}$	:	spatial acceleration of link i.
f_i	$\in R^{6\times1}$	:	composite force of link i.
f_i^*	$\in R^{6\times1}$	:	spatial force of link i.
$\mathcal{L}_i$	$\in R^{6\times1}$	:	composite momentum of link i.
$\mathcal{L}_i^*$	$\in R^{6\times1}$	:	spatial momentum of link i.
$\boldsymbol{I}_i^C$	$\in R^{6\times1}$	:	composite inertia matrix of link i.

Table 2. Main Symbols in Spatial Notation

4.1 Spatial Notation

In the spatial notation, linear and angular components are dealt with in a unified framework and results in a concise form (*e.g.* 6×1 vector or 6×6 matrix). In this expression, a spatial velocity, $\mathbf{v}_i$, and a spatial force, f_i, of link i are defined as :

$$\mathbf{v}_i = \begin{bmatrix} \boldsymbol{v}_i \\ \boldsymbol{\omega}_i \end{bmatrix} \text{ and } f_i = \begin{bmatrix} \mathcal{F}_i \\ \mathcal{T}_i \end{bmatrix},$$

where $\boldsymbol{v}_i$, $\boldsymbol{\omega}_i$, $\mathcal{F}_i$, and $\mathcal{T}_i$ represent the 3×1 linear and angular velocity, the force and moment in terms of link i in frame i, respectively.

The simple joint model, $\boldsymbol{S}_i \in R^{6\times1}$, for prismatic and rotational joint is defined, so that 1 is assigned along the prismatic or rotational axis : *e.g.*

$$\boldsymbol{S}_i = [\,0\,0\,1\,0\,0\,0\,]^T \text{ for prismatic joint at z axis}$$

and

$$\boldsymbol{S}_i = [\,0\,0\,0\,0\,0\,1\,]^T \text{ for rotational joint at z axis.}$$

More complex multi-degrees-of-freedom joint is introduced in (Featherstone, 1987; Lilly, 1992).

The spatial inertia matrix of link i in frame c_i, $\boldsymbol{I}_{c_i}$, is a symmetric positive definite matrix as:

$$\boldsymbol{I}_{c_i} = \begin{bmatrix} m_i\boldsymbol{E}_3 & \mathbf{0} \\ \mathbf{0} & \mathbf{I}_{c_i} \end{bmatrix} \in R^{6\times6},$$

where m_i is the mass of link i, and $\mathbf{I}_{c_i} \in R^{3\times3}$ is the inertia matrix around the center of mass of the link i in frame c_i. $\boldsymbol{E}_3$ stands for the 3×3 identity matrix.

The 6×6 spatial transformation matrix, ${}^h_i\boldsymbol{X}$, transforms a spatial quantity from frame i to frame h as:

$${}^h_i\boldsymbol{X} = \begin{bmatrix} {}^h_i\boldsymbol{R} & \mathbf{0} \\ \widetilde{{}^h\boldsymbol{r}_i}\,{}^h_i\boldsymbol{R} & {}^h_i\boldsymbol{R} \end{bmatrix} \in R^{6\times6}. \tag{14}$$

where ${}^h_i\boldsymbol{R} \in R^{3\times3}$ is a rotation matrix and ${}^h\boldsymbol{r}_i \in R^{3\times1}$ is a position vector from the origin of frame h to that of frame i expressed in frame h. $\widetilde{\{\cdot\}}$ denotes the skew-symmetric matrix. Unlike

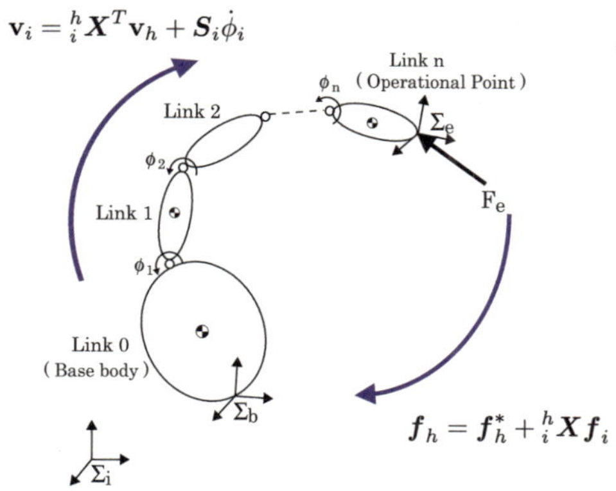

Fig. 3. Forward Chain Approach

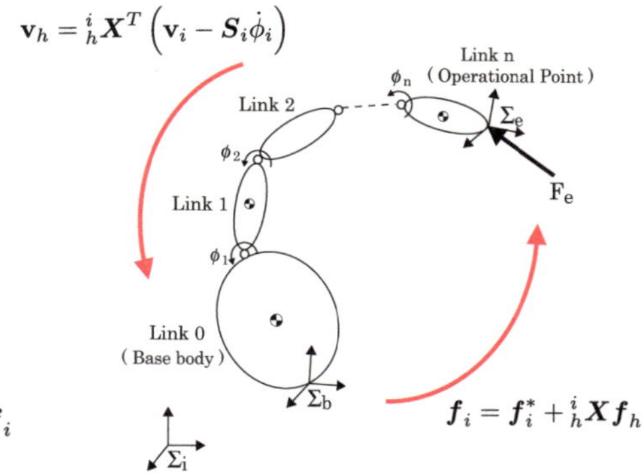

Fig. 4. Inverted Chain Approach

the conventional 3×3 rotational matrix, the spatial transformation matrix is not orthogonal. The transformation matrix from frame h to frame i, ${}^i_h\boldsymbol{X}$, is expressed as:

$$ {}^i_h\boldsymbol{X} = \begin{bmatrix} {}^h_i\boldsymbol{R}^T & \mathbf{0} \\ \left(\widetilde{{}^h\boldsymbol{r}_i}\,{}^h_i\boldsymbol{R}\right)^T & {}^h_i\boldsymbol{R}^T \end{bmatrix} \in R^{6\times 6}. \tag{15} $$

4.2 Spatial Quantity

The most of the spatial quantities including the spatial velocity, acceleration, and the spatial force are iteratively calculated. To carry out the iterative calculation, two approaches can be found, which are dependent on the direction of the iteration. In the first approach, the kinematic quantities such as the velocity and the acceleration are computed by outward recursion from the link 0 toward the operational points as shown in Fig. 3 and the spatial force is obtained by inward recursion from the operational points toward the link 0. This approach is termed *forward chain approach*. In the second approach, the direction of the recursive calculation is opposed to the first one, so that the kinematic quantities are computed by inward recursion and the spatial force is obtained by outward recursion as shown in Fig. 4. This approach is termed *inverted chain approach*.

4.2.1 Forward Chain Approach

In the forward chain approach, the spatial velocity of link i is computed by the spatial velocity of its parent link and its joint velocity as:

$$ \mathbf{v}_i = {}^h_i\boldsymbol{X}^T\mathbf{v}_h + \boldsymbol{S}_i\dot{\phi}_i, \;\; (\mathbf{v}_i = \mathbf{v}_0 \text{ at } i = 0)\,. \tag{16} $$

The spatial acceleration can be calculated similarly in the following form:

$$ \mathbf{a}_i = {}^h_i\boldsymbol{X}^T\mathbf{a}_h + \mathbf{v}_i \widehat{\times} \boldsymbol{S}_i\dot{\phi}_i + \boldsymbol{S}_i\ddot{\phi}_i, \;\; (\mathbf{a}_i = \mathbf{a}_0 \text{ at } i = 0)\,. \tag{17} $$

where $\widehat{\times}$ denotes the spatial cross-product operator associated with a spatial vector $\left[\boldsymbol{a}^T, \boldsymbol{b}^T\right]^T$ defined as:

$$ \begin{bmatrix} \boldsymbol{a} \\ \boldsymbol{b} \end{bmatrix} \widehat{\times} = \begin{bmatrix} \widetilde{\boldsymbol{b}} & \widetilde{\boldsymbol{a}} \\ \mathbf{0} & \widetilde{\boldsymbol{b}} \end{bmatrix} \in R^{6\times 6}. $$

The iterative calculation of the composite force acted on the link h is defined as:

$$f_h = f_h^* + {}^h_i X f_i\,, \quad (f_n = f_n^*)\,, \tag{18}$$

where

$$f_i^* = I_i \mathbf{a}_i + \mathbf{v}_i \widehat{\times} I_i \mathbf{v}_i\,. \tag{19}$$

I_i is the spatial inertia matrix of link i in frame i expressed as:

$$I_i = {}^i_{C_i} X I_{C_i} {}^i_{C_i} X^T\,,$$

where I_i is a symmetric positive definite matrix. The integral of eq. (19) represents the spatial momentum of link h and the composite momentum of link h can be derived as :

$$\mathcal{L}_h = I_h \mathbf{v}_h + {}^h_i X \mathcal{L}_i\,, \ (\mathcal{L}_n = I_n \mathbf{v}_n)\,. \tag{20}$$

In the above recursive calculation of the momentum, the composite rigid-body inertia of link h can be obtained as follows, which is the summation of inertia matrices of link h and its children links (Lilly, 1992):

$$I_h^C = I_h + {}^h_i X I_i^C {}^h_i X^T, \ (I_n^C = I_n)\,. \tag{21}$$

Note that the velocity and the acceleration of link 0 are arbitrary, not only zero but also non-zero values are acceptable since our main focus is on the free-flying or free-floating space robots. We assume that those values are measurable or can be estimated.

4.2.2 Inverted Chain Approach

In the inverted chain approach, the transformation matrix, ${}^i_h X$, is used to obtain each spatial quantity. As mentioned before, the direction to calculate each quantity is opposed to that of the forward chain approach.
The spatial velocity is expressed as:

$$\mathbf{v}_h = {}^i_h X^T (\mathbf{v}_i - S_i \dot{\phi}_i)\,, \ (\mathbf{v}_h = \mathbf{v}_n \text{ at } h = n)\,. \tag{22}$$

The spatial acceleration is described as:

$$\mathbf{a}_h = {}^i_h X^T \left(\mathbf{a}_i - \mathbf{v}_i \widehat{\times} S_i \dot{\phi}_i - S_i \ddot{\phi}_i\right), \ (\mathbf{a}_h = \mathbf{a}_n \text{ at } h = n)\,. \tag{23}$$

The composite force is derived as:

$$f_i = f_i^* + {}^i_h X f_h\,, \ (f_0 = f_0^*)\,. \tag{24}$$

Under the same conditions, the results of the forward chain approach and the inverted chain approach are consistent.

5. Recursive Computation of Generalized Jacobian Matrix

This section presents efficient recursive calculations of the generalized Jacobian matrices introduced in (3) and (5). We introduce here the recursive algorithms in the framework of the spatial notation. Yokokohji proposed the recursive calculation of the generalized Jacobian matrix in (Yokokohji et al., 1993). However, by using the spatial notation, the recursive calculations can be improved to simpler and faster methods than one proposed in (Yokokohji et al., 1993).

5.1 Generalized Jacobian Matrix of Base Body (Link 0)

In the recursive expression, the total linear and angular momentum around the origin of frame 0 is expressed from eqs. (16) and (20) as follows :

$$\begin{aligned}\mathcal{L}_0 &= \sum_{k=0}^{n} {}^0_k\boldsymbol{X}\boldsymbol{I}_k\mathbf{v}_k \\ &= \boldsymbol{I}_0\mathbf{v}_0 + \sum_{k=1}^{n}\left[{}^0_k\boldsymbol{X}\boldsymbol{I}_k\left({}^0_k\boldsymbol{X}^T\mathbf{v}_0 + \boldsymbol{S}_k\dot{\phi}_k\right)\right] .\end{aligned} \tag{25}$$

where $\mathcal{M}_0 =^I_0 X\mathcal{L}_0$. From eq. (25), the velocity of the base can be expressed as a function of the velocity of the joint as :

$$\mathbf{v}_0 = -\left(\boldsymbol{I}_0^C\right)^{-1}\sum_{k=1}^{n} {}^0_k\boldsymbol{X}\boldsymbol{I}_k^C\boldsymbol{S}_k\dot{\phi}_k + \left(\boldsymbol{I}_0^C\right)^{-1}\mathcal{L}_0 , \tag{26}$$

where $\boldsymbol{I}_0^C$ and $\boldsymbol{I}_k^C$ denote composite rigid-body inertia matrix of the base and for link k computed by (21), respectively. The coefficient of the first term on the righthand side corresponds to the generalized Jacobian matrix of the base (link 0) in the frame 0 :

$${}^0\boldsymbol{J}_b^* = -\left(\boldsymbol{I}_0^C\right)^{-1}\sum_k {}^0_k\boldsymbol{X}^T\boldsymbol{I}_k^C\boldsymbol{S}_k . \tag{27}$$

Transformed into the inertial frame, eq. (27) equals to the expression (3) :

$$\boldsymbol{J}_b^* = {}^I_0\boldsymbol{X}\,{}^0\boldsymbol{J}_b^*. \tag{28}$$

5.2 Generalized Jacobian Matrix of Operational Point

Once ${}^0\boldsymbol{J}_b^*$ is obtained, the generalized Jacobian matrix of the operational point is straightforwardly derived. The spatial velocity of the operational point e_i can be expressed as:

$$\mathbf{v}_{e_i} = {}^n_{e_i}\boldsymbol{X}^T\mathbf{v}_n, \tag{29}$$

where $\mathbf{v}_n$ is the velocity of the link n, on which the operational point is determined, and is obtained from eq. (16). By substituting (26) into (29), the generalized Jacobian matrix of the operational point in the frame e_i can be derived as:

$${}^{e_i}\boldsymbol{J}_{m_i}^* = \sum_{k=1}^{n}\left({}^k_{e_i}\boldsymbol{X}^T\boldsymbol{S}_k - {}^0_{e_i}\boldsymbol{X}^T\left(\boldsymbol{I}_0^C\right)^{-1}\boldsymbol{I}_k^C\boldsymbol{S}_k\right) \tag{30}$$

Consequently, the generalized Jacobian matrix in the inertial frame, (5), can be obtained as follows:

$$\boldsymbol{J}_{m_i}^* = {}^I_{e_i}\boldsymbol{X}^{e_i}\boldsymbol{J}_{m_i}^*. \tag{31}$$

6. Efficient Algorithms of Operational Space Dynamics

This section describes recursive algorithms of the operational space dynamics, eq. (8). We recall here the operational space dynamics :

$$\mathbf{\Gamma}_e \ddot{\mathbf{x}}_e + \boldsymbol{\mu}_e = \mathcal{F}_e \,, \quad \begin{bmatrix} \mathcal{F}_b \\ \boldsymbol{\tau} \end{bmatrix} = \boldsymbol{J}_e^T \mathcal{F}_e \,. \tag{32}$$

A main focus is on developing computational efficient algorithms of $\mathbf{\Gamma}_e$ and $\boldsymbol{\mu}_e$ of a n-link, p-operational-point branching space robot system. The derivation of $\boldsymbol{J}_e$ is omitted in this chapter since its algorithm is well-known. The algorithms of $\mathbf{\Gamma}_e$ and $\boldsymbol{\mu}_e$ are developed with the concept of the articulated body dynamics (Featherstone, 1987; Lilly, 1992). Firstly, the operational space formulation of a single operational point on the space robot is developed by using the inverted chain approach. Then, the algorithms of the multi-operational point system are further developed.

6.1 Single Operational Point in Space Robot

As mentioned in Section 1, a unique characteristic of the space robot is that the base satellite is inertially free and the system is invertible in its modeling unlike the ground-based robot system. Based on this characteristic, the operational space dynamics of the space robot is derived in the framework of the articulated-body dynamics. In the conventional articulated-body dynamics, the articulated-body inertia and its associated bias force are calculated inward from the operational point to the base body (link 0). When the system is inverted, the articulated-body dynamics is calculated in the opposed direction, namely outward from the base body to the operational point. This approach introduced as the *inertia propagation method* in (Lilly, 1992). We make use of the inverted chain approach for the iterative calculation of the inertia matrix, $\mathbf{\Gamma}_e$, and the bias force vector, $\boldsymbol{\mu}_e$ in the operational space.

6.1.1 Operational space inertia matrix

Operational space inertia matrix corresponds to the articulated-body inertia calculated from the base body to the operational point with the initial condition, $\boldsymbol{I}_0^A = \boldsymbol{I}_0$:

$$\boldsymbol{I}_i^A = \boldsymbol{I}_i + {}^i\boldsymbol{I}_h^{A\,i}{}_h\boldsymbol{L} \,, \; (i = 0 \cdots n) \,, \tag{33}$$

where

$${}^i_h\boldsymbol{L} = \boldsymbol{E}_6 - \frac{\boldsymbol{S}_i \boldsymbol{S}_i^{T\,i} \boldsymbol{I}_h^A}{\alpha_i} \,,$$

$${}^i\boldsymbol{I}_h^A = {}^i_h\boldsymbol{X} \boldsymbol{I}_h^{A\,i}{}_h\boldsymbol{X}^T \,, \quad \alpha_i = \boldsymbol{S}_i^{T\,i} \boldsymbol{I}_h^A \boldsymbol{S}_i .$$

$\boldsymbol{E}_6$ represents the 6×6 identity matrix. The superscript i at left side of the symbols describes the quantities of link h expressed in the frame i.

The symbols without the superscript at the left side expresses the quantities represented in their own frame. Consequently, the inertia matrix, $\mathbf{\Gamma}_e$, in the inertial frame is obtained by using the following spatial transformation.

$$\mathbf{\Gamma}_e = {}^I_e\boldsymbol{X} \boldsymbol{I}_e^{A\,I}{}_e\boldsymbol{X}^T . \tag{34}$$

6.1.2 Operational space bias force vector

Likely the operational space inertia matrix, the associated bias force, $\boldsymbol{\mu}_e$, is calculated in the outward recursive manner. The bias force in the inverted chain approach is calculated as the following algorithm with the initial condition, $\boldsymbol{p}_0^A = \boldsymbol{p}_0 = \mathbf{v}_0 \hat{\times} \boldsymbol{I}_0 \mathbf{v}_0$:

$$\boldsymbol{p}_i^A = \boldsymbol{p}_i + {}^i\boldsymbol{p}_h^A - {}_h^i\boldsymbol{L}\boldsymbol{c}_i - \frac{{}^i\boldsymbol{I}_h^A \mathbf{S}_i \mathbf{S}_i^T {}^i\boldsymbol{p}_h^A}{\alpha_i}, \tag{35}$$

where

$${}^i\boldsymbol{p}_h^A = {}_h^i\boldsymbol{X}\boldsymbol{p}_h^A, \quad \boldsymbol{p}_i = \mathbf{v}_i \hat{\times} \boldsymbol{I}_i \mathbf{v}_i, \quad \boldsymbol{c}_i = \mathbf{v}_i \hat{\times} \mathbf{S}_i \dot{\boldsymbol{\phi}}_i.$$

Finally, the operational space bias force in the inertial frame, $\boldsymbol{\mu}_e$, is obtained as follows:

$$\boldsymbol{\mu}_e = {}_e^I\boldsymbol{X}\boldsymbol{p}_e^A + \boldsymbol{\Gamma}_e \begin{bmatrix} v_e \times \omega_e \\ \mathbf{0}_3 \end{bmatrix}, \tag{36}$$

where the relationship between the spatial acceleration, $\mathbf{a}_i$, and the conventional acceleration of a point fixed in a rigid body, $\ddot{x}_i$, is used here, *i.e.* $\ddot{x}_i = \mathbf{a}_i - \left((v_i \times \omega_i)^T, \mathbf{0}_3^T \right)^T$, since the spatial acceleration, $\mathbf{a}_i$, differs from the conventional acceleration of a point fixed in a rigid body, $\ddot{x}_i$ (Featherstone, 1987). The vector $\mathbf{0}_3$ represents the 3×1 zero vector.

6.2 Multi-Operational Points in Space Robot

The lack of the fixed base and the dynamic coupling between each operational point lead to the computational complexity in the multi-operational points on the space robot. However, no fixed base provides an arbitrary choice of link 0 in the modeling of the kinematic connectivity. In addition, the proposed algorithms enable to formulate the dynamics of arbitrary operational points, not only the end-effectors in the real system but also the base-satellite or other controlled points in operational space. For instance, if both base-satellite and one end-effector are operated simultaneously, these two points are determined as operational points. In this subsection, the operational space inertia matrix and the bias force vector are derived based on the forward chain approach. Note that the direction of the recursive calculation is opposed to the approach in the previous subsection although the same symbols are used.

6.2.1 Operational space inertia matrix

The inverse of the operational space inertia matrix, $\boldsymbol{\Gamma}_e^{-1}$, consists of diagonal matrices, $\boldsymbol{\Gamma}_{e_i,e_i}^{-1}$ and off-diagonal matrices, $\boldsymbol{\Gamma}_{e_i,e_j}^{-1}$ as expressed in the following equation:

$$\boldsymbol{\Gamma}_e^{-1} = \begin{bmatrix} \boldsymbol{\Gamma}_{e_1,e_1}^{-1} & \cdots & \boldsymbol{\Gamma}_{e_1,e_p}^{-1} \\ \vdots & \boldsymbol{\Gamma}_{e_i,e_j}^{-1} & \vdots \\ \boldsymbol{\Gamma}_{e_p,e_1}^{-1} & \cdots & \boldsymbol{\Gamma}_{e_p,e_p}^{-1} \end{bmatrix}. \tag{37}$$

The inertial quantity $\boldsymbol{\Gamma}_{e_i,e_i}^{-1}$ describes the inertia of link i if the force is applied to only i-th operational point, and the inertial quantity $\boldsymbol{\Gamma}_{e_i,e_j}^{-1}$ describes the cross-coupling inertia matrix, which expresses the dynamic influence of the i-th operational point due to the force of the j-th operational point. Once the inverse of the operational space inertia matrix is calculated, the operational space inertia matrix is obtained as the following recursive calculation. Figure 5 shows the process of the calculation of the inertia matrix. In the figure, arrows indicate the direction of the calculation.

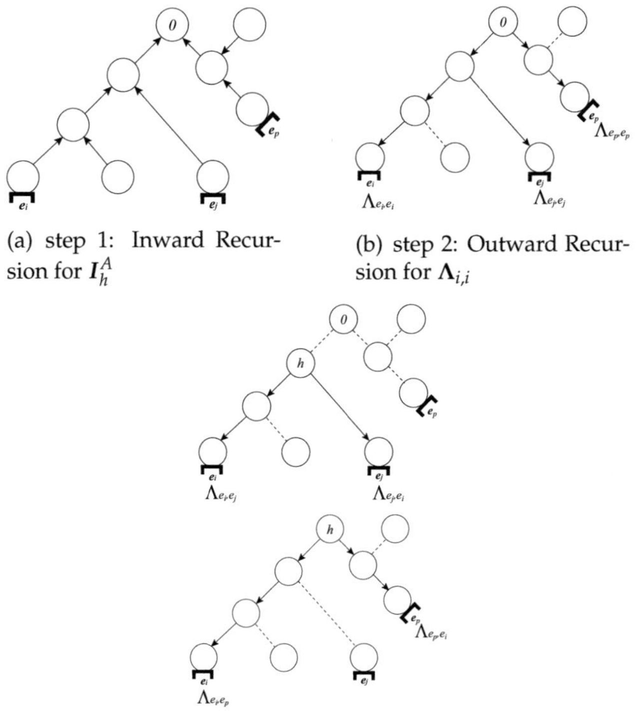

(a) step 1: Inward Recursion for $\boldsymbol{I}_h^A$

(b) step 2: Outward Recursion for $\boldsymbol{\Lambda}_{i,i}$

(c) step 3: Outward Recursion for $\boldsymbol{\Lambda}_{i,j}$

Fig. 5. Recursive process of the inertia matrix

1. **Inward Recursion**: Compute the articulated-body inertia of link i in the forward chain approach:

$$\boldsymbol{I}_h^A = \boldsymbol{I}_h + {}^h_i\boldsymbol{L}\boldsymbol{I}_i^A {}^h_i\boldsymbol{X}^T, \quad (\boldsymbol{I}_n^A = \boldsymbol{I}_n) \tag{38}$$

$${}^h_i\boldsymbol{L} = {}^h_i\boldsymbol{X}\left(\boldsymbol{E}_6 - \frac{\boldsymbol{I}_i^A\boldsymbol{S}_i\boldsymbol{S}_i^T}{\beta_i}\right), \quad \beta_i = \boldsymbol{S}_i^T\boldsymbol{I}_i^A\boldsymbol{S}_i.$$

2. **Outward Recursion**: Compute the block diagonal matrices:

$$\boldsymbol{\Lambda}_{i,i} = \frac{\boldsymbol{S}_i\boldsymbol{S}_i^T}{\beta_i} + {}^{h_i}_i\boldsymbol{L}^T\boldsymbol{\Lambda}_{h_i,h_i}{}^{h_i}_i\boldsymbol{L}, \quad \left(\boldsymbol{\Lambda}_{0,0} = \left(\boldsymbol{I}_0^A\right)^{-1}\right) \tag{39}$$

3. **Outward Recursion**: Compute the cross-coupling inertia matrices:

$$\boldsymbol{\Lambda}_{i,j} = \begin{cases} {}^{h_i}_i\boldsymbol{L}^T\boldsymbol{\Lambda}_{j,h_i}^T & if\ j = h \\ \boldsymbol{\Lambda}_{i,h_j}{}^{h_j}_j\boldsymbol{L} & if\ i = h \end{cases} \tag{40}$$

4. **Spatial Transformation**: Compute the inverse inertia matrices in the inertial frame:

$$\boldsymbol{\Gamma}_{e_i,e_j}^{-1} = {}^I_{e_i}\boldsymbol{X}\,\boldsymbol{\Lambda}_{e_i,e_j}\,{}^I_{e_j}\boldsymbol{X}^T \tag{41}$$

5. **Matrix Inversion:** Compute the operational space inertia matrix, $\boldsymbol{\Gamma}_e$, for the space robot by inverting $\boldsymbol{\Gamma}_e^{-1}$.

6.2.2 Operational space bias force vector

To derive the bias force vector in operational space, the bias acceleration vector of each operational point is firstly derived in a recursive manner. The multiplication of the operational space inertia matrix $\boldsymbol{\Gamma}_e$ and the bias acceleration provides the bias force. The recursive algorithm of the bias force vector is shown in the following.

1. **Inward Recursion:** Compute the bias force of link i in the forward chain approach from the operational points to the link 0:

$$\boldsymbol{p}_h^A = \boldsymbol{p}_h + \mathbf{X}_i^h \boldsymbol{p}_i^A + {}_i^h\boldsymbol{L}\boldsymbol{I}_i^A \boldsymbol{c}_i - \frac{{}_i^h\mathbf{X}\boldsymbol{I}_i^A \mathbf{S}_i \mathbf{S}_i^T \boldsymbol{p}_i^A}{\beta_i}, \quad (\boldsymbol{p}_n^A = \boldsymbol{p}_n) \tag{42}$$

where

$${}^h\boldsymbol{p}_i^A = {}_i^h\mathbf{X}\boldsymbol{p}_i^A, \quad \boldsymbol{p}_i = \mathbf{v}_i \widehat{\times} \boldsymbol{I}_i \mathbf{v}_i, \quad \boldsymbol{c}_i = \mathbf{v}_i \widehat{\times} \mathbf{S}_i \dot{\boldsymbol{\phi}}_i.$$

2. **Outward Recursion:** Compute the bias acceleration of each operational point:

$$\boldsymbol{a}_i = {}_i^h\mathbf{X}^T \boldsymbol{a}_h + \boldsymbol{c}_i + \mathbf{S}_i \boldsymbol{b}_i, \quad \left(\boldsymbol{a}_0 = \left(\boldsymbol{I}_0^A\right)^{-1} \boldsymbol{p}_0^A\right) \tag{43}$$

where

$$\boldsymbol{b}_i = -\frac{\mathbf{S}_i^T \left[\boldsymbol{I}_i^A \left(\mathbf{X}_i^h \boldsymbol{a}_h + \boldsymbol{c}_i\right) + \boldsymbol{p}_i\right]}{\beta_i}.$$

3. **Spatial Transformation:** Compute the bias acceleration of each operational point with respect to the inertial frame:

$${}^I\boldsymbol{a}_{e_i} = {}_{e_i}^I\boldsymbol{X}\boldsymbol{a}_{e_i} \tag{44}$$

4. **Matrix Multiplication:** Compute the bias force of the operational points with respect to the inertial frame:

$$\boldsymbol{\mu}_e = \boldsymbol{\Gamma}_e \left(\begin{bmatrix} \vdots \\ {}^I\boldsymbol{a}_{e_i} \\ \vdots \end{bmatrix} - \begin{bmatrix} \vdots \\ \boldsymbol{v}_{e_i} \times \boldsymbol{\omega}_{e_i} \\ \mathbf{0}_3 \\ \vdots \end{bmatrix} \right), \quad (i = 1 \cdots p), \tag{45}$$

where the relationship between the conventional acceleration and the spatial acceleration, $\ddot{\boldsymbol{x}}_i = \boldsymbol{a}_i - \left((\boldsymbol{v}_i \times \boldsymbol{\omega}_i)^T, \mathbf{0}_3^T\right)^T$, is used.

6.3 Free-Floating Space Robot

In the previous subsections, we derive the inertia matrix and the bias force of eq. (8). In the free-floating space robot, one needs to refer to eq. (12). Since the operational inertia matrix $\left(\boldsymbol{\Lambda}^{-1} + \boldsymbol{\Lambda}_b^{-1}\right)^{-1} = \boldsymbol{\Gamma}_e$, the bias force $\boldsymbol{\Gamma}_e \boldsymbol{\mu} = \boldsymbol{\mu}_e$ in eq. (12) and $\boldsymbol{J}_m^*$ is calculated in Section 5, one need to derive only the matrix $\boldsymbol{\Lambda}$. The inertia matrix $\boldsymbol{\Lambda}$ can be easily obtained by $\boldsymbol{\Lambda} = \left(\boldsymbol{\Gamma}_e^{-1} - \boldsymbol{\Lambda}_b^{-1}\right)^{-1}$, where the inertia matrix $\boldsymbol{\Lambda}_b$ corresponds to the composite inertia matrix of the base $\boldsymbol{\Lambda}_b = {}_0^I\boldsymbol{X}\boldsymbol{I}_0^C$.

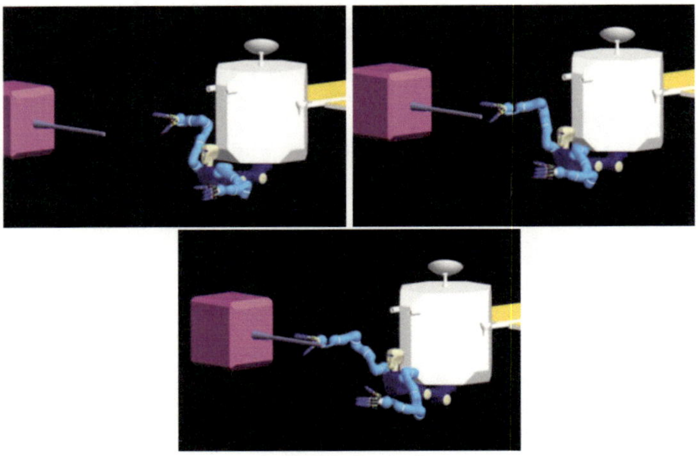

Fig. 6. Target grasping sequence, wait, approach, and grasp

7. Operational Space Task

The operational space task is illustrated by using the proposed algorithms. The robotic model considered here has total 24 degrees-of-freedom including two 7-degrees-of-freedom arms, a 4-degrees-of-freedom torso, and a 6-degrees-of-freedom base body. In practice, the redundancy of the system provides the null-space motion in the joint space. To obtain the null space motion, the classical joint space inverse dynamics is simply applied.The null space motion performs the self-motion in the joint space such as self-collision avoidance, posture behavior while the operational space motion is carried out.

Figure 6 shows the motion sequence for target grasping by a chaser-robot, consisting of waiting phase, approaching phase, and grasping phase. In those processes, the end-effector of the right hand, the head of the robot and the base body are determined as three operational points. In the simulation, it is demonstrated that the right arm approaches to the target while the rest of active joints are operated to keep the orientation of the head and the orientation of the base-satellite constant.

8. Conclusions

This chapter proposed efficient recursive algorithms of the operational space dynamics for the free-flying and the free-floating space robots. In the space robot, the operational space formulation is more complex than that of the ground-based robot due to the lack of the fixed base. However, by virtue of no fixed base, the space robot can be switched around in its modeling. By making use of this unique characteristic, firstly the operational space formulation of a single-serial-arm space robot has been developed. Then, the efficient algorithm of the operational space formulation of the branching-arms space robot has been proposed by using the concept of the articulated-body system. The realistic simulation with 24 DOF space robot system was illustrated to verify the efficiency of the proposed algorithms.

9. References

Chang, K.-S. & Khatib, O. (1999). Efficient algorithm for extended operational space inertia matrix, *Proc. of the IEEE/RSJ International Conference on Intelligent Robots and Systems*, Vol. 1, Kyongju, Korea, pp. 350 – 355.

Chang, K.-S. & Khatib, O. (2000). Operational space dynamics: Efficient algorithms for modeling and control of branching mechanisms, *Proc. of the 2000 IEEE International Conference on Robotics and Automation*, San Francisco, CA, pp. 850 – 856.

Featherstone, R. (1987). *Robot Dynamics Algorithms*, Kluwer Academic Publishers.

Khatib, O. (1987). A Unified Approach for Motion and Force control of Robot Manipulators: The Operational Space Formulation, *IEEE Journal of Robotics and Automation* **RA-3**(1): 43 – 53.

Lilly, K. W. (1992). *Efficient Dynamic Simulation of Robotic Mechanisms*, Kluwer Academic Publishers.

Russakow, J., Khatib, O. & Rock, S. M. (1995). Extended operational space formulation for serial-to-parallel chain (branching) manipulators, *Proc. of IEEE International Conference on Robotics and Automation*, Nagoya, Japan, pp. 1056 – 1061.

Umetani, Y. & Yoshida, K. (1989). Resolved motion rate control of space manipulators with generalized jacobian matrix, *IEEE Transactions on Robotics and Automation* **5**(3): 303 – 314.

Xu, Y. & Kanade, T. (eds) (1993). *Space Robotics: Dynamics and Control*, Kluwer Academic Publishers.

Yokokohji, Y., Toyoshima, T. & Yoshikawa, T. (1993). Efficient computational algorithms for trajectory control of free-flying space robots with multiple arms, *IEEE Transactions on Robotics and Automation* **9**(5): 571 – 580.

3

Modeling and Control of Mechanical Systems in Terms of Quasi-Velocities

Farhad Aghili
Canadian Space Agency, Department of Space Technologies, Saint-Hubert, Quebec Canada

1. Introduction

Multi-body systems' (MBS) dynamics are often described by the second-order nonlinear equations parameterized by a configuration-dependent inertia matrix and the nonlinear vector containing the Coriolis and centrifugal terms. Since these equations are the cornerstone for simulation and control of robotic manipulators, many researchers have attempted to develop efficient modelling techniques to derive the equations of motion of multi-body systems in novel forms. A unifying idea for most modeling techniques is to describe the equations of motion in terms of general coordinates and their time–derivatives. In classical mechanics of constrained systems, a generalized velocity is taken to be an element of tangential space of configuration manifold, and a generalized force is taken to be the cotangent space. However, neither does space possess a natural metric as the generalized coordinates or the constrains may have a combination of rotational and translational components. As a result, the corresponding dynamic formulation in not invariant and a solution depends on measure units or a weighting matrix selected Aghili (2005); Angeles (2003); Lipkin and Duffy (1988); Luca and Manes (1994); Manes (1992). There also exist other techniques to describe the equations of motion in terms of *quasi–velocities*, i.e., a vector whose Euclidean norm is proportional to the square root of the system's kinetic energy, which can lead to simplification of these equations Aghili (2008; 2007); Bedrossian (1992); Gu (2000); Gu and Loh (1987); Herman (2005); Herman and Kozlowski (2006); Jain and Rodriguez (1995); Junkins and Schaub (1997); Kodischeck (1985); Kozlowski (1998); Loduha and Ravani (1995); Papastavridis (1998); Rodriguez and Kertutz-Delgado (1992); Sinclair *et al.* (2006); Spong (1992). A recent survey on some of these techniques can be found in Herman and Kozlowski (2006). In short, the square–root factorization of mass matrix is used as a transformation to obtain the quasi–velocities, which are a linear combination of the velocity and the generalized coordinates Herman and Kozlowski (2006); Papastavridis (1998).

It was shown by Kodistchek Kodischeck (1985) that if the square–root factorization of the inertia matrix is integrable, then the robot dynamics can be significantly simplified. In such a case, transforming the generalized coordinates to quasi–coordinates by making use of the integrable factorization modifies the robot dynamics to a system of double integrator. Then, the cumbersome derivation of the Coriolis and centrifugal terms is not required. It was later realized by Gu *et al.* Gu and Loh (1987) that such a transformation is a canonical transformation because it satisfies Hamilton's equations. Rather than deriving the mass matrix of MBS first and then obtaining its factorization, Rodriguez *et al.* Rodriguez and Kertutz-Delgado (1992)

derived the closed–form expressions of the mass matrix factorization of an MBS and its inverse directly from the link geometric and inertial parameters. This eliminates the need for the matrix inversion required to compute the forward dynamics.
The interesting question of when the factorization of the inertia matrix is integrable, i.e., the factorization being the Jacobian of some quasi–coordinates, was addressed independently in Spong (1992) and Bedrossian (1992). Using the notion that the inertia matrix defines a metric tensor on the configuration manifold, Spong Spong (1992) showed that the necessary and sufficient condition for the existence of an integrable factorization of the inertia matrix is that the metric tensor is a Euclidean metric tensor.[1] It turned out that for most of the practical robot systems, the condition is not satisfied meaning that integration of the quasi–velocities does not produce any quasi–coordinates. Jain and Rodriguez Jain and Rodriguez (1995) described quasi–velocities are obtained as a result of diagonalizing the inertia matrix. Instead of diagonalizing globally in configuration space, they look at a diagonalizing transformation in the velocity space. The transformation replaces generalized velocity with the quasi–velocities, without replacing the configuration variables. The concept of quasi–velocities has also been used for the set-point control of manipulators Herman (2005); Herman and Kozlowski (2001); Jain and Rodriguez (1995); Kozlowski (1998); Kozolowski and Herman (2000). However, the problem of the tracking control of manipulators using quasi–velocities feedback still remains unsolved owing to unintegrability of the quasi–velocities.
The goal of this chapter is to extend the concept of quasi–velocities for an gauge-invariant formulation of constrained MBS that can be used for simulation, analysis, and control purposes Aghili (2009). The main focus of previous works on modeling of constrained or unconstrained mechanical systems using the notion of quasi–velocities, e.g., Junkins and Schaub (1997); Loduha and Ravani (1995), has been decoupling of the equations of motion, which yields a dynamical system with an identity mass matrix. Analysis of constraint force has not been considered in the previous works. In this paper, we took advantages of the fact that quasi–velocities are not unique but they are related by unitary transformations and found a particular transformation which allows to decouple the equations of motions and the equation of constraints in such a way that separate control inputs are associated to each set of equations. This facilitates motion/force control of constrained systems such as robotic manipulators. Moreover, unlike other approaches Aghili (2005); Doty *et al.* (1993*a*); Luca and Manes (1994); Schutter and Bruyuinckx (1996), this formulation does not require any weighting matrix to achieve gauge-invariance when both translational and rotational components are involved in the generalized coordinates or in the constraint equations. Some properties of the quasi–velocities dynamic formulation are presented that could be useful for control purposes. Finally, the dynamic model is used for developing tracking control of constrained MBS based on a combination of feedbacks on the vector of reduced quasi–velocities and the vectors of configuration-variables Aghili (2009).

[1] A manifold with a Euclidean metric is said to be "flat" and the curvature associated with it is identically zero Jain and Rodriguez (1995).

2. Quasi-Variables Transformation

2.1 Square-Root Factorization of the Mass Matrix

Dynamics of MBS with kinetic energy, T, and potential energy, P, obey the standard *Euler–Lagrange* (EL) equations, which are given as

$$\frac{\mathrm{d}}{\mathrm{d}t}\left(\frac{\partial T}{\partial \dot{\boldsymbol{q}}}\right) - \frac{\partial T}{\partial \boldsymbol{q}} = \boldsymbol{f}, \tag{1}$$

where $\boldsymbol{q} \in \mathbb{R}^n$ is the vector of configuration-variables[2] used to define the configuration of the system, and f is the generalized forces acting on the system. The generalized forces $\boldsymbol{f} = \boldsymbol{f}_p + \boldsymbol{f}_a$ contain all possible external forces including the conservative forces $\boldsymbol{f}_p = -\partial P/\partial \boldsymbol{q}$ owing to gravitational energy plus all active and dissipative forces represented by $\boldsymbol{f}_a$. The system kinetic energy is in the following quadratic form:

$$T(\boldsymbol{q}, \dot{\boldsymbol{q}}) = \frac{1}{2}\dot{\boldsymbol{q}}^T \boldsymbol{M}(\boldsymbol{q})\dot{\boldsymbol{q}}, \tag{2}$$

where the generalized inertia matrix $\boldsymbol{M}(\boldsymbol{q})$ is *symmetric* and *positive definite* for all $\boldsymbol{q}$. It is well known that any symmetric positive-definite matrix $\boldsymbol{M}$ can be decomposed as

$$\boldsymbol{M} = \boldsymbol{W}\boldsymbol{W}^T, \tag{3}$$

where $\boldsymbol{W}$ is the *square root factorization* of $\boldsymbol{M}$, e.g. the Cholesky decomposition; see Appendix A.
Considering the transformation

$$\bar{\boldsymbol{W}} = \boldsymbol{W}\boldsymbol{V},$$

where $\boldsymbol{V}$ is an orthogonal matrix, i.e., $\boldsymbol{V}\boldsymbol{V}^T = \boldsymbol{V}^T\boldsymbol{V} = \boldsymbol{I}$, one can trivially verify that $\bar{\boldsymbol{W}}\bar{\boldsymbol{W}}^T = \boldsymbol{M}$. Thus, we get the following remark

Remark 1. *The square–root factorization* (3) *is not unique, rather they are related by unitary transformations.*

Now, substituting (3) into (2) and then applying the EL formulation yields

$$\begin{aligned} \boldsymbol{f} &= \frac{\mathrm{d}}{\mathrm{d}t}\Big(\boldsymbol{W}\boldsymbol{W}^T\dot{\boldsymbol{q}}\Big) - \frac{1}{2}\Big(\frac{\partial}{\partial \boldsymbol{q}}\|\boldsymbol{W}^T(\boldsymbol{q})\dot{\boldsymbol{q}}\|^2\Big)^T \\ &= \boldsymbol{W}\frac{\mathrm{d}}{\mathrm{d}t}(\boldsymbol{W}^T\dot{\boldsymbol{q}}) + \Big(\dot{\boldsymbol{W}} - \frac{\partial(\boldsymbol{W}^T(\boldsymbol{q})\dot{\boldsymbol{q}})^T}{\partial \boldsymbol{q}}\Big)\boldsymbol{W}^T\dot{\boldsymbol{q}} \end{aligned} \tag{4}$$

Note that (4) is obtained using the property that for any vector field $a(q)$, we have

$$\frac{\partial}{\partial \boldsymbol{q}}\|\boldsymbol{a}(\boldsymbol{q})\|^2 = 2\boldsymbol{a}^T\frac{\partial \boldsymbol{a}}{\partial \boldsymbol{q}}. \tag{5}$$

Define

$$\boldsymbol{v} \triangleq \boldsymbol{W}^T(\boldsymbol{q})\dot{\boldsymbol{q}} \tag{6a}$$

$$\boldsymbol{u} \triangleq \boldsymbol{W}^{-1}(\boldsymbol{q})\boldsymbol{f}, \tag{6b}$$

[2] also known as generalized coordinates

which are called here as the vectors of *quasi–velocities* and *quasi–forces*, respectively. It should be pointed out that in analytical dynamics, quasi-velocities are broadly defined as any linear combination of velocities Baruh (1999); Corben and Stehle (1960); Meirovitch (1970). Since $\det \boldsymbol{W} = \sqrt{\det \boldsymbol{M}} \neq 0$, $\boldsymbol{W}^{-1}$ is well–defined and hence the reciprocals of relations (6) always exist. Pre-multiplying (4) by $\boldsymbol{W}^{-1}$ and the substituting (6) into the resultant equation, we arrive at the equations of mechanical systems expressed by the quasi–variables:

$$\dot{\boldsymbol{v}} + \boldsymbol{\Gamma}\boldsymbol{v} = \boldsymbol{u}, \tag{7a}$$

where

$$\boldsymbol{\Gamma} \triangleq \boldsymbol{W}^{-1}\Big(\dot{\boldsymbol{W}} - \frac{\partial \boldsymbol{v}^T}{\partial \boldsymbol{q}}\Big) \tag{7b}$$

is the Coriolis term associated with the quasi–velocities. Note that the quasi–velocities is factored out in the derivation of the Coriolis term (7b) that is different from the previous derivation Jain and Rodriguez (1995). As will be seen in the following section, this representation is useful when the formulation is extended for constrained MBS.

2.2 Changing Coordinates by Unitary Transformations

Remark 1 states that the quasi–velocities (and also quasi–forces) can not be uniquely determined. Rather, the following variables:

$$\bar{\boldsymbol{v}} = \boldsymbol{V}^T\boldsymbol{v} \qquad \text{and} \qquad \bar{\boldsymbol{u}} = \boldsymbol{V}^T\boldsymbol{u}, \tag{8}$$

obtained by any unitary transformation $\boldsymbol{V}$, are also valid choices for the new quasi–velocities and quasi–forces. Now we are interested to derive the equations of motion expressed by the new quasi–variables $\bar{\boldsymbol{v}}$. To this end, using the reciprocal of (8), i.e., $\boldsymbol{v} = \boldsymbol{V}\bar{\boldsymbol{v}}$ and $\boldsymbol{f} = \boldsymbol{V}\bar{\boldsymbol{f}}$, into (7a) and then multiplying the resultant equation by $\boldsymbol{V}$, we arrive at

$$\dot{\bar{\boldsymbol{v}}} + \boldsymbol{V}^T\dot{\boldsymbol{V}}\bar{\boldsymbol{v}} + \boldsymbol{V}^T\boldsymbol{\Gamma}\boldsymbol{V}\bar{\boldsymbol{v}} = \bar{\boldsymbol{u}} \tag{9}$$

Analogous to the rotation transformation in the three-dimensional Euclidean space, consider matrix $\boldsymbol{V}$ as a transformation in the n-dimensional space. Then, it is known that the time–derivative of a differentiable orthogonal matrix V satisfies a differential equation of this form Schaub *et al.* (1995)

$$\dot{\boldsymbol{V}} = -\boldsymbol{\Omega}\boldsymbol{V}, \tag{10}$$

where $\boldsymbol{\Omega}$ is a *skew symmetric matrix* representing the angular rates in $n - D$ space Bar-Itzhack (1989). The elements of $\boldsymbol{\Omega}$ can be interpreted as a generalized eigenvector axis angular velocity Junkins and Schaub (1997). It is worth noting that in the three-dimensional space, the angular rate matrix can be obtained from the vector of angular velocity by $\boldsymbol{\Omega} = [\boldsymbol{\omega}\times]$. For the n–dimensional case, the method for computing the elements of matrix $\boldsymbol{\Omega}$ can be found in Junkins and Schaub (1997); Oshman and Bar-Itzhack (1985); Schaub *et al.* (1995). Finally, by replacing (10) in (9), we can show that the latter equation is equivalent to

$$\dot{\bar{\boldsymbol{v}}} + \bar{\boldsymbol{\Gamma}}\bar{\boldsymbol{v}} = \bar{\boldsymbol{u}},$$

where

$$\bar{\boldsymbol{\Gamma}} = \boldsymbol{V}^T(\boldsymbol{\Gamma} - \boldsymbol{\Omega})\boldsymbol{V}. \tag{11}$$

2.3 Conservation of Kinetic Energy

The kinetic energy expressed by the quasi–velocities is trivially

$$T = \frac{1}{2}\|\boldsymbol{v}\|^2. \tag{12}$$

Note that in the following, $\|\cdot\|$ denotes either the matrix 2-norm or the Euclidean vector norm. In the absence of any external force, the principle of conservation of kinetic energy dictates that the kinetic energy of mechanical system is bound to be constant, i.e., $\boldsymbol{u} = \boldsymbol{0} \Longrightarrow \dot{T} = 0$. On the other hand, the zero-input response of a mechanical system is $\dot{\boldsymbol{v}} = -\boldsymbol{\Gamma}\boldsymbol{v}$. Substituting the latter equation in the time-derivative of (12) gives

$$\boldsymbol{v}^T\boldsymbol{\Gamma}\boldsymbol{v} = \boldsymbol{0}, \tag{13}$$

which is consistent with the earlier result reported by Jain *et al.* Jain and Rodriguez (1995) that the Coriolis term associated with quasi–velocities does no mechanical work. Note that (13) is a necessary but not a sufficient condition for $\boldsymbol{\Gamma}$ to be a *skew-symmetric* matrix.

2.4 State-Space Model

It should be pointed out that despite of the one-to-one correspondence between velocity coordinate $\dot{\boldsymbol{q}}$ and the quasi–velocity $\boldsymbol{v}$, they are not synonymous. This is because the integration of the former variable leads to the generalized coordinate, while that of the latter variable does not always lead to a meaningful vector describing the configuration of the mechanical system. Defining a matrix $\boldsymbol{R} = \boldsymbol{W}\boldsymbol{\Gamma}$, we can calculate its ijth element from (7b) through the following equations

$$R_{ij} = \sum_k \Big(\frac{\partial W_{ij}}{\partial q_k} - \frac{\partial W_{kj}}{\partial q_i}\Big)\dot{q}_k. \tag{14}$$

Here W_{ij} and $\dot{q}_k$ are the (i,j)th entry of matrix $\boldsymbol{W}$ and the kth element of vector $\dot{\boldsymbol{q}}$, respectively. Now let us assume that $\dot{\boldsymbol{\xi}} = \boldsymbol{v}$, where $\boldsymbol{\xi}$ is called *quasi–coordinates*. For $\boldsymbol{\xi}$ to be an explicit function of $\boldsymbol{q}$, i.e., $\boldsymbol{\xi} = \boldsymbol{\xi}(\boldsymbol{q})$, it must be the gradient of a scalar function meaning that $\boldsymbol{\xi}$ is a *conservative vector field*. In that case, (6a) implies that $\boldsymbol{W}^T(\boldsymbol{q})$ is actually a Jacobian as $W_{ij} = \partial\xi_j/\partial q_i$. Since the Jacobian is an invertible matrix, $\boldsymbol{\xi}(\boldsymbol{q})$ must be an invertible function meaning that there is a one-to-one correspondence between $\boldsymbol{\xi}$ and $\boldsymbol{q}$. Under this circumstance, $\boldsymbol{\xi}$ and $\boldsymbol{v}$ are indeed alternative possibilities for generalized coordinates and generalized velocities and that can fundamentally simplify the equations of motion Bedrossian (1992); Gu and Loh (1987); Kodischeck (1985); Spong (1992). It can be also seen from (7b) that if $\boldsymbol{\xi}(\boldsymbol{q})$ exists and it is a smooth function, then the expression in the parenthesis of the right-hand side of (14) vanishes, i.e.,

$$\frac{\partial W_{ij}}{\partial q_k} - \frac{\partial W_{kj}}{\partial q_i} = \frac{\partial^2\xi_j}{\partial q_i\partial q_k} - \frac{\partial^2\xi_j}{\partial q_k\partial q_i} = 0,$$

because of the equality of mixed partials. Thus, $\boldsymbol{\Gamma} \equiv \boldsymbol{0}$ and the equations of motion become a simple integrator system.

Technically speaking, a necessary and sufficient condition for the existence of the quasi–coordinates, $\boldsymbol{\xi}$, is that the Riemannian manifold defined by the robot inertia matrix $\boldsymbol{M}(\boldsymbol{q})$ be locally flat[3]. However, that has been proved to be a very stringent condition Bedrossian

[3] By definition, a Riemannian manifold that is locally isometric to Euclidean manifold is called a locally flat manifold Spong (1992).

(1992). Nevertheless, vector $\boldsymbol{x}^T = [\boldsymbol{q}^T \;\; \boldsymbol{v}^T]$ is sufficient to describe completely the states of MBS. Hence, similar to Jain and Rodriguez (1995), we look at the transformation only in the velocity space. That is, only the velocity coordinate is replaced with the quasi–velocity whereas the generalized coordinate remains. Setting (6a) and (7a) in state space form gives

$$\frac{\mathrm{d}}{\mathrm{d}t}\begin{bmatrix}\boldsymbol{q}\\ \boldsymbol{v}\end{bmatrix} = \begin{bmatrix}\boldsymbol{W}^{-T}\\ -\boldsymbol{\Gamma}\end{bmatrix}\boldsymbol{v} + \begin{bmatrix}\boldsymbol{0}\\ \boldsymbol{I}\end{bmatrix}\boldsymbol{u}. \tag{15}$$

It is interesting to note that dynamics system (15) is in the form of the so-called *second-order kinematic model* of constrained mechanism, which appears in kinematics of nonholonomic systems. This is the manifestation of the fact that the integration of quasi–velocities, in general, does not lead to quasi–coordinates.

3. Constrained Mechanical Systems

3.1 Equations of Motion

In this section, we extend the notion of the quasi–velocity for modeling of constrained mechanical systems where the coordinates are related by a set of m algebraic equations $\boldsymbol{\Phi}(\boldsymbol{q}) = \boldsymbol{0}$. The constraints can be written in the Pfaffian form as

$$\boldsymbol{A}(\boldsymbol{q})\dot{\boldsymbol{q}} = \boldsymbol{0} \tag{16}$$

where Jacobian $\boldsymbol{A} = \partial\boldsymbol{\Phi}/\partial\boldsymbol{q} \in \mathbb{R}^{m\times n}$ is not necessarily a full-rank matrix because of the possible redundant constraints. The EL equations of the constrained MBS with kinetic energy T are

$$\frac{\mathrm{d}}{\mathrm{d}t}\left(\frac{\partial T}{\partial \dot{\boldsymbol{q}}}\right) - \frac{\partial T}{\partial \boldsymbol{q}} = \boldsymbol{f} - \boldsymbol{A}^T\boldsymbol{\lambda}, \tag{17}$$

where $\boldsymbol{\lambda} \in \mathbb{R}^m$ are the *generalized Lagrangian multipliers*.

Using any form of the square–root factorizations in a development similar to (6)-(7), we can show that (17) is equivalent to

$$\dot{\boldsymbol{v}} + \boldsymbol{\Gamma}\boldsymbol{v} = \boldsymbol{u} - \boldsymbol{\Lambda}^T\boldsymbol{\lambda}, \tag{18}$$

where

$$\boldsymbol{\Lambda} \triangleq \boldsymbol{A}\boldsymbol{W}^{-T}. \tag{19}$$

It can be verified that the quasi–velocities satisfy the following Paraffin constraint equation:

$$\boldsymbol{\Lambda}\boldsymbol{v} = \boldsymbol{0}. \tag{20}$$

Also, (20) may suggest that $\boldsymbol{\Lambda}$ be taken as the Jacobian of the constraint with respect to the quasi–coordinates. However, this is true only if the quasi–coordinates ever exist. This means that, in general, system (18) together with (20) most likely constitutes a non-holonomic system even though the configuration–variables q satisfies a holonomic constraint equation.

Since $\boldsymbol{W}$ is a full-rank matrix, we can say $\mathrm{rank}(\boldsymbol{\Lambda}) = \mathrm{rank}(\boldsymbol{A}) = r$, where $r \leq m$ is the number of independent constraints. Then, according to the *singular value decomposition* (SVD) there exist unitary (orthogonal) matrices $\boldsymbol{U} = [\boldsymbol{U}_1 \;\; \boldsymbol{U}_2] \in \mathbb{R}^{m\times m}$ and $\boldsymbol{V} = [\boldsymbol{V}_1 \;\; \boldsymbol{V}_2] \in \mathbb{R}^{n\times n}$ (i.e., $\boldsymbol{U}^T\boldsymbol{U} = \boldsymbol{I}_m$ and $\boldsymbol{V}^T\boldsymbol{V} = \boldsymbol{I}_n$) such that

$$\boldsymbol{\Lambda} = \boldsymbol{U}\boldsymbol{\Sigma}\boldsymbol{V}^T \quad \text{where} \quad \boldsymbol{\Sigma} = \begin{bmatrix}\boldsymbol{S} & \boldsymbol{0}\\ \boldsymbol{0} & \boldsymbol{0}\end{bmatrix} \tag{21}$$

and $\boldsymbol{S} = \text{diag}(\sigma_1, \cdots, \sigma_r)$ with $\sigma_1 \geq \cdots \geq \sigma_r > 0$ being the singular values Klema and Laub (1980); Press *et al.* (1988). The unitary matrices are partitioned so that the dimensions of the submatrices $\boldsymbol{U}_1$ and $\boldsymbol{V}_1$ are consistent with those of $\boldsymbol{S}$. That is the columns of $\boldsymbol{U}_1$ and $\boldsymbol{V}_2$ are the corresponding sets of orthonormal eigenvalues which span the range space and the null space of $\boldsymbol{\Lambda}$, respectively Golub and Loan (1996). As will be seen in the following, derivation of the equations of motion hinges on computing a basis for the kernel of matrix $\boldsymbol{\Lambda}$, which constitute the columns of $\boldsymbol{V}_2$. Fortunately, there are many powerful symbolic algorithms and even commercial softwares to do that Anton (2003).
Now, we take advantage of the arbitration in choosing the square–root factorization to find a particular one that leads to decoupling of the equations of motion and those of constrained force. Consider the unitary transformation (8) where the orthogonal matrix $\boldsymbol{V}$ corresponds to decomposition (21). Then, the equations of motion expressed in terms of the new quasi–variables become

$$\dot{\bar{\boldsymbol{v}}} + \bar{\boldsymbol{\Gamma}}\bar{\boldsymbol{v}} = \bar{\boldsymbol{u}} - \bar{\boldsymbol{\Lambda}}^T\boldsymbol{\lambda}, \tag{22}$$

where $\bar{\boldsymbol{\Lambda}} \triangleq \boldsymbol{\Lambda}\boldsymbol{V}$ and $\bar{\boldsymbol{\Gamma}}$ has been already defined in (11). Again, it can be easily verified that the new quasi–velocities satisfy the following Pfaffian constraints:

$$\bar{\boldsymbol{\Lambda}}\bar{\boldsymbol{v}} = \boldsymbol{0}. \tag{23}$$

At the first glance, the transformed system (22)–(23) reassembles (18)–(20) without gaining any simplification. However, it is the structure of $\bar{\boldsymbol{\Lambda}}$ that will result in further simplification. Using (21) in the definition of $\bar{\boldsymbol{\Lambda}}$ gives

$$\bar{\boldsymbol{\Lambda}} = \begin{bmatrix}\boldsymbol{\Lambda}_r & \boldsymbol{0}_{m\times(n-r)}\end{bmatrix} \quad \text{where} \quad \boldsymbol{\Lambda}_r \triangleq \boldsymbol{U}_1\boldsymbol{S}. \tag{24}$$

Since $\boldsymbol{\Lambda}_r \in \mathbb{R}^{m\times r}$ is a full-rank matrix, it can be inferred from (23) that the first rth elements of the transformed quasi–velocity $\bar{\boldsymbol{v}}$ must be zero. That is,

$$\bar{\boldsymbol{v}} = \begin{bmatrix}\boldsymbol{0}_{r\times 1} \\ \boldsymbol{v}_r\end{bmatrix}, \tag{25}$$

where $\boldsymbol{v}_r \in \mathbb{R}^{n-r}$ represents a set of *reduced quasi–velocities*– in the following, the subscript $_r$ denotes variables associated with the reduced-order variables. Clearly, the zero components of the transformed quasi–velocities are due to the r–independent constraints. It can be verified that (25) is equivalent to

$$\boldsymbol{V}_2^T\boldsymbol{v} = \boldsymbol{v}_r. \tag{26}$$

Now, by using (26) in the reciprocal of relation (6a), we can show that there is a one-to-one correspondence between $\boldsymbol{v}$ and $\dot{\boldsymbol{q}}$ as

$$\dot{\boldsymbol{q}} = \boldsymbol{W}^{-T}\boldsymbol{V}_2\boldsymbol{v}_r, \quad \text{and} \quad \boldsymbol{v}_r = \boldsymbol{V}_2^T\boldsymbol{W}^T\dot{\boldsymbol{q}}. \tag{27}$$

Moreover, by virtue of (25), we partition the quasi–forces accordingly as

$$\bar{\boldsymbol{u}} = \begin{bmatrix}\boldsymbol{u}_o \\ \boldsymbol{u}_r\end{bmatrix}, \quad \text{where} \quad \begin{array}{ll}\boldsymbol{u}_o & \triangleq \boldsymbol{V}_1^T\boldsymbol{W}^{-1}\boldsymbol{f} \\ \boldsymbol{u}_r & \triangleq \boldsymbol{V}_2^T\boldsymbol{W}^{-1}\boldsymbol{f}\end{array}. \tag{28}$$

In addition, we assume that matrix $\bar{\boldsymbol{\Gamma}}$ is divided into four block matrices

$$\bar{\boldsymbol{\Gamma}}_{ij} = \boldsymbol{V}_i^T(\boldsymbol{\Gamma} - \boldsymbol{\Omega})\boldsymbol{V}_j, \qquad i,j = 1,2, \tag{29}$$

and then define

$$\boldsymbol{\Gamma}_r \triangleq \bar{\boldsymbol{\Gamma}}_{22}, \quad \text{and} \quad \boldsymbol{\Gamma}_o \triangleq \bar{\boldsymbol{\Gamma}}_{12}. \tag{30}$$

Now, substituting (25) into (22) and then using definitions (28) and (29), we arrive at

$$\dot{\boldsymbol{v}}_r + \boldsymbol{\Gamma}_r \boldsymbol{v}_r = \boldsymbol{u}_r, \tag{31a}$$

and

$$\boldsymbol{\Lambda}_r^T \boldsymbol{\lambda} + \boldsymbol{\Gamma}_o \boldsymbol{v}_r = \boldsymbol{u}_o \tag{31b}$$

Apparently, (31a) and (31b) represent the equations of motion and those of constraint force which are completely decoupled from each other. Note that the partitioned components of the quasi–forces, i.e., $\bar{\boldsymbol{u}}_r$ and $\bar{\boldsymbol{u}}_o$, contribute exclusively to the motion system and the constraint force system, respectively. Now, we are ready to combine (31a) and (27) into the state–space form:

$$\frac{\mathrm{d}}{\mathrm{d}t}\begin{bmatrix} \boldsymbol{q} \\ \boldsymbol{v}_r \end{bmatrix} = \begin{bmatrix} \boldsymbol{W}^{-T}\boldsymbol{V}_2 \\ -\boldsymbol{\Gamma}_r \end{bmatrix} \boldsymbol{v}_r + \begin{bmatrix} \boldsymbol{0} \\ \boldsymbol{I} \end{bmatrix} \boldsymbol{u}_r. \tag{32}$$

The above equation can be viewed as the special case of Kane's equations Kane (1961); Kane and Levinson (1985) where all particles have unit mass.

The Lagrangian multipliers can be uniquely obtained from (31b) through matrix inversion only if $r = m$, i.e., in the presence of no redundant constraints. Otherwise, there are fewer equations than unknowns, and hence there is no unique solution to (31b). Nevertheless, the minimum norm solution can be found by

$$\min \|\boldsymbol{\lambda}\| \quad \leftarrow \quad \boldsymbol{\lambda} = \boldsymbol{U}_1 \boldsymbol{S}^{-1}(\boldsymbol{u}_o - \boldsymbol{\Gamma}_o \boldsymbol{v}_r). \tag{33}$$

3.2 Calculating the Coriolis Term

The Coriolis force term $\boldsymbol{\Gamma}_r$ itself characterized completely the motion dynamics of a constrained mechanical system expressed by reduced quasi–velocities. In this section, we describe $\boldsymbol{\Gamma}_r$ expressed in terms of $\boldsymbol{v}_r$ that appears to be simpler than (30). First, in view of (5) and the facts that $\boldsymbol{v} = \boldsymbol{V}_2 \boldsymbol{v}_r$ and $\|\boldsymbol{v}_r\| = \|\boldsymbol{v}\|$, one can verify that

$$\frac{\partial \boldsymbol{v}_r}{\partial \boldsymbol{q}} = \frac{\partial \boldsymbol{v}}{\partial \boldsymbol{q}} \boldsymbol{V}_2. \tag{34}$$

Now, consider the relation between $\boldsymbol{v}_r$ and $\dot{\boldsymbol{q}}$ as

$$\boldsymbol{v}_r = \boldsymbol{W}_r^T(\boldsymbol{q})\dot{\boldsymbol{q}},$$

where $\boldsymbol{W}_r = \boldsymbol{W}\boldsymbol{V}_2$. Then, from (7b), (10), (30), and (35) we obtain

$$\begin{aligned} \boldsymbol{\Gamma}_r &= \boldsymbol{V}_2^T \boldsymbol{W}^{-1}(\dot{\boldsymbol{W}} - \frac{\partial \boldsymbol{v}^T}{\partial \boldsymbol{q}})\boldsymbol{V}_2 + \boldsymbol{V}_2 \dot{\boldsymbol{V}}_2 \\ &= \boldsymbol{V}_2^T \boldsymbol{W}^{-1}(\dot{\boldsymbol{W}}_r - \boldsymbol{W}\dot{\boldsymbol{V}}_2 - \frac{\partial \boldsymbol{v}^T}{\partial \boldsymbol{q}}\boldsymbol{V}_2) + \boldsymbol{V}_2^T \dot{\boldsymbol{V}}_2 \\ &= \boldsymbol{V}_2^T \boldsymbol{W}^{-1}(\dot{\boldsymbol{W}}_r - \frac{\partial \boldsymbol{v}_r^T}{\partial \boldsymbol{q}}). \end{aligned} \tag{35}$$

Finally, by noting that $\boldsymbol{V}_2^T \boldsymbol{W}^{-1} = \boldsymbol{W}_r^+$ is a left inverse of $\boldsymbol{W}_r$, that is, $\boldsymbol{W}_r^+ \boldsymbol{W}_r = \boldsymbol{I}$, we can express (35) by

$$\boldsymbol{\Gamma}_r = \boldsymbol{W}_r^+ (\dot{\boldsymbol{W}}_r - \frac{\partial \boldsymbol{v}_r^T}{\partial \boldsymbol{q}}), \tag{36}$$

which closely resembles the Coriolis term of unconstrained mechanical systems in (7b). It is interesting to note that $\boldsymbol{W}_r \in \mathbb{R}^{n\times(n-r)}$ can be thought of as the factorization of the semi-positive "mass matrix" $\boldsymbol{M}_r = \boldsymbol{W}_r\boldsymbol{W}_r^T = \boldsymbol{W}\boldsymbol{P}\boldsymbol{W}^T$, where $\boldsymbol{P} = \boldsymbol{V}_2\boldsymbol{V}_2^T$ is a projection matrix which projects vectors from $\mathbb{R}^n$ to the null space of system (20). A comparison between systems (31a)–(36) and (7) reveals that the formulation of constrained mechanical systems remains essentially similar to that of unconstrained mechanical systems if the quasi–velocity is simply replaced by a reduced quasi–velocity.
Finally, a development similar to (35) shows that

$$\boldsymbol{\Gamma}_o = \boldsymbol{W}_r^-\big(\dot{\boldsymbol{W}}_r - \frac{\partial \boldsymbol{v}_r^T}{\partial \boldsymbol{q}}\big),$$

where $\boldsymbol{W}_r^- = \boldsymbol{V}_1^T\boldsymbol{W}^{-1}$ is an annihilator for $\boldsymbol{W}_r$, i.e., $\boldsymbol{W}_r^-\boldsymbol{W}_r = \boldsymbol{0}$.

4. Force/Motion Control

In general, it should be always possible to choose a minimal set of independent velocity coordinate, equal in number of the degrees-of-freedom (DOF) exhibited by the mechanical system. However, a minimal set of independent generalized coordinates may not exist; a well-known example is the orientation configuration of a rigid-body that can not be expressed by a three-dimensional vector. However, the conventional control of constrained mechanical system relies on the existence of a minimal set of parameters defining the configuration of a constrained MBS. In this section, we provide velocity and position feedbacks from (reduced) quasi–velocities and (dependent) configuration variables, respectively, for tracking control and regulating a constrained MBS. Interestingly enough, the control challenge, then, becomes similar to that of non-holonomic systems, as the configuration of MBS can not be represented by any quasi–coordinates.

4.1 Properties

First, we explore some properties of system (31) that will be useful in control design purposes.

Remark 2. *Using* (13) *and the fact that* $\boldsymbol{\Omega}$ *is a skew-symmetric matrix in definition* (29), *we can say*

$$\boldsymbol{v}_r^T\boldsymbol{\Gamma}_r\boldsymbol{v}_r = 0.$$

4.2 Tracking Control

Due to presence of only r independent constraints, the actual number of degrees of freedom of the system is reduced to $n-r$. Thus, in principle, there must be $n-r$ independent variables $\boldsymbol{\theta}(\boldsymbol{q}) \in \mathbb{R}^{n-r}$, which is also called a *minimal set of generalized coordinates*. In view of the time-derivative of the minimal set of generalized coordinates, $\frac{\mathrm{d}}{\mathrm{d}t}\boldsymbol{\theta}(\boldsymbol{q})$, and (27), we get

$$\dot{\boldsymbol{\theta}} = \boldsymbol{B}(\boldsymbol{\theta})\boldsymbol{v}_r, \qquad \text{where} \qquad \boldsymbol{B} \triangleq \frac{\partial \boldsymbol{\theta}}{\partial \boldsymbol{q}}\boldsymbol{W}^{-T}(\boldsymbol{q})\boldsymbol{V}_2. \tag{37}$$

Since both velocities variables $\boldsymbol{v}_r$ and $\dot{\boldsymbol{\theta}}$ are with the same dimension, the reciprocal of mapping (37) must uniquely exist, i.e., $\boldsymbol{v}_r = \boldsymbol{B}^{-1}(\boldsymbol{\theta})\dot{\boldsymbol{\theta}}$.
We adopt a Lyapunov-based control scheme (Canudas de Wit *et al.*, 1996, p. 74) for designing a feedback control in terms of quasi–velocities. Define the composite error

$$\boldsymbol{\epsilon} \triangleq \tilde{\boldsymbol{v}}_r + \boldsymbol{B}^{-1}(\boldsymbol{\theta})\boldsymbol{K}_p\tilde{\boldsymbol{\theta}}, \tag{38}$$

where $\boldsymbol{K}_p > 0$, $\tilde{\boldsymbol{v}}_r = \boldsymbol{v}_r - \boldsymbol{v}_{r_d}$, and $\tilde{\boldsymbol{\theta}} = \boldsymbol{\theta} - \boldsymbol{\theta}_d$. Also, define the new variable as $\boldsymbol{s} = \boldsymbol{v}_{r_d} - \boldsymbol{B}^{-1}\boldsymbol{K}_p\tilde{\boldsymbol{\theta}}$, which is used in the following control law:

$$\boldsymbol{u}_r = \dot{\boldsymbol{s}} + \boldsymbol{\Gamma}_r\boldsymbol{s} - \boldsymbol{K}_d\boldsymbol{\epsilon}, \tag{39}$$

where $\boldsymbol{K}_d > 0$. Applying control law (39) to system (31a) gives the dynamics of the error ϵ in terms of the first-order differential equation:

$$\dot{\boldsymbol{\epsilon}} = -(\boldsymbol{\Gamma}_r + \boldsymbol{K}_d)\boldsymbol{\epsilon}. \tag{40}$$

As shown in Appendix B, the solution of (40) is bounded by

$$\|\boldsymbol{\epsilon}\| \leq \|\boldsymbol{\epsilon}(0)\| e^{-\eta_1 t}, \tag{41}$$

where $\eta_1 = 2\lambda_{\min}(\boldsymbol{K}_d)$, and hence the composite error $\boldsymbol{\epsilon}$ is exponentially stable.
Pre-multiplying both sides of (38) by $\boldsymbol{B}(\boldsymbol{\theta})$, the resultant equation can be rearranged to the following differential equation

$$\dot{\tilde{\boldsymbol{\theta}}} = -\boldsymbol{K}_p\tilde{\boldsymbol{\theta}} + \big(\boldsymbol{B}(\boldsymbol{\theta}) - \boldsymbol{B}(\boldsymbol{\theta}_d)\big)\boldsymbol{v}_d + \boldsymbol{B}(\boldsymbol{\theta})\boldsymbol{\epsilon}. \tag{42}$$

Now, it remains to show that the solution of the above non-autonomous system converges to zero. We assume that the matrix function $\boldsymbol{B}(\boldsymbol{\theta})$ is bounded and sufficiently smooth so that it satisfies the Lipschitz condition, i.e., there exists a finite scalar $c_l > 0$ such that

$$\|\boldsymbol{B}(\boldsymbol{\theta}) - \boldsymbol{B}(\boldsymbol{\theta}^*)\| \leq c_l\|\boldsymbol{\theta} - \boldsymbol{\theta}^*\| \qquad \forall \boldsymbol{\theta}, \boldsymbol{\theta}^* \in \mathbb{R}^{n-r}. \tag{43}$$

Furthermore, there exists scalar $c_b > 0$ such that

$$\boldsymbol{B}(\boldsymbol{\theta}) \leq c_b\boldsymbol{I}. \tag{44}$$

Assuming that the command velocity is bounded, i.e., $\|\boldsymbol{v}_{r_d}\| \leq c_v$, we can show that the solution of the above differential equation satisfies

$$\|\tilde{\boldsymbol{\theta}}\| \leq \|\tilde{\boldsymbol{\theta}}(0)\| e^{-\eta_2 t} + \frac{c_b}{c_l c_v}\|\boldsymbol{\epsilon}(0)\| \big(e^{-\eta_1 t} - e^{-\eta_2 t}\big), \tag{45}$$

where $\eta_2 = \lambda_{\min}(\boldsymbol{K}_p) - c_l c_v$; see Appendix C for details. Equation (45) implies exponential stability of error $\|\tilde{\boldsymbol{\theta}}\|$ provided that $\eta_2 > 0$, i.e.,

$$\lambda_{\min}(\boldsymbol{K}_p) > c_l c_v. \tag{46}$$

The above development can be summarized in the following theorem.

Theorem 1. *Assume that the mass matrix factorization is a smooth function satisfying the Lipschitz condition and that $\|\boldsymbol{v}_d\|$ is bounded. Then, for a sufficiently large position gain, i.e., (46) is satisfied, the error trajectories of the configuration-variables and quasi–velocities of a constrained MBS under control law* (38)–(39) *exponentially converge to zero.*

Tracking of the desired constraint force $\boldsymbol{\lambda}_d$ can be achieved simply by compensating for the velocity perturbation term in (31b), i.e.,

$$\boldsymbol{u}_o = \boldsymbol{\Lambda}_r^T\boldsymbol{\lambda}_d + \boldsymbol{\Gamma}_o\boldsymbol{v}_r. \tag{47}$$

It is worth noting that in view of the norm identity $\|\boldsymbol{u}\|^2 = \|\boldsymbol{u}_r\|^2 + \|\boldsymbol{u}_o\|^2$, we cay say that $\|\boldsymbol{u}\|$ is minimum if $\boldsymbol{u}_0 \equiv \boldsymbol{0}$. That is tantamount to minimization of weighted norm of the generalized forces where the weight matrix is the inertia matrix because $\|\boldsymbol{u}\|^2 = \boldsymbol{f}^T\boldsymbol{M}(\boldsymbol{q})\mathbf{f}$. Interestingly enough, not additional weighing matrix is required even if the elements of the generalized contains both force and torque components.

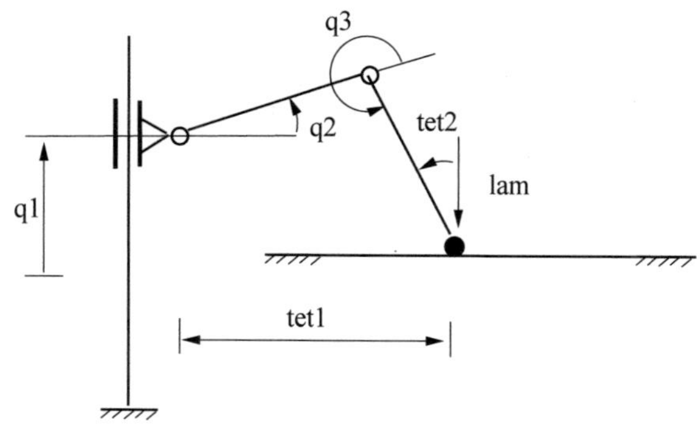

Fig. 1. A constrained 3-PRR manipulator.

4.3 Gauge Invariant

A problem that often arises in robotics, namely hybrid control or minimum solution to joint rate or force, is that generalized coordinate $\boldsymbol{q}$ may have a combination of rotational and translational components that can be even compounded by having combination of rotational and translational constraints Doty *et al.* (1993*b*). This may lead to inconsistent results, i.e., results that are invariant with respect to changes in dimensional units unless adequate weighting matrixes are used Aghili (2005); Doty *et al.* (1993*b*); Featherstone and Fijany (1999); Featherstone *et al.* (1999); Manes (1992). For example, the minimum joint rate rates, min $\|\dot{\boldsymbol{q}}\|$, or minimum norm force, min $\|\boldsymbol{f}\|$, are not meaningful quantities if the robot has both revolute and prismatic joints Doty *et al.* (1993*b*).

An important property of the reduced quasi–velocities and quasi–forces is that they always have homogenous units. As a matter of fact, since

$$\|\boldsymbol{v}_r\| = \|\boldsymbol{v}\| = \sqrt{2T},$$

we can say that all elements of the vector of quasi–velocity $\boldsymbol{v}$ or $\boldsymbol{v}_r$ must have a homogenous unit $[\sqrt{\text{kg}}\text{m/s}]$. This is true even if the vector of the generalized coordinate or the constraints have combinations of rotational and translational components. Similarly, one can argue that the elements of the quasi–forces have always identical unit $[\sqrt{\text{kg}}\text{m/s}^2]$, regardless of the units of the generalized force of the constraint wrench. Therefore, minimization of $\|\boldsymbol{v}\|$ or min $\|\boldsymbol{u}\|$ is legitimate because the latter vectors have always homogeneous units. Moreover, the selection matrices which are often needed in hybrid position-force control of manipulators when both translational and rotational constraints are involved between its end effector and its environment Featherstone *et al.* (1999)—and that yields a problem with gauge invariance—becomes a non-issue here.

5. Analytical Example

Fig. 1 illustrates a PRR manipulator, with one prismatic and two revolute joints. The vertical motion of the manipulator tip-point is constrained by a solid surface. The prismatic joint provides the vertical motion of the robot base, which is with mass of m. Clearly, the vectors of generalized coordinates, $\boldsymbol{q}$, and generalized force, $\boldsymbol{f}$, have inhomogeneous components. We assume that each link is uniform with length of l and mass of m. Then, the constraint Jacobian can be expressed by

$$\boldsymbol{A}(\boldsymbol{q}) = \begin{bmatrix} 1 & l(c_{23} + c_2) & lc_{23} \end{bmatrix},$$

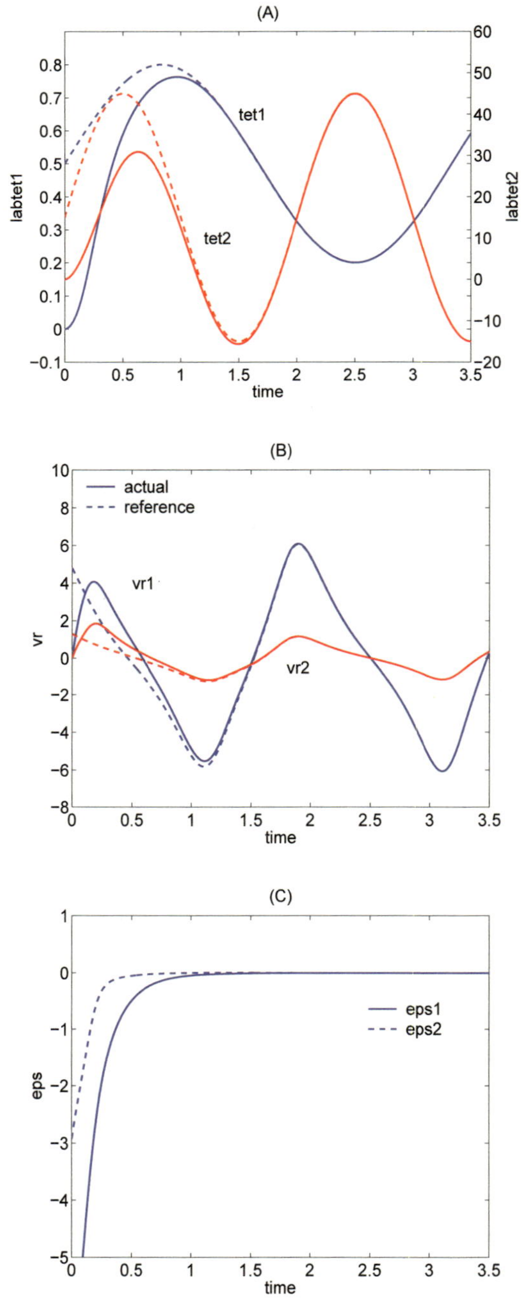

Fig. 2. Simulated motion tracking.

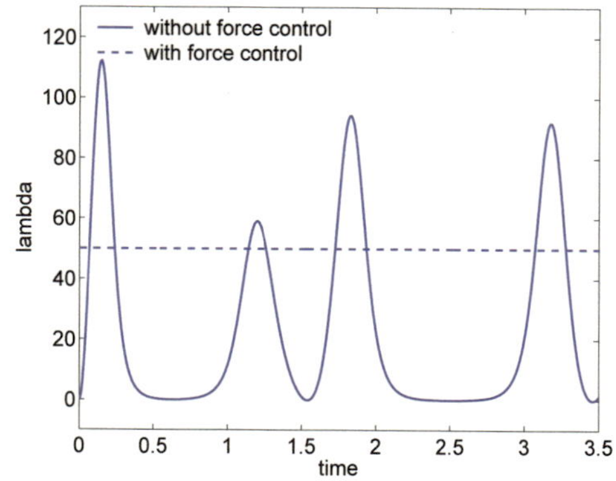

Fig. 3. Simulated constrained force.

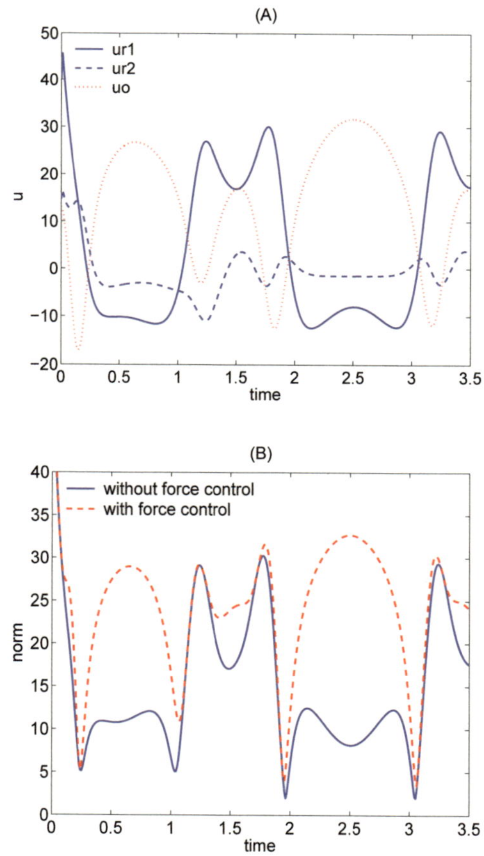

Fig. 4. Trajectories of the quasi–forces.

where $c_{23} = \cos(q_2 + q_3)$ and $c_2 = \cos(q_2)$.
Let us define the minimal set generalized coordinates as $\boldsymbol{\theta} = [\theta_1 \;\; \theta_2]^T$ with θ_1 and θ_2 being the horizontal location of the tip and the angle of the last link with respect to the vertical line, respectively; see Fig. 1. Then from the kinematics, we get

$$\boldsymbol{\theta}(\boldsymbol{q}) = \begin{bmatrix} lc_{23} + lc_2 \\ q_2 + q_3 - \frac{3\pi}{2} \end{bmatrix}$$

Now assume that the control objective is to track the following desired trajectories

$$\theta_{1_d}(t) = 0.3\sin(0.6\pi t) + 0.5 \quad \text{(m)}$$
$$\theta_{2_d}(t) = \frac{\pi}{6}\sin(\pi t) + \frac{\pi}{12} \quad \text{(rad)}$$

Figs. 2A and 2B show the actual and desired trajectories of the position and quasi-velocities when the quasi-velocity feedback (38)-(39) is applied for the following parameters: $m = 5$ kg, $l = 1$ m, $\boldsymbol{K}_p = 3\boldsymbol{I}$, and $\boldsymbol{K}_d = 5\boldsymbol{I}$. The time-history of the composite error, $\boldsymbol{\epsilon}$, shown in Fig. 2C demonstrates tracking of the reference motion trajectory. Fig.3 illustrates trajectories of the constraint force, λ, for two cases: i) no force control is applies, ii) the force control law (47) is applied to achieve

$$\lambda_d = 50 \qquad \text{(N)}.$$

Trajectories of the corresponding motion input, $\boldsymbol{u}_r$, and force input, u_o, components of the quasi–forces are shown in Fig. 4A. Trajectories of the Euclidean norm the quasi–forces with and without force control are illustrated in the bottom of Fig. 4B. Clearly, the quasi–forces norm is automatically minimized norm if the force control input, u_o is set to zero. It is worth noting that the norm of quasi-forces is an invariant quantity even though the vector of generalized force has both force and torque components.

Appendix A

According to the *Cholesky decomposition*, a symmetric and positive-definite matrix $\boldsymbol{M}$ can be decomposed efficiently into $\boldsymbol{M} = \boldsymbol{L}\boldsymbol{L}^T$, where $\boldsymbol{L}$ is a lower–triangular matrix with strictly positive–diagonal elements; $\boldsymbol{L}$ is also called the *Cholesky triangle*. The Cholesky decomposition is a particular case of the well–known LU decomposition for symmetric matrices. Nevertheless, the Cholesky decomposition is twice as efficient as the LU decomposition. The following formula can be used to obtain the Cholesky triangle through some elementary operations

$$l_{ii} = \big(m_{ii} - \sum_{k=1}^{i-1} l_{ik}^2\big)^{1/2} \qquad \forall i = 1, \cdots, n \tag{48}$$
$$l_{ji} = \big(m_{ji} - \sum_{k=1}^{i-1} l_{jk} l_{ik}\big)/l_{ii} \qquad \forall j = i+1, \cdots, n$$

Since $\boldsymbol{L}$ is a lower-triangular matrix, its inverse can be simply computed by the back substitution technique.

Appendix B

Consider the following positive–definite function:

$$V = \frac{1}{2}\|\boldsymbol{\epsilon}\|^2$$

In view of Remark 2, the time-derivative of the above function along the error trajectory (40) is obtained as

$$\begin{aligned}\dot{V} &= -\boldsymbol{\epsilon}^T\boldsymbol{\Gamma}_r\boldsymbol{\epsilon} - \boldsymbol{\epsilon}^T\boldsymbol{K}_d\boldsymbol{\epsilon}\\ &= -\boldsymbol{\epsilon}^T\boldsymbol{K}_d\boldsymbol{\epsilon}\end{aligned}$$

which gives

$$\dot{V} \leq -2\lambda_{\min}(\boldsymbol{K}_d)V.$$

Thus

$$V \leq V(0)e^{-2\lambda_{\min}(K_d)t},$$

which is equivalent to (41).

Appendix C

Consider the following positive–definite function

$$V = \frac{1}{2}\|\tilde{\boldsymbol{\theta}}\|^2, \tag{49}$$

whose time–derivative along (42) gives

$$\dot{V} = -\tilde{\boldsymbol{\theta}}^T\boldsymbol{K}_p\tilde{\boldsymbol{\theta}} + \tilde{\boldsymbol{\theta}}^T\boldsymbol{B}(\boldsymbol{\theta})\boldsymbol{\epsilon} + \tilde{\boldsymbol{\theta}}^T\big(\boldsymbol{B}(\boldsymbol{\theta}) - \boldsymbol{B}(\boldsymbol{\theta}_d)\big)\boldsymbol{v}_d.$$

From (44) and (43), we can find a bound on $\dot{V}$ as

$$\begin{aligned}\dot{V} &\leq -\lambda_{\min}(\boldsymbol{K}_p)\|\tilde{\boldsymbol{\theta}}\|^2 + c_b\|\tilde{\boldsymbol{\theta}}\|\,\|\boldsymbol{\epsilon}\| + c_l c_v\|\tilde{\boldsymbol{\theta}}\|^2\\ &\leq -2\eta_2 V + c_b\sqrt{2V}\|\boldsymbol{\epsilon}\|,\end{aligned} \tag{50}$$

which is in the form of a Bernoulli differential inequality. The above nonlinear inequality can be linearized by the following change of variable $U = \sqrt{V}$, i.e.,

$$\dot{U} \leq -\eta_2 U + \frac{c_b}{\sqrt{2}}\|\boldsymbol{\epsilon}(0)\|e^{-\eta_1 t} \tag{51}$$

In view of the comparison lemma (Khalil, 1992, p. 222) and (41), one can show that the solution of (51) must satisfy

$$U \leq U(0)e^{-\eta_2 t} + \frac{c_b\|\boldsymbol{\epsilon}(0)\|}{\sqrt{2}}\int_0^t e^{-\eta_2(t-\tau)-\eta_1\tau}d\tau,$$

which is equivalent to (45).

6. References

Aghili, F. (2005). A unified approach for inverse and direct dynamics of constrained multibody systems based on linear projection operator: Applications to control and simulation. *IEEE Trans. on Robotics* **21**(5), 834–849.

Aghili, F. (2008). A gauge-invariant formulation for constrained robotic systems using square-root factorization and unitary transformation. In: *IEEE/RSJ Int. Conf. on Intelligent Robots and Systems*. Nice, France. pp. 2814–2821.

Aghili, F. (2009). A gauge-invariant formulation for constrained mechanical systems using square-root factorization and unitary transformation. *ASME Journal of Computational and Nonlinear Dynamics*, Vol. 4, No. 3.

Aghili, Farhad (2007). Simplified lagrangian mechanical systems with constraints using square-root factorization. In: *Multibody Dynamics 2007, ECCOMAS Thematic Conference*. Milano, Italy.

Angeles, J. (2003). A methodology for the optimum dimensioning of robotic manipulators. Memoria del 5o. Congreso Mexicano de Robótica. UASLP, San Luis Potosí, SLP, México. pp. 190–203.

Anton, H. (2003). *Contemporary Linear Algebra with Maple Manual Set*. Wiley.

Bar-Itzhack, I. Y. (1989). Extension of the euler's theorem to n-dimensional spaces. *IEEE Trans. on Aerospace and Electronics* **25**(6), 903–909.

Baruh, H. (1999). *Analytical Dynamics*. McGraw Hill. London.

Bedrossian, N. S. (1992). Linearizing coordinate transformation and riemann curvature. In: *IEEE Int. Conf. on Decision and Control*. Tucson, Arizona. pp. 80–4.

Canudas de Wit, Carlos, Siciliano, Bruno and Bastin, Georges, Eds. (1996). *Theory of Robot Control*. Springer. London, Great Britain.

Corben, H. C. and P. Stehle (1960). *Classical Mechanics*. John Wiley & Sons.

Doty, K. L., C. Melchiorri and C. Bonivento (1993*a*). A theory of generalized inverses applied to robotics. *The International Journal of Robotics Research* **12**(1), 1–19.

Doty, Keith L., Claudio Melchiorri and Claudio Bonivento (1993*b*). A theory of generalized inverses applied to robotics. *The International Journal of Robotics Research* **12**(1), 1–19.

Featherstone, R. and A. Fijany (1999). A technique for analyzing constrained rigid-body systems, and its application to constraint force algorithm. *IEEE Trans. on Robotics & Automation* **15**(6), 1140–1144.

Featherstone, R., S.S. Thiebaut and O. Khatib (1999). A general contact model for dynamically-decoupled force-motion control. In: *IEEE International Conference on Robotics & Automation*. Detroit, Michigan. pp. 3281–3286.

Golub, G. H. and C. F. Van Loan (1996). *Matrix Computations*. The Johns Hopkins University Press. Baltimore and London.

Gu, E. Y. L. (2000). A configuration manifold embedding model for dynamic control of redundant robots. *The Int. Journal of Robotics Research* **19**(3), 289–304.

Gu, Y. L. and N. K. Loh (1987). Control system modeling by using of a canonical transformation. In: *IEEE Conference on Decision and Control*. Lauderdale, FL. pp. 1–4.

Herman, P. (2005). PD controller for manipulator with kinetic energy term. *Journal of Intelligent Robotic Systems* **44**, 101–121.

Herman, P. and K. Kozlowski (2001). Some remarks on two quasi-velocities approaches in PD joint space control. In: *IEEE/RSJ Int. Conf. On Intelligent Robots and Systems*. Maui, Hawaii, USA. pp. 1888–1893.

Herman, P. and K. Kozlowski (2006). A survey of equations of motion in terms of inertial quasi-velocities for serial manipulators. *Arch. Appl. Mech.* **76**, 579–614.

Jain, A. and G. Rodriguez (1995). Diagonalized lagrangian robot dynamics. *IEEE Trans. on Robotics & Automation* **11**(4), 571–584.

Junkins, John L. and Hanspeter Schaub (1997). An instantaneous eigenstructure quasivelocity formulation for nonlinear multibody dunamics. *The Journal of Astronautical Sciences* **45**(3), 279–295.

Kane, T. R. (1961). Dynamics of nonholonomic systems. *Transactions of the ASME Journal of Applied Mechanics* **28**, 574–578.

Kane, Thomas R. and David A. Levinson (1985). *Dynamics: Theory and Applications*. McGraw-Hill series in Mechanical Engineering. McGraw-Hill Book Company. New York, NY.

Khalil, Hassan K. (1992). *Nonlinear Systems*. Macmillan Publishing Company. New-York.

Klema, V. C. and A. J. Laub (1980). The singular value decomposition: its computation and some applictions. *IEEE Trans. on Automatic Control* **25**(2), 164–176.

Kodischeck, D. E. (1985). Robot kinematics and coordinate transformation. In: *IEEE Int. Conf. on Decision and Control*. Lauderdale, FL. pp. 1–4.

Kozlowski, K. (1998). *Modelling and Identification in Robotics*. Springer–Verlag. London.

Kozolowski, K. and P. Herman (2000). A comparision of conrtol algorithm for serial manipulators in terms of quasi-velocity. In: *IEEE/RSJ Int. Conf. on Intelligent Robots & Systems*. Takamatsu, Japan. pp. 1540–1545.

Lipkin, H. and J. Duffy (1988). Hybrid twist and wrench control for a robotic manipulator. *Trans. ASME J. of Mechanism, Transmission, and Automation in Design* **110**, 138–144.

Loduha, T. A. and B. Ravani (1995). On first-order decoupling of equatrions of motion for constrained dynamical systems. *ASME Journal of Applied Mechanics* **62**, 216–222.

Luca, A. De and C. Manes (1994). Modeling of robots in contact with a dynamic environment. *IEEE Trans. on Robotics & Automation* **10**(4), 542–548.

Manes, C. (1992). Recovering model consistence for force and velocity measures in robot hybrid control. In: *IEEE Int. Conference on Robotics & Automation*. Nice, France. pp. 1276–1281.

Meirovitch, L. (1970). Rigid body dynamics. *Methods of Analytical Dynamics* pp. 157–160.

Oshman, Y. and I. Bar-Itzhack (1985). Eigenfactor solution of the matrix riccati equation–a continous square root algorithm. *IEEE Trans. on Automatic Control* **AC-30**(10), 971–978.

Papastavridis, J. G. (1998). A panoramic overview of the principles and equations of motion of advanced engineering dynamics. *ASME Applied Mechanics Reviews* **51**(4), 239–265.

Press, W. H., B. P. Flannery, S. A. Teukolsky and W. T. Vetterling (1988). *Numerical Recipies in C: The Art of Scientific Computing*. Cambridge University Press. NY.

Rodriguez, G. and K. Kertutz-Delgado (1992). Spatial operator facorization and inversion of manipulator mass matrix. *IEEE Trans. on Robotics & Automation* **8**(1), 65–76.

Schaub, Hanspeter, Panagiotis Tsiotras and John L. Junkins (1995). Principal rotation representations of proper NxN orthogonal matrices. *International Journal of Engineering Sciences.* **33**(2), 2277–2295.

Schutter, J. De and H. Bruyuinckx (1996). *The Control Handbook*. Chap. Force Control of Robot Manipulators, pp. 1351–1358. CRC. New York.

Sinclair, A. J., J. E. Hurtado and J. L. Junkins (2006). Linear feedback control using quasi velocities. *Journal of Guidance, Control, and Dynamics* **29**(6), 1309–1314.

Spong, M. W. (1992). Remarks on robot dynamics: Canonical transformations and riemannian geometry. In: *IEEE Int. Conference on Robotics & Automation*. Nice, France. pp. 554–559.

4

Practical Model-based and Robust Control of Parallel Manipulators Using Passivity and Sliding Mode Theory

Houssem Abdellatif, Jens Kotlarski, Tobias Ortmaier and Bodo Heimann
houssem.abdellatif@imes.uni-hannover.de
jens.kotlarski@imes.uni-hannover.de
tobias.ortmaier@imes.uni-hannover.de
bodo.heimann@imes.uni-hannover.de
Institute of Mechatronic Systems, Leibniz University of Hannover, Appelstr. 11a, D-30167 Hannover, Germany

Abstract
This chapter provides a practical strategy to realize accurate and robust control for 6 DOFs (degrees of freedom) parallel robots. The presented approach consists in two parts. The first basic part is based on the the compensation of the desired dynamics in combination with controller/observer for the single actuators. The passivity formalism offers an excellent framework to design and to tune the closed-loop dynamics, such that the desired behavior is obtained. The basic algorithm is proved to be locally robust towards uncertainties. The second part of the control strategy consists in a sliding mode controller. To keep the practical and computational efficient implementation, the proposed switching control considers explicitly only the friction model. Here we opt for the so called model-decomposition paradigm and we use additional integral action to improve robustness. The proposed approach is substantiated with experimental results demonstrating the effectiveness and success of the strategy that keeps control setup simple and intuitive.

Keywords parallel manipulators, robust control, passivity formalism, sliding mode control, desired dynamics compensation, velocity observer

1. Introduction

Due to their complexity, the practical control of parallel kinematic manipulators is challenging. The missing of appropriate control strategies plays a key role such that the promising potentials of such machines, like high dynamics and high accuracy could not be exploited satisfactorily in practice. Speaking about practical is speaking about control approaches that respect computational limitation of conventional control systems and do not require additional hardware setups, like external sensors or additional actuators.

The proposed chapter presents a complete control strategy that is suitable for parallel manipulators and that is robust to different sources of uncertainties. The issue of robust control in robotics is not quite new and has been addressed by different authors since more than two

decades. Fundamental works have been presented in (Abdallah *et al.*, 1991; Qu and Dawson, 1996). For the present study two families of robust controller are interesting: linear high gain controller, known to provide local uniform ultimate boundedness Berghuis and Nijmeijer (1994); Egeland (1987); Qu and Dawson (1996) and nonlinear structure variable or switching controller that can guarantee global stability Liu and Goldenberg (1996); Spong (1992). Even if the fundamentals of robust control have been already elaborated, their practical implementation in the industrial field has been barely considered. This is especially the case for 6 DOFs parallel robots, that are more complex and suffer especially from uncertainties due to the high coupled structure Kim *et al.* (2000). For that reason we try to close this practical gap by proposing a closed concept for the robust control of parallel manipulators.
The core of the scheme consists in the feedforward compensation of the inverse dynamics. Such type of compensation is preferable then the feedback one, since the latter requires the measurement or at least the precise knowledge of the endeffector's pose, translational and angular velocities, which is not easy to manage without additional and expensive sensors (Abdellatif and Heimann, 2007; Abdellatif *et al.*, 2005; Burdet *et al.*, 2000; Kim *et al.*, 2005; Wang and Ghorbel, 2006). The feedback controllers of the single actuators are kept linear and simple to avoid additional computational effort. The necessity of velocity error feedback for the typical stabilizing control of robotic systems is avoided by using observers of actuator's velocities Berghuis (1993); Burdet *et al.* (2000). The latter are also kept linear. The simultaneous design of controller/observer pairs is achieved by means of the passivity formalism. Both elements are tuned up, such that the closed-loop is robust against parametric uncertainties of the implemented inverse dynamics model and against the use of feedforward compensation as such, that introduces systematic errors into the control loop.
We demonstrate in this paper that the combination of desired dynamics compensation and linear robust control provides exponential ultimate boundedness. Nevertheless such approach remains conservative in the way that it demands higher feedback the more uncertainties affect the system. High feedback is limited in practice by the actuation constraints. We propose therefore to keep this basic scheme to encounter systematic or small parametric uncertainties, like those of the rigid-body model and to augment the scheme with sliding mode control. To keep the practical and computational efficient implementation, the proposed switching control considers explicitly only the friction model that is known to be more affected by time-varying uncertainties. Here we opt for the so called model-decomposition paradigm and we use additional integral action to improve robustness (Liu and Goldenberg, 1996).
The control approach is substantiated by a multitude of experimental results achieved on a directly actuated 6-DOFs parallel manipulator and by using a commercial control system. It is shown that the proposed strategy is highly appropriate to achieve high tracking accuracy at high dynamics, exploiting therefore the benefits of parallel manipulators in a practical way.
The chapter is organized as follows. Section 2 provides the reader with a preliminary discussion on the challenges that faces the control of complex parallel manipulators. Section 3 is dedicated to the passivity-based design of the control algorithm. Afterwards and in section 4, this algorithm is augmented with a sliding mode part to enhance robustness and accuracy. Section 5 provides experimental results, that substantiate the proposed control strategy.

2. Motivation and Preliminaries

2.1 Motivation for the proposed controller

Classically, the majority of model-based controller in robotics have been derived based on the famous equations of motion for Euler-Lagrangian mechanical systems:

$$\boldsymbol{\tau} = \boldsymbol{M}(\boldsymbol{z})\ddot{\boldsymbol{s}} + \boldsymbol{C}(\boldsymbol{z}, \dot{\boldsymbol{s}})\dot{\boldsymbol{s}} + \boldsymbol{g}(\boldsymbol{z}), \tag{1}$$

with $\boldsymbol{\tau}$, $\boldsymbol{z}$, $\dot{\boldsymbol{s}}$ and $\ddot{\boldsymbol{s}}$ being the generalized forces, coordinates, velocities and accelerations, respectively. $\boldsymbol{M}$, $\boldsymbol{C}$ and $\boldsymbol{g}$ consist in the positive definite and symmetric mass matrix, the Coriolis and centrifugal-forces matrix and the gravity vector, respectively. Notice that the generalized forces do not necessarily match with the actuating forces $\boldsymbol{Q}_{\mathrm{a}}$ that correspond to the actuation variables or actuator displacements $\boldsymbol{q}_{\mathrm{a}}$. From the actuation and sensing point of view, both $\boldsymbol{Q}_{\mathrm{a}}$ and $\boldsymbol{q}_{\mathrm{a}}$ are the only available physical interfaces to the robotic system.
The well known approaches in robotics like the computed-torque or the non-adaptive basic controller in (Slotine and Li, 1991) provide a control law that is composed of a nonlinear compensating part $\boldsymbol{u}_{\mathrm{C}}$ and a stabilizing part $\boldsymbol{u}_{\mathrm{a}}$, such that the actuation input is provided as

$$\boldsymbol{u} = \boldsymbol{u}_{\mathrm{C}} + \boldsymbol{u}_{\mathrm{a}}. \tag{2}$$

Classically, the first part $\boldsymbol{u}_{\mathrm{C}}$ compensates for the nonlinear dynamics corresponding to the actual configuration $(\boldsymbol{z}, \dot{\boldsymbol{s}})$ of the robot and according to the model given by (1) or a similar variation of it. In such manner, the closed-loop dynamics is approximately linearized and could be stabilized by achieving feedback control via $\boldsymbol{u}_{\mathrm{a}}$. Mostly, the latter is realized as a linear control (e.g. PD or PID) of the actuators corresponding to their respective tracking errors $\boldsymbol{e} = \boldsymbol{q}_{\mathrm{a}} - \boldsymbol{q}_{\mathrm{a,d}}$, with $\boldsymbol{q}_{\mathrm{a,d}}$ being the desired displacements of the actuators. As it is well documented in the text book of (Qu and Dawson, 1996) and proven by a series of journal publications (Abdallah *et al.*, 1991; Berghuis and Nijmeijer, 1994; Egeland, 1987; Qu and Dorsey, 1991*a*;*b*), the robustness of classic model-based strategies has been demonstrated. As long as the feedback action is strong enough, the closed loop is robust against uncertainties.
The realization of any model-based control is formally straightforward for the classic open-chain robot, since the actuation or control variables coincide with the generalized ones: $\boldsymbol{z} \triangleq \boldsymbol{q}_{\mathrm{a}}$ and $\dot{\boldsymbol{s}} \triangleq \dot{\boldsymbol{q}}_{\mathrm{a}} = \frac{d}{dt}\boldsymbol{q}_{\mathrm{a}}$. The dynamics given by (1) can be re-written in the very well known form

$$\boldsymbol{Q}_{\mathrm{a}} = \boldsymbol{M}(\boldsymbol{q}_{\mathrm{a}})\ddot{\boldsymbol{q}}_{\mathrm{a}} + \boldsymbol{C}(\boldsymbol{q}_{\mathrm{a}}, \dot{\boldsymbol{q}}_{\mathrm{a}})\dot{\boldsymbol{q}}_{\mathrm{a}} + \boldsymbol{g}(\boldsymbol{q}_{\mathrm{a}}). \tag{3}$$

Since the configuration and the actuation space are the same, no mapping between both is necessary. The dynamics and therefore the control law can be calculated and derived directly from the knowledge of the actuation variables. The latter are practically always available and are provided by the actuation sensors, such incremental encoder or motor current. This advantageous case is not given for high mobility parallel manipulators. The configuration of such systems are defined with respect to the end-effector pose, velocities and accelerations $\boldsymbol{x}$, $\boldsymbol{v}$ and $\boldsymbol{a}$ such like (1) becomes

$$\boldsymbol{\tau} = \boldsymbol{M}(\boldsymbol{x})\boldsymbol{a} + \boldsymbol{C}(\boldsymbol{x}, \boldsymbol{v})\boldsymbol{v} + \boldsymbol{g}(\boldsymbol{x}). \tag{4}$$

The computation of the nonlinear part $\boldsymbol{u}_{\mathrm{C}}$ needs consequently the additional knowledge of the end-effector motion, which is not available in practice. It is therefore mandatory to have a mapping that provides the necessary but non measurable configuration variables from the measurable actuation ones. For this reason and as it is brilliantly discussed by Wang and

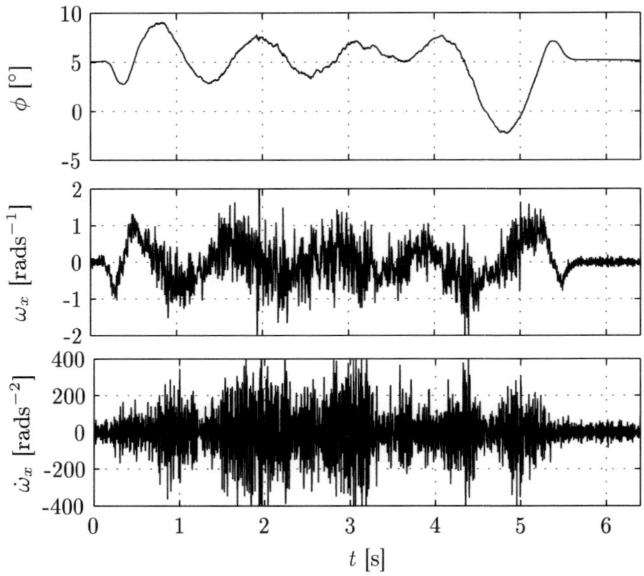

Fig. 1. Results of the direct kinematics for the first rotational DOF. Top: orientation angle, middle: angular velocity, bottom: angular acceleration.

Ghorbel (2006), a common point to model-based control schemes for parallel robots is that the direct or the forward kinematics problem (i.e. the determination of the end-effector motion given the measured joint positions) needs to be solved in real-time in order to compute the dynamics compensating term $\boldsymbol{u}_{\mathrm{C}}$ (Burdet *et al.*, 2000; Cheng *et al.*, 2003; Kim *et al.*, 2005; Ting *et al.*, 2004). In general such operation do not have a closed-form solution and is achieved in an iterative numerical manner. This does not only cause a severe computational problem but yields high noisy estimation of the velocities and accelerations of the end-effector, even for very small termination conditions of the direct kinematics and especially for the rotational DOFs. The use of the Jacobian and its time derivative to calculate the velocities and accelerations may yield modest or acceptable results for lower-mobility manipulators, like reported in Cheng *et al.* (2003); Pietsch *et al.* (2005); Ren *et al.* (2005) and Vivas and Poignet (2005). The results are however not acceptable for accurate tracking of 6 DOFs mechanisms. Figure 1 depicts an experimental example for the first rotational DOF of a spacial parallel manipulator (see system description in section 5.1). It is obvious that both velocities and the accelerations are not suitable for providing reliable dynamics and control inputs.
A second crucial issue for the control of parallel robots is the high complexity of their dynamics, that compromises the real-time implementation of $\boldsymbol{u}_{\mathrm{C}}$. Thus, many researchers suggest the simplification of the dynamics model (Caccavale *et al.*, 2003; Lee *et al.*, 2003; Pietsch *et al.*, 2005; Vivas and Poignet, 2005; Wang *et al.*, 2007) to ensure real-time ability. This will increase the uncertainties to be counteracted by using higher feedback action. Due to the limitation of actuator energy, it is not always possible to implement high-gain control. Model approximation leads in the most of cases to a significant deterioration of the tracking quality (Abdellatif and Heimann, 2007; Denkena *et al.*, 2006). This is especially the case for the range of high accelerations and velocities. The recommended and practical choice is to concentrate all computationally intensive terms in $\boldsymbol{u}_{\mathrm{C}}$ and to keep the controller $\boldsymbol{u}_{\mathrm{a}}$ linear and as simple as possible.

A third point to be considered is common to all robotic systems that are governed by (1) and stabilized by velocity feedback. The quality of actuator's velocity signal affects directly the possible range of the robust high-gain feedback (Berghuis, 1993; Slotine and Li, 1991). Since the direct measurement of actuator's velocities is not practical, the numerical calculation is highly noisy and causal filtering introduces delay, it is recommended to use observation techniques. This has been suggested in many works and in a variety of complexity (Berghuis, 1993; Celani, 2006; de Wit and Slotine, 1991). A discussion concerning this subject in relationship to fully parallel manipulators is though still missing in literature. The presented approach will contribute to close such gap.

2.2 Preliminary analysis

This section is dedicated to the analysis of different properties that are useful for the comprehension of subsequent development.

2.2.1 Passivity of parallel robots with respect to the actuation space

Rigid multi-body systems with dynamics described by (1) are known to be passive from the generalized forces $\boldsymbol{\tau}$ to the generalized velocities $\dot{\boldsymbol{s}}$ by satisfying following property (Ortega *et al.*, 1998)

$$\exists\, 0 < \beta < \infty, \quad \int_0^t \dot{\boldsymbol{s}}^{\mathrm{T}}(t)\boldsymbol{\tau}(t)dt \geq -\beta \quad \forall\, t \geq 0. \tag{5}$$

As proved in (Berghuis, 1993; Ortega *et al.*, 1998; Qu and Dawson, 1996), the passivity property results directly from the nature of the Christoffels symbols constituting the generalized Coriolis and centrifugal forces, such that the matrix $\dot{\boldsymbol{M}}(\boldsymbol{z}) - 2\boldsymbol{C}(\boldsymbol{z}, \dot{\boldsymbol{s}})$ is skew symmetric $\forall\, t$. In that sense and by substituting $\dot{\boldsymbol{s}}$ and $\boldsymbol{\tau}$ with their corresponding values, an open-chain robot is passive from $\boldsymbol{Q}_{\mathrm{a}}$ to $\dot{\boldsymbol{q}}_{\mathrm{a}}$ (the most studied case (Berghuis, 1993; Ortega *et al.*, 1998; Slotine and Li, 1991)) and a 6 DOFs parallel robot is passive from $\boldsymbol{\tau}$ to $\boldsymbol{v}$. The latter is not directly relevant for control design, since we need the passivity of parallel robots also with respect to the actuated space from the actuation input or forces $\boldsymbol{Q}_{\mathrm{a}}$ to the velocities of the active joints $\dot{\boldsymbol{q}}_{\mathrm{a}}$. To investigate such passivity, the equations of motion (4) are transformed into the actuation space

$$\boldsymbol{Q}_{\mathrm{a}} = \boldsymbol{M}_{\mathrm{a}}(\boldsymbol{x})\ddot{\boldsymbol{q}}_{\mathrm{a}} + \boldsymbol{C}_{\mathrm{a}}(\boldsymbol{x}, \boldsymbol{v})\dot{\boldsymbol{q}}_{\mathrm{a}} + \boldsymbol{g}_{\mathrm{a}}(\boldsymbol{x}) \tag{6}$$

with

$$\begin{aligned}
\boldsymbol{M}_{\mathrm{a}}(\boldsymbol{x}) &= \boldsymbol{J}^{\mathrm{T}}(\boldsymbol{x})\boldsymbol{M}_{\mathrm{a}}(\boldsymbol{x})\boldsymbol{J}(\boldsymbol{x}) \\
\boldsymbol{C}_{\mathrm{a}}(\boldsymbol{x}, \boldsymbol{v}) &= \boldsymbol{J}^{\mathrm{T}}(\boldsymbol{x})\boldsymbol{C}(\boldsymbol{x}, \boldsymbol{v})\boldsymbol{J}(\boldsymbol{x}) + \boldsymbol{J}^{\mathrm{T}}(\boldsymbol{x})\boldsymbol{M}(\boldsymbol{z})\dot{\boldsymbol{J}}(\boldsymbol{x}), \\
\boldsymbol{g}_{\mathrm{a}}(\boldsymbol{x}) &= \boldsymbol{J}^{\mathrm{T}}(\boldsymbol{x})\boldsymbol{g}(\boldsymbol{x})
\end{aligned}$$

and $\boldsymbol{J}(\boldsymbol{x}) = \frac{\partial v}{\partial \dot{\boldsymbol{q}}_{\mathrm{a}}}$ being the jacobian matrix of the robot. For non-singular jacobian[1] the mass matrix $\boldsymbol{M}_{\mathrm{a}}$ is also positive definite. Due to the transformation, the term $\boldsymbol{C}_{\mathrm{a}}(\boldsymbol{x}, \boldsymbol{v})$ does not satisfy the properties of the Christoffel's symbols, such that the skew symmetry of $\dot{\boldsymbol{M}}_{\mathrm{a}} - 2\boldsymbol{C}_{\mathrm{a}}$ is not evident anymore. However the relevant skew-symmetric property

$$\boldsymbol{u}^{\mathrm{T}}\left(\dot{\boldsymbol{M}}_{\mathrm{a}} - 2\boldsymbol{C}_{\mathrm{a}}\right)\boldsymbol{u} = 0 \quad \forall\, \boldsymbol{u} \in \mathbb{R}^6 \tag{7}$$

[1] The parallel manipulator is assumed to be mechanically designed, such that a singularity in the workspace is avoided

can be proven (see Appendix). The availability of the fundamental requirement (7) allows to demonstrate the passivity of 6 DOFs parallel manipulator from $\boldsymbol{Q}_{\mathrm{a}}$ to $\dot{\boldsymbol{q}}_{\mathrm{a}}$ in a straightforward manner.

It is important to point out, that even if the dynamics equations (6) are available with respect to the actuation space, the non-measurable variables $\boldsymbol{x}$ and $\boldsymbol{v}$ are still necessary to calculate the different equation parts. Furthermore, the term $\boldsymbol{C}_{\mathrm{a}}(\boldsymbol{x},\boldsymbol{v})\,\dot{\boldsymbol{q}}_{\mathrm{a}}$ decreases the flexibility and variability of the control design in contrast to the case of serial robots. For the latter the Coriolis and centrifugal forces $\boldsymbol{C}_{\mathrm{a}}(\boldsymbol{q}_{\mathrm{a}},\dot{\boldsymbol{q}}_{\mathrm{a}})\dot{\boldsymbol{q}}_{\mathrm{a}}$ (see eq. (3)) allows a variable interchanging of desired and actual velocities in the corresponding control term to shape the energy of the closed-loop system in a more sophisticated way (Berghuis, 1993; Slotine and Li, 1991; Wen and Bayard, 1988). Due to this fact and since Coriolis and centrifugal forces are directly responsible for the global stability of Euler-Lagrange systems, the conditions on the control parameters for parallel manipulators are more conservative than those of classic open-chain systems (see section 3 for discussion).

Before proposing the control design, it is necessary to recall that the different components of the dynamics equations are bounded, that is

$$0 < \underline{m} \leq \|\boldsymbol{M}_{\mathrm{a}}(\boldsymbol{x})\| \leq \overline{m} \quad \forall \boldsymbol{x} \tag{8}$$

$$\|\boldsymbol{C}_{\mathrm{a}}(\boldsymbol{x},\boldsymbol{v})\| \leq \overline{c}\|\dot{\boldsymbol{q}}_{\mathrm{a}}\| \quad \forall \boldsymbol{x},\, \boldsymbol{v}. \tag{9}$$

with $\|\cdot\|$ being the euclidean norm and where $\underline{x}$ and $\overline{x}$ denote generally the minimal and maximal eigenvalue of a Matrix $\boldsymbol{X}$, respectively. Finally the dynamics of a robotic parallel manipulator is available in a linear form with respect to a minimal set of parameters $\boldsymbol{p}$:

$$\boldsymbol{M}_{\mathrm{a}}(\boldsymbol{x})\ddot{\boldsymbol{q}}_{\mathrm{a}} + \boldsymbol{C}_{\mathrm{a}}(\boldsymbol{x},\boldsymbol{v})\dot{\boldsymbol{q}}_{\mathrm{a}} + \boldsymbol{g}_{\mathrm{a}}(\boldsymbol{x}) \;\equiv\; \boldsymbol{A}\,(\boldsymbol{x},\boldsymbol{v},\boldsymbol{a})\,\boldsymbol{p} \tag{10}$$

which is known to be the computationally most efficient (Abdellatif *et al.*, 2005).

2.2.2 Impact of desired dynamics compensation

The desired dynamics compensation is achieved by the choice

$$\boldsymbol{u}_{\mathrm{C}} \triangleq \boldsymbol{M}_{\mathrm{a}}(\boldsymbol{x}_{\mathrm{d}})\ddot{\boldsymbol{q}}_{\mathrm{a,d}} + \boldsymbol{C}_{\mathrm{a}}(\boldsymbol{x}_{\mathrm{d}},\boldsymbol{v}_{\mathrm{d}})\dot{\boldsymbol{q}}_{\mathrm{a,d}} + \boldsymbol{g}_{\mathrm{a}}(\boldsymbol{x}_{\mathrm{d}}) \;=\; \boldsymbol{A}\,(\boldsymbol{x}_{\mathrm{d}},\boldsymbol{v}_{\mathrm{d}},\boldsymbol{a}_{\mathrm{d}})\,\boldsymbol{p} \tag{11}$$

where $'\mathrm{d}'$ being the subscript that distinguishes desired variables. By considering (11), (2) and (6) and by assuming - at this stage of analysis - a perfect model knowledge the following equation

$$\boldsymbol{M}_{\mathrm{a}}(\boldsymbol{x})\ddot{e} + \boldsymbol{C}_{\mathrm{a}}(\boldsymbol{x},\boldsymbol{v})\dot{e} - \boldsymbol{u}_{\mathrm{a}} - \underline{\underline{\boldsymbol{\Delta}}} = \boldsymbol{0}. \tag{12}$$

results for the closed-loop dynamics. The term $\underline{\underline{\boldsymbol{\Delta}}}$ is equal to

$$\begin{aligned}\underline{\underline{\boldsymbol{\Delta}}} = {} & \left(\boldsymbol{M}_{\mathrm{a}}(\boldsymbol{x}_{\mathrm{d}}) - \boldsymbol{M}_{\mathrm{a}}(\boldsymbol{x})\right)\ddot{\boldsymbol{q}}_{\mathrm{a,d}} + \left(\boldsymbol{C}_{\mathrm{a}}(\boldsymbol{x}_{\mathrm{d}},\boldsymbol{v}_{\mathrm{d}}) - \boldsymbol{C}_{\mathrm{a}}(\boldsymbol{x},\boldsymbol{v})\right)\dot{\boldsymbol{q}}_{\mathrm{a,d}} \\ & + \left(\boldsymbol{g}_{\mathrm{a}}(\boldsymbol{x}_{\mathrm{d}}) - \boldsymbol{g}_{\mathrm{a}}(\boldsymbol{x})\right)\end{aligned} \tag{13}$$

and corresponds to the systematic errors introduced by using feedforward or desired dynamics compensation instead of feedback- or actual dynamics compensation. With help of the dynamics properties (8,9) it can be demonstrated that such term is bounded (Burdet *et al.*, 2000; Qu and Dawson, 1996)

$$\|\underline{\underline{\boldsymbol{\Delta}}}\| \leq \bar{\alpha}\|\boldsymbol{e}\| + \bar{c}v_{+}\|\dot{e}\| \;\forall t, \boldsymbol{x}, \boldsymbol{v} \text{ and } \boldsymbol{a} \tag{14}$$

with $\bar{\alpha}$ being a strict positive constant and $v_{+} = \sup_t \|\dot{\boldsymbol{q}}_{\mathrm{a,d}}(t)\|$. The boundedness of the systematic error norm $\|\underline{\underline{\boldsymbol{\Delta}}}\|$ is a fundamental property for the design of control schemes with desired dynamics compensation.

3. Robust Control with Desired Dynamics Compensation

The proposed control scheme is basically composed of the compensation term given by (11) and underlying independent control loops for the single actuators. These are linear control/observer combinations that are to be tuned according to the passivity formalism. This proposed basic control scheme is the adaptation of the approach proposed by (Berghuis and Nijmeijer, 1994) to the case of desired dynamics compensation. The same idea was also briefly studied by (Burdet *et al.*, 2000) but remained without successful experimental implementation.

3.1 Proposed control scheme

Based on the works in (Berghuis, 1993; Qu and Dawson, 1996) we propose for a 6 DOFs parallel manipulator the following robust and computationally high efficient controller

$$\boldsymbol{u} = \underbrace{\boldsymbol{A}\left(\boldsymbol{x}_{\mathrm{d}}, \boldsymbol{v}_{\mathrm{d}}, \boldsymbol{a}_{\mathrm{d}}\right)\hat{\boldsymbol{p}}}_{\boldsymbol{u}_{\mathrm{C}}} \underbrace{-\boldsymbol{K}_{\mathrm{D}}\left(\boldsymbol{s}_1 - \boldsymbol{s}_2\right)}_{\boldsymbol{u}_{\mathrm{a}}}, \tag{15}$$

where $\hat{\boldsymbol{p}}$ is the estimate of the real parameters. The matrix $\boldsymbol{K}_{\mathrm{D}}$ is positive definite. The control variables are defined as follows

$$\begin{aligned} \boldsymbol{s}_1 &= \dot{\boldsymbol{e}} + \boldsymbol{\Lambda}_1 \boldsymbol{e} \\ \boldsymbol{s}_2 &= \dot{\hat{\boldsymbol{e}}} + \boldsymbol{\Lambda}_2 \hat{\boldsymbol{e}}, \end{aligned}$$

where both $\boldsymbol{\Lambda}_1$ and $\boldsymbol{\Lambda}_2$ are positive definite matrices, $\boldsymbol{e} = \boldsymbol{q}_{\mathrm{a}} - \boldsymbol{q}_{\mathrm{a,d}}$ and $\hat{\boldsymbol{e}} = \boldsymbol{q}_{\mathrm{a}} - \hat{\boldsymbol{q}}_{\mathrm{a}}$ denote the tracking and observer errors, respectively. The vectors $\boldsymbol{s}_1$ and $\boldsymbol{s}_2$ correspond to first order sliding tracking and observer variables, respectively (Slotine and Li, 1991). It is here important to notice that due to the assumed absence of the actuator velocity signals $\dot{\boldsymbol{q}}_{\mathrm{a}}$ either $\boldsymbol{s}_1$ nor $\boldsymbol{s}_2$ can be calculated separately. However, their difference is obtainable for the feedback term $\boldsymbol{u}_{\mathrm{a}}$ from the available signals. It is straightforward to prove that

$$\boldsymbol{s}_1 - \boldsymbol{s}_2 = \dot{\hat{\boldsymbol{q}}}_{\mathrm{a}} - \boldsymbol{\Lambda}_2\left(\boldsymbol{q}_{\mathrm{a}} - \hat{\boldsymbol{q}}_{\mathrm{a}}\right) - \dot{\boldsymbol{q}}_{\mathrm{a,d}} + \boldsymbol{\Lambda}_1\left(\boldsymbol{q}_{\mathrm{a}} - \boldsymbol{q}_{\mathrm{a,d}}\right)$$

contains only available signals. The velocity observer is proposed as suggested by Berghuis and Nijmeijer (1994)

$$\begin{array}{rcl} \dot{\hat{\boldsymbol{q}}}_{\mathrm{a}} &=& \boldsymbol{z}_{\mathrm{o}} + \boldsymbol{L}_{\mathrm{D}}\left(\boldsymbol{q}_{\mathrm{a}} - \hat{\boldsymbol{q}}_{\mathrm{a}}\right) \\ \dot{\boldsymbol{z}}_{\mathrm{o}} &=& \ddot{\boldsymbol{q}}_{\mathrm{a,d}} + \boldsymbol{L}_{\mathrm{P}}\left(\boldsymbol{q}_{\mathrm{a}} - \hat{\boldsymbol{q}}_{\mathrm{a}}\right) \end{array} \tag{16}$$

with $\boldsymbol{z}_{\mathrm{o}}$ being the internal observer variable, $\boldsymbol{L}_{\mathrm{D}} = l_{\mathrm{D}}\boldsymbol{I} + \boldsymbol{\Lambda}_2$, $\boldsymbol{L}_{\mathrm{P}} = l_{\mathrm{D}}\boldsymbol{\Lambda}_2$ being symmetric positive definite and $l_{\mathrm{D}} > 0$ is a strict positive scalar quantity. The observer error dynamics are obtained from (16)

$$\ddot{e} = \ddot{\hat{e}} + \boldsymbol{L}_{\mathrm{D}}\dot{\hat{e}} + \boldsymbol{L}_{\mathrm{P}}\hat{e}$$

yielding

$$\ddot{e} = \ddot{\hat{e}} + (l_{\mathrm{D}}\boldsymbol{I} + \boldsymbol{\Lambda}_2)\dot{\hat{e}} + l_{\mathrm{D}}\boldsymbol{\Lambda}_2\hat{e} = \dot{\boldsymbol{s}}_2 + l_{\mathrm{D}}\boldsymbol{s}_2. \tag{17}$$

The control error dynamics are obtained by combining (6) and (15)

$$\boldsymbol{M}_{\mathrm{a}}(\boldsymbol{x})\ddot{e} + \boldsymbol{C}_{\mathrm{a}}(\boldsymbol{x}, \boldsymbol{v})\dot{e} + \boldsymbol{K}_{\mathrm{D}}(\boldsymbol{s}_1 - \boldsymbol{s}_2) - \underline{\underline{\boldsymbol{\Delta}}} - \underline{\boldsymbol{\Delta}} = \boldsymbol{0}. \tag{18}$$

Besides the systematic errors $\underline{\underline{\boldsymbol{\Delta}}}$ introduced by the desired dynamics compensation (see section 2.2.2), the term $\underline{\boldsymbol{\Delta}}$ arises in the last equation. It results from model uncertainties that may be the consequence of biased parameter estimate $\hat{\boldsymbol{p}}$ or unmodeled dynamics $\boldsymbol{Q}_{\mathrm{dist}}$. By considering the realistic assumption of bounded disturbances $\|\boldsymbol{Q}_{\mathrm{dist}}\| < \overline{Q}\ \forall t > 0$ we obtain an upper bound on $\underline{\boldsymbol{\Delta}}$

$$\|\underline{\boldsymbol{\Delta}}\| \leq \|\boldsymbol{A}\left(\boldsymbol{x}_{\mathrm{d}}, \boldsymbol{v}_{\mathrm{d}}, \boldsymbol{a}_{\mathrm{d}}\right) \overline{\Delta \boldsymbol{p}}\| + \overline{Q}. \tag{19}$$

where the parameter uncertainties $\overline{\Delta \boldsymbol{p}}$ can be calculated by evaluating the confidence intervals of the estimate as known from the identification theory (Abdellatif *et al.*, 2008).
Considering both control and observer dynamics (17) and (18) the following closed-loop dynamics are obtained[2]

$$\boldsymbol{M}_{\mathrm{a}}\ddot{\boldsymbol{e}}+\boldsymbol{C}_{\mathrm{a}}\boldsymbol{s}_1 = -\boldsymbol{K}_{\mathrm{D}}\boldsymbol{s}_1+\boldsymbol{K}_{\mathrm{D}}\boldsymbol{s}_2+\boldsymbol{C}_{\mathrm{a}}\boldsymbol{\Lambda}_1\boldsymbol{e}+\underline{\underline{\boldsymbol{\Delta}}}+\underline{\boldsymbol{\Delta}}, \tag{20}$$

$$\begin{aligned}\boldsymbol{M}_{\mathrm{a}}\dot{\boldsymbol{s}}_2+\boldsymbol{C}_{\mathrm{a}}\boldsymbol{s}_2 = &-\boldsymbol{K}_{\mathrm{D}}\boldsymbol{s}_1+\left(-l_{\mathrm{D}}\boldsymbol{M}_{\mathrm{a}}+\boldsymbol{K}_{\mathrm{D}}+\boldsymbol{C}_{\mathrm{a}}\right)\boldsymbol{s}_2\\ &-\boldsymbol{C}_{\mathrm{a}}\dot{\boldsymbol{e}}+\underline{\underline{\boldsymbol{\Delta}}}+\underline{\boldsymbol{\Delta}}.\end{aligned} \tag{21}$$

For the obtained nonautonomous nonlinear system the energy function

$$V = H_1 = \frac{1}{2}\boldsymbol{s}_1^{\mathrm{T}}\boldsymbol{M}_{\mathrm{a}}\boldsymbol{s}_1+\frac{1}{2}\boldsymbol{e}^{\mathrm{T}}\boldsymbol{K}_1\boldsymbol{e}+\frac{1}{2}\boldsymbol{s}_2^{\mathrm{T}}\boldsymbol{M}_{\mathrm{a}}\boldsymbol{s}_2+\frac{1}{2}\hat{\boldsymbol{e}}^{\mathrm{T}}\boldsymbol{K}_2\hat{\boldsymbol{e}} \tag{22}$$

is a Lyapunov-function (Berghuis, 1993; Qu and Dawson, 1996), with

$$\boldsymbol{K}_1 = \boldsymbol{\Lambda}_1(2\boldsymbol{\Lambda}_1^{-1}\boldsymbol{K}_{\mathrm{D}} - \boldsymbol{M}_{\mathrm{a}})\boldsymbol{\Lambda}_1 \quad \text{and} \quad \boldsymbol{K}_2 = 2\boldsymbol{\Lambda}_2^{-1}\boldsymbol{K}_{\mathrm{D}}.$$

By defining the error state vector

$$\boldsymbol{z}_{\mathrm{e}} = \left[\dot{\boldsymbol{e}}^{\mathrm{T}}\left(\boldsymbol{\Lambda}_1\boldsymbol{e}\right)^{\mathrm{T}}\dot{\hat{\boldsymbol{e}}}^{\mathrm{T}}\left(\boldsymbol{\Lambda}_2\hat{\boldsymbol{e}}\right)^{\mathrm{T}}\right]^{\mathrm{T}}$$

we obtain

$$V = \frac{1}{2}\boldsymbol{z}_{\mathrm{e}}^{\mathrm{T}}\boldsymbol{P}\boldsymbol{z}_{\mathrm{e}} \tag{23}$$

with

$$\boldsymbol{P}_1 = \begin{bmatrix} \begin{bmatrix} \boldsymbol{M}_{\mathrm{a}} & \boldsymbol{M}_{\mathrm{a}} \\ \boldsymbol{M}_{\mathrm{a}} & 2\boldsymbol{\Lambda}_1^{-1}\boldsymbol{K}_{\mathrm{D}} \end{bmatrix} & \boldsymbol{0} \\ \boldsymbol{0} & \begin{bmatrix} \boldsymbol{M}_{\mathrm{a}} & \boldsymbol{M}_{\mathrm{a}} \\ \boldsymbol{M}_{\mathrm{a}} & \boldsymbol{M}_{\mathrm{a}}+2\boldsymbol{\Lambda}_2^{-1}\boldsymbol{K}_{\mathrm{D}} \end{bmatrix} \end{bmatrix}.$$

Using (20), (21) and the skew symmetric property of $\dot{\boldsymbol{M}}_{\mathrm{a}} - 2\boldsymbol{C}_{\mathrm{a}}$ (see Appendix) the time derivative of V results in

$$\begin{aligned}\dot{V}(\boldsymbol{z}_{\mathrm{e}}, t) = &-\boldsymbol{z}_{\mathrm{e}}^{\mathrm{T}}\boldsymbol{Q}\boldsymbol{z}_{\mathrm{e}} - \boldsymbol{s}_2^{\mathrm{T}}\left[l_{\mathrm{D}}\boldsymbol{M}_{\mathrm{a}}(\boldsymbol{x}) - 2\boldsymbol{K}_{\mathrm{D}} - \boldsymbol{C}_{\mathrm{a}}(\boldsymbol{x}, \boldsymbol{v})\right]\boldsymbol{s}_2\\ &+\dot{\boldsymbol{e}}^{\mathrm{T}}\boldsymbol{C}_{\mathrm{a}}(\boldsymbol{x}, \boldsymbol{v})\boldsymbol{\Lambda}_1\boldsymbol{e} - \boldsymbol{s}_2^{\mathrm{T}}\boldsymbol{C}_{\mathrm{a}}(\boldsymbol{x}, \boldsymbol{v})\dot{\boldsymbol{e}}\\ &+\left(\boldsymbol{s}_1^{\mathrm{T}} + \boldsymbol{s}_2^{\mathrm{T}}\right)\left(\underline{\underline{\boldsymbol{\Delta}}} + \underline{\boldsymbol{\Delta}}\right)\end{aligned} \tag{24}$$

[2] The arguments $\boldsymbol{x}$ and $\boldsymbol{v}$ were omitted for convenience

with

$$Q = \begin{bmatrix} \begin{bmatrix} K_{\mathrm{D}} - \Lambda_1 M_{\mathrm{a}} & 0 \\ 0 & K_{\mathrm{D}} \end{bmatrix} & 0 \\ 0 & \begin{bmatrix} K_{\mathrm{D}} & 0 \\ 0 & K_{\mathrm{D}} \end{bmatrix} \end{bmatrix}.$$

Furthermore and according to the property (8), the time derivative of the Lyapunov-function is bounded (Qu and Dawson, 1996)

$$\dot{V}(z_{\mathrm{e}}, t) \leq \phi_0 \|z_{\mathrm{e}}(t)\| - \phi_1 \|z_{\mathrm{e}}(t)\|^2 + \phi_2 \|z_{\mathrm{e}}(t)\|^3$$

with

$$\phi_0 = 2\sqrt{2}\left(\|A(x_{\mathrm{d}}, v_{\mathrm{d}}, a_{\mathrm{d}})\,\overline{\Delta p}\| + \overline{Q}\right), \tag{25}$$

$$\phi_1 = \underline{k_{\mathrm{D}}} - \overline{\lambda_1}\overline{m} - (1+3\sqrt{2})\bar{c}v_+ - 2\sqrt{2}\,\overline{\lambda_1}^{-1}\bar{\alpha}, \tag{26}$$

$$\phi_2 = (1+3\sqrt{2})\bar{c} \tag{27}$$

resulting from the error dynamics of the here studied case. For given initial error $z_{\mathrm{e}}(0)$ the closed-loop system is locally uniformly and ultimately bounded when the following inequalities are fulfilled

$$\begin{aligned} \phi_1 &> 2\sqrt{\phi_0\,\phi_2}, \\ \phi_1^2 + \phi_1\sqrt{\phi_1^2 - 4\phi_0\phi_2} &> 2\phi_0\phi_2\left(1 + \sqrt{\tfrac{p_{\mathrm{M}}}{p_{\mathrm{m}}}}\right), \\ \phi_1 + \sqrt{\phi_1^2 - 4\phi_0\phi_2} &> 2\phi_2\|z_{\mathrm{e}}(0)\|\sqrt{\tfrac{p_{\mathrm{M}}}{p_{\mathrm{m}}}}. \end{aligned} \tag{28}$$

The variables p_{m} and p_{M} can be obtained from the eigenvalues of P_1, as given by Berghuis and Nijmeijer (1994)

$$\begin{aligned} p_{\mathrm{m}} &= \tfrac{1}{3}\underline{p} &&= \tfrac{1}{3}\min\{\underline{m}, 2\overline{\lambda_2}^{-1}\overline{\lambda_1}\overline{m}\} \\ p_{\mathrm{M}} &= 3\overline{p} &&= 3\max\{2\underline{\lambda_1}^{-1}\overline{k_{\mathrm{D}}}, 2\underline{\lambda_2}^{-1}\overline{k_{\mathrm{D}}}\} \end{aligned}. \tag{29}$$

It is then possible to chose the matrices K_{D}, Λ_1 and Λ_2 such that their eigenvalues satisfy (28,29), i.e.

$$l_{\mathrm{D}} > \underline{m}^{-1}\left[2\,\overline{k_{\mathrm{D}}} + \bar{c}v_+\right], \tag{30}$$

$$\underline{k_{\mathrm{D}}} > \phi_1 + \overline{\lambda_1}\overline{m} + (1+3\sqrt{2})\bar{c}v_+ + 2\sqrt{2}\,\overline{\lambda_1}^{-1}\bar{\alpha} \tag{31}$$

which finishes the control design procedure. An additional benefit is the analytical availability of the radius R of the region of final error convergence

$$R = \frac{2\phi_0}{\phi_1 + \sqrt{\phi_1^2 - 4\phi_0\phi_2}}. \tag{32}$$

It is straightforward to deduce that the theoretical case of a perfect model ($\phi_0 = 0$) provides semi-global exponential stability under the regarded controller/observer combination (15,16). In contrast to the work Berghuis (1993) for serial manipulators, two major differences can be stressed out. First, the presented robust control scheme for 6 DOFs parallel manipulators uses consequently the compensation of desired dynamics. Second, the necessity of x and v for the calculation of the inverse dynamics and especially the Coriolis and centrifugal terms shrinks the theoretically possible region of attraction. Both effects yield more conservative conditions on boundedness and therefore on stability, i.e. the terms $2\sqrt{2}\,\overline{\lambda_1}^{-1}\bar{\alpha}$ and $(1+3\sqrt{2})\bar{c}v_+$ in (26).

3.2 Considering friction

For the sake of simplicity, friction has not been regarded in the above discussed design. This is not associated with a loss of generality, since friction preserves the passivity of the system (Berghuis, 1993; Slotine and Li, 1991). For the exemplarily case of the classic modeling approach of a superposition of coulomb friction and viscous damping, we obtain for every passive or active joint i

$$Q_{\mathrm{f}_i} = f_{1i}\,\mathrm{sgn}(\dot{q}_i) + f_{2i}\dot{q}_i. \tag{33}$$

The overall friction that occurs in each actuator is obtained from (33) by means of kinematic transformation (Abdelllatif *et al.*, 2007):

$$\boldsymbol{F}_{\mathrm{a}} = \left(\frac{\partial \dot{\boldsymbol{q}}}{\partial \dot{\boldsymbol{q}}_{\mathrm{a}}}\right)^{\mathrm{T}} \boldsymbol{Q}_{\mathrm{f}} = \boldsymbol{A}_{\mathrm{f}}\left(\boldsymbol{x},\boldsymbol{v},\boldsymbol{a}\right)\boldsymbol{p}_{\mathrm{f}}.$$

The resulting model is also linear with respect to the parameter vector $\boldsymbol{p}_{\mathrm{f}}$ that groups all friction coefficients f_{1i} and f_{2i}. Consequently, the compensating term (11) can be updated by a friction part:

$$\boldsymbol{u}_{\mathrm{C}} = \boldsymbol{A}\left(\boldsymbol{x}_{\mathrm{d}},\boldsymbol{v}_{\mathrm{d}},\boldsymbol{a}_{\mathrm{d}}\right)\hat{\boldsymbol{p}} + \boldsymbol{A}_{\mathrm{f}}\left(\boldsymbol{x}_{\mathrm{d}},\boldsymbol{v}_{\mathrm{d}},\boldsymbol{a}_{\mathrm{d}}\right)\hat{\boldsymbol{p}}_{\mathrm{f}}.$$

The parametric uncertainties is consequently updated by the friction parameter estimate bias and (19) may be re-written to

$$\|\underline{\boldsymbol{\Delta}}\| \leq \|\boldsymbol{A}\left(\boldsymbol{x}_{\mathrm{d}},\boldsymbol{v}_{\mathrm{d}},\boldsymbol{a}_{\mathrm{d}}\right)\overline{\Delta\boldsymbol{p}}\| + \|\boldsymbol{A}_{\mathrm{f}}\left(\boldsymbol{x}_{\mathrm{d}},\boldsymbol{v}_{\mathrm{d}},\boldsymbol{a}_{\mathrm{d}}\right)\overline{\Delta p_{\mathrm{f}}}\| + \overline{Q}.$$

As it is known for the Lyapunov-based design, the additional uncertainties yield more conservative bounds and therefore more conservative choice of the controller parameters. This is especially the case for parallel robots, since the friction forces discussed here depend on the system's configuration. As demonstrated in (Abdelllatif *et al.*, 2007), the resulting friction that is to be counteracted by an actuator j is expressed as:

$$F_{\mathrm{a}_j} = r_{1j}(\boldsymbol{x})\,\mathrm{sgn}(\dot{q}_{\mathrm{a}_j}) + r_{2j}(\boldsymbol{x})\dot{q}_{\mathrm{a}_j} \tag{34}$$

and is not only dependent on the actuator velocity but also on the pose $\boldsymbol{x}$ of the manipulator. The upper bounds of $\|\underline{\boldsymbol{\Delta}}\|$ can be extended to

$$\|\underline{\boldsymbol{\Delta}}\| \leq \overline{r_2} + \overline{\alpha}\|e\| + (\overline{r_1} + \overline{c}v_+)\|\dot{e}\|$$

and integrated in the design procedure. Since r_1 and r_2 are widely varying over the workspace, their upper estimates $\overline{r_1}$ and $\overline{r_2}$ increase the conservatism of the control design, in comparison e.g. to open-chain robots, where the friction forces depend only on the actuator's velocity. The interested reader is referred to the article (Abdelllatif *et al.*, 2007) for deeper insight into the consideration of friction for parallel manipulators.

4. Augmenting the Scheme with Sliding Mode Control

We demonstrate in the previous section that the combination of desired dynamics compensation and linear feedback provides robustness in the sense of local exponential ultimate boundedness. Such approach remains conservative in the way that the robustness is achieved primarily through higher feedback. High feedback is limited in practice by the actuation constraints. Alternatively, nonlinear sliding mode (or switching) control strategy could provide robustness in a global manner (Slotine and Li, 1991). Therefore, we propose to extend the basic algorithm given by (2) to

$$\boldsymbol{u} = \boldsymbol{u}_{\mathrm{C}} + \boldsymbol{u}_{\mathrm{a}} + \boldsymbol{u}_{\mathrm{R}}, \tag{35}$$

with $\boldsymbol{u}_{\mathrm{R}}$ being the robustifying switching control term. The basic scheme is kept to encounter systematic or parametric uncertainties, like those resulting from biased estimated rigid-body model parameters. It is now extended with the new term $\boldsymbol{u}_{\mathrm{R}}$. Inspired by our long experience with parallel manipulators and in order to keep the practical and computational efficient implementation, the proposed switching control considers explicitly only the friction model that is more affected by time varying uncertainties and not only by constant bias.
The proposed control approach combines and merges a multitude of schemes, that have been proposed in early years for serial manipulators. Primarily we use the parameter-based saturation principle as proposed among others by Spong (1992) and we opt for the so called model-decomposition paradigm of Liu and Goldenberg (1996) to limit the robust action on the friction part. Our contributions are: first, to extend such strategies to the case of desired dynamics compensation; second, to consider the observer dynamics within the control law and last, to implement the control for the case of complex spatial parallel manipulators.

4.1 Proposed scheme with sliding mode control

The proposed extended controller is the following:

$$\begin{aligned}\boldsymbol{u} = &\underbrace{\boldsymbol{A}(\boldsymbol{x}_{\mathrm{d}}, \boldsymbol{v}_{\mathrm{d}}, \boldsymbol{a}_{\mathrm{d}})\,\hat{\boldsymbol{p}} + \boldsymbol{A}_{\mathrm{f}}(\boldsymbol{x}_{\mathrm{d}}, \boldsymbol{v}_{\mathrm{d}}, \boldsymbol{a}_{\mathrm{d}})\,\hat{\boldsymbol{p}}_{\mathrm{f}}}_{\boldsymbol{u}_{\mathrm{C}}} \underbrace{-\boldsymbol{K}_{\mathrm{D}}(\boldsymbol{s}_1 - \boldsymbol{s}_2)}_{\boldsymbol{u}_{\mathrm{a}}} \\ &+ \underbrace{\boldsymbol{A}_{\mathrm{f}}(\boldsymbol{x}_{\mathrm{d}}, \boldsymbol{v}_{\mathrm{d}}, \boldsymbol{a}_{\mathrm{d}})\,\boldsymbol{u}_{\mathrm{f}}}_{\boldsymbol{u}_{\mathrm{R}}},\end{aligned} \tag{36}$$

with $\boldsymbol{u}_{\mathrm{f}}$ being dimensionally equal to the friction parameter vector $\boldsymbol{p}_{\mathrm{f}}$: $(\dim(\boldsymbol{u}_{\mathrm{f}}) = n_{\mathrm{f}})$ and is a robust parametric correction vector:

$$u_{\mathrm{f},k} = \begin{cases} -\rho_k \dfrac{\mathbf{Y}_{\mathrm{f},k}}{\|\mathbf{Y}_{\mathrm{f},k}\|} & \text{if } \|\mathbf{Y}_{\mathrm{f},k}\| > \epsilon_k \\ -\frac{\rho_k}{\epsilon_k}\mathbf{Y}_{\mathrm{f},k} - K_{\mathrm{I},k}\int_{t_0}^{t}\mathbf{Y}_{\mathrm{f},k}d\tau & \text{if } \|\mathbf{Y}_{\mathrm{f},k}\| \le \epsilon_k \end{cases}, \quad \text{for } k = 1\dots n_{\mathrm{f}}. \tag{37}$$

and

$$\boldsymbol{Y}_{\mathrm{f}} = \boldsymbol{A}_{\mathrm{f}}^{\mathrm{T}}(\boldsymbol{x}_{\mathrm{d}}, \boldsymbol{v}_{\mathrm{d}}, \boldsymbol{a}_{\mathrm{d}})\,(\boldsymbol{s}_1 - \boldsymbol{s}_2)\,. \tag{38}$$

Both parameters ρ_k and ϵ_k can be adjusted individually for each friction parameter $p_{\mathrm{f},k}$. They correspond in the sense of saturation control to the uncertainty bounds and to the width of the boundary layers, respectively. The parameter ρ_k depends on the modeling and the parameter estimate precision, whereas ϵ_k is a positive control parameter, that have to be chosen with respect to the control goals. In classical approaches the boundary layer is shaped as thin as possible by using very small ϵ to guarantee high tracking accuracy. This implies high feedback action with its all related disadvantages. Therefore, Liu and Goldenberg has proposed the integral control action, given above in the second equation (37). This is motivated by the fact, that some aspects of parametric uncertainties like estimate bias affect the system as an offset in the parameter domain. Thus, integral action in the same domain is the adequate method to counteract such type of uncertainties. To avoid windups due to large initial errors, the integral action is restricted to the case when the errors are small and are within the boundary layer.

The integral term helps enlarging the boundary layer ϵ and therefore decreasing the feedback action by keeping the same tracking accuracy. It is highly recommended for use in practice. The proof of uniform ultimate boundedness can follow by combining the method shown in the previous section 2.2.1 and the procedure demonstrated in (Liu and Goldenberg, 1996). The proof - although straightforward - is too long to be put here. It shall be noticed that the Lyapunov function candidate remains the same as (22) for the region outside the boudary layer, and is extended with the term $\frac{1}{2}\boldsymbol{\xi}^{\mathrm{T}}\boldsymbol{K}_{\mathrm{I}}\boldsymbol{\xi}$ in the contrary case. Here

$$\boldsymbol{\xi} = \boldsymbol{K}_{\mathrm{I}}^{-1}\Delta\boldsymbol{p}_{\mathrm{f}} - \int_{t_0}^{t} \boldsymbol{Y}_{\mathrm{f}} d\tau. \tag{39}$$

The next section provides a discussion on the provided sliding mode controller as well as its comparison to alternative approaches from literature.

4.2 Discussion

As previously mentioned, the proposed control given by (37-38) results from merging classic approaches provided for serial manipulators and their adaptation to the case of desired dynamics compensation with additional consideration of observer dynamics. In the original approaches Liu and Goldenberg (1996); Spong (1992), the vector $\boldsymbol{Y}_{\mathrm{f}}$ resulted by using only the sliding variable $\boldsymbol{s}_1$ (or similar variations of it), which is in general noisy[3] and cause the shrinking of control band width. Using the smoother variable $(\boldsymbol{s}_1 - \boldsymbol{s}_2)$ as well as the noise-free desired values $\boldsymbol{x}_{\mathrm{d}}$, $\boldsymbol{v}_{\mathrm{d}}$ and $\boldsymbol{a}_{\mathrm{d}}$ allows for more freedom when tuning the feedback gain or adjusting the boundary layer parameters $\boldsymbol{\rho}$ and ϵ. Even if the theory provides global uniform and ultimate boundedness, practically the use of sliding observer component plays a key role in the amelioration of tracking accuracy. It is believed, that the experimental studies in many publications would provide better result, if an observer has been implemented.
Even if the control laws (37-38) appear complicated, they do not cause any major losing of computational efficiency. This is due to the fact, that we consciously limited the switching control to the friction model. The related part $\boldsymbol{A}_{\mathrm{f}}$ is very simple to obtain by 36 additions and 54 multiplications (Abdellatif *et al.*, 2005). Extending the switching control to the rigid-body part is not very efficient, since the rigid-body model requires about twenty times more computational effort. It is questionable to spend so much effort to counteract uncertainties of rigid-body parameters, that and in exception of playloads are not affected by important uncertainties. This is an additional important difference between our algorithms and other alternative robust switching controllers developed for parallel manipulators (Kim *et al.*, 2000; 2005; 2001).

5. Experimental Study

This section is dedicated to the experimental implementation of the proposed control strategy on a 6 DOFs spatial parallel manipulator, which will be presented briefly in the first subsection.

5.1 Case study hexapod

All proposed approaches are substantiated on the hexapod PaLiDA (see Fig. 2), that has been designed and constructed by the institute of production engineering and machine tools of the university of Hannover. The machine is equipped with fast direct drives variable in length and

[3] since the velocity errors are calculated by numerical differentiation

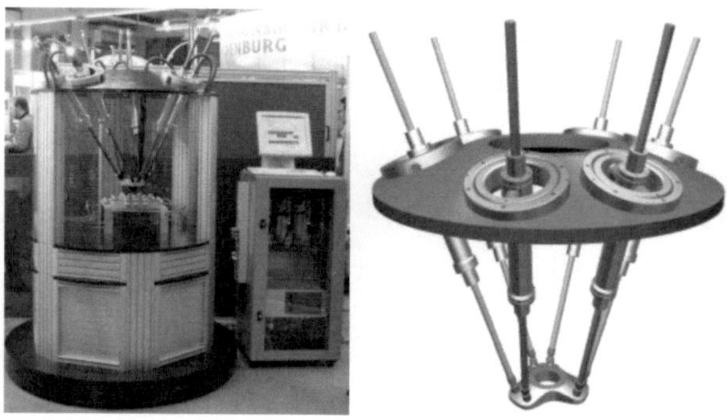

Fig. 2. Case study: hexapod PaLiDA. Left: presentation at the Hannover industrial fair, right: CAD model

has been designed to be a mixture of a high-speed manipulator and a tool machine (Denkena *et al.*, 2006). The application area covers fast handling and light cutting machining tasks with low process forces. Central requirement is therefore to ensure acceptable tracking errors at highest possible velocities and accelerations at the presence of disturbances. The maximal actuation force is about 230N, whereas actuator accelerations of about 2-3 g could be achieved. The internal hall sensors are affected with significant measurement noise, such that any feedback strategy of numerically differentiated variables is challenging. The control system consists in a commercial dSPACE Power-PC 604e single processor unit (333 MHz). The sample time is 0.5 ms. The proposed control approach requires (including path-planning) about 0.15 ms of computational time, which demonstrates its efficiency.

k	p	unit	$\hat{p}$
1	f_{1_α}	[Nm]	0.654
2	f_{1_β}	[Nm]	0.675
3	f_{1_1}	[N]	8.148
4	f_{1_2}	[N]	5.288
5	f_{1_3}	[N]	16.574
6	f_{1_4}	[N]	7.743
7	f_{1_5}	[N]	6.295
8	f_{1_6}	[N]	9.525
9	f_{2_1}	[Nsm^{-1}]	18.774
10	f_{2_1}	[Nsm^{-1}]	16.092
11	f_{2_1}	[Nsm^{-1}]	4.428
12	f_{2_1}	[Nsm^{-1}]	17.444
13	f_{2_1}	[Nsm^{-1}]	17.915
14	f_{2_1}	[Nsm^{-1}]	3.454

Table 1. friction parameters p_k with corresponding estimated values $\hat{p}_k$

As it has been derived and deeply discussed in former publications (see e.g. (Abdellatif and Heimann, 2007; Abdellatif *et al.*, 2005)) the dynamics model used for the feedforward controller $\boldsymbol{u}_C$ contains 24 minimal parameters. The rigid body model part corresponds to a set of

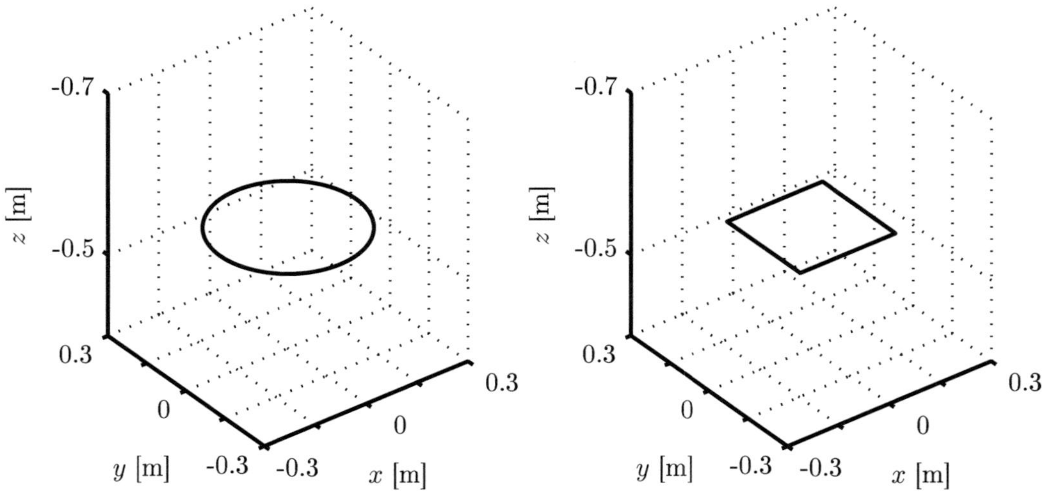

Fig. 3. Investigated motions for the experimental study, left: motion 1, right motion 2.

10 minimal parameters $\boldsymbol{p}$. As given by the modeling approach (33), the friction of each of the 6 actuators j is modeled by a dry friction coefficient f_{1_j} and a viscous dumping coefficient f_{2_j}. Friction in the passive joints is modeled only as dry friction with a common parameter for all α_j (the first revolute joint of each strut) and another one for all β_j-joints (the second revolute joint of each strut). The friction model contains therefore 14 different parameters (see Table 1 for an overview and (Abdelllatif *et al.*, 2007) for more details).

For the experimental validation of the proposed control, two exemplarily test motions (see Fig. 3) are chosen. Both of them are demanding in point of view of achieved dynamics and velocities. The circular motion ($\varnothing$ = 20mm) allows for highest actuation forces of about 230 N. The quadratic one (edge length= 28mm) allows for highest possible actuator velocities of about $1.5\,\mathrm{ms}^{-1}$. To keep the influence of kinematic accuracy the same over the experiments, we chose both test motions in the middle of the workspace and at the same hight. The experimental comparison focuses on the two proposed schemes: the passivity based approach presented in section 3, which is denoted in the following by (P-FF) and the scheme with additional Sliding mode control (P-SM). Additionally, both proposed schemes are compared to the classic feedforward computed-force technique (CF-FF), that consists in forwarding the dynamic model by keeping the actuators controlled by standard PID (or PD) controllers (see Abdellatif *et al.* (2005)). The common control parameters for P-FF and P-SM are set equal for a meaningful comparison. For the classic CF-FF approachh, the PID parameters are tuned heuristically, but the derivative part could not has set as high as for the passivity-based schemes, due to the absence of a velocity observer. This can be stated already as an improvement of the proposed control, that allows for higher control bandwidth thanks to the integration of the observer. All different control parameters used in the following experimental study are given in Table 2.

5.2 Experimental results

The first experiment consists in comparing the three control strategies CF-FF, P-FF und P-SM. For this purpose, the identification of the model parameters has been achieved to obtain reliable estimate for $\hat{\boldsymbol{p}}$ and $\hat{\boldsymbol{p}}_\mathrm{f}$ (see (Abdellatif *et al.*, 2008) and results in Table 1). To examine the robustness towards parametric uncertainties, different friction parameters $\bar{\boldsymbol{p}}_\mathrm{f} = \frac{1}{2}\hat{\boldsymbol{p}}_\mathrm{f}$ has

	K_P	k_I	K_D	Λ_1	Λ_2	l_D	ρ	ϵ	$K_I[10^3]$
CF-FF	38000	85000	1000	-	-	-	-	-	-
PF-FF	-	-	1400	42	42	180	-	-	-
PF-SM	-	-	1400	42	42	180	[6.32 7.15 99.39 74.43 129.05 72.67 104.74 140.78 210.72 148.49 228.80 151.94 178.42 221.57]	[0.06 0.06 0.16 0.16 0.16 0.16 0.16 0.16 0.16 0.16 0.16 0.16 0.16 0.16]	[0.31 0.35 49.69 37.21 64.52 36.33 52.37 70.39 105.36 74.24 114.40 75.97 89.21 110.78]

Table 2. control parameters for the experimental study

been used to compute the feedforward controller u_C. As demonstrated by our experience with the system, such choice is very realistic. Friction parameters could even vary more then here assumed.

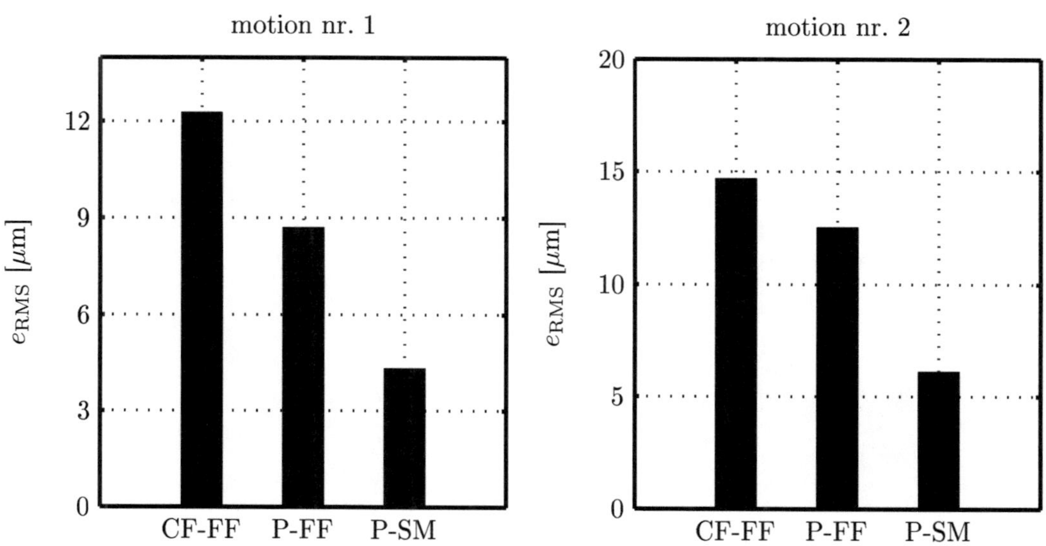

Fig. 4. Comparing the resulting root mean squares errors resulting from the two test motions and by using three different controllers.

Comparing the root mean squares of the tracking errors with respect to the actuators yields the results in Fig. 4. As expected the classic approach CF-FF shows the lowest tracking accuracy. This is due to the fact, that the highly noisy velocity error signal inhibits increasing

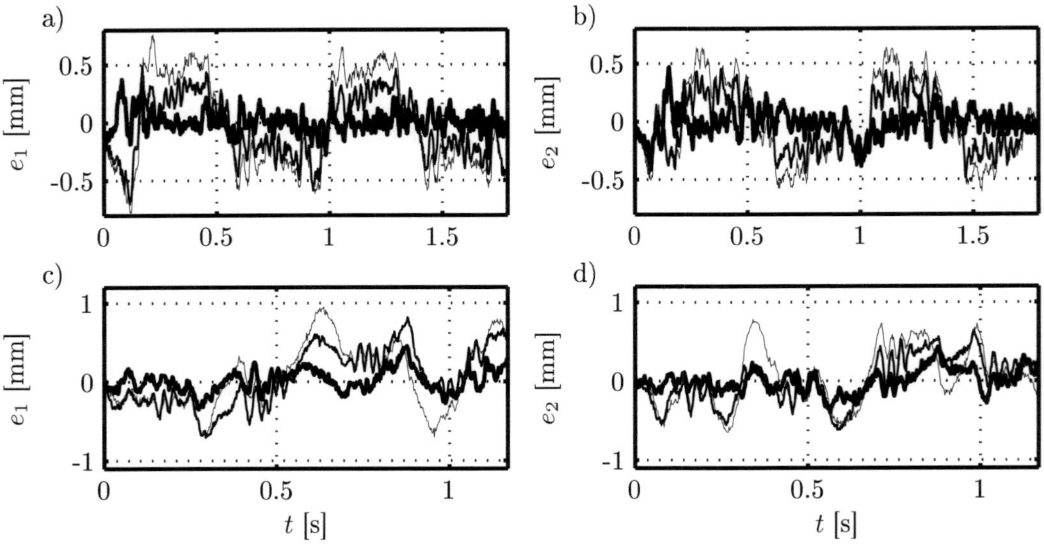

Fig. 5. Comparing the tracking errors of exemplarily chosen actuators resulting for the two test motions (top: motion nr. 1, bottom: motion nr. 2) and by using three different controllers: Thick line: P-SM, medium line: P-FF, thin line: CF-FF

the feedback action significantly without compromising the stability of the system. The use of well tuned observer is a key issue to improve the tracking performance. It is also clear that the controller augmented with the sliding mode component outperforms the other two. This can be concluded by examining the time histories of the tracking errors (see some examples depicted in 5). The approach P-FF is able to counteract the effects of the parameter uncertainty by increasing the feedback. Such operation has to be performed by the operator or by the control engineer. The feedback action should then be tuned in relationship to the estimate upper bound of the uncertainty. The switching mode robust controller is able to react autonomously on the deviation of the nominal model. Its operating mode can be illustrated for the present case by examining the robust corrections performed for the first and for the 10th friction parameters (see Fig. 6 and Fig. 7, respectively).

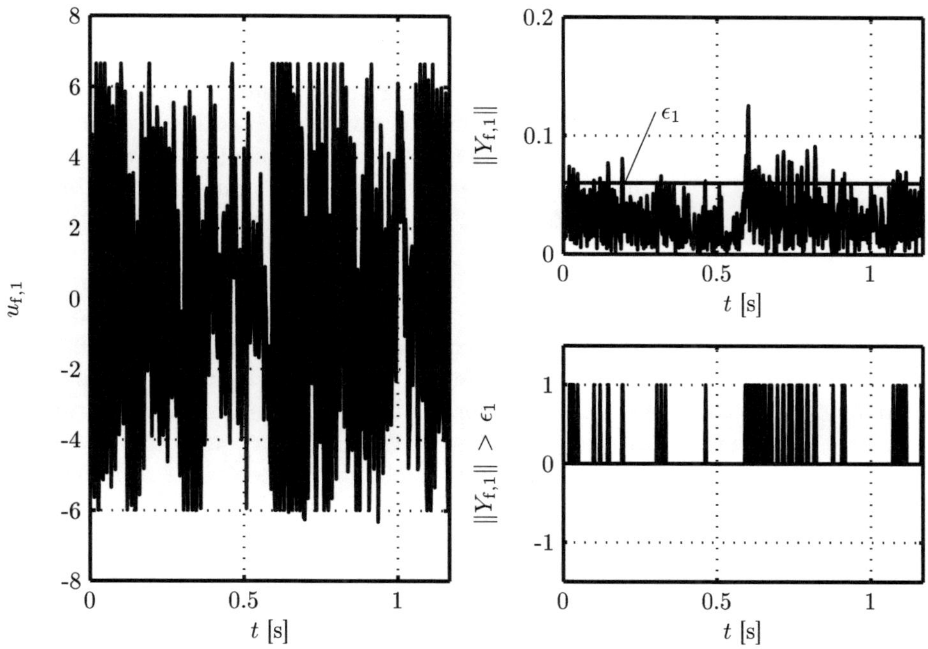

Fig. 6. Robut correction for the first entry of the input vector $\boldsymbol{u}_\mathrm{f}$ corresponding to the first friction parameters f_{1_α}

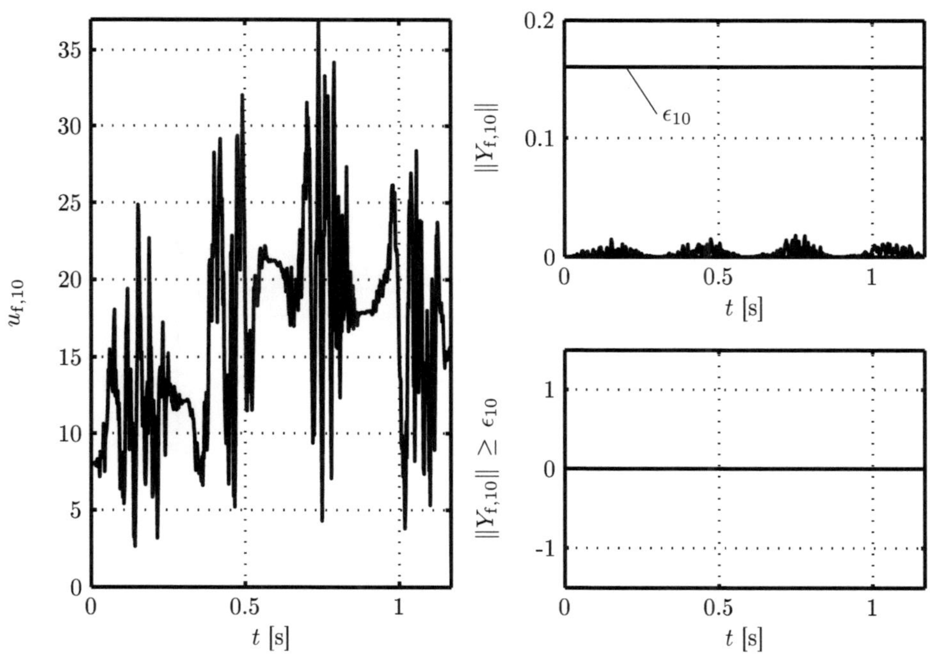

Fig. 7. Robut correction for the first entry of the input vector $\boldsymbol{u}_\mathrm{f}$ corresponding to the 10th friction parameters f_{2_2}

As given mathematically in (36-38) the entries of the switching robust controller $\boldsymbol{u}_{\mathrm{R}}$ depend on the corresponding state $\|\boldsymbol{Y}_{\mathrm{f},k}\| > \epsilon_k$ (see right, bottom of the figures) . If the latter is fulfilled then the boundary layer is violated and the controller switches to pull back $\boldsymbol{Y}_{\mathrm{f},k}$ within the layer. This is the case, when high uncertainty is given like it is the case for the friction parameter f_{1_α} of the passive joint. A higher switching action occurs (see left plot of Fig. 6). In the contrary case, like for $u_{f,10}$ the algorithms reacts on the error dynamics $\boldsymbol{s}_1 - \boldsymbol{s}_2$ by adjusting the output. As it can be better observed in Fig. 7, the output $u_{f,10}$ is highly correlated with the corresponding $\boldsymbol{Y}_{\mathrm{f},k}$.
The case of significant initial errors showed however some drawbacks of the control approach P-SM. The most important one is that in such case the switching controller is too aggressive and leads very quickly to the violation of the actuator constraints. The design of the controller has been made without any consideration of input constraints which explains such undesirable phenomena. This issue is left for future work and for future improvement. Figure 8 shows corresponding experimental results achieved by driving the quadratic motion by significant initial errors. The exemplarily depicted tracking errors for the second actuator demonstrates that the P-SM controller yields the biggest overshoot. Additionally, it exhibits lower tracking convergence quality. Notice that we used the region of final convergence R (see eq. (32)) in Fig. 7 only for illustration purpose and in order to improve the understanding of the results. To remedy the bad behavior of the sliding mode controller in presence of high initial errors, another mild tuning of the parameters is required, e.g. increasing $\boldsymbol{K}_{\mathrm{I}}$ and decreasing $\boldsymbol{\epsilon}$ by 100 times and 4 times, respectively. The corresponding experimental results are denoted by P-SM′. It is clear that a tradeoff should be met between robustness and tracking accuracy, which is the classic problem in control practice.

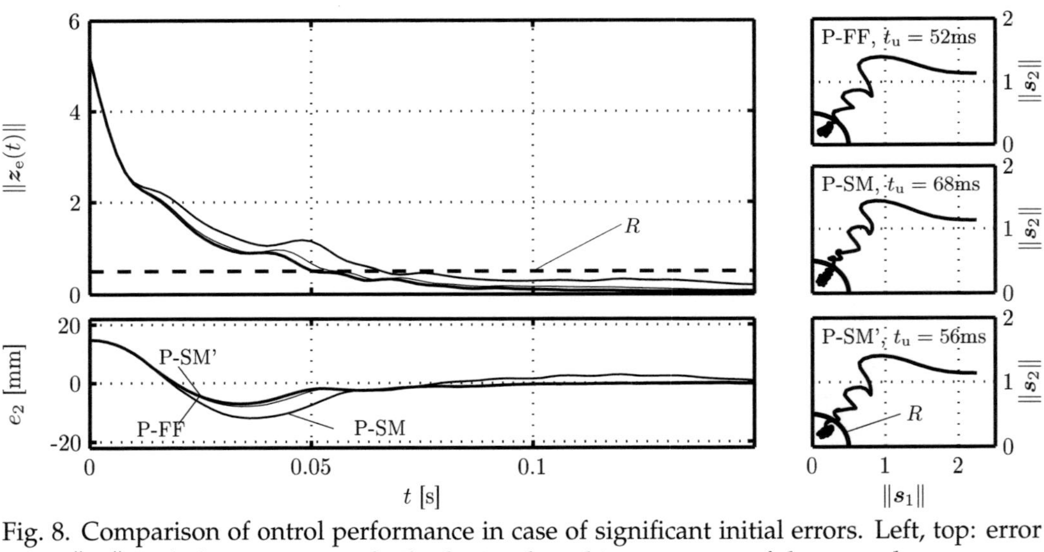

Fig. 8. Comparison of ontrol performance in case of significant initial errors. Left, top: error norm $\|\boldsymbol{z}_{\mathrm{e}}\|$, Left, bottom: exemplarily depicted tracking accuracy of the second actuator, on the right side: convergence of the control and observer sliding errors for the compared control approaches and with resulting convergence time t_{u}.

The final and concluding experiment compares the accuracy of the three control approaches in the cartesian space. Figure 9 shows the tracking performance of the circle as well as that of a corner achieved by the algorithms CF-FF, P-FF and P-SM. In this experiment we switched back to the case of zero initial errors. The sliding mode approach outperforms the other two controllers.

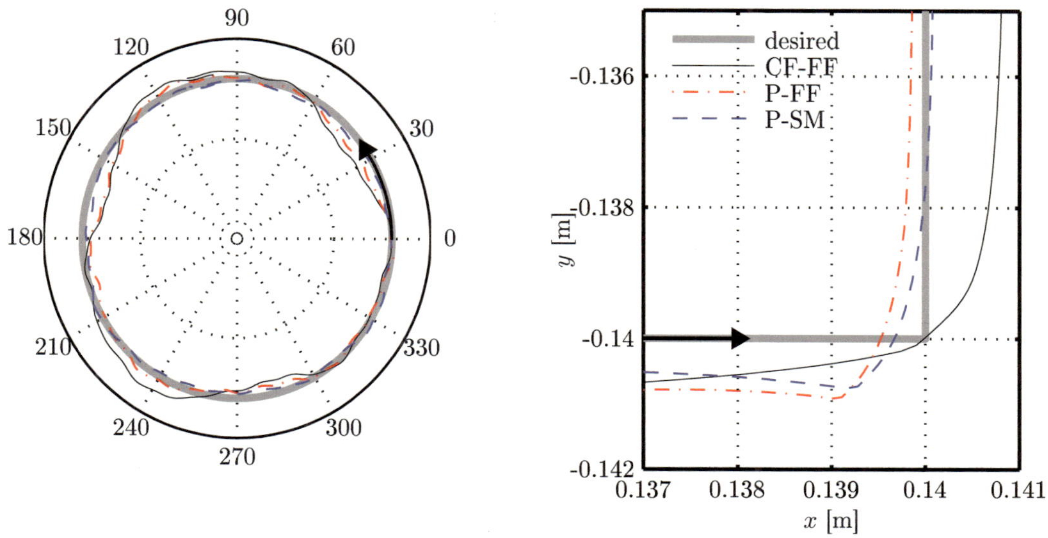

Fig. 9. Peroformance of tracking a circle (left) and a sharp corner (right) achieved by the three investigated control approaches

6. Conclusions

In this paper an experimentally approved practical methodology for the robust control of 6 DOFs parallel manipulators has been presented. A discussion on the key issues for a successful control strategy for such systems has been provided. The computational efficiency of the control has two aspects: the first one is the calculation of the complex dynamics model and the second is the determination of the end-effector motion. Both aspects can be solved by feedforward desired dynamics compensation, that is in this sense more appropriate then the feedback dynamics compensation. The use of observers for the actuator velocities allows to increase the control bandwidth. We used the passivity paradigm to develop an approach that merges the feedforward compensation technique with the observer-based feedback to provide a first basic controller, that is locally uniformly and ultimately bounded.

In a second step the basic algorithm is extended with a robust switching term. This has been designed to harmonize with the basic algorithm by consequently using desired dynamics and the consideration of the observer dynamics. The practicability of the approach is improved by restricting the robust term on the friction part of the model, which is classically affected by important and time-varying uncertainty.

The presented methodology has been investigated and substantiated by a set of experiments. It has been demonstrated that the algorithm augmented with the switching term exhibits the best performance. Nevertheless, a tradeoff between accuracy and stability should be met while tuning the controller, especially with respect to significant initial errors.

7. Appendix

Proof of the skew symmetric property of $\dot{M}_a - 2C_a$. Given the skew symmetry of $\dot{M} - 2C$ (Ortega *et al.*, 1998) following transformations hold

$$\begin{aligned} M_a &= J^T M_a J, \\ C_a &= J^T M_a \dot{J} + J^T C J, \end{aligned}$$

substitution yields

$$\begin{aligned} \dot{M}_a - 2C_a &= \frac{d}{dt}\left(J^T M_a J\right) - 2J^T M_a \dot{J} - 2J^T C J, \\ &= J^T \left(\dot{M}_a - 2C_a\right) J + \dot{J} M J - J M \dot{J}. \end{aligned}$$

Let $u \in \mathbb{R}^6$. It results

$$\begin{aligned} u^T \left(\dot{M}_a - 2C_a\right) u &= u^T J^T \left(\dot{M}_a - 2C_a\right) J u \\ &\quad + u^T \dot{J} M J u - u^T J M \dot{J} u, \\ &= u^T \dot{J} M J u - u^T J M \dot{J} u, \\ &= 0 \; \forall u \in \mathbb{R}^6 \end{aligned}$$

since $u^T \dot{J} M J u$ scalar, it results

$$u^T \dot{J} M J u = \left(u^T \dot{J} M J u\right)^T = u^T J M \dot{J} u.$$

yielding that

$$u^T \left(\dot{M}_a - 2C_a\right) u = 0 \quad \forall u \in \mathbb{R}^6$$

which completes the proof.

8. References

Abdallah, C., D. Dawson, P. Dorato and M. Jamshidi (1991). Survey of robust control for rigid robots. *IEEE Control Systems Magazine* **11**(2), 24–30.

Abdellatif, Houssem and Bodo Heimann (2007). *Industrial Robotics: Theory, Modeling and Control*. Chap. Model-Based Control for Industrial Robots: Uniform Approaches for Serial and Parallel Structures, pp. 523–556. Pro-Literatur Verlag.

Abdellatif, Houssem, Bodo Heimann and Jens Kotlarski (2008). *Parallel Manipulators: New Developments*. Chap. On the Robust Dynamics Identification of Parallel Manipulators: Methodology and Experiments, pp. 1–20. I-Tech Education and Publishing. Vienna, Austria.

Abdellatif, Houssem, Martin Grotjahn and Bodo Heimann (2005). High efficient dynamics calculation approach for computed-force control of robots with parallel structures. In: *Proc. of the 44th IEEE Conference on Decision and Control and the 2005 European Control Conference, CDC-ECC05*. Seville, Spain. pp. 2024–2029.

Abdelllatif, Houssem, Martin Grotjahn and Bodo Heimann (2007). Independent identification of friction characteristics for parallel manipulators. *ASME Journal of Dynamic Systems, Measurment and Control* **129**, 294–302.

Berghuis, Harry (1993). Model-based Robot Control: from Theory to Practice. PhD thesis. Universiteit Twente. Enschede, The Netherlands.

Berghuis, Harry and Henk Nijmeijer (1994). Robust control of robots via linear estimated state feedback. *IEEE Transactions on Automatic Control* **39**(10), 2159–2162.

Burdet, E., M. Honegger and A. Codourey (2000). Controllers with desired dynamic compensation and their implementation of a 6dof parallel manipulator. In: *Proc. of the 2000 IEEE/RSJ Int. Conf. on Intelligent Robots and Systems*. Takamatsu, Japan. pp. 39–45.

Caccavale, Fabrizio, Bruno Siciliano and Luigi Villani (2003). The tricept robot: Dynamics and impedance control. *IEEE/ASME Transactions on Mechatronics* **8**(2), 263–268.

Celani, Fabio (2006). A luenberger-style observer for robot manipulators with position measurements. In: *Proc. of 14th Mediterranean Conference on Control and Automation*. Ancona, Italy.

Cheng, Hui, Yiu Kuen and Zexiang Li (2003). Dynamics and control of redundantly actuated parallel manipulators. *IEEE/ASME Transactions on Mechatronics* **8**(4), 483–491.

de Wit, C. Canudas and J.-J. E. Slotine (1991). Sliding observers for robot manipulators. *Automatica* **27**(5), 859–864.

Denkena, Berend, Christian Holz and Houssem Abdellatif (2006). Model-based control of a hexapod with linear direct drives. *Int. Journal of Computer Integrated Manufacturing* **19**(5), 463–472.

Egeland, Olav (1987). On the robustness of the computed torque technique in manipulator control. *Modeling, Identification and Control* **8**(3), 149–158.

Kim, Dong Hwan, Ji-Yoon Kang and Kyo-IL Lee (2000). Robust tracking control design for a 6 dof parallel manipulator. *Journal of Robotic Systems* **17**(10), 527–547.

Kim, Hag Seong, Young Man Cho and Kyo-IL Lee (2005). Robust nonlinear task space control for a 6 dof parallel manipulator. *Automatica* **41**, 1591–1600.

Kim, Jong-Phil, Sung-Guang Kim and Jeha Ryu (2001). Robust adaptive control of a 6 hexaslide type parallel manipulator. *ICASE: The institute of Control, Automation and Sysetems Engineers, KOREA* **3**(4), 262–267.

Lee, Se-Han, Jae-Bok Song, Woo-Chun Choi and Daehie Hong (2003). Position control of a stewart platform using inverse dynamics control with approximate dynamics. *Mechatronics* **13**, 605–619.

Liu, Guangjun and Andrew A. Goldenberg (1996). Uncertainty decomposition-based robust control of robot manipulators. *IEEE Transactions on Control Systems Technology* **4**(4), 384–393.

Ortega, R., Loría, A., Nicklasson, P.J. and Sira-Ramirez, H.J., Eds. (1998). *Passivity-based Control of Euler-Lagrange Systems*. Communications and Control Engineering. Springer-Verlag.

Pietsch, Ingo T., Mathias Kreft, Oliver T. Becker, Carlos C. Bier and Jüergen Hesselbach (2005). How to reach the dynamic limits of parallel robots? an autonomous control approach. *IEEE Transactions on Automation Science and Engineering* **2**(4), 369–380.

Qu, Zhihua and Darren M. Dawson (1996). *Robust Tracking Control of Robot Manipulators*. IEEE Press. Piscataway, USA.

Qu, Zhihua and John Dorsey (1991*a*). Robust pid control of robots. *Int. Journal of Robotics and Automation* **6**(4), 228–235.

Qu, Zhihua and John Dorsey (1991*b*). Robust tracking control of robots by a linear feedback law. *IEEE Transactions on Automatic Control* **36**(9), 1081–1084.

Ren, Lu, James K. Mills and Dong Sun (2005). Controller design applied to planar parallel manipulators for trajectory tracking control. In: *Proc. of the 2005 IEEE Int. Conference on Robotics and Automation*. Barcelona, Spain. pp. 980–985.

Slotine, Jean-Jeaques E. and Weiping Li (1991). *Applied Nonlinear Control*. Patience Hall. Engelwood Cliffs, NJ.

Spong, Mark W. (1992). On the robust control of robot manipulators. *IEEE Transactions on Automatic Control* **37**(11), 1782–1786.

Ting, Yung, Yu-Shin Chen and Ho-Chin Jar (2004). Modeling and control for a gough-stewart platform cnc-machine. *Journal of Robotic Systems* **21**(11), 609–623.

Vivas, Andres and Philippe Poignet (2005). Predictive functional control of a parallel robot. *Control Engineering Practice* (13), 863–874.

Wang, Jinsong, Jun Wu, Liping Wang and Tiemin Li (2007). Simplified strategy of the dynamic model of 6-ups parallel kinematic machine for real-time control. *Mechanism and Machine Theory* **42**, 1119–1140.

Wang, Zhiyong and Fathi H. Ghorbel (2006). Control of closed kinematic chains using a singularly perturbed dynamics model. *ASME Journal of Dynamic Systems, Measurement and Control* **128**, 142–151.

Wen, John T. and David S. Bayard (1988). New class of control laws for robotic manipulators, part 1. non-adaptive case. *Int. Journal of Control* **47**(5), 1361–1385.

5

Reinforcement learning approach to object-contact motion with estimation of low-dimensional submanifold and mode-boundary

Yuichi Kobayashi and Shigeyuki Hosoe
Tokyo University of Agriculture and Technology, RIKEN RTC
Japan

1. Introduction

One of the challenging issues for robotic motion learning is the acquisition of object-contact motions. It is known that reinforcement learning is an effective approach to robot motion learning. For example, multi-legged locomotion (Kimura et al., 2001), biped locomotion (Nakanishi et al., 2004), and stand-up motion of a link robot (Morimoto & Doya, 2001) are known as successful applications of reinforcement learning to robot motion learning. However, it is difficult to realize object-contact motion learning by simply applying reinforcement leaning. The object-contact motions include complex dynamics caused by mode-switching of contact/non-contact or stick/slip. The object-contact motions include both continuous dynamics and discrete events, which has been investigated as hybrid systems control problems (van der Schaft & Schumacher, 2000). In robotic applications, such mode switching is an important issue for the development of learning methods but hitherto not so much attention has been paid from reinforcement learning viewpoint. Apart from reinforcement learning researches, mode switching has been actively discussed in robotic literature. For example, (Schlegl et al., 2002) analysed multi-fingered grasping problem as a hybrid system and proposed a hybrid control architecture with grasping and regrasping. (Yashima et al., 2003) proposed a planning strategy for multi-fingered robotic hand with contact mode switching. There are also general approaches for systematic modelling, planning and control such as Mixed Logic Dynamical model (e.g. (Yin et al., 2007)). Most of these approaches, however, assume that the dynamics of the system is completely known. From the viewpoint of increasing abilities of learning robots, it is very important to consider how to deal with unknown factors in hybrid control problems. In this chapter two different learning schemes are proposed for the acquisition of contact-motion control, incorporating some specific characteristics of the contact motion dynamics. The first approach utilizes the characteristics that the actual motion of the system is restricted to a lower-dimensional submanifold by holonomic constraint when a contact mode is maintained (Fig.1). This helps for narrowing the learning space. The other one depends on the characteristics of object contact motions that the dynamics of the system changes with mode switching of

contact/non-contact or stick/slip. By explicitly approximating the boundary of the mode switching, efficiency of learning can be improved.
In the first learning scheme, an acquisition method of mappings with a parameterization of the low-dimensional submanifold and the approximation method of the boundary of a discontinuous reward function have been proposed. The method is evaluated in an object rotating task. The effectiveness of the proposed learning method was verified by making comparisons with ordinal Q-learning and Dyna.
In the second learning scheme, object manipulation with stick/ slip mode switching is realized by a hierarchical structure. The method is applicable to tasks when the desired configuration of the object are specified so that they can be attained only when the switching between stick and slip modes is utilized. The upper layer determines a global trajectory and a sequence of mode switching by reinforcement learning while giving the lower layer a command for small motion not accompanying any (sliding/sticking) mode changes. The lower layer realizes a desired small regional motion using the estimated boundary information about mode switches by Support vector machine (SVM). The proposed learning architecture is evaluated in simulation of an object control task with stick/ slip mode switching.
The remainder of the chapter is composed as follows. In section 2, learning of manipulation with constraint is discussed, followed by evaluation by simulation in section 3. The second manipulation task with stick/ slip mode switching is described in section 4 including the proposal of hierarchical learning architecture. The learning method is evaluated by simulation in section 5. Section 6 gives discussion and conclusions.

2. Learning of Manipulation with Constraint

2.1 Problem assumptions and overview of learning method

The reinforcement learning problem discussed in this section is classified as a control problem with continuous state space $X \subset \Re^{N_1}$ and discrete state set A. The learning agent decides actions by $\pi : X \rightarrow A$ and receives reward $R_t = R(x_t, a_t), x_t \in X, a_t \in A$ with state transition from x_t to x_{t+1}, where $a_t = \pi(x_t)$ and $R(x,a)$ is unknown by the agent. The objective of the agent is to obtain the optimal policy π^* that maximizes cumulated expectation of discounted reward $E_{\pi^*}\left[\Sigma_t \gamma^t R_t\right]$ with discount factor $\gamma(0 < \gamma \leq 1)$.

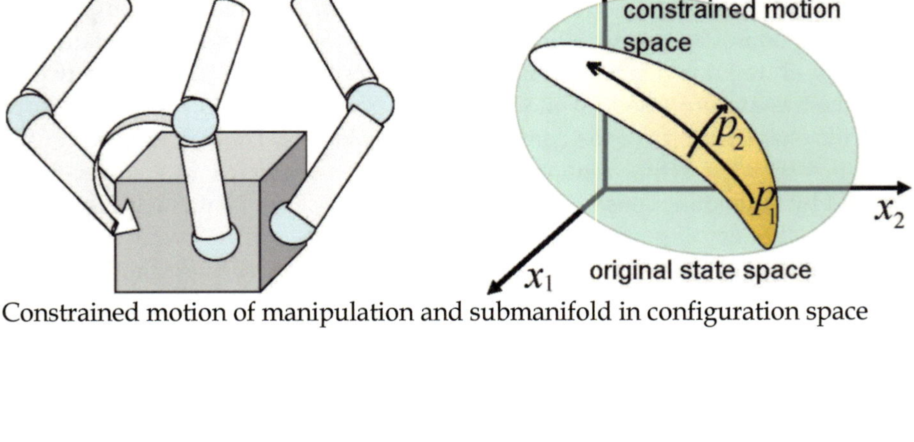

Fig. 1. Constrained motion of manipulation and submanifold in configuration space

In addition to these ordinal formulations, we assume the followings.

(a) The state variable x e X can be expressed as

$$\mathbf{x} = \begin{bmatrix} \mathbf{z} \\ \dot{\mathbf{z}} \end{bmatrix}, \mathbf{z} \in \Re^{N}, 2N = N_1. \tag{1}$$

(b) The motion of z is constrained to a lower-dimensional submanifold P of the configuration space Z as long as the agent is successfully keeping the motion without any failure.

(c) Reward $R^{G}(p, \dot{p}, a)$ is given for reaching the goal region where $p \in P$.

(d)When the agent fails the task, a trial is finished and a new trial is started from an initial state.

The first assumption is a standard framework for dynamic systems. The second assumption is valid for the problems with holonomic constraints. For simplicity, it will be assumed that N=2 i.e., Z is a plane. This implies that P must be a curve in the plane since it is a real submanifold in Z. Therefore, the motion of $\mathbf{z}$ can be expressed by one-dimensional parameter $p \in \Re$ and state of the system is expressed by $[p, \dot{p}] \in X \subset \Re$ as far as the constraint is maintained.

Failing in the fourth assumption means that the agent cannot continue the task because of some fatal error. When the failure happens, it is assumed that the agent can recognize the situation by getting some negative reward. That is, failure is expressed as

$$R_a^F \begin{cases} R_{fail} < 0 & \text{if fails with action a} \\ 0 & \text{otherwise} \end{cases} \tag{2}$$

The boundary between the region where R_a^F takes 0 and the region where R_a^F takes $R_{fail}$is assumed to be continuous in S as indicated in the right hand of Fig. 2. In the application to reinforcement learning framework, the value function is calculated using the sum of reward as $R = R^{G}(p, \dot{p}, a) + R^{F}(p, \dot{p}, a)$.

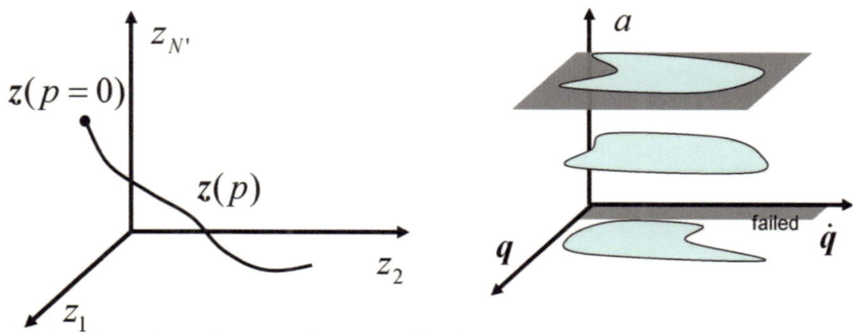

Fig. 2. Constrained motion of manipulation and submanifold in configuration space.

Based on the above assumptions, a learning scheme composed of the following three basic elements will be proposed.

(a) A mapping from Z to one-dimensional parameter space $P \subset \Re$.

(b) Model-based reinforcement learning with system dynamics and reward function approximation

(c) Approximated representation for reward function R_a^F with discontinuous jump as indicated by Eqn. (2)
The reward function approximation in (c) mainly focuses on the reward R^F for failing the task. The function approximation of system dynamics and reward function are presented based on the mapping in (a).

2.2 Mapping to one-dimensional space

It is assumed that z belongs to a plane, i.e., N=2 in (1) and z(t) moves along a one-dimensional submanifold (a curve) as long as no failure happens. In on-line learning, the agent constructs an estimated representation of the curve using observed sampled data of z(t). For this, we can use various kind of curve fitting theories such as Spline, Bezier or Ferguson curve (Farin, 2001). Here, we will apply the second order polynomial interpolation method for simplicity. In on-line learning, z(t) is observed at every time step with interval Δt and new samples are added if necessary. Let Q denote the set of sample points. We add new sample points when

$$Q_L < \min_{q_i e Q} \left\| q_i - z(t) \right\|, \tag{3}$$

where Q_L denotes the threshold length used for judgment. The threshold value Q_L should be sufficiently small for the precise representation of the nonlinear curve.

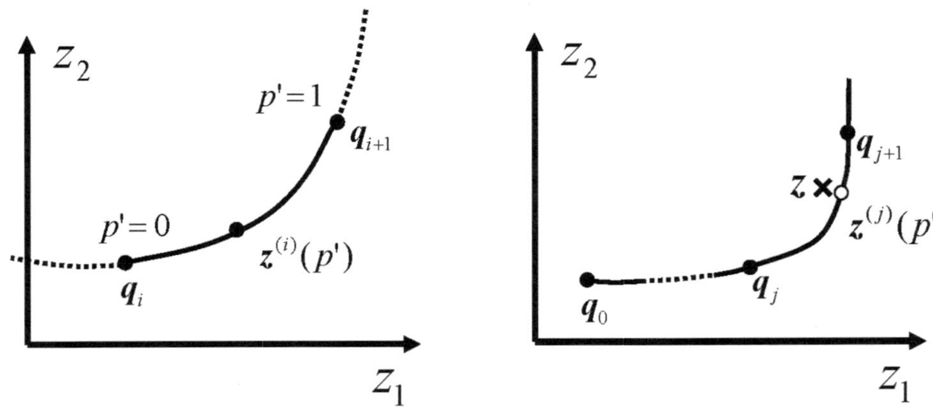

Fig. 3. Mapping to 1D submanifold and finding nearest point on curve

When given n sampled points $q_i \in Z, i = 0, \cdots, n-1$ of **z**(t), i-th interval on the one-dimensional curve can be approximated by

$$\mathbf{z}^{(i)}(p') = \mathbf{q}_i + \mathbf{v}_{1i}p' + \mathbf{v}_{2i}p'^2, p' \in [0,1], \tag{4}$$

where $z^{(i)}$ denotes i-th interval of the curve, as shown in the right hand of Fig. 3. $\mathbf{v}_{1i}, \mathbf{v}_{2i} \in \Re^{N'}$ are coefficient vectors which specify the shape of the curve in the i-th interval. When we differentiate this equation by p', we have

$$\frac{\partial \mathbf{z}^{(i)}}{\partial p'}(p') = \mathbf{v}_{1i} + 2\mathbf{v}_{2i}p', \tag{5}$$

which gives the tangential vector of the curve at $\mathbf{z}^{(i)}(p')$. Considering the continuity of i-th and (i+1)-th intervals ($\mathbf{z}^{(i)}(1) = \mathbf{z}^{(i+1)}(0)$), Eqn. (4) has to satisfy

$$\mathbf{q}_i + \mathbf{v}_{1i} + \mathbf{v}_{2i} = \mathbf{q}_{i+1}, \quad i = 0, \cdots, n-1. \tag{6}$$

Similarly considering the continuity of the tangents ($\frac{\partial \mathbf{z}^{(i)}}{\partial p}(1) = \frac{\partial \mathbf{z}^{(i)}}{\partial p}(0)$) the followings have to hold:

$$\mathbf{v}_{1i} + 2\mathbf{v}_{2i} = \mathbf{v}_{1i+1}, \quad i = 0, \cdots, n-2. \tag{7}$$

Let $I_N \in \Re^{N \times N}$ denote the identity matrix. By letting

$$A_1 = \begin{bmatrix} I_N I_N & & & O \\ & I_N I_N & & \\ & & \ddots & \\ O & & & I_N I_N \end{bmatrix} \in \Re^{nN \times 2nN}, \tag{8}$$

$$A_2 = \begin{bmatrix} I_N & 2I_N & -I_N & & & & O \\ & & I_N & 2I_N & -I_N & & \\ & & & & \ddots & & \\ O & & & & I_N & 2I_N & -I_N \end{bmatrix} \in \Re^{(n-1)N \times 2nN}, \tag{9}$$

and

$$V = \begin{bmatrix} \mathbf{v}_{10} \\ \mathbf{v}_{20} \\ \vdots \\ \mathbf{v}_{1n-1} \\ \mathbf{v}_{2n-1} \end{bmatrix} \in \Re^{2nN}, \quad Q_1 = \begin{bmatrix} \mathbf{q}_1 - \mathbf{q}_0 \\ \vdots \\ \mathbf{q}_n - \mathbf{q}_{n-1} \end{bmatrix} \in \Re^{nN}, \tag{10}$$

we can rewrite (6) and (7) in the matrix form as

$$AV = Q, \; A = \begin{bmatrix} A_1 \\ A_2 \end{bmatrix}, Q = \begin{bmatrix} Q_1 \\ O_{(n-1)N \times 1} \end{bmatrix}. \tag{11}$$

V can be calculated using pseudo-inverse matrix as

$$V = A^T \left(AA^T\right)^{-1} Q. \tag{12}$$

Thus, we can decide the parameters $\mathbf{v}_{1i}$ and $\mathbf{v}_{2i}$ from sample points $\mathbf{q}_i$. Using V, we can specify p for observed z by finding the nearest point on curve. The squared error for i-th interval on the curve is defined as

$$E_i(p', z) = \left\| \mathbf{q}_i + V_{1i}p' + V_{2i}p'^2 - z \right\|^2, p' \in [0,1], i = 0, \ldots, n-1. \tag{13}$$

Parameter p and I giving the minimum will be represented as

$$[i_{min}, p_{min}] = \arg\min_{i,p'} E_i(p', \mathbf{z}). \tag{14}$$

The minimum of $E_i(p, z)$ can be obtained by solving the cubic equation derived from the local minimum condition $\frac{\partial E_i(p,z)}{\partial p}=0$. Eqn. (14) is calculated at every time step t to obtain $i_{min}(t)$ and $p_{min}(t)$. By connecting parameterization (4) for subintervals, the whole curve can be parameterized by:

$$p(t)=i_{min}(t)+p_{min}(t) \tag{15}$$

$$\dot{p}(t) = (p(t) - p(t-\Delta t))/\Delta t. \tag{16}$$

Since p_{min} and i_{min} are determined from **z** by Eqn(14), p_{min}, i_{min} and therefore p is described as a function of **z**. We interpret this as Eqn. (14) and Eqn. (15) defines a mapping from **z**(t) to p(t). When the minimum in Eqn. (14) is not attained in any intervals, that is, **z** lies in the outside of the curve, p(t) is calculated by

$$p(t)=\begin{cases} -\|q_0-z\|/Q_L & if\ \|q_0 - z(t)\| < \|q_{n-1} - z(t)\| \\ n+\|q_{n-1}-z\|/Q_L & otherwise \end{cases} \tag{17}$$

instead of Eqn. (15). We define $\mathbf{s}(t)\equiv[p(t),\dot{p}(t)]^T$ for later discussion. In reinforcement learning application, the sample points are obtained successively while exploration.

2.3 Approximation of Reward Function

Now, based on the obtained mapping as above, the reward function can be represented by using lower dimensional parameter as $R_a(\mathbf{s})$ instead of $R(\mathbf{x},a)$. Based on the assumption for the reward function, two components of $R_a(\mathbf{s})$ are to be approximately constructed through learning.

$$\tilde{R}_a(s)=\tilde{R}_a^G(s)+\tilde{R}_a^F(s),\ \forall a\in A \tag{18}$$

The approximate learning of $\tilde{R}_a^F(\mathbf{s})$ is done by following procedures. Suppose the reward for failure R_{fail} is given at time t_{fail}. Sample is added at this moment as shown below.

$$M_a \leftarrow M_a \cup \{u^a\},\ u^a=\left[s\left(t_{fail}\right)^T, R_{fail}\right]^T, \tag{19}$$

where $a=a(t_{fail})$. Similarly, observed data at the previous time step of t_{fail} is also added by

$$M_a \leftarrow M_a \cup \{u^a\},\ u^a=\left[s\left(t_{fail}-\Delta t\right)^t, 0\right]^T. \tag{20}$$

Using the sample set M_a, the approximate function for $\tilde{R}_a^F(\mathbf{s})$ is constructed by the nearest neighbor method. First, we find $\mathbf{u}_{min}(s)$ as

$$u_{min}(s)=\arg\min_{u^a\in M_a}\left\|s-\left[u_1^a,u_2^a\right]^T\right\| \tag{21}$$

Using above, the approximation function is defined by

$$\tilde{R}_a^F(s)=u_{min3}^a(s), \tag{22}$$

where u_i^a denotes i-th component of vector $\mathbf{u}^a$. Left hand of Fig. 4. shows an example of approximating $\tilde{R}_a^F(\mathbf{s})$ using nearest neighbour method, which corresponds to Voronoi division of the state space. A circle in the figure corresponds to $\mathbf{u}_3^a=0$ and a cross corresponds to $\mathbf{u}_3^a = R_{fail}$.

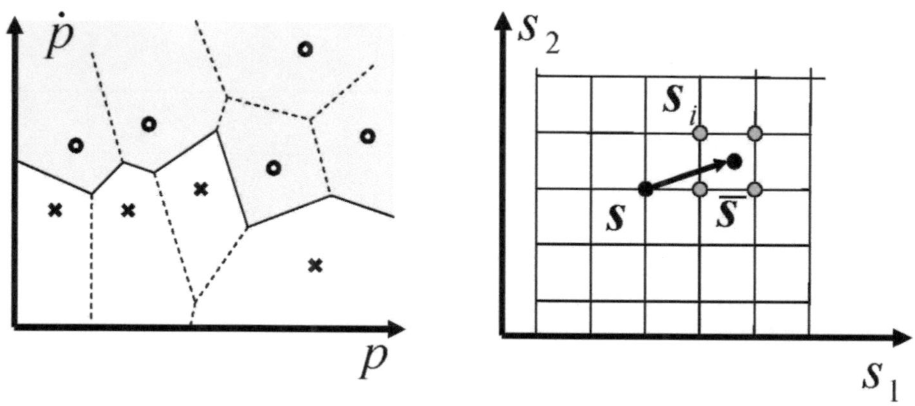

Fig. 4. Voronoi division of state space and state transition approximation

2.4 Reinforcement learning of Controlling Constrained Motion

The basic idea of the proposed learning method is based on the model-based learning method, which is known as Dyna (Sutton, 1991). The main procedures for model-based reinforcement learning are as follows:

- state transition and reward function are approximated by online sampling,
- value function V(s) is updated using these approximations, and
- policy π is improved successively using V(s).

State transition can be denoted as a function of state variables and action.

$$\dot{s}(t) = f(s(t), a(t)). \tag{23}$$

The state transition function is approximated by the same framework of the nearest-neighbor method described in section 2.3. Value function V(s) is updated with respect to each grid as

$$V(\mathbf{s}) = \max_{a \in A}\left[\gamma \sum_{i=1}^{d} \rho(\mathbf{s}_i | \mathbf{s}, a) V(\mathbf{s}_i) + \tilde{R}_a(\mathbf{s})\right] \tag{24}$$

as discussed in optimal control (e.g. (Munos & Moore, 2001)). The state transition probability ρ is calculated for the neighbors in the two-dimensional grid as

$$\rho(\mathbf{s}_i | \mathbf{s}, a) = \frac{\exp\left(-\frac{\|\mathbf{s}_i - \bar{\mathbf{s}}\|^2}{\sigma^2}\right)}{\sum_{k=1}^{d} \exp\left(-\frac{\|\mathbf{s}_k - \bar{\mathbf{s}}\|^2}{\sigma^2}\right)}, \quad \bar{\mathbf{s}} = \mathbf{s} + \tilde{f}(\mathbf{s}, a)\Delta t, \tag{25}$$

where $\tilde{f}$ denotes the approximated state transition function. The right hand of Fig.4. shows an example of $\bar{\mathbf{s}}$ and its neighbor grids in the two-dimensional case. In the ε-greedy policy, action is chosen randomly with a small probability e and decided by

$$\pi(\mathbf{p}) = \arg\max_{a \in A}\left[\gamma \sum_{i=1}^{d} \rho(\mathbf{p}_i | \mathbf{p}, a) V(\mathbf{p}_i) + \tilde{R}_a(\mathbf{p})\right], \tag{26}$$

with probability 1 − ε.
The total learning algorithm proposed for the constrained motion acquisition including successive addition of sample $\mathbf{q}_i$ is described as Table 1.

1.	$Q \leftarrow \{\mathbf{q}_0\}, \mathbf{q}_0 = [q_1(t_0), q_2(t_0)]^T$
2.	Initialize state $t = 0, \mathbf{x}(0) = \mathbf{x}_0$
3.	Observe $\mathbf{z}(t)$
4.	If Eqn. (3) holds, add sample data $\mathbf{q}_j = \mathbf{q}(t), j = \lvert Q \rvert + 1, Q \leftarrow Q \cup \{\mathbf{q}_j\}$
5.	Obtain p(t) by the mapping using Eqn. (15)
6.	Decide control input based on ε-greedy: (a) Random action selection at probability ε (b) Otherwise, $a = \pi(\mathbf{s}(t))$.
7.	State transition with reward R_t and judgement for failing the task (whether $R_t = R_{fail}$ or not)
8.	Data sampling for identification of state transition and reward functions.
9.	If trial completed or failed, update $V(\mathbf{s})$ and $\pi(\mathbf{s})$ and go to 2.
10.	$t \leftarrow t+\Delta t$ and go to 3.

Table 1. Learning algorithm for constrained motion

3. Simulation and evaluation of Constrained Motion Learning

3.1 Dynamics of the model

A two-link manipulator and an object with one-DOF rotational movement are shown in Fig. 5. The contact between the link and the object is point contact without rolling. The objective of the task is to stabilize the object around some posture by the link manipulator.

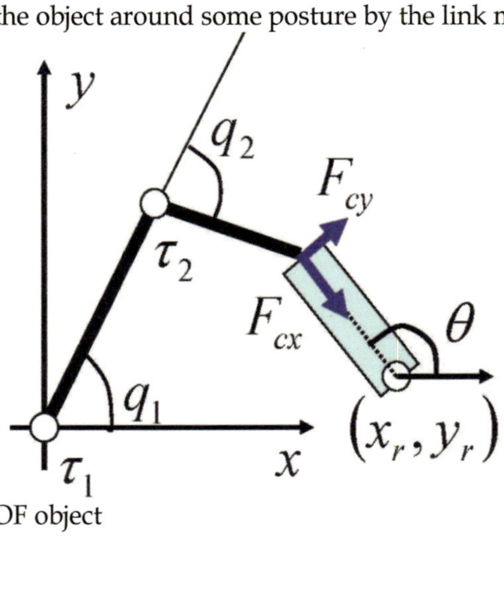

Fig. 5. Rotation of one-DOF object

Let q_1 and q_2 be the angles of the link manipulator and θ be angle of the object. τ_1 and τ_2 are the torques for joints 1 and 2, respectively. Because of the constraint, the actual motion of this problem has one DOF as long as the manipulator remains in contact with the object. Letting $q = [q_1, q_2]^T$, the position constraint while the end of link 2 contacts with the object can be described as

$$l_1 \cos q_1 + l_2 \cos(q_1 + q_2) = x_r L \cos\theta \tag{27}$$

$$l_1 \sin q_1 + l_2 \sin(q_1 + q_2) = y_r L \sin\theta, \tag{28}$$

where L, l_1 and l_2 are the lengths of the object, link 1 and link 2, respectively, and $[x_r, y_r]^T$ is the center of object rotation. The motion equation of the manipulator can be written as

$$M_q(q)\ddot{q} + h_q(q,\dot{q}) + J_q^T F = [\tau_1, \tau_2]^T, \tag{29}$$

where $F \in \Re^2$ denotes the contact force between the link and the object. Let m and I_0 denote the mass and inertia of the object, respectively. μ_R denotes the viscosity resistance coefficient. The dynamics of the object can be written as

$$I_o\ddot{\theta} + \mu_R\dot{\theta} + \frac{1}{2}Lmg\cos\theta - J_\theta^T F = 0\ . \tag{30}$$

We define h_θ as

$$h_\theta(\theta,\dot{\theta}) = \mu_R\dot{\theta} + \frac{1}{2}\mathrm{Lmg}\cos\theta. \tag{31}$$

The Jacobian matrices J_q and J_θ in Eqn. (29) and (30) are obtained by differentiating the constraints (27) and (28) by time. Using above-mentioned notations, the integrated motion equation may be written as

$$\begin{bmatrix} M_q(q) & 0 \\ & 0 \\ 0 & 0 & I_0 \end{bmatrix} \begin{bmatrix} \ddot{q}_1 \\ \ddot{q}_2 \\ \ddot{\theta} \end{bmatrix} + \begin{bmatrix} h_q(q,\dot{q}) \\ h_\theta(\theta,\dot{\theta}) \end{bmatrix} + \begin{bmatrix} J_q^T \\ J_\theta^T \end{bmatrix} F = \begin{bmatrix} \tau_1 \\ \tau_2 \\ 0 \end{bmatrix}, \tag{32}$$

or equivalently as

$$M(z)\ddot{z} + h(z,\dot{z}) + J^T F = \tau, \tag{33}$$

where $z = [q_1, q_2, \theta]^T$ and $\tau = [\tau_1, \tau_2, 0]^T$. The contact force F can be written as

$$F = (JM^{-1}J^T)^{-1}\{JM^{-1}(\tau - h) + \dot{J}\dot{z}\}. \tag{34}$$

When we define F_c in the object coordinate

$$F_C = \begin{bmatrix} F_{CX} \\ F_{CY} \end{bmatrix} = R_\theta F \tag{35}$$

by rotating F by θ, the conditions for maintaining contact with the object are written as

$$F_{CY} > 0 \tag{36}$$

$$\mu F_{CY} > F_{CX}, \tag{37}$$

where μ denotes the static friction coefficient. Eqn. (37) gives the condition for avoiding slipping. If either of the conditions is broken, a trial is aborted and the agent restarts the trial from the initial state z_0. These conditions for maintaining contact give the boundary of reward function for maintaining constraints.

3.2 Assumption for learning agent

It is assumed that the agent

- observes q_1 and q_2 and their velocities $\dot{q}_1$ and $\dot{q}_2$
- he force F_c and the object angle θ, but receives the reward for reaching goal region and the reward for failing to maintain contact with the object.

In addition to these assumptions for agent observation, the agent utilizes the knowledge described in section 3.1 through the proposed mapping method and reward function approximation.

3.3 Simulation Conditions

We evaluate the proposed learning method in the problem described in section 3.1. Although we show the effectiveness of the proposed learning method through a problem where analytical solutions can be easily found, it does not mean this method is restricted to such problems. The method can be applied to other problems where we can not easily derive analytical solutions, e.g., manipulation problems with non-spherical fingertips or with moving joints structures, which can be seen in human arms.

Physical parameters are set as $l_1 = 2, l_2 = 2, L = 1/2$ [m], $m_0 = 0.8$[kg], μ= 0.8. $[x_r, y_r] = [2.5, 0]$ and the initial state is set as $\mathbf{z}_0 = [3/\pi, -2/3\pi, \pi]^T$. Sampling time for the control is 0.25[sec] and is equivalent to one step in a trial. We have 4 x 4 actions by discretizing τ_1 and τ_2 into [60, 30, 0,-60][Nm]. One trial is finished after 1,000 steps or when either of conditions (27) or (28) is broken. If either $\theta(t)$ or $\dot{\theta}(t)$ goes out of the interval $[\theta_{min}, \theta_{max}] = [0, \pi]$ or $[\dot{\theta}_{min}, \dot{\theta}_{max}] = [-5, 5]$, a trial is also aborted. The reward function is given as

$$R(x,a) = R^1(x,a) + R^2(x), \tag{38}$$

where each component is given by

$$R^1(x,a) = \begin{cases} 10 & if \ |\theta - \theta_d| < \pi/10 \\ -|\theta - \theta_d| & otherwise \end{cases} \tag{39}$$

and

$$R^2(\boldsymbol{x}) = \begin{cases} 0 & \text{if (27) and (28) hold} \\ -100 & \text{otherwise} \end{cases} \tag{40}$$

The desired posture of the object is $\theta_d = \pi/2$. The threshold length for adding new samples in the mapping construction is set as Q_L =0.05. The state space constructed by $s \in \Re^2$ is divided into 40x40 grids with the the regions $[p_{min}, p_{max}]$ = [0, 5] and $[\dot{p}_{min}, \dot{p}_{max}]$ = [−5, 5]. The parameters for reinforcement learning are set set as ε=0.1 and γ=0.95

The proposed reinforcement learning method is compared with two candidates.

- Model-based reinforcement learning without mapping F_Q using $[q_1, q_2, \dot{q}_1, \dot{q}_2]$ as state variables.
- Ordinal Q-learning with state space constructed by the state variables $s = [p, \dot{p}]$

The first method is applied to evaluate the effect of introducing the mapping to one-dimensional space. The second method is applied to see that the explicit approximation of discontinous reward function can accelerate learning.

3.4 Simulation Results

The obtained mapping is depicted in the left hand of Fig. 6. The bottom circle corresponds to the initial state with $\mathbf{z}_0$ and each circle in the figure denotes a sample. The right hand of Fig. 6. shows the reward profiles obtained through trials. We can see that performance is not always sufficiently good even after many trials. This is caused by the ε-greedy policy and the nature of the problem. When the agent executes random action based on the ε-greedy policy, it can easily fail to maintain contact with the object even after it acquired a sufficiently good policy not to fail.

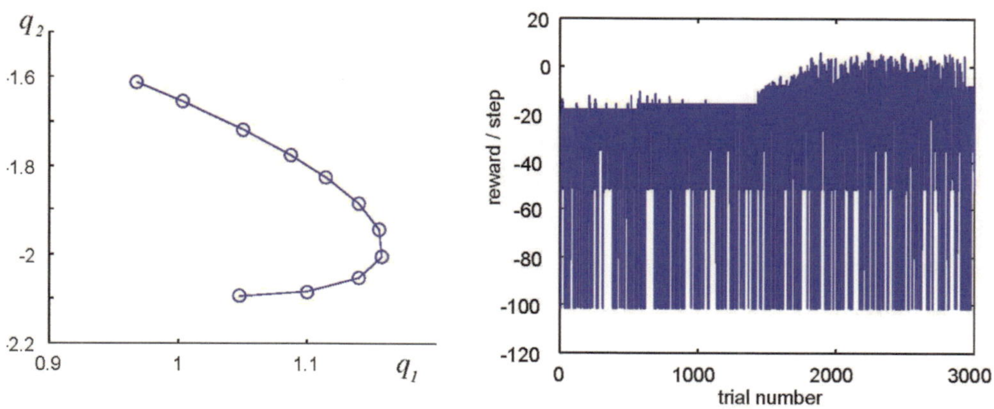

Fig. 6. Obtained 1-D mapping and learning curve obtained by the proposed method

The left hand of Fig.7 shows the state value function $V(s)$. It can be seen that the result of exploration in the parameterized state space is reflected in the figure where the state value is non-zero. The positive state value means that it was possible to reach the desired configuration through trials. The right hand of Fig.7 shows the learning result with Q-learning as a comparison. In the Q-learning case, the object did not reach the desired goal region within 3,000 trials. With four-dimensional model-based learning, it was possible to reach the goal region. Table 2 shows comparisons between the proposed method and the model-based learning method without lower-dimensional mapping. The performances of the obtained controllers after 3,000 trials learning are evaluated without random exploration (that is, ε=0) with ten test sets. The average performance of the proposed method was higher. This is caused by the fact that the controller obtained by the learning method without the mapping failed to keep contact between the arm and the object at earlier stages of the rotating task in many cases, which resulted in smaller cumulated rewards. Additionally in the case of the method without the mapping, calculation time for the control was three times as long as the proposed method case.

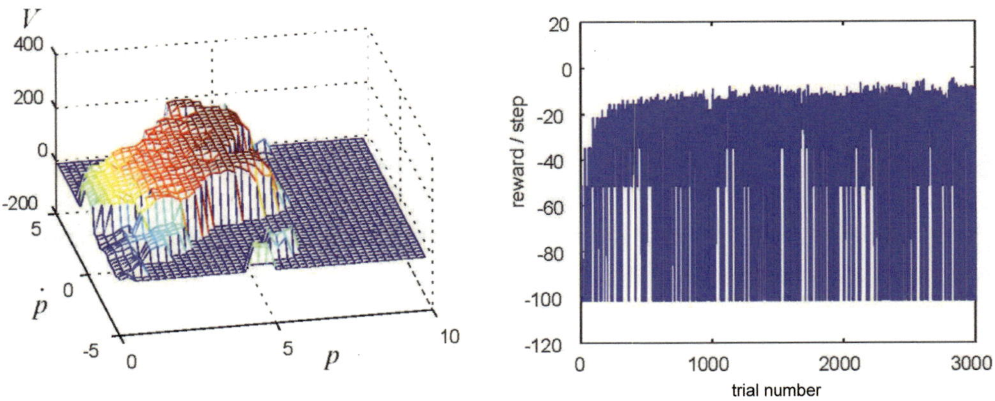

Fig. 7. State value function and learning curve obtained by Q-learning

	Proposed method	Without mapping
Average reward	-3.87	-90.7
Standard deviation	8.53	34.8

Table 2. Comparison with model-based reinforcement learning without mapping

The examples of the sampled data for reward approximation are shown in Fig. 8. Circles in the left hand figure denote $u_3^a = 0$ and the crosses denote $v_3^a = R_{fail}$. The reward functions $\widetilde{R}_{13}^{F}(\mathbf{s})$ approximated using corresponding sample data are also shown in the figure. Fig. 9 shows an example of the trajectories realized by the obtained policy $\pi(s)$ without random action decisions in the parameterized state space and in the physical space, respectively.

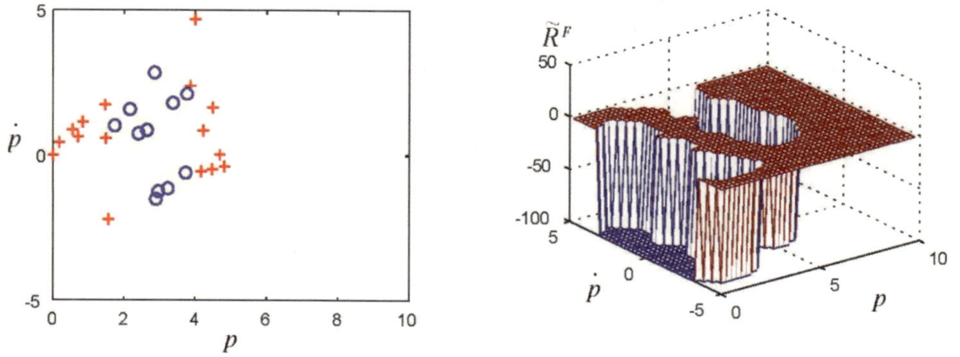

Fig. 8. Sampled data for reward estimation (a=13) and approximated reward $\tilde{R}_{13}^F(s)$

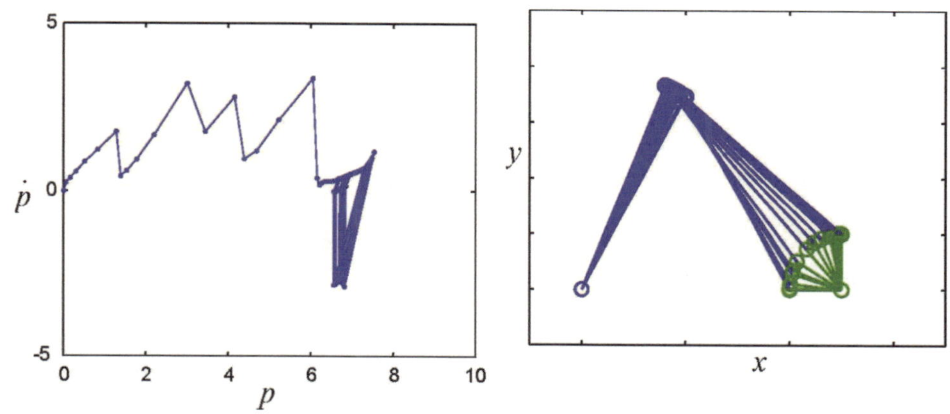

Fig. 9. Trajectory in the parameterized state space and trajectory of links and object

3.5 Discussion

The result of simulation showed that the reinforcement learning approach effectively worked for the manipulation task. Through comparison between Q-learning and model-based reinforcement learning without the proposed mapping, we saw that the proposed mapping and reward function approximation improved the learning performance including calculation time. Some parameter settings should be adjusted to make the problem more realistic, e.g., friction coefficient, which may require more trials to obtain a sufficient policy by learning. For the purpose of focusing on the state space construction, we assumed discrete actions in the learning method. In the example of this manipulation task, however, the continuous control of input torques plays an important role in realizing more dexterous manipulation. It is also useful for the approximation of reward to consider the continuity of actions. The proposed function approximation with low-dimensional mapping is expected to be a base for such extensions.

4. Learning of Manipulation with Stick/Slip contact mode switching

4.1 Object Manipulation Task with Mode Switching

This section presents a description of an object manipulation task and a method for simulating motions with mode switching. Note that mathematical information described in this section is not used by the learning agent. Thus, the agent can not predict mode switching using equations described in this section. Instead, it estimates the mode boundary by directly observing actual transitions (off-line).

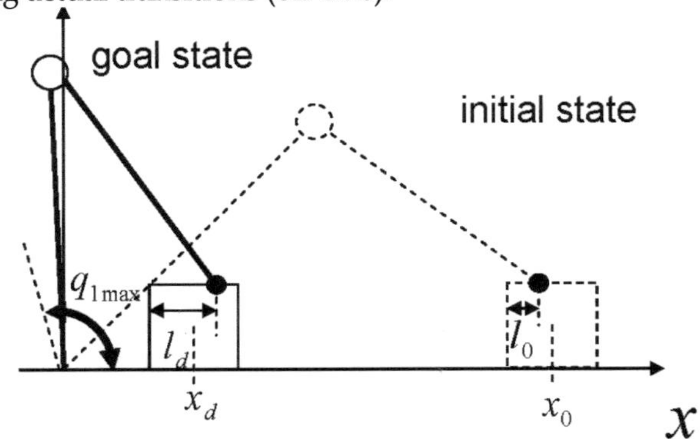

Fig. 10. Manipulation of an object with mode switching

An object manipulation task is shown in Fig.10. The objective of the task is to move the object from initial configuration to a desired configuration. Here, it is postulated that this has to be realized by putting robot hand onto the object and moving it forward and backward by utilizing friction between the hand and the object as shown in the figure. Note that, due to the limited working ranges of joint angles, mode change (switching contact conditions between the hand and the object from slipping mode to stick mode and vice versa) is generally indispensable to achieve the task. For example, to move the object close to the manipulator, it is necessary once to slide the hand further (from the initial position) on the object so that the contact point becomes closer to point B in Fig.11.

Physical parameters are as described in Fig.11. The followings are assumed about physical conditions for the manipulation:

- The friction is Coulomb type frictions and the coefficient of static friction is equal to the coefficient of kinetic friction
- The torque of the manipulator is restricted to $\tau_{1\min} \le \tau_1 \le \tau_{1\max}$ and $\tau_{2\min} \le \tau_2 \le \tau_{2\max}$.
- The joint angles have limitations of $q_{1\min} \le q_1 \le q_{1\max}$ and $q_{2\min} \le q_2 \le q_{2\max}$.
- The object and the floor contact at a point and the object does not do rotational motion.
- A mode where both contact points (hand and object / object and floor) are slipping is omitted (Controller avoids such mode).

In what follows the contact point between the hand and the object will be referred as point 1 and the contact point between the object and the floor as point 2. It is assumed that the agent can observe at each control sampling time the joint angles of the manipulator and their velocities and also

- position and velocity of the object and the ones of contact point 1.
- contact mode at contact point 1 and 2 (stick/slip to positive direction of x axis/slip to negative direction of x axis/apart).

Concerning the learning problem, the agent is assumed to know or not know the following factors: It knows the basic dynamics of the manipulator, i.e., gravity compensation and Jacobian matrix are known (they correspond to g_q and J_q in Eqn. (41)). On the other hand, the agent does not know conditions for the mode switching. That is, friction conditions are unknown including friction coefficients. The agent also does not know the limitation of joint angles and sizes (vertical and horizontal lengths) of the object.

From the viewpoint of application to the real robot, it might be not easy to measure the contact mode precisely, because 1) it is difficult to detect small displacement of the object (e.g. assuming visual sensor) and 2) the slipping phenomenon could be stochastic. In the real application, estimation of mode boundary might require further techniques such as noise reduction.

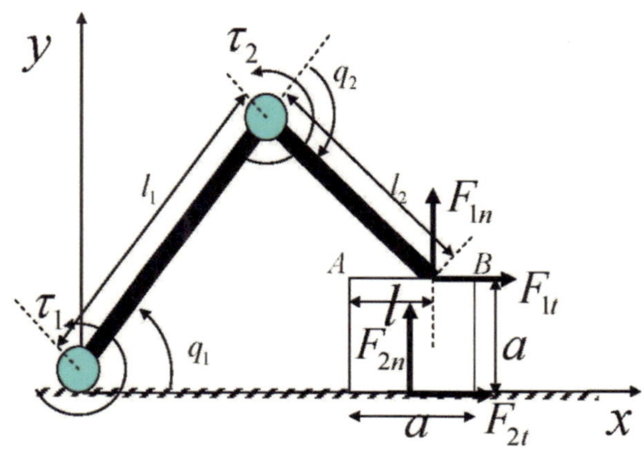

Fig. 11. Manipulator and a rectangular object

4.2 System Dynamics and Physical Simulation

Motion equation of the manipulator is expressed by

$$M_q(q)\ddot{q}+h_q(q,\dot{q})=[\tau_1,\tau_2]^T+J_t^TF_t+J_n^TF_n, \tag{41}$$

where $q=[q_1,q_2]^T$, $F_t=[F_{1t},F_{2t}]^T$, $F_n=[F_{1n},F_{2n}]^T$, $J_t=\left[J_{1t}^T,0_{2\times1}\right]^T$, $J_n=J_n=\left[J_{1n}^T,0_{2\times1}\right]^T$ and $J_q=\left[J_{1t}^T,J_{1n}^T\right]^T$ is Jacobian matrix of the manipulator. F_{it} and F_{in} denote tangential and normal force at point i, respectively. Zero vectors in J_t and J_n denote that the contact forces at point 2 do not affect the dynamics of the manipulator. Letting $\boldsymbol{\varphi}=[x,y]^T$, motion equation of the object is expressed by

$$M_O\ddot{\varphi}+g_O=W_tF_t+W_nF_n, \tag{42}$$

where $g_o=[0,mg]^T$ and

$$W_t = \begin{bmatrix} -1 & 1 \\ 0 & 0 \end{bmatrix}, \ W_n = \begin{bmatrix} 0 & 0 \\ -1 & 1 \end{bmatrix}. \quad \cdot \tag{43}$$

δ_i denotes contact mode at contact point i and defined as

$$\delta_i(t+\Delta t) = \begin{cases} 0(stick) & \text{if } v_{it}(t+\Delta t) = 0 \\ 1(\text{slip to +direction}) & \text{if } v_{it}(t+\Delta t) > 0 \\ -1(\text{slip to -direction}) & \text{if } v_{it}(t+\Delta t) < 0 \end{cases}, \tag{44}$$

where v_{it} denotes relative (tangential) velocity at contact point i. At each contact point, normal and tangential forces satisfy the following relation based on Coulomb friction law.

$$\mu F_n - |F_t| \geq 0. \tag{45}$$

Relative velocities of the hand and the object at contact point 1 are written as

$$v_n = J_n \dot{q} + W_n^T \dot{\varphi}, \ v_t = J_t \dot{q} + W_t^T \dot{\varphi}. \tag{46}$$

By differentiating and substituting Eqns. (41) and (42), the relation between relative acceleration and contact force can be obtained as

$$a = AF + a^0, \ a = [a_n \ \ a_t]^T, \ F = [F_n \ \ F_t]^T, \tag{47}$$

where

$$A = JMJ^T, M = \begin{bmatrix} M_q^{-1} & \\ & M_O^{-1} \end{bmatrix}, J = \begin{bmatrix} J_n & W_n^T \\ J_t & W_t^T \end{bmatrix}, a^0 = \dot{J}\begin{bmatrix} \dot{q} \\ \dot{\varphi} \end{bmatrix} + JM\begin{bmatrix} \tau - h_q \\ g_O \end{bmatrix}. \tag{48}$$

By applying Euler integration to (47) with time interval Δt, relation between relative velocity and the contact force can be obtained as

$$\boldsymbol{v}(t+\Delta t) = KF + \boldsymbol{b}, \quad K \equiv A\Delta t, \quad \boldsymbol{b} \equiv \boldsymbol{a}^0 \Delta t + \boldsymbol{v}(t). \tag{49}$$

On the other hand, normal components of contact force and relative velocity have the following relation.

$$F_{in} \geq 0, \tag{50}$$

$$v_{in} \geq 0, \tag{51}$$

$$F_{in} = 0 \ \text{ or } \ v_{in} = 0. \tag{52}$$

This relation is known as linear complementarity. By solving (49) under conditions of (45) and (50)-(52), contact forces and relative velocities at next time step can be calculated. In this chapter, projected Gauss-Seidel method (Nakaoka, 2007) is applied to solve this problem.

4.3 Hierarchical Architecture for Manipulation Learning

The upper layer deals with global motion planning in x-l plane using reinforcement learning. Unknown factors on this planning level are 1) limitation of state space of x-l plane caused by the limitation of joint angles and 2) reachability of each small displacement by lower layer. The lower layer deals with local control which realizes small displacement given by the upper layer as command. The estimated boundary between modes by SVM is used for control input (torque) generation.

Fig.12 shows an overview of the proposed learning architecture. Configuration of the system is given to the upper layer after discretization and interpretation as discrete states. Actions in the upper layer are defined as transition to adjacent discrete states. Policy defined by reinforcement learning framework gives action a as an output. The lower layer gives control input τ using state variables and action command a. Physical relation between two layers is explained in Fig.4. Discrete state transition in the upper layer corresponds to small displacement in x-l plane. When an action is given as command, the lower layer generates control inputs that realizes the displacement by repeating small motions for small time period Δt until finally s' is reached. In this example in the figure, l is constant during state transition.

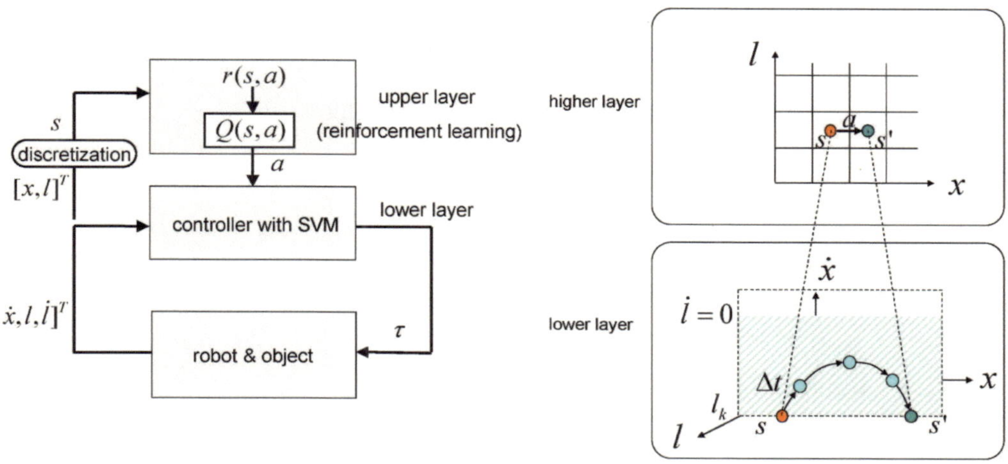

Fig. 12. Hierarchical learning structure

4.4 Upper layer learning for Trajectory Generation

For simplicity and easiness of implementation, Q-learning (Sutton, 1998) is applied in the upper layer. The action value function is updated by the following TD-learning rule:

$$Q(s,a) \leftarrow Q(s,a) + \alpha\left[r + \gamma \max_{a} Q(s',a) - Q(s,a)\right] \tag{53}$$

The action is decided by the ε-greedy method. That is, a random action is selected by small probability ε and otherwise the action is selected as $a=\mathrm{argmax}Q(s,a)$. The actual state transition is achieved by the lower layer. The reward is given to the upper layer depending on the state transition.

4.5 Lower Controller Layer with SVM Mode-Boundary Learning

When current state $X(t)=\left[x(t),l(t),\dot{x}(t),\dot{l}(t)\right]^T$ control input $\tau(t)$ are given, contact mode at next time $\delta(t+\Delta t)$ can be calculated by projected Gauss-Seidel method. This relation between X, u and δ can be learned as a classification problem in X-u space. A nonlinear Support Vector Machine is used in our approach to learn the classification problem. Thus, mode transition data are collected off-line by changing $x, 1, \dot{x}, \dot{1}, \tau_1, \tau_2$. Let m_s denote training set size and $\mathbf{d} \in \Re^{m_s}$ denote a vector with plus or minus ones, where plus and minus correspond respectively to different two modes. In non-linear SVM with Gaussian kernel, by introducing kernel function K (with query point v) as

$$K(v,\mu_i) = \exp\left(-\frac{\|v-\mu_i\|^2}{\sigma^2}\right), \tag{54}$$

where $\mu_i (=[x, l, \dot{x}, \dot{l}, \tau_1, \tau_2]^T)$ denotes i-th data for mode boundary estimation and σ denotes a width parameter for the Gaussian kernel, separation surface between two classes is expressed as

$$\sum_{i=1}^{m_s} d_i w_i K(v,\mu_i) = 0, \tag{55}$$

where w is a solution of the following optimization problem:

$$\min_w \left\{ \frac{1}{2} w^T Q w - e^T w \right\}, \tag{56}$$

where Q is given by

$$Q = \frac{1}{v} + HH^T, \;\; H=D[\mu_0 - e], \; v > 0 \tag{57}$$

and $\mathbf{e} \in \Re^{n_s}$ denote the vector of ones. $D = diag\left[d_1, \ldots, d_{m_s}\right], \mu_0 = \left[\mu_1, \ldots, \mu_{m_s}\right]^T$ and v is a parameter for the optimization problem. Note that matrix D gives labels of modes. For implementation of optimization in (56), Lagrangian SVM (Mangasarian & Musicant, 2001) is used. After collecting data set of D and $\boldsymbol{\mu}_0$ and calculating SVM parameter w, (55) can be used to judge the mode at next time step when $X = \left[x(t), 1(t), \dot{x}(t), \dot{1}(t)\right]^T$ is given.

When the action command a is given by the upper layer, the lower layer generates control input by combining PD control and mode boundary estimation by SVM. Let $\Delta\chi(a) = [\Delta x, \Delta l]^T$ denote displacement in x-l space which corresponds to action a (notice that here χ is different from X because velocities are not necessary in the upper layer). When Δl = 0, the command a means that the modes should be maintained as $\delta_1 \neq 0$ and $\delta_2 = 0$. When Δl = 0 on the other hand, it is required that the modes should be $\delta_1 = 0$ and $\delta_2 \neq 0$. Thus, the desired mode can be decided depending on the command $\Delta\chi(a)$. First, PD control input u_{PD} is calculated as

$$u_{PD} = -K_P J_q^T [\Delta x + \Delta 1, 0]^T - K_D \dot{q} + g_q - J_q^T F_d, \tag{58}$$

where F_d is desired contact force and K_P , K_D are PD gain matrices. In order to realize the desired mode retainment, $\mathbf{u}_{PD}$ is verified by (55). If it is confirmed that $\mathbf{u}_{PD}$ maintains the desired mode, $\mathbf{u}_{PD}$ is used as control input. If it is found that $\mathbf{u}_{PD}$ is not desirable, a searching algorithm for finding u is applied until a desirable control input is found. $\tau_1 - \tau_2$ space is discretized into small grids. The grid points are tested one by one using (55) until the desirable condition is satisfied. The total learning algorithm is described in Table 3.

1. Lower layer: Estimate mode switching boundary by sampling mode transition data off-line.
2. Environment: Initialize state.
3. Observe X(t) and obtain discrete state s.
4. Upper layer: Action selection. Action is given to the lower layer as a subgoal ($\Delta\chi$).
5. Lower layer: Mode retainment/transition operation corresponding action a using mode boundary estimation:
   a) Calculate PD control input $\mathbf{u}_{PD}$ using (58).
   b) Judge whether desired mode retainment is possible using (55).
   c) If it is possible to realize desired motion, use $\mathbf{u}_{PD}$ as control input. Otherwise, explore and find an appropriate input u using SVM.
   d) Environment: State transition to $X(t+\Delta t)$
   e) Repeat a)-d) until subtotal is reached. If mode retainment is failed or the hand went out of limitation, negative reward is given and state is initialized.
6. Environment: State transition and reward r(s,a) is given to the upper layer.
7. Upper layer: Update Q(s, a) using (53)
8. Repeat 3.-7. until goal is reached or a fixed number of steps are repeated.
9. Go to 2.

Table 3. Algorithm for hierarchical learning of stick/ slip switching motion control

5. Simulation results of Stick/Slip Switching Motion Learning

Physical parameters for simulation are set as followings:

- Lengths of links and sizes of the object: $1_1 = 1.0, 1_2 = 1.0, a = 0{,}336[m]$ (Object is a square.)
- Masses of the links and the object: $m_1 = 1.0, m_2 = 1.0$ [kg]
- Time interval for one cycle of simulation and control: Δt =0.02[sec]
- Coefficients of static (and kinetic) friction: $\mu_1 = 0.6, \mu_2 = 0.2$
- Joint angle limitation is set as $q_{1\min} = 0, q_{1\max} = 1{,}6[rad]$ (No limitation forq_2).
- Torque limitations are set as $\tau_{1\min} = -5, \tau_{1\max} = 20$ and $\tau_{2\min} = -20, \tau_{2\max} = 5$.

Initial states of the manipulator and the object are set as $\left[x_0, 1_0, \dot{x}_0 \dot{1}_0\right]^T = [1.440, 0.1090, 0, 0]^T$. Corresponding initial conditions for the manipulator are $[q_1, q_2, \dot{q}_1, \dot{q}_2]^T = [\pi/3, -\pi/2, 0, 0]^T$. Goal state is given as $[x_d, l_d, \dot{x}_d, \dot{1}_d] = [0.620, 0.3362, 0, 0]^T$ (as indicated in Fig.10)

Parameters for Q-learning algorithm are set as $\gamma = 0.95$, $\alpha = 0.5$ and $\varepsilon = 0.1$. The state space is defined as $0.620 < x < 1.440$, $0 < l < 0.336(= a)$ and x and l axes are discretized into 6. Thus total number of discrete states is 36. There are four actions in the upper layer Q-learning, each corresponds to the transition to adjacent state in x, l space. Reward is defined as r(s, a) $= r^1(\mathbf{s},a) + r^2(\mathbf{s},a)$ and r^1 and r^2 are specified as followings. Let $\mathbf{s}_d$ denote the goal state in discrete state space and r^1 is given as

$$r^1(s,a) = \begin{cases} 10 & if \quad s = s_d \\ 0 & \text{otherwise} \end{cases} \tag{1}$$

r^2 is given as $r^2(\mathbf{s},a) = -1$ when constraints are broken or the hand moves out of the state space.

5.1 Mode boundary estimation by SVM

Before applying reinforcement learning, mode transition data are collected and used for mode boundary estimation by SVM. Data are sampled for grid points in X, by discretizing $\lfloor x, l, \dot{x}, \dot{l}, \tau_1, \tau_2 \rfloor$ by [5, 10, 10, 10, 10, 10]. Two graphs in Fig. 13. show examples of mode boundary estimation. In the left hand, $x - \dot{x}$ plane is shown by fixing other variables as l = 0.183 and $\boldsymbol{\tau} = [-1,5]^T$ by setting $\dot{l} = 0$. The curve in the figure shows the region where mode 'stick' for contact point 1 and mode 'slip to negative direction of x-axis' for contact point 2 are maintained. In the left hand, $l - \dot{l}$ plane is shown by fixing other variables as l = 0.966 and $\boldsymbol{\tau} = [5.5, 2.5]^T$ by setting $\dot{x} = 0$. The curve shows the region where mode 'slip to positive direction of x-axis' for contact point 1 and mode `stick' for contact point 2 are maintained.

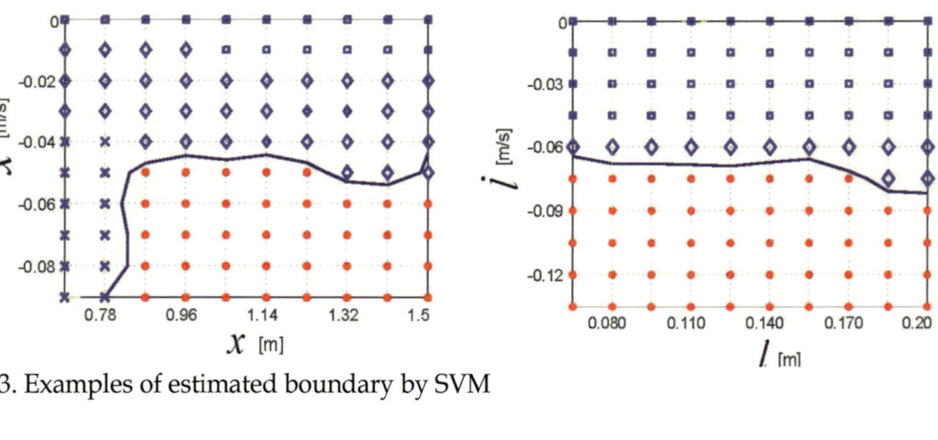

Fig. 13. Examples of estimated boundary by SVM

5.2 Learning of manipulation

The profile of reward per step (average) is shown in the left hand of Fig.14. Trajectories from initial configuration to the desired one were obtained after 200 trials. It takes value of around 6 or 7 because it is an average of one trial, in which reward of -1 is obtained at the beginning and later reward of 10 is obtained, as far as it stays at the desired configuration. The right hand of Fig.14 shows state value function V(s), which is calculated from action value function by $\max_a Q(s,a)$ (s_1 and s_2 correspond to discretization of l and x, respectively). It can be seen that the value of the desired state is the highest in the state space. 500 steps trials are tested for 20 times. For all cases, it was possible to achieve the control to the desired state, though numbers of trials required to achieve learning are different (around several hundred trials).

The left hand of Fig.15 shows a trajectory obtained by the hierarchical controller with the greedy policy. Totally five mode switching are operated to achieve desired configuration. The right hand of Fig.15 shows the profiles of joint torques. Continuous torques are calculated by the lower layer.

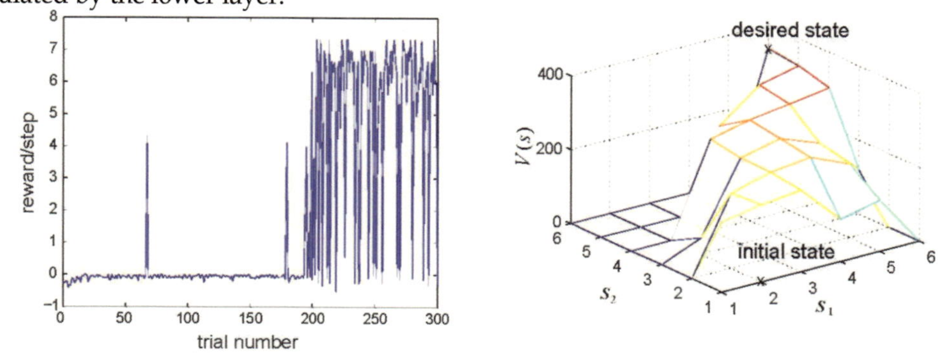

Fig. 14. Learning profile and obtained state value function

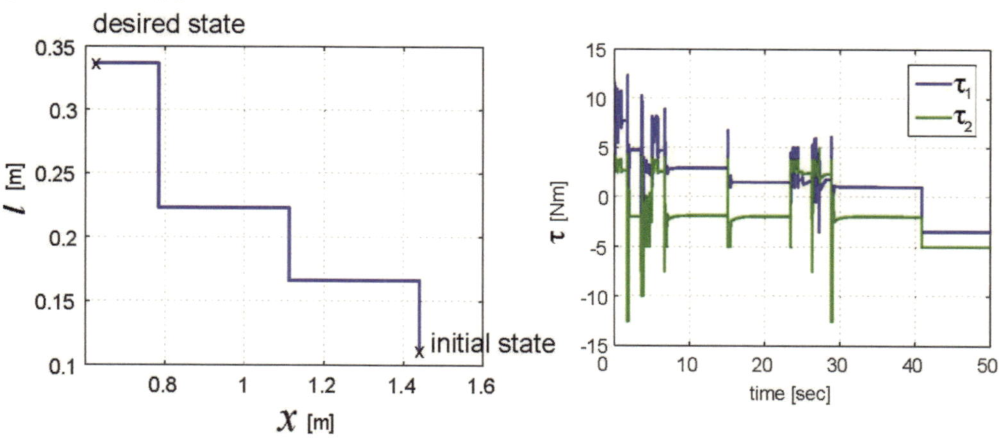

Fig. 15.Trajectory on l-x plane and joint torque profiles

Fig.16. shows contact modes δ for contact point 1 and 2. By comparing two figures, it can be seen that when $\delta_1 = 1$ (contact point 1 is slipping and the hand is moving to right), $\delta_2 = 0$ (contact point 2 is stick mode and the object is stopping) holds. On the contrary, when $\delta_1 = 0$ (contact point 1 is stick mode), $\delta_2 = -1$ (contact point 2 is slipping and the object is moving to left) holds. That is, the hand is moving together with the object. Thus, the manipulator is switching `slipping the hand on the object to right' mode and `moving the object to left' mode. Note that there are instances when both contact modes becomes stick, that is, $\delta_1 = \delta_2 = 0$. This is caused by the learning architecture, which requires stopping at the end of each action of Q-learning. If there is no stop for each action, the total motion would be much more smooth and faster.

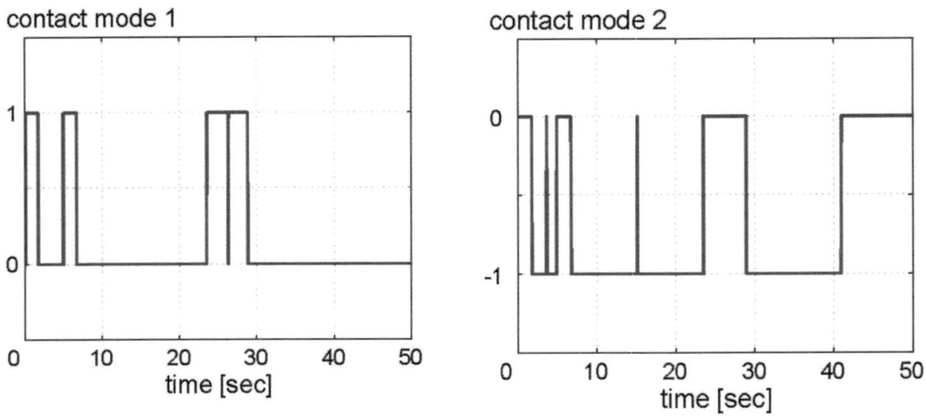

Fig. 16. Examples of estimated boundary by SVM

5.3 Discussion

The lower layer controller achieved local control of the manipulator using SVM boundary obtained off-line sampling. On-line data sampling and on-line boundary estimation of the mode boundaries will be one of our future works. On the other hand, there were some cases where the lower layer controller could not find appropriate torques to realize desired mode. Improvement of the lower layer controller will realize faster learning in the upper layer. One might think that it would be much easier to learn mode boundary in $F_{it} - F_{in}$ space using measurement of contact force F_i for contact point i, because the boundary can be expressed by simple linear relation in contact force space. There are two reasons for applying boundary estimation in the torque space: 1) In more general cases, it is not appropriate to assume that contact forces can be always measured. E.g., in whole body manipulation (Yoshida et al., 2006), it is difficult to measure contact force because contact can happen at any point on the arm. 2) From the viewpoint of developing learning ability, it is also an important learning problem to find an appropriate transformation of coordinate systems so that boundaries between modes can be simply expressed. This will be also one of our future works.

In order to extend the proposed framework to more useful applications such as multi-finger object manipulation, a higher-dimensional state space should be considered. If dimension of the state space is higher, the boundary estimation problem by SVM will require more computational load. The problem 2) mentioned above will be a key-technique to realize a compact and effective boundary estimation to the high-dimensional problems. The dimension of state space for the reinforcement learning should remain low enough so that the learning approach is applicable. Otherwise, other planning techniques might be better to be applied.

6. Conclusion

In this chapter, we proposed two reinforcement learning approaches for object contact robotic motion. The first approach realized a holonomic constrained motion control by making use of a function giving a map from the general motion space to the constrained lower dimensional one and the reward function approximation. This mapping can be regarded as giving function approximation for the extraction of nonlinear lower dimensional parameters. By comparing the proposed method with the ordinal reinforcement learning method, the superiority of the proposed learning method was confirmed. From a more general perspective, we are investigating multidimensional mapping for broader applications. In addition, it is important to consider the continuity of action (force control input) in the manipulation task.

In the second approach, a hierarchical approach of mode switching control learning was proposed. In the upper layer, reinforcement learning was applied for global motion planning. In the lower layer, SVM was applied to learn the boundaries between contact modes and utilized to generate control input which realized mode retainment control. In simulation, it was shown that an appropriate trajectory was obtained by reinforcement learning with mode switching of stick/slip. For further development, fast learning of mode boundaries will be required.

7. References

Andrew G. Barto, Steven J. Bradke & Satinder P. Singh: Learning to Act using Real-Time Dynamic Programming, Artificial Intelligence, Special Volume: Computational Research on Interaction and Agency, 72, 1995, pp. 81-138.Gerald Farin: Curves and Surfaces for CAGD, Morgan Kaufmann Publishers, 2001.

Z. Gabor, Z. Kalmar, & C. Szesvari: Multi-criteria reinforcement learning, Proc. of the 15thInt. Conf. on Machine Learning, pp. 197-205, 1998.

Peter Geibel: Reinforcement Learning with Bounded Risk, Proc. of 18th Int. Conf. on Machine Learning, pp. 162-169, 2001.

H. Kimura, T. Yamashita and S. Kobayashi, Reinforcement Learning of Walking Behaviorfor a Four-Legged Robot, Proc. of IEEE Conf. on Decision and Control, 411-416,2001.

Cheng-Peng Kuan & Kuu-Young Young: Reinforcement Learning and Robust Control for Robot Compliance Tasks, Journal of Intelligent and Robotic Systems, 23, pp.165-182,1998.

O. L. Mangasarian and David R. Musicant, Lagrangian Support Vector Machines, Journal of Machine Learning Research, 1, 161-177, 2001.

H. Miyamoto, J. Morimoto, K. Doya and M. Kawato: Reinforcement learning with via-pointrepresentation, Neural Networks, 17, 3, 299-305, 2004.

Saleem Mohideen & Vladimir Cherkassky, On recursive calculation of the generalized inverse of a matrix, ACM Transactions on Mathematical Software 17, Issue 1, pp.130 - 147, 1991

J. Morimoto and K. Doya, Acquisition of stand-up behavior by a real robot using hierarchical reinforcement learning. Robotics and Autonomous Systems 36 (1): 37-51, 2001.

R. Munos, A. Moore, Variable Resolution Discretization in Optimal Control, Machine Learning, No.1, pp.1-31,2001.

J. Nakanishi, J. Morimoto, G. Endo, G. Cheng, S. Schaal, M. Kawato, Learning from demonstration and adaptation of biped locomotion. Robotics and AutonomousSystems 47(2-3): 79-91, 2004

S. Nakaoka, S. Hattori, F. Kanehiro, S. Kajita and H. Hirukawa, Constraint-based Dynamics Simulator for Humanoid Robots with Shock Absorbing Mechanisms, The 2007 IEEE/RSJ International Conference on Intelligent Robots and Systems, 2007

A. van der Schaft & H. Schumacher: An Introduction to Hybrid Dynamical Systems, Springer, 2000.

Richard S. Sutton: Dyna, an Integrated Architecture for Learning, Planning, and Reacting, Proc. of the 7th Int. Conf. on Machine Learning, pp. 216-224, 1991.

Richard S. Sutton: Learning to Predict by the Methods of Temporal Differences, Machine Learning, 1988, 3, 9-44.

T. Schlegl, M. Buss, and G. Schmidt, Hybrid Control of Multi-fingered Dextrous Robotic Hands, S. Engell. G. Frehse, E. Schnieder (Eds.): Modelling, Analysis and Design of Hybrid Systems, LNCIS 279, 437-465, 2002.

V. N. Vapnik, The Nature of Statistical Learning Theory, Springer, 1995.M. Yashima, Y. Shiina and H. Yamaguchi, Randomized Manipulation Planning for A Multi-Fingered Hand by Switching Contact Modes, Proc. 2003 IEEE Int. Conf. on Robotics and Automation, 2003.

Y. Yin, S. Hosoe, and Z. Luo, A Mixed Logic Dynamical Modelling Formulation and Optimal Control of Intelligent Robots, Optimization Engineering, Vol.8, 321-340,2007.

E. Yoshida, P. Blazevic, V. Hugel, K. Yokoi, and K. Harada, Pivoting a Large Object: Whole-body Manipulation by a Humanoid Robot, Applied Bionics and Biomechanics, vol. 3, no. 3, 227-235, 2006

Ball Control in High-speed Throwing Motion Based on Kinetic Chain Approach

Taku Senoo*1, Akio Namiki*2 and Masatoshi Ishikawa*1
*1 *University of Tokyo*, *2 *Chiba University*
Japan

1. Introduction

In recent years, many robotic manipulation systems have been developed. However such systems were designed with a primary goal of the emulation of human capabilities, and less attention to pursuing of the upper limit in terms of speed for mechanical systems. In terms of motor performance, there are few robots equipped with quickness. Fast movement for robot systems provides not only improvement in operating efficiency but also new robotic skills based on the features peculiar to high-speed motion. For example some previous studies have been reported such as dynamic regrasping (Furukawa et al., 2006), high-speed batting (Senoo et al., 2006) and so on. However, there is little previous work where high-speed hand-arm coordination manipulation is achieved.

In this paper we report on experiments on the robotic throwing motion using a hand-arm system as shown in Fig.1. First a strategy for arm control is proposed based on the "kinetic chain" which is observed in human throwing motion. This strategy produces efficient high-speed motion using base functions of two types derived from approximate dynamics. Next the method of release control with a robotic hand is represented based on analysis related to contact state during a fast swing. The release method employs features so that the apparent force, almost all of which is generated by high-speed motion, plays a roll in robust control of the ball direction. Finally our high-speed manipulation system is described and experimental results are shown.

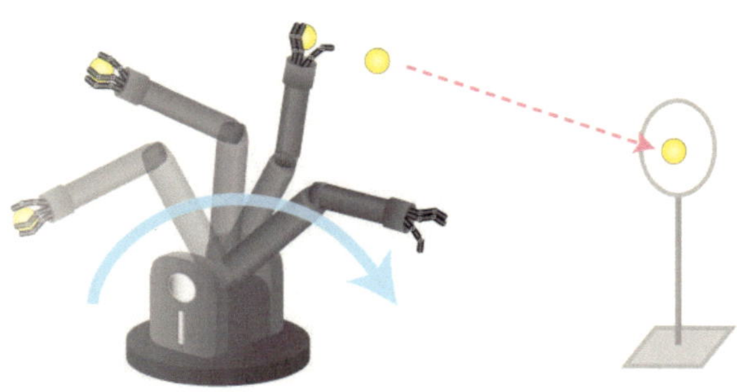

Fig. 1. Throwing motion using a hand-arm system

2. Speeding Up Swing Motion

In this section we extract a motional framework for producing high-speed movement based on human swing motion. The proposed swing model can be adapted to any two-link robot arm with rotational joints. It is possible to convert model-based motion into motion corresponding to actual equipment even if there are differences in kinematics between them.

2.1 Human Swing Motion

We see human motion at tremendous speeds in sports. For example the speed of an elbow joint in the throwing motion can be up to 40 [rad/s] (Werner, 1993). However the observed torque of triceps brachii, which generates the elbow motion, is remarkably low considering the speed. Focusing on the speed or power in distal upper extremity, their values are dramatically increased just before release time. This is because a human has a mechanism to explosively radiate kinetic energy accumulated from the early stages of a swing motion. This mechanism is called "kinetic chain" and it achieves high-speed swing motion efficiently. Two factors are particularly important. One is two-dimensional kinetic chain, which means power transmission (Putnam, 1993). This motion has characteristics so that the peak of the velocity waveform is continuously moved from the body trunk to the distal part. The other is three-dimensional kinetic chain. This means the effect of motion generated by rotation about axes, the directions of which are different from each other like gyro (Mochiduki et al., 1993).

2.2 Swing Model and Its Dynamics

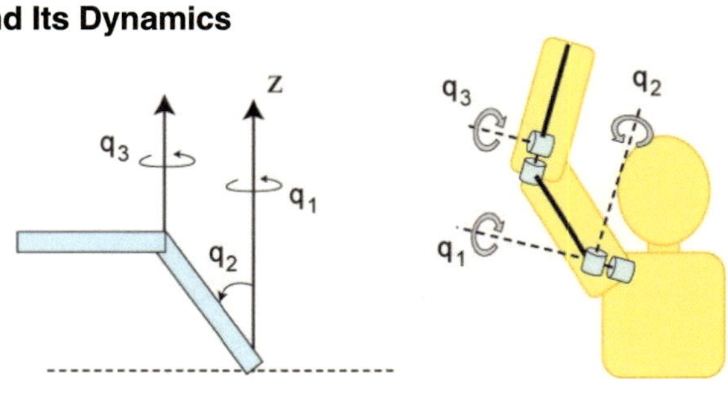

Fig. 2. Swing Model

We propose a swing model to constitute the framework of kinetic chain. Figure 2 shows the swing model composed of the upper arm and the lower arm. This model has a total of 3-DOF and consists of two revolution joints q_1, q_3 at the shoulder and the elbow respectively and a coupled bending joint q_2 to keep the lower arm horizontal. This model corresponds to the above two factors of kinetic chain as described below.

- Axis-1 and axis-3 are parallel. The state with the elbow in extension ($q_2 = \pi/2$) results in two-dimensional planar model.
- Axis-2 is perpendicular to other axes. The rotation about axis-2 produces 3D interferential action.

Next we derive the equation of motion. To simplify the problem, we ignore the moment of inertia. Gravity is also ignored to clarify the effect of interaction between joints. Because of this assumption the dynamics does not depend on the choice of coordinate system, so the model can accommodate various types of throwing such as overhand pitch, sidearm pitch and underarm pitch. The torque $\boldsymbol{\tau}$ is computed as

$$\boldsymbol{\tau} = \boldsymbol{M}(\boldsymbol{q})\ddot{\boldsymbol{q}} + \boldsymbol{h}(\boldsymbol{q}, \dot{\boldsymbol{q}}) \ , \tag{1}$$

where $M(q)$ is the inertia matrix, $h(q, \dot{q})$ is the coriolis and centrifugal force term. Those elements are represented as

$$\begin{aligned} &M_{22} = J_1 + 4J_2C_3^2 + 4A_{12}S_2C_3 + B_2 \ , M_{23} = M_{32} = -A_{12}C_2S_3 \ , \ M_{33} = J_2 \ , \\ &h_2 = -\frac{1}{2}(J_1 + B_2)\sin(2q_2)\dot{q}_1^2 - 4J_2\sin(2q_3)\dot{q}_2\dot{q}_3 - A_{12}C_2C_3(2\dot{q}_1\dot{q}_3 + \dot{q}_3^2 + 2\dot{q}_2^2 + \dot{q}_1^2) - 4A_{12}S_2S_3\dot{q}_2\dot{q}_3 \ , \\ &h_3 = 2A_{12}C_2C_3\dot{q}_1\dot{q}_2 + A_{12}S_2S_3(\dot{q}_2^2 + \dot{q}_1^2) + 2J_2\sin(2q_3)\dot{q}_2^2 \ , \end{aligned} \tag{2}$$

where $S_i \equiv \sin q_i$ and $C_i \equiv \cos q_i$. Other constant parameters are defined as follows;

$$J_1 = m_1 L_{1g}^2 \ , \ J_2 = m_2 L_{2g}^2 \ , A_{12} = m_2 L_1 L_{2g} \ , \ B_2 = m_2 L_1^2 \ , \tag{3}$$

where m_i, L_i, L_{ig} corresponding to i-th link means the mass, the entire length and the length to center of gravity respectively. We assume that the parameter including m_1 is larger than other parameters because the upper arm is generally heavier than the lower arm;

$$J_1 \gg J_2 \ , \ A_{12} \ , \ B_2 \ . \tag{4}$$

2.3 Decomposition into Base Functions

The essence of the kinetic chain approach is efficient transmission of power from body trunk to distal part. Because in this model joint-1 represents the source of power, the motion driven by the interaction is desirable behavior for joint-2 and joint-3. Therefore we set $\boldsymbol{\tau} \simeq \boldsymbol{0}$ except for joint-1;

$$\boldsymbol{M}(\boldsymbol{q})\ddot{\boldsymbol{q}} + \boldsymbol{h}(\boldsymbol{q}, \dot{\boldsymbol{q}}) \simeq \boldsymbol{0} \ . \tag{5}$$

Suppose that joint-1 can output higher power than the other joints and achieve steady state high-speed rotation instantaneously;

$$\dot{q}_1 \gg \dot{q}_i \ \ (i \neq 1) \ , \ \dot{q}_1 = \text{constant}, \ \ \ddot{q}_1 = 0 \ . \tag{6}$$

To obtain motion in joint-2, the vector $\ddot{q} = [0, \ddot{q}_2, 0]^T$ is substituted for Eq.(5) thereby setting joint-3 in continuous uniform motion $\ddot{q}_3 = 0$. The dynamics of joint-2 is approximated using the above assumption;

$$(J_1 + B_2)\ddot{q}_2 - \frac{1}{2}(J_1 + B_2)\dot{q}_1{}^2 \sin(2q_2) \simeq 0 \ . \tag{7}$$

If we express $\sin 2q_2 \simeq 2q_2$ using the first-order approximation of the Taylor series, this equation becomes a second order differential equation for q_2. The solution is

$$q_2 \simeq \alpha_2 \exp\left(\omega_2 t + \phi_2\right) \ , \tag{8}$$

where ω_2, ϕ_2 and α_2 are frequency, phase, and amplitude respectively. That is an exp-type base function representing a three-dimensional interaction of inertia force.

To obtain motion of joint-3, the vector $\ddot{q} = [0, 0, \ddot{q}_3]^T$ is substituted for Eq.(5) thereby setting joint-2 in continuous uniform motion $\ddot{q}_2 = 0$. The dynamics of joint-3 is approximated using the above assumption;

$$J_2\ddot{q}_3 + A_{12}\dot{q}_1{}^2 S_2 S_3 \simeq 0 \ . \tag{9}$$

If we express $S_3 \simeq q_3$ using the first-order approximation of the Taylor series and consider S_2 as constant, this equation becomes a second order differential equation for q_3. The solution is

$$q_3 \simeq \alpha_3 \sin(\omega_3 t + \phi_3) \ , \tag{10}$$

where ω_3, ϕ_3 and α_3 are frequency, phase, and amplitude respectively. That is a sin-type base function representing the two-dimensional kinetic chain.

Although we can calculate the frequency parameters $\boldsymbol{\omega}$ by Eqs.(7) and (9), these equations are rough approximations of the dynamics. Therefore we calculate the frequency parameters as well as other unknown parameters $\boldsymbol{\phi}, \boldsymbol{\alpha}$ in the next section. In addition we also set the trajectory of joint-1 as a sin-type base function. We express parameters concerning the base function as

$$\boldsymbol{\xi}^T = [\ \boldsymbol{\omega}^T \quad \boldsymbol{\phi}^T \quad \boldsymbol{\alpha}^T\ ] \ . \tag{11}$$

The constant term $\vec{c}$ is added to the base functions to ensure continuity of motion. Then the trajectory of joint angles is

$$\boldsymbol{q}(\boldsymbol{\xi}, t) = \begin{bmatrix} \alpha_1 \sin(\omega_1 t + \phi_1) + c_1 \\ \alpha_2 \exp(\omega_2 t + \phi_2) + c_2 \\ \alpha_3 \sin(\omega_3 t + \phi_3) + c_3 \end{bmatrix} . \tag{12}$$

The trajectory of velocity and acceleration is the same type of function. Because the behavior of exp-type functions is monotonically increasing (or decreasing), we change the sign of ω_2 at a certain time to decelerate swing motion.

2.4 Superposition of Base Functions

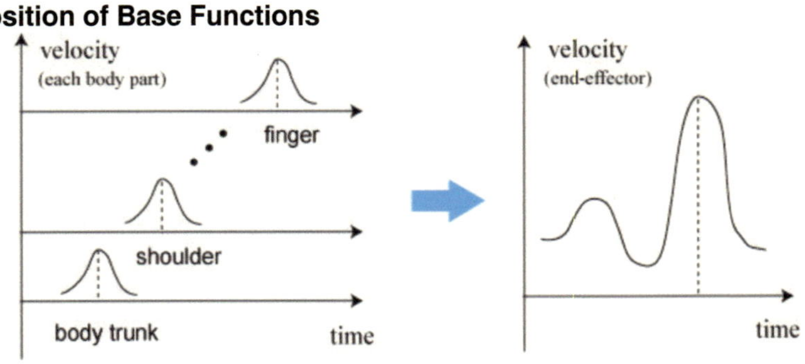

Fig. 3. Superposition of base functions

We synthesize the decomposed base functions so that the speed of the end-effector is increased as shown in Fig.3. The velocity of the end-effector $\dot{r}_E$ is represented as a function f of base function parameters and the time variable:

$$\dot{\boldsymbol{r}}_E = \boldsymbol{f}(\boldsymbol{\xi}, t) \ . \tag{13}$$

In each joint the start time t_{si} and the termination time t_{ei} of the swing are defined by Eq.(12) so that the joint velocity equals zero;

$$t_{si} = \frac{-\phi_i - \pi/2}{\omega_i}, \quad t_{ei} = \frac{-\phi_i + \pi/2}{\omega_i} \ . \tag{14}$$

Although the joint velocity for exp-type functions does not go to zero exactly, the time is defined in a similar way.

Under the kinematics constraint and dynamics constraint, the parameter $\boldsymbol{\xi}$ is set so as to maximize the following evaluation function:

$$\max_{\boldsymbol{\xi},\, t} \dot{\boldsymbol{r}}_E^T K_r \dot{\boldsymbol{r}}_E \tag{15}$$

subuject to $\boldsymbol{q}_{\min} \leq \boldsymbol{q} \leq \boldsymbol{q}_{\max}$, $\dot{\boldsymbol{q}}_{\min} \leq \dot{\boldsymbol{q}} \leq \dot{\boldsymbol{q}}_{\max}$, $\boldsymbol{\tau}_{\min} \leq \boldsymbol{\tau} \leq \boldsymbol{\tau}_{\max}$, $t_s \leq t \leq t_e$,
where K_r is a positive definite matrix, $t_s = \min t_{si}$, $t_e = \max t_{ei}$, and a suffix max or min means the maximum and minimum of the variable respectively. This computation is equivalent to maximizing the translational kinetic energy of the end-effector motion when $K_r = [\frac{1}{2}m_e, \frac{1}{2}m_e, \frac{1}{2}m_e]$, where m_e is the effective mass of the end-effector.

2.5 Simulation

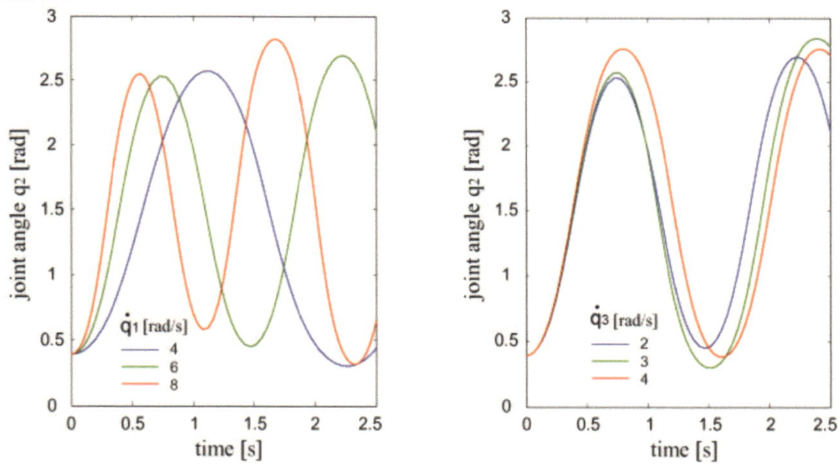

Fig. 4. Change of joint angle q2 with respect to angular velocity

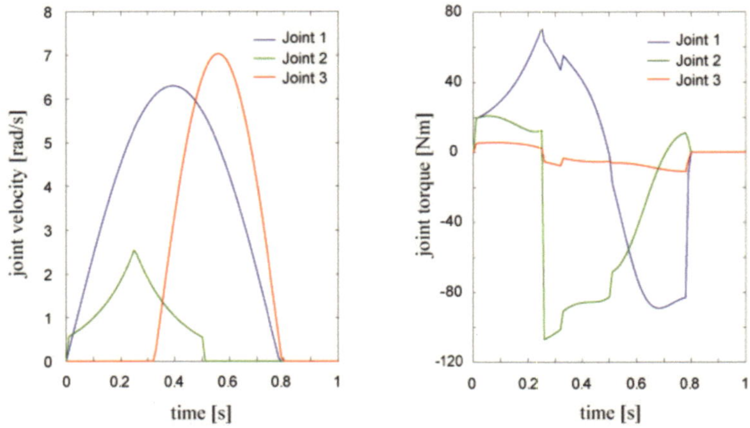

Fig. 5. Time response of joint velocity and joint torque

For optimization, we set the constraint using data from the barrett arm, which is described in section 4.1. The SQP method was used as the optimized calculation.
The change in joint angle q_2 with respect to joint velocities $\dot{q}_1, \dot{q}_3$ is shown in Fig.4. The left figure shows the case where $\dot{q}_1$ is constant and $\dot{q}_3$ varies. The right figure shows the reverse case. In those figures the frequency of joint-2 depends not on $\dot{q}_3$ but on $\dot{q}_1$. Because this result corresponds to the behavior of Eq.(7), it proves the validity of the approximation of the dynamics. In addition it turns out that joint-2 oscillates around $q_2 = \pi/2$. This means that the apparent force acts so that it extends the arm as in a human throwing motion. The oscillation is attributable to the mechanism of rotational joints. Figure 5 represents the time response of joint velocity and torque. In the left figure it turns out that the trajectories of both joint-1 and joint-3 are expressed by a sin-type function and the trajectory of joint-2 is expressed by an exp-type function. In the sin-type trajectories we can observe the transition of peak time as a typical characteristic of planar kinetic chain. The torque of joint-1 increases sharply from the start time $t = 0$. On the other hand the torque of joint-2 gradually goes to zero although joint-2 also gets into motion. This is as expected for an exp-type function. At $t = 0.26$ the torque of joint-2 decreases dramatically. This is because the motion of joint-2 switches to decreasing speed due to the constraint of joint angles against the force heading to $q_2 = \pi/2$. Moreover joint-3 moves fast despite the low torque of joint-3 during its motion. This is as expected for a sin-type function.

2.6 Analogous Swing Model

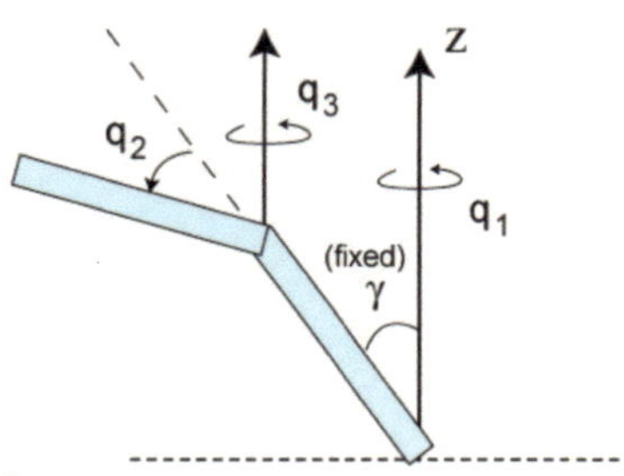

Fig. 6. Analogous Swing Model

Let's consider the analogous swing model shown in Fig.6. The difference between the two models is the placement of joint-2. The angle γ between axis-1 and the upper arm is fixed. We calculate the dynamics of this model and simulate swing motion in a similar way.
The dynamics of joint-2 is approximated as

$$J_2\ddot{q}_2 - \frac{1}{2}J_2\dot{q}_1^2\sin(2(\gamma + q_2)) \simeq 0 \ . \tag{16}$$

This equation is solved with an exp-type function. The motion of joint-2 oscillates around $q_2 + \gamma = \pi/2$. The dynamics of joint-3 is approximated as

$$J_2S_{\gamma 2}\ddot{q}_3 + A_{12}\dot{q}_1^2S_\gamma S_3 \simeq 0 \ . \tag{17}$$

This equation is solved with a sin-type function.
The analogy between the two models shows the following behavior of high-speed swing motion. The rotations about parallel axes produce a planar sin-type kinetic chain. Moreover rotation about the axis which is perpendicular to the parallel axes produces an exp-type kinetic chain. This motion oscillates around the plane that is perpendicular to the sin-type axes.
We adopt the latter model in the experiment for the following reasons. First, the speed is limited to some extent in the former model because of the constraint of joint angle q_2. Second, the latter model requires less compensating torque for gravity than the former one.

3. Control of a thrown ball

In this section we analyze the model of the contact state between a hand and a ball. Next the method for ball control is described.

3.1 Modeling of Contact State

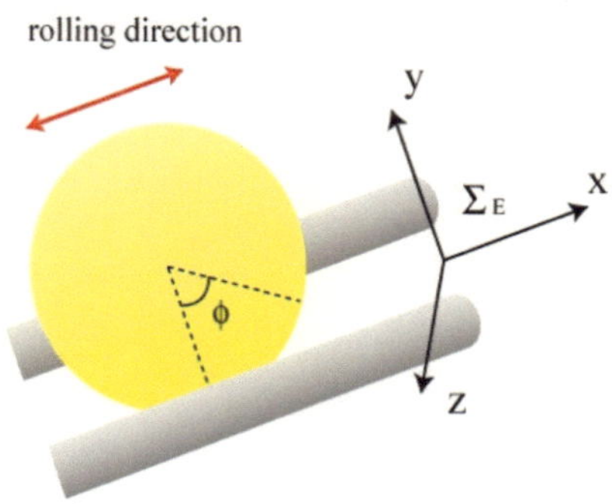

Fig. 7. Contact model between a hand and a ball

Suppose that a hand and a ball are both rigid bodies. The equation of ball motion in the coordinate Σ_E moving in translational acceleration α_0 and angular velocity ω_0 against standard coordinate Σ_0 is expressed as

$$m\ddot{\boldsymbol{r}}' = \boldsymbol{F}' + \boldsymbol{g}' - m(\boldsymbol{\alpha}_0' + \dot{\boldsymbol{\omega}}_0' \times \boldsymbol{r}') + 2m\dot{\boldsymbol{r}}' \times \boldsymbol{\omega}_0' + m(\boldsymbol{\omega}_0' \times \boldsymbol{r}') \times \boldsymbol{\omega}_0' \ , \tag{18}$$

where m is mass of the ball, $\boldsymbol{r}$ is ball position, $\boldsymbol{F}$ is contact force, $\boldsymbol{g}$ is gravity force, and a suffix $'$ means that the variable is expressed in the coordinate Σ_E. The third term is the inertia force due to accelerated motion, the fourth term is the Coriolis force, and the fifth term is the centrifugal force. Since the throwing motion includes three dimensional fast rotatory motion, Eq.(18) contains the considerable effect of the apparent force peculiar to a non-inertial system.

The motion of the hand is set to throw a ball by following three steps.

i. The hand grasps the ball with three fingers so as not to drop it until a given time.
ii. The contact state is switched to two-finger contact by releasing one finger at a certain time.
iii. The two fingers move in the same way and release the ball while the ball rolls along the fingers.

Suppose that the x-axis is set along the two fingers for release, the y-axis is perpendicular to the plane of the two fingers, and the z-axis is set so that the three axes constitute orthogonal coordinates as shown in Fig.7.

The condition where the ball with radius a and moment of inertia $I = 2ma^2/5$ rolls at angles ϕ without sliding is represented as

$$I\ddot{\phi} = aF'_x; \qquad x' = a\phi \ . \tag{19}$$

Since the ball cannot be moved in the y-axis and z-axis until release time, the following conditions are satisfied:

$$y' = \dot{y}' = \ddot{y}' = 0, \qquad z' = \dot{z}' = \ddot{z}' = 0 \ . \tag{20}$$

The dynamics of rolling motion is calculated as follows when Eqs.(19) and (20) are substituted for Eq.(18):

$$\ddot{x}' = \frac{5}{7}\left\{-\alpha'_{0x} + g'_x + ({\omega'_{0y}}^2 + {\omega'_{0z}}^2)x'\right\} \ . \tag{21}$$

Similarly, the normal force F'_y is computed as

$$F'_y = m\left\{\alpha'_{0y} - g'_y + 2\omega'_{0z}\dot{x}' + (\dot{\omega}'_{0z} + {\omega'_{0x}}^2)x'\right\} \ . \tag{22}$$

The ball is released from the fingers when the normal force is reduced to zero or less. Then we analyzed the relation between the normal force and the motion of rolling.

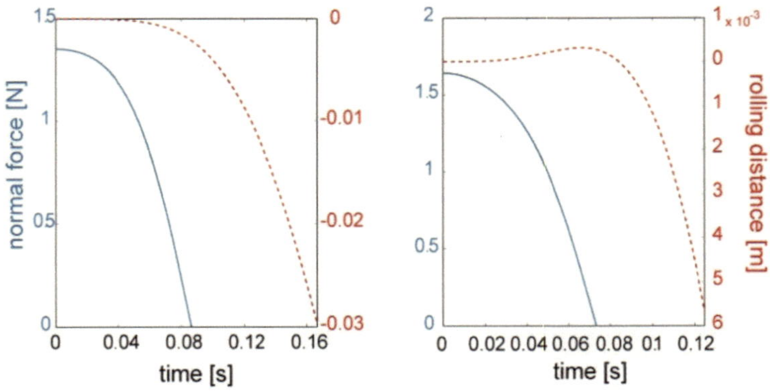

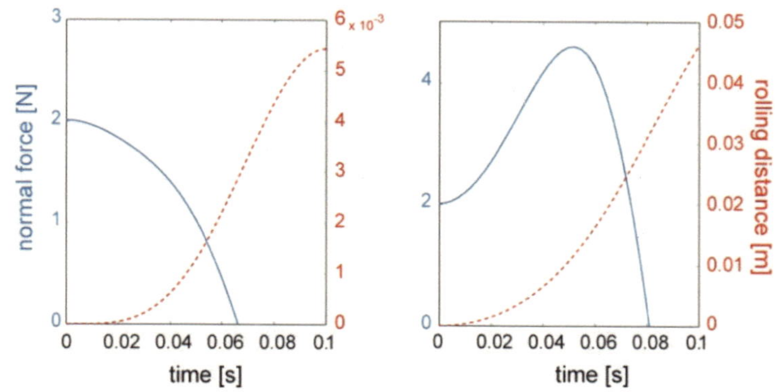

Fig. 8. Relation between normal force and rolling distance

Figure 8 shows the rolling distance and the normal force under varying conditions. Depending on the conditions of the acceleration and the posture of the manipulator, the ball can roll in either the positive or negative direction, or the hand sometimes releases the ball without the effect of a normal force as soon as the grasping finger is released from the ball. The faster the manipulator moves, the longer generally the rolling distance is. However the rolling distance cannot be dramatically increased. It is only a few centimeters. This is because the motion time becomes short until release of the ball, while fast motion brings fast speed of rolling. Therefore we assume that the ball is released before it rolls to the fingertip.

3.2 Released Ball Motion

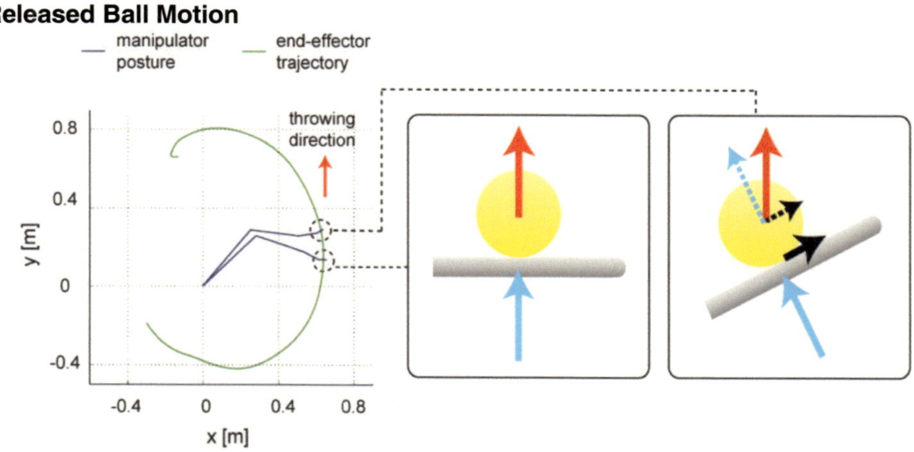

Fig. 9. Rolling effect for control of throwing direction

The velocity of the ball expressed in standard coordinates is calculated as follows:

$$v = v_0 + R\dot{r}' + \omega \times Rr' \ , \tag{23}$$

where v_0 is translational velocity of the end-effector, and R is a rotation matrix from Σ_0 to Σ_E. If the rolling distance is short, the above equation can be approximated as $v \simeq v_0 + R\dot{r}'$. That is, throwing direction depends mainly on the translational velocity of the end-effector and rolling velocity of the ball.

3.3 Strategy for Ball Control

Based on the above results, two elements are important to ball control. One is control of release timing and the other is control of the direction of release. These two elements are greatly affected by the translational acceleration and the translational velocity respectively. However it is difficult to control these variables in parallel with high accuracy. In addition it is difficult to positively apply the rolling motion to directional control. Therefore we propose a strategy for ball control in which the rolling motion acts secondarily to maintain the throwing direction.

We control the hand so that release of grasping finger and release of the ball are performed at the same time. Because both rolling distant and rolling velocity are zero at release time, the normal force is represented as $F'_y = m(\alpha'_{0y} - g'_y)$ by Eq.(22). Therefore the grasping finger is controlled to release the ball when the following condition is satisfied;

$$\alpha'_{0y} < g'_y \ . \tag{24}$$

We control the hand so that the ball is released in the tangential direction of the end-effector trajectory at release point. This condition of control is represented as

$$v'_{0x} = 0, \qquad \alpha'_{0x} < 0 \ . \tag{25}$$

The rolling acceleration near release point is expressed as $\ddot{x}' = \frac{5}{7}(-\alpha'_{0x} + g'_x)$ by Eq.(21). Therefore the ball rolls toward distal direction of the finger if $\alpha'_{0x} < g'_x$. On the other hand the velocity of the finger just after release is $v'_{0x} < 0$ by Eq.(25). This result means that if the ball is not released at the desired release time due to some errors, the rolling of the ball affects the release direction of the ball toward the target as shown in Fig.9. As a result robust control of the ball is achieved.

4. Experiments

4.1 System Configuration

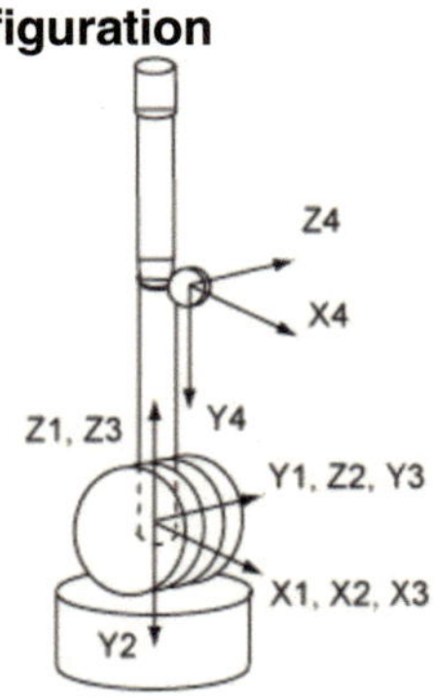

(a) Arm kinematics

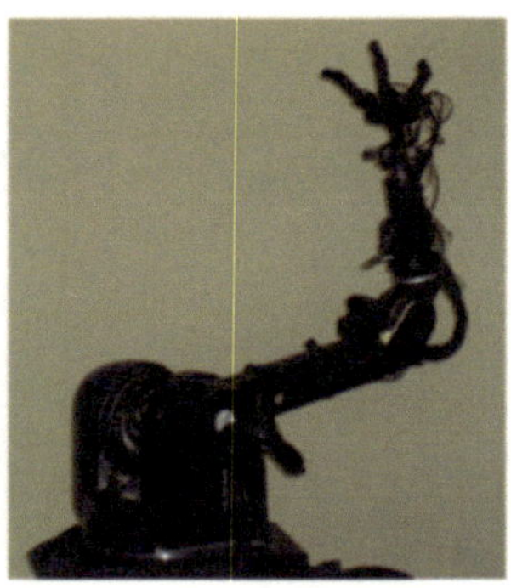

(b) Hand-Arm system

Fig. 10. High-speed manipulator system

The arm is a wire-drive manipulator (Barrett Technology Inc.). The kinematics of the manipulator is shown in Fig.10(a). The manipulator has 4-DOF consisting of alternately revolution and bending motion. High-speed movement with maximum velocity of the end-effector of 6 [m/s] and maximum acceleration of 58 [m/s^2] is achieved.

The hand consists of three fingers and a wrist. It has 10-DOF in total. A small harmonic drive gear and a high-power mini actuator are fitted in each finger link (Namiki et al., 2003). The design of this actuator is based on the new concept that maximum power output, rather than rated power output, should be improved. The hand can close its joints at 180 [deg] per 0.1 [s]. Its maximum velocity is 300 [rpm], and the maximum output is 12 [N].
Figure 10(b) shows the hand-arm system.

4.2 Experimental Setting

The manipulator threw a urethane ball with radius 5 [cm] towards the target from 3 [m] distance. The target is a net with radius 10 [cm]. The dynamics of the swing model is transformed to the one corresponding to the arm shown in Fig.10(a) due to the difference between mechanisms. In addition we adopted a gravity-compensated PD controller so as to enable accurate control of high-speed movement. Because the swing motion is generated so that Eq.(5) is satisfied, this simple controller plays an equivalent role as well as following the computed torque method;

$$\boldsymbol{\tau} = \boldsymbol{g}(\boldsymbol{q}_d) + K_p\Delta\boldsymbol{q} + K_v\Delta\dot{\boldsymbol{q}} \simeq M\ddot{\boldsymbol{q}}_d + \boldsymbol{h}(\boldsymbol{q}_d, \dot{\boldsymbol{q}}_d) + \boldsymbol{g}(\boldsymbol{q}_d) + K_p\Delta\boldsymbol{q} + K_v\Delta\dot{\boldsymbol{q}} \ . \tag{26}$$

4.3 Experimental Result

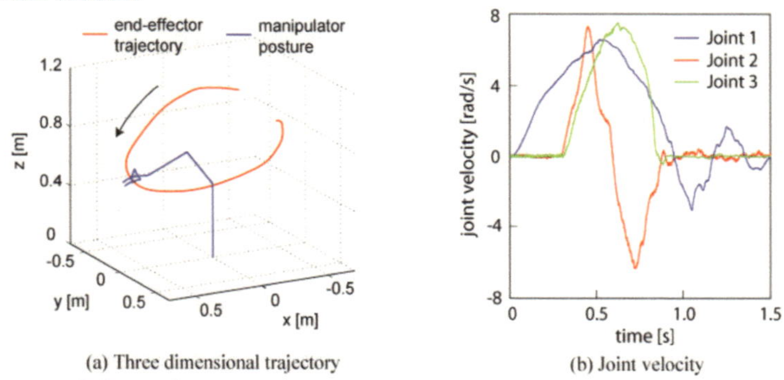

Fig. 11. Experimental data of manipulator motion

Figure 11(a) shows the trajectory of the end-effector. It turns out that the motion in the z-axis direction oscillates around the height of the elbow. This is caused by the exp-type function. The time response of joint velocities is shown in Fig.10(b). It turns out that both joint-1 and joint-3 correspond to a sin-type mode while joint-2 corresponds to an exp-type mode. The figure also indicates that accurate control on fast swing motion is achieved even by the simple controller except the overshoot of joint-1.
In Fig.12and Fig.13, the change in contact state and the trajectory of a thrown ball are shown as a continuous sequence of pictures taken at intervals of 100 [ms] and 66 [ms] respectively. It turns out that grasping state with three fingers is switched to release state with two fingers, and then the ball is thrown toward the target.
The success rate was about 40 %. Failure was caused mainly by uncertainty in initial calibration of the grasping state and nonuniform deformation of a ball due to grasping force. Accordingly, a tactile or force feedback control would improve the success rate. These experimental results are shown as a movie on the following web site.

(*http://www.k2.t.u-tokyo.ac.jp/fusion/HighspeedThrowing/*)

Fig. 12. Serial photographs of the change in contact state

Fig. 13. Serial photographs of a ball thrown toward a target

5. Discussion

(a) Radially distributed model (b) Deformation

Fig. 14. Elastic ball model

In this experiment the manipulator throws a urethane ball, which consists of elastic elements. To examine the effects on release motion, we analyzed the difference between the rigid model and the elastic model. We assume that the ball consists of isotropic elastic elements which are radially distributed as shown in Fig.14(a). Because the ball symmetrically deforms relative to the tangential direction as shown in Fig.14(b), tangential force caused by the elasticity is totally zero. The normal force F_n is calculated as

$$F_n = \int E\, dS/a = \pi E(1 - \frac{y}{a})y^2, \tag{27}$$

where a is radius of the ball, E is Young's modulus, and y is the deformation volume of a ball respectively.

For a certain throwing situation, the time response of the normal force and the displacement with respect to Young's modulus E is shown in Fig.15. It turns out that the release time is lengthened and the deformation velocity becomes faster when the Young's modulus E becomes smaller. This is because elastic potential energy can be accumulated in an elastic ball. The figure indicates that the delay of release timeg caused by the elastic elements is a few milliseconds.

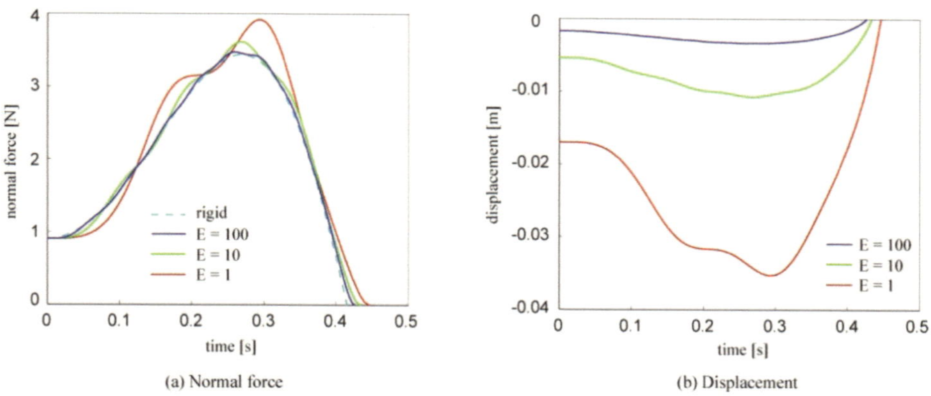

Fig. 15. Time response of normal force and displacement

6. Conclusion

In this paper we presented a strategy for high-speed swing motion based on the kinetic chain approach. In addition, we analyzed the contact model for ball control. From this analysis, robust control of release direction was proposed. As a result, the ball control in high-speed throwing motion was achieved using the hand-arm system.

Our future work will concentrate on a detailed robustness evaluation of ball control and an adoption of tactile or force feedback control. Moreover we plan to control other high-speed dexterous manipulations with hand-arm coordination.

7. References

Furukawa, N.; Namiki, A.; Senoo, T & Ishikawa, M. (2006). Dynamic Regrasping Using a High-speed Multifingered Hand and a High-speed Vision System, *Proceedings of IEEE Int. Conf. on Robotics and Automation*, pp. 181-187, 2006.

Mochiduki, Y.; Matsuo, T.; Inokuchi, S. & Omura, K. (1993). Dynamics analysis for the effect of centrifugal and coriolis forces in swing a bat, *Int. symp. on Biomechanics in Sports*, pp. 393-396.

Namiki, A.; Imai, Y.; Ishikawa, M. & Kaneko, M. (2003). Development of a High-speed Multifingered Hand System and Its Application to Catching, *IEEE/RSJ Proc. of Int. Conf. on Intelligent Robots and Systems*, pp. 2666-2671.

Putnam, C. (1993) Sequential motions of body segments in striking and throwing skills, *J. of Biomech.*, pp. 125-135.

Senoo, T.; Namiki, A & Ishikawa, M. (2006). Ball Control in High-speed Batting Motion using Hybrid Trajectory Generator, *Proceedings of IEEE Int. Conf. on Robotics and Automation*, pp. 1762-1767, 2006.

Werner, S. (1993). Biomechanics of the elbow during baseball pitching, *Journal of Orthop. Sports Phys.*, vol.17, No.6, pp. 274-278.

7

Shape Classification using Tactile Information in Rotation Manipulation by Universal Robot Hand

Hiroyuki Nakamoto†, Futoshi Kobayashi‡ and Fumio Kojima‡
† *Hyogo Prefectural Institute of Technology (Japan)*
‡ *Kobe University (Japan)*

1. Introduction

Recently, a variety of humanoid robots have been developed to help humans by working in human environments (Kaneko, 2002). However, these robots can not work as well as humans, because many robotic elements still require advanced research and development. One of these elements is a dexterous robot hand. For humanoid robots to be able to do various jobs in environments designed for humans and to use the same tools that humans use, those hands must have the same size and number of fingers and a tactile sense similar to those of the human hand. It is necessary to develop dexterous robot hands. Examples of multi-fingered robot hands developed includes the Utah/MIT Dexterous Hand (Hollerbach & Jacobsen, 1996), which has four four-jointed fingers driven by pneumatic actuators with tendon cables and a full tactile sensing suite using a capacitance sensor suitable for contact control (Johnston et al., 1996). Other examples are the Gifu Hand (Mouri et al., 2002) having 20 joints with 16 degrees of freedom (DOF) and the KH Hand type S (Mouri et al., 2005) having 20 joints with 15 DOF. These lightweight hands have six-axis force sensors and distributed tactile sensors consisting of a pressure sensor sheet with conductive ink. Despite the obvious importance of tactile sensors, they have not been actively used to recognize objects. Among the few studies achieving effective dexterous movement using tactile sensors, Pribadi (Pribadi et al., 1989) and Allen (Allen & Mechelman, 1990) focused on the sensing of global object shapes and fitting these shapes to object models. Okamura proposed detecting features based on definitions using a hemispherical fingertip having a tactile sensor (Okamura & Cutkosky, 2001). Despite their important results, these methods require extensive investigation but do not use distributed tactile sensor very effectively. By the way, Klatzky researched haptic exploratory procedures by the human hand (Klatzky & Lederman, 1990). In this research, exploratory procedures are classified into eight motions, i.e., lateral motion, pressure, static contact. They are important to obtain object's attributes. Especially "contour following", which can obtain object's outer shape in a manipulation, is necessary to manipulate an object dexterously. However, this procedure has not been realized in a robot hand's manipulation.

To obtain object's shape while a robot hand manipulates an object, we propose a shape classification. It uses the five-fingered robot hand, called Universal Robot Hand, which has distributed tactile sensors. Universal Robot Hand rotates an object, and tactile sensors measure pressure distribution once every 10 ms. A kurtosis is calculated from each pressure distribution, and dynamic programming (DP) matches the time-series kurtosis pattern with reference patterns and outputs valuated values. The evaluated value is classified based on a given threshold. Experiments demonstrated that three symmetrical objects are classified accurately.

2. Universal Robot Hand and Rotational Manipulation

2.1 Universal Robot Hand

Each of Universal Robot Hand's five four-jointed 3-DOF fingers is 333.7 mm long (Fig. 1) and Table 1. Each Joint is driven by a miniaturized DC motor and a reduction gear (Harmonic Drive Systems, Inc.) built into each link, and also has a joint torque sensor and a rotary encoder. Joints 3 and 4 drive at equal ratios, like a human finger. Tactile sensors are attached to the ventral sides of links.

Universal Robot Hand's three-layer tactile sensor consists of urethane gel (Exseal Co.), pressure sensitive rubber (Inaba Rubber Co., Ltd.) and an electrode pattern sheet (Fig. 2). This simple arrangement provides high productivity and easy maintenance. The urethane gel has a hardness of ASKER-C 15 and 2.5 mm thick. When an object comes in contact with this sensor, the force and shape of the contact surface deform the urethane gel. This deformation creates a pressure distribution in the pressure sensitive rubber. The rubber has about 0.5 mm thick and contains electrically conductive particles. When the pressed rubber

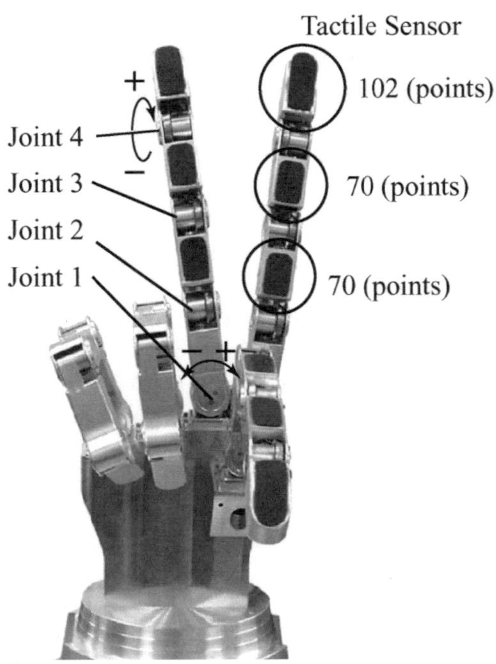

Fig. 1. Universal Robot Hand

Joints	4
Actuators	3
Actuator driver	DC micro motor
Angle range (deg)	$-10 \le$ Joint 1 ≤ 10 $0 \le$ Joint 2-4 ≤ 90
Maximum joint velocity (rpm)	29
Maximum joint torque (Nmm)	110

Table 1. Finger specifications

becomes thin, resistance in that region drops. Measuring this electrical resistance enables to be calculated the pressure in the region. Measuring the electrical resistance of the rubber requires two electrodes. One of the two electrodes sends voltage and the other receives it. We designed these electrodes on a flexible multi-layer polyimide board to simplify the sensor structure. Each measurement point is 3.4 × 1.8 mm and individual points are spaced at 0.2 mm. Each measurement point detects a force range of 0 to 1 N. As shown in Fig. 1, a finger has three tactile sensors. There are 102 measurement points on each fingertip, and the other two parts of the finger have 70 points each. The tactile sensor takes 10 ms to measure one pressure distribution.

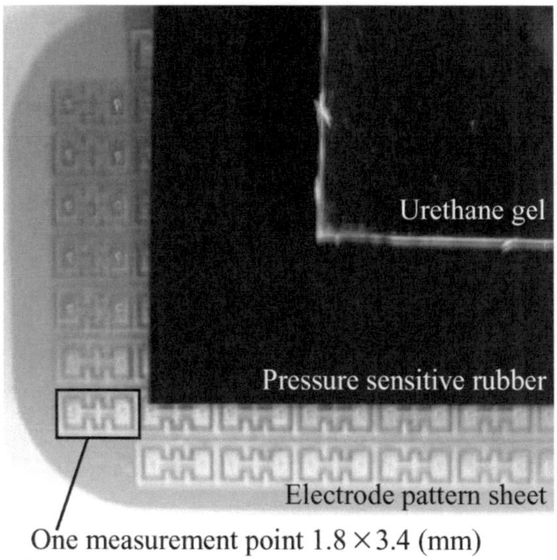

Fig. 2. Structure of Tactile Sensor

2.2 Rotational Manipulation

Universal Robot Hand uses thumb and index finger to rotate an object. Two fingers pinch an object at the position to classify the shape. The thumb moves forward and the index finger moves backward simultaneously and vice versa, while the tactile sensor on the index finger measures pressure distributions on the contact surface between the object and fingertip. Assuming that the object does not slip and the object's radius is known, its rotational angle is converted to positioning fingertip and joint angles calculated using the Newton method. Repeating this process generates a rotational manipulation trajectory up to target angles of the object. The joint angle is used to calculate joint torque with proportional derivative (PD) control. Stiffness control adjusts the torque so that contact force remains almost constant. Universal Robot Hand repeats rotational manipulation and classifies the area's shape repetitively. We call this manipulation repetitive rotational manipulation. Repetitive rotational manipulation is shown in Fig. 3. Fig. 3 (1)-(3) show the thumb moves forward and the index finger moves backward, then the octagonal cylinder is rotated clockwise.

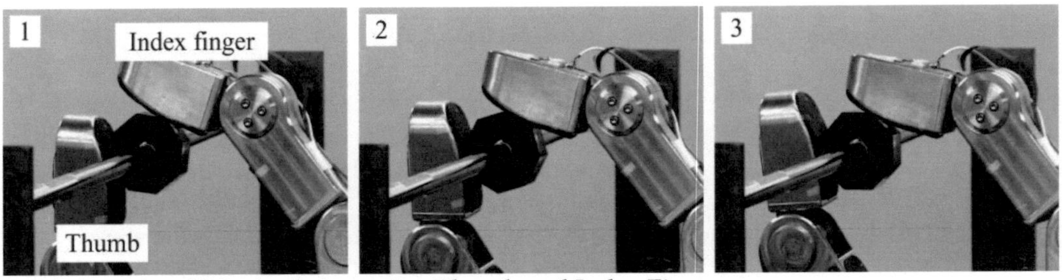

Fig. 3. Rotational Manipulation using Thumb and Index Finger

3. Pressure Distribution Quantification

We propose a pressure distribution quantification prior to shape classification. A tactile sensor measures the pressure distribution based on the contact shape and total contact force (Fig. 4). Environmental conditions such as temperature and humidity are assumed to be constant. When the total contact force is constant, the pressure distribution depends entirely on the contact shape, i.e., if it is convex, pressure distribution exhibits sharp convexity and, if flat, gentle convexity.

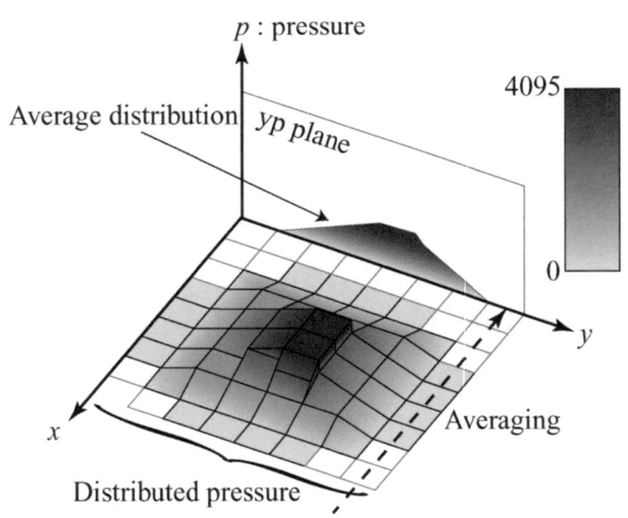

Fig. 4. Pressure Distribution Quantification

Based on these assumptions, the contact surface is quantified as follows: As shown in Fig. 4, the Axes of coordinates are set to the pressure distribution, with average distribution defined as average x axis pressure for the yp-plane. Kurtosis γ is calculated from average distribution as follows:

$$\gamma = \sum_{n=1}^{N} \frac{p_i (c_i - \mu)^4}{N\sigma^4}$$

where $p_i(i=1..N)$ is the average distribution pressure, $c_i(i=1..N)$ pressure weighting coefficients, N average distribution length, μ average, and σ standard deviation. Kurtosis is a statistical parameter quantifying distribution sharpness. If distribution becomes sharp, kurtosis becomes high, and vice versa. Kurtosis is suitable for contact shape classification because it expresses pressure distribution sharpness in a single value.

Although kurtosis can quantify the static contact shape, single kurtosis cannot classify shapes changing during rotational manipulation, which requires time-series pressure distribution. We therefore applied our work to time-series pressure distributions in rotational manipulation.

4. Shape Classification

4.1 Proposal Method

Our proposed method of shape classification is shown in Fig. 5. Universal Robot Hand obtains regular pressure distributions. Each pressure distribution is averaged in the direction of finger width to yield an average distribution. Kurtosis is calculated repeatedly, yielding time series kurtosis (Fig. 5). For a prism, kurtosis is high at an edge and low at a face, therefore time-series kurtosis has extreme values for feature contacts. Our method extracts patterns from a time-series kurtosis between the extreme values, as shown in Fig. 5.

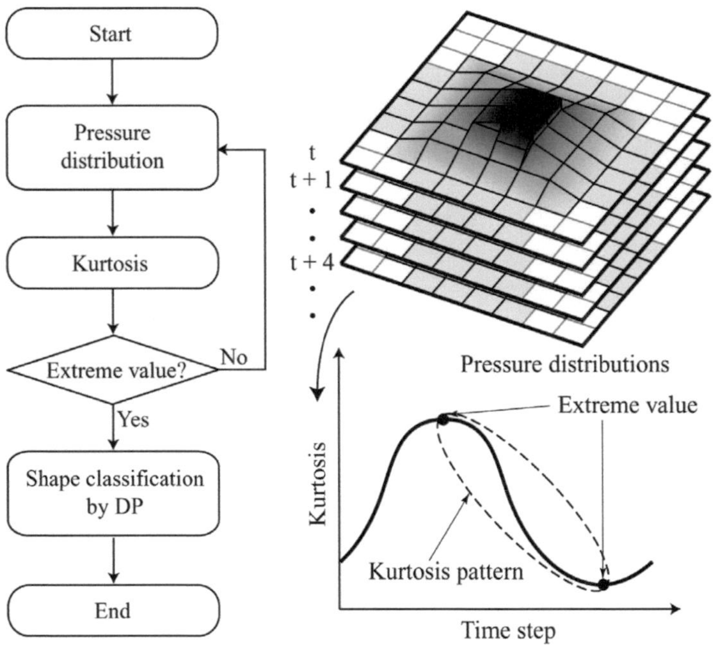

Fig. 5 Shape Classification

We call this extracted pattern, kurtosis patterns. DP matching (Uesaka & Ozeki, 1990) evaluates the similarity between a kurtosis pattern and a reference pattern to classify the contact shape. The reference pattern is the kurtosis pattern for a known shape. DP can shrink or stretch patterns, even if the pattern length is different. This allows to change rotational velocities. DP selects the optimal local path for the minimum accumulated distance between former two patterns, then matches and evaluates the two patterns.

4.2 Dynamic Programming

In DP matching, Pattern Z is expressed as a reference pattern:

$$Z = \{z(j) \mid 1 \le j \le J\}$$

where j is the number of elements and J reference pattern length. Pattern U is expressed as a kurtosis pattern:

$$U = \{u(i) \mid 1 \le i \le I\}$$

where i is the number of elements and I kurtosis pattern length. The local distance between $u(i)$ and $z(j)$ is denoted by $d(i,j)$, define as follows:

$$d(i,j) = (u(i) - z(j))^2$$

Distance accumulated between the reference and input pattern at point (i, j) is denoted by $S(i, j)$. With DP matching, $S(i, j)$ is updated using the following iterative equations:
Initial Conditions (i=1, j=1):

$$S(1,1) = d(1,1)$$

(i=1,2 $\le j \le$ J):

$$S(1,j) = \infty$$

Iterative Equation ($2 \le i \le I, j = 1$):

$$S(i,1) = d(i,1) + S(i-1,1)$$

($2 \le i \le I, j = 2$):

$$S(i,2) = d(i,2) + \min\begin{cases} S(i-1,2) \\ S(i-1,1) \end{cases}$$

($2 \le i \le I, 3 \le j \le J$):

$$S(i,j) = d(i,j) + \min\begin{cases} S(i-1,j) \\ S(i-1,j-1) \\ S(i-1,j-2) \end{cases}$$

These equations select the optimal local path for the minimum accumulated distance, Meaning $S(i, j)$ as the minimum accumulated distance between the reference and input pattern shrinks or stretches from 1/2 to 2 times along the time axis. DP output A is defined as follows:

$$A = \frac{1}{T} S(I, J)$$

This normalizes output with the sum of weights. T is the length of matched patterns. We use the evaluated value of similarity between the reference and kurtosis patterns with output A.

5. Experiments

5.1 Experiments at a constant velocity

Conditions for experiments confirming that time-series kurtosis indicated contact shape and classified highly accurately during repetitive rotational manipulations assumed that rotated objects are a cylinder, hexagonal prism, and octagonal prism (Fig. 5). Objects are 15 mm thick, and the inscribed circles of the cross-sectional outer-edge on each object are 35 mm in diameter. Universal Robot Hand rotated the cylinder by 0.63 radian, the hexagonal and octagonal prisms from edge to face. The rotational velocity was constant, 0.15 rad/s.

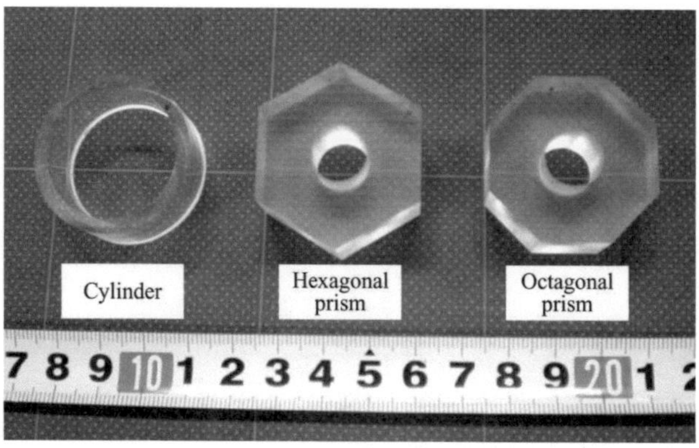

Fig. 6. Objects

The relationship between kurtosis and time step is shown in Fig. 7. Cylinder kurtosis is almost constant at 2.5. Hexagonal and octagonal prisms kurtosis, however, is a time-series waveform. Local maxima kurtosis of the hexagonal prism are higher than those of the octagonal prism, showing edge sharpness because the hexagonal prism has edges sharper than the octagonal prism. Local minima kurtosis of the hexagonal prism are lower than that of the octagonal prism. These results reflect the hexagonal prism plane length for 20.2 mm and that for the octagonal prism of 14.5 mm. The time-series kurtosis acquired in rotational manipulation therefore reflects object shape accurately.

We calculated the evaluated values using DP. Kurtosis patterns of the hexagonal and octagonal prisms, which range from the local maximum to next minimum values, are extracted from the time-series kurtosis. For the cylinder, kurtosis patterns for 0.63 radian rotation are extracted. The number of kurtosis patterns of each object is 20. The evaluated values between the kurtosis and reference patterns provided beforehand are calculated. Averages and standard deviations of the evaluated values are shown in Table 2, which shows that the lower the evaluated value, the greater is the similarity between that the kurtosis pattern and the reference pattern. For different objects, the evaluated value averages, as well as the standard deviations, are high. The cause of these high averages is that the kurtosis patterns are not similar to the reference patterns. On the other hand, for the same object, averages and standard deviations are small, because the kurtosis patterns are similar to reference patterns. If threshold is 0.01, the classification results are shown in Table 3. The classification accuracy for the hexagonal and octagonal prisms is 90%, and that for the cylinder is 95%. These classification results are very high, confirming the effectiveness of the proposed classification.

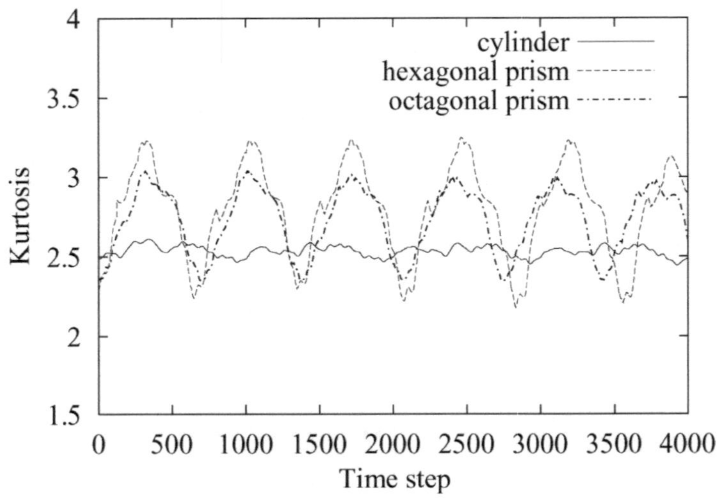

Fig. 7. Kurtosis vs. Time Step

		reference		
		hexagon*	octagon*	cylinder
hexagon*	AVE	0.00121	0.01262	0.07642
	SD	0.00106	0.00285	0.03566
octagon*	AVE	0.1179	0.00241	0.03566
	SD	0.00571	0.00227	0.00433
cylinder	AVE	0.10297	0.06107	0.00545
	SD	0.01434	0.01093	0.00259

* Prism shape

Table 2. Averages and Standard Deviations of Evaluated Value

	Hexagon*	octagon*	cylinder
Classification Accuracy (%)	90	90	95

* Prism shape

Table 3. Classification Accuracy

5.2 Experiments using Five Different Velocities

The rotational velocity was changed by 0.5, 0.75, 1.5 times, and double original speed, giving 0.075, 0.113, 0.225, and 0.3 rad/s, respectively. The objects shown in Fig. 6 were used. Universal Robot Hand rotated the cylinder by 0.63 rad, the hexagonal and octagonal prisms from edge to face. The relationship between kurtosis and time step of the hexagonal prism is shown in Fig. 8. Corresponding to rotational velocities, time-series kurtosis are shrunk or stretched, i.e., the time-series kurtosis's period of 0.3 rad/s is half that of 0.15 rad/s. Local maxima and minima kurtosis are constant, even if rotational velocities are different. These results mean that time-series kurtosis can reflect the object's shape.

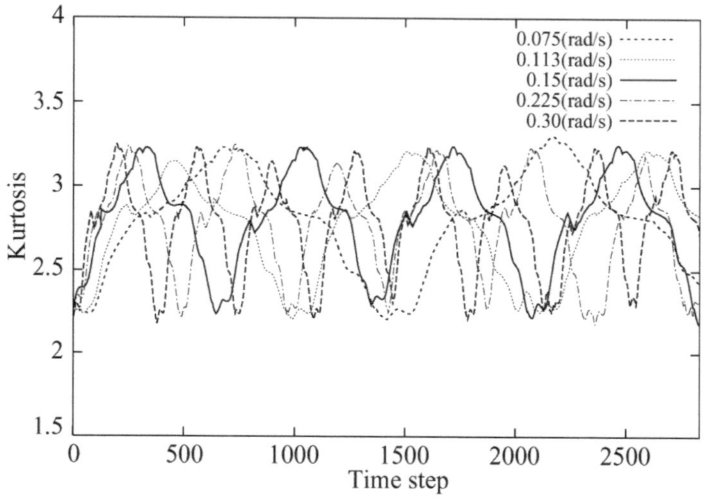

Fig. 8. Time-series Kurtosis of Hexagonal Prism at five different velocities

By the same described in 4.1 and using the same reference patterns, we calculated classification accuracies from 20 kurtosis patterns of each velocity. The results including in 4.1 are shown in Table 4. The classification accuracies of each rotational velocity are high, despite the reference patterns of a rotational velocity 0.15 rad/s. These results show that our proposed method is robust accommodating changes in rotational velocity. Shape classification does not require reference patterns for each rotational velocity, and is confirmed its effectiveness.

	Rotational Velocity (rad/s)				
	0.075	0.113	0.15	0.225	0.3
hexagonal prism(%)	85	90	90	90	85
octagonal prism(%)	85	90	90	85	80
cylinder(%)	90	95	95	95	95

Table 4. Classification Accuracies using Five Different Velocity

6. Conclusion

We proposed a shape classification in rotation manipulation. In this classification, time-series kurtosis is calculated from pressure distributions, and kurtosis patterns are extracted from time-series kurtosis. DP calculates evaluated values between kurtosis pattern and

reference patterns to determine whether to classify a contact shape if the evaluated value is falls below a given threshold. Experiments demonstrated the effectiveness while Universal Robot Hand rotates objects repetitively.
The whole outer shape classification we also proposed in continuous rotation manipulation (Nakamoto et al. 2009) is applicable to manipulation where the robot hand continuously rotates an object in one direction. We plan to combine repetitive and continuous classification that uses repetitive classification in regular processing continuous classification when it cannot classify a shape. This combination is expected to classify shapes robustly. We also plan to downsize the robot hand and improve the tactile sensor.

7. References

P. K. Allen & P. Michelman (1990). Acquisition and interpretation of 3-d sensor data from touch, *IEEE Transactions on Robotics and Automation*, Vol.6, No.4, pp.397-404.

John M. Hollerbach & Stephen C. Jacobsen (1996). Anthropomorphic robots and human interactions, *Proceedings of 1st International Symposium on Humanoid Robots* pp.83-91.

D. Johnston, P. Zhang, J. Hollerbach and S. Jacobsen (1996). A Full Tactile Sensing Suite for Dextrous Robot Hands and Use in Contact Force Control, *Proceedings of the 1996 IEEE International Conference on Robotics and Automation*, pp.661-666.

K. Kaneko, F. Kanehiro, S. Kajita, K. Yokoyama, K. Akachi, T. Kawasaki, S. Ota, and T. Isozumi (2002). Design of Prototype Humanoid Robotics Platform for HRP, *Proceedings of the 2002 IEEE/RSJ International Conference on Intelligent Robots and Systems*, pp.2431-2436.

R. Klatzky & S. Lederman (1990). Intelligent Exploration by the Human Hand, *Dextrous Robot hands*, pp.66-81, Springer.

T. Mouri, H. Kawasaki, K. Yoshikawa, J. Takai and S. Ito (2002). Anthropomorphic Robot Hand: Gifu Hand III, *Proceedings of International Conference ICCAS2002*, pp.1288-1293.

H. Nakamoto, F. Kobayashi, N. Imamura, H. Shirasawa, and F. Kojima (2009). Shape Classification in Continuous Rotation Manipulation by Universal Robot Hand, *Journal of Advanced Computational Intelligence and Intelligent Informatics*, Vol.13, No.3, pp.178-184, ISSN 1343-0130

M. Okamura & R. Cutkosky (2001). Feature Detection for Haptic Exploration with Robotic Fingers, *The International Journal of Robotics Research*, Vol.20, No.12, pp.925-938.

K. Pribadi, J. S. Bay and H. Hemami (1989). Exploration and dynamic shape estimation by a robotic probe, *IEEE Transaction on Systems, Man, and Cybernetics*, Vol.19, No.4, pp.840-846.

Y. Uesaka & K. Ozeki (1990). DP Matching, *Algorithm of Pattern Recognition and Learning (in Japanese)*, pp.91-108, Bun-ichi Sogo Shyuppan.

8

Biologically Inspired Robot Arm Control Using Neural Oscillators

Woosung Yang[1], Nak Young Chong[2] and Bum Jae You[1]
[1]*Korea Institute of Science and Technology, Korea*
[2]*Japan Advanced Institute of Science and Technology, Japan*

1. Introduction

Humans or animals exhibit natural adaptive motions against unexpected disturbances or environment changes. This is because that, in general, the neural oscillator based circuits on the spinal cord known as Central Pattern Generators (CPGs) might contribute to efficient motor movement and novel stability properties in biological motions of animal and human. Based on the CPGs, most animals locomote stably using inherent rhythmic movements adapted to the natural frequency of their body dynamics in spite of differences in their sensors and actuators.

For such reasons, studies on human-like movement of robot arms have been paid increasing attention. In particular, human rhythmic movements such as turning a steering wheel, rotating a crank, etc. are self-organized through the interaction of the musculoskeletal system and neural oscillators. In the musculoskeletal system, limb segments connected to each other with tendons are activated like a mechanical spring by neural signals. Thus neural oscillators may offer a reliable and cost efficient solution for rhythmic movement of robot arms. Incorporating a network of neural oscillators, we expect to realize human nervous and musculoskeletal systems in various types of robots.

The mathematical description of a neural oscillator was presented in Matsuoka's works (Matsuoka, 1985). He proved that neurons generate the rhythmic patterned output and analyzed the conditions necessary for the steady state oscillations. He also investigated the mutual inhibition networks to control the frequency and pattern (Matsuoka, 1987), but did not include the effect of the feedback on the neural oscillator performance. Employing Matsuoka's neural oscillator model, Taga *et al.* investigated the sensory signal from the joint angles of a biped robot as feedback signals (Taga et al., 1991), showing that neural oscillators made the robot robust to the perturbation through entrainment (Taga, 1995). This approach was applied later to various locomotion systems (Miyakoshi et al., 1998), (Fukuoka et al., 2003), (Endo et al., 2005), (Yang et al., 2008).

Besides the examples of locomotion, various efforts have been made to strengthen the capability of robots from biological inspiration. Williamson created a humanoid arm motion based on postural primitives. The spring-like joint actuators allowed the arm to safely deal with unexpected collisions sustaining cyclic motions (Williamson, 1996). And the neuro-mechanical system coupled with the neural oscillator for controlling rhythmic arm motions

was proposed (Williamson, 1998). Arsenio suggested the multiple-input describing function technique to control multivariable systems connected to multiple neural oscillators (Arsenio, 2000).
Even though natural adaptive motions were accomplished by the coupling between the arm joints and neural oscillators, the correctness of the desired motion was not guaranteed. Specifically, robot arms are required to exhibit complex behaviors or to trace a trajectory for certain type of tasks, where the substantial difficulty of parameter tuning emerges. The authors have presented encouraging simulation results in controlling the arm trajectory incorporating neural oscillators (Yang et al., 2007 & 2008). This chapter addresses how to control the trajectory of a real robot arm whose joints are coupled to neural oscillators for a desired task. For achieving this, real-time feedback from sensory information is implemented to exploit the entrainment feature of neural oscillators against unknown disturbances.
In the following section, a neural controller is briefly explained. An optimization procedure is described in Section 3 to design the parameters of the neural oscillator for a desired task. Details of dynamic responses and simulation and experimental verification of the proposed method are discussed in Section 4 and 5, respectively. Finally, conclusions are drawn in Section 6.

2. Rhythmic Movement Using a Neural Oscillator

2.1 Matsuoka's neural oscillator

Our work is motivated by studies and facts of biologically inspired locomotion control employing oscillators. Especially, the basic motor pattern generated by the CPG of inner body of human or animal is usually modified by sensory signals from motor information to deal with environmental disturbances. The CPGs drive the antagonistic muscles that are reciprocally innervated to form an intrinsic rhythm generating mechanism around each joint. Hence, adapting this mechanism actuated by the CPGs which consists of neural oscillator network, we can design a new type of biologically inspired robots that can accommodate unknown interactions with the environments by controlling internal loading (or force) of the body.
For implementing this, we use Matsuoka's neural oscillator consisting of two simulated neurons arranged in mutual inhibition as shown in Fig. 1. If gains are properly tuned, the system exhibits limit cycle behaviors. Now we propose the control method for dynamic systems that closely interacts with the environment exploiting the natural dynamics of Matsuoka's oscillator.

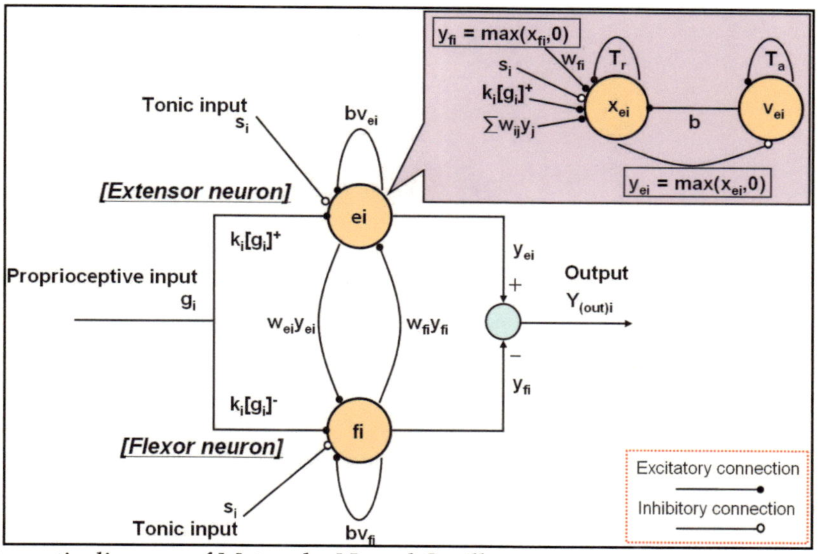

Fig. 1. Schematic diagram of Matsuoka Neural Oscillator

$$
\begin{aligned}
T_r\dot{x}_{ei} + x_{ei} &= -w_{fi}y_{fi} - \sum_{j=1}^{n} w_{ij}y_j - bv_{ei} - \sum k_i[g_i]^+ + s_i \\
T_a\dot{v}_{ei} + v_{ei} &= y_{ei} \\
y_{ei} &= [x_{ei}]^+ = \max(x_{ei},\ 0) \\
T_r\dot{x}_{fi} + x_{fi} &= -w_{ei}y_{ei} - \sum_{j=1}^{n} w_{ij}y_j - bv_{fi} - \sum k_i[g_i]^- + s_i \\
T_a\dot{v}_{fi} + v_{fi} &= y_{fi} \\
y_{fi} &= [x_{fi}]^+ = \max(x_{fi},\ 0),\ (i = 1,\ 2,\ \cdots,\ n)
\end{aligned}
\tag{1}
$$

where x_{ei} and x_{fi} indicate the inner state of the i-th neuron for $i=1{\sim}n$, which represents the firing rate. Here, the subscripts 'e' and 'f' denote the extensor and flexor neurons, respectively. $v_{e(f)i}$ represents the degree of adaptation and b is the adaptation constant or self-inhibition effect of the i-th neuron. The output of each neuron $y_{e(f)i}$ is taken as the positive part of x_i and the output of the oscillator is the difference in the output between the extensor and flexor neurons. w_{ij} is a connecting weight from the j-th neuron to the i-th neuron: w_{ij} are 0 for $i{\neq}j$ and 1 for $i{=}j$. $w_{ij}y_i$ represents the total input from the neurons arranged to excite one neuron and to inhibit the other, respectively. Those inputs are scaled by the gain k_i. T_r and T_a are the time constants of the inner state and the adaptation effect, respectively, and s_i is an external input with a constant rate. $w_{e(f)i}$ is a weight of the extensor neuron or the flexor neuron and g_i indicates a sensory input from the coupled system.

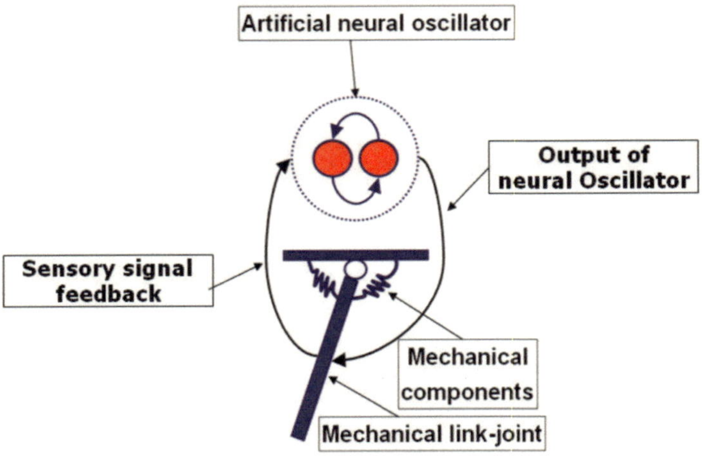

Fig. 2. Mechanical system coupled to the neural oscillator

Figure 2 conceptually shows the control method exploiting the natural dynamics of the oscillator coupled to the dynamic system that closely interacts with environments. This method enables a robot to adapt to changing conditions. For simplicity, we employ a general 2nd order mechanical system connected to the neural oscillator as seen in Fig. 4. The desired torque signal to the joint can be given by

$$\tau_i = k_i(\theta_{vi} - \theta_i) - b_i\dot{\theta}_i, \tag{2}$$

where k_i is the stiffness of the joint, b_i the damping coefficient, θ_i the joint angle, and θ_{vi} is the output of the neural oscillator that produces rhythmic commands of the *i*-th joint. The neural oscillator follows the sensory signal from the joints, thus the output of the neural oscillator may change corresponding to the sensory input. This is what is called "entrainment" that can be considered as the tracking of sensory feedback signals so that the mechanical system can exhibit adaptive behavior interacting with the environment.

2.2 Entrainment property of the neural oscillator

Generally, it has been known that the Matsuoka's neural oscillator exhibits the following properties: the natural frequency of the output signal increases in proportion to $1/T_r$. The magnitude of the output signal also increases as the tonic input increases. T_r and T_a have an effect on the control of the delay time and the adaptation time of the entrained signal, respectively. Thus, as these parameters decrease, the input signal is well entrained. And the minimum gain k_i of the input signal enlarges the entrainment capability, because the minimum input signal is needed to be entrained appropriately in the range of the natural frequency of an input signal. In this case, regardless of the generated natural frequency of the neural oscillator and the natural frequency of an input signal, the output signal of the neural oscillator locks onto an input signal well in a wide range.
Figure 3 illustrates the entrainment procedure of the neural oscillator. If we properly tune the parameters of the neural oscillator, the oscillator exhibits the stable limit cycle behaviors.

In Figure 1, the gain k of the sensory feedback was sequentially set as 0.02, 0.2 and 0.53 such as Figure 3 (a), (b) and (c). When k is 0.02, the output of the neural oscillator can't entrain the sensory signal input as shown in Figure 3 (a). The result of Figure 3 (b) indicates the signal partially entrained. If the gain k is properly set as 0.53, the neural oscillator produces the fully entrained signal as illustrated in Figure 3 (c) in contrast to the result of Figure 3 (b).

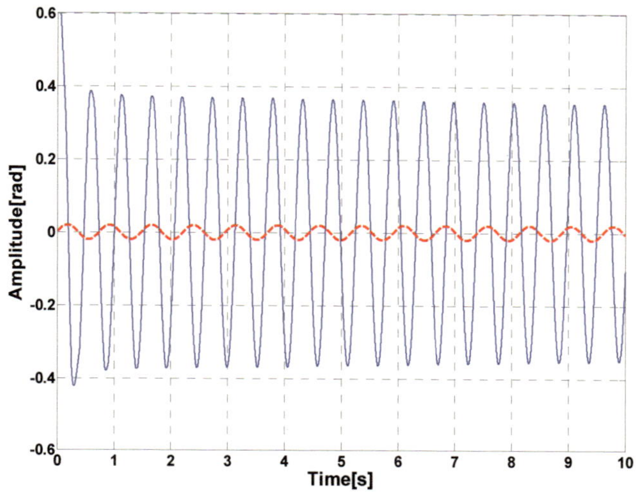

(a)

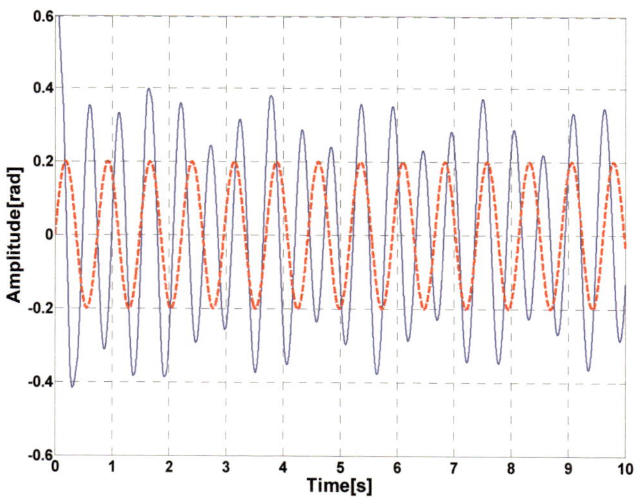

(b)

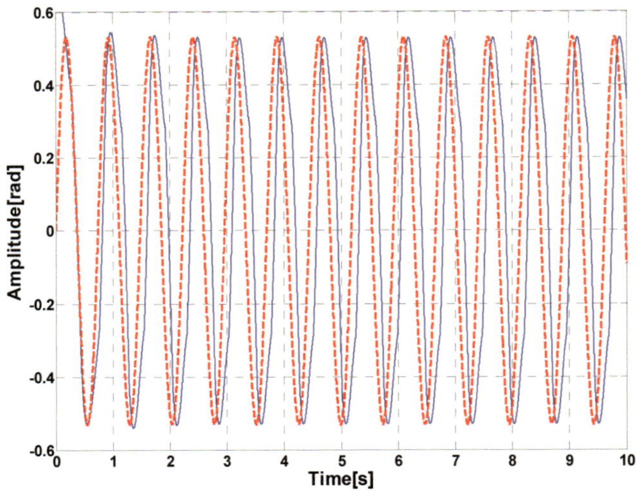

(c)

Fig. 3. Simulation results on the entrainment property of the neural oscillator. The solid line is the output of the neural oscillator and the dashed line indicates the sensory signal input.

3. Optimization of Neural Oscillator Parameters

The neural oscillator is a non-linear system, thus it is generally difficult to analyze the dynamic system when the oscillator is connected to it. Therefore a graphical approach known as the describing function analysis has been proposed earlier (Slotine & Li, 1991). The main idea is to plot the system response in the complex plane and find the intersection points between two Nyquist plots of the dynamic system and the neural oscillator. The intersection points indicate limit cycle solutions. However, even if a rhythmic motion of the dynamic system is generated by the neural oscillator, it is usually difficult to obtain the desired motion required by the task. This is because many oscillator parameters need to be tuned, and different responses occur according to the inter-oscillator network. Hence, we describe below how to determine the parameters of the neural oscillator using the *Metropolis* method (Yang et al., 2007 & 2008) based on simulated annealing (SA) (Kirkpatrick, 1983), which guarantees convergence to the global extremum (Geman & Geman, 1984).

For the process of minimizing some cost function E, $X=[T_r, T_a, w, s, \cdots]^T$ is selected as the parameters of the neural oscillator to be optimized; the initial temperature T_0 is the starting parameter; the learning rate ν is the step size for X. Specifically, the parameters are replaced by a random number N in the range [-1,1] given by;

$$X_i = X_{i-1} + \nu \cdot N \tag{3}$$

If the change in the cost function ΔE is less than zero, the new state X_i is accepted and stored at the i-th iteration. Otherwise, another state is drawn with the transition probability, $Prob_i(E)$ given by

$$\Pr ob_i(E) = (\frac{1}{Z(T)})\exp(-\frac{\Delta E}{c}) > \gamma, \quad (4)$$

where γ is a random value uniformly distributed between 0 and 1. The temperature cooling schedule is $c_i=k \cdot c_{i-1}$ (k is the Boltzmann constant or effective annealing gain) and $Z(T)$ is a temperature-dependant normalization factor. If ΔE is positive and $Prob_i(E)$ is less than γ or equal to zero, the new state X_i is rejected. Here the lower cost function value and large difference of ΔE indicate that X_i is the better solution. If temperature approaches zero, the optimization process terminates.

Even though SA has several potential advantages over conventional algorithms, it may be faced with a crucial problem. When searching for optimal parameters, it is not known whether the desired task is performed correctly with the selected parameters or not. We therefore added the task completion judgment and cost function comparison steps as shown in Fig. 4 by thick-lined boxes. If the desired task fails, the algorithm reloads previously stored parameters and selects the parameters that give the lowest cost function value. Then the optimization process is restarted with the selected parameters until it finds the parameters of the lowest cost function that allow the task to be done correctly.

4. Crank Rotation of Two-link Planar Arm

To validate the proposed control scheme, we evaluate the crank rotation task with a two-link planar arm whose joints are coupled to neural oscillators as shown in Fig. 5. The inter-oscillator network is not established, because the initial condition of the same sign will be equivalent to the excitatory connection between two oscillators. We focus on the entrainment property of the arm.

The crank rotation is modeled by generating kinematic constraints and an appropriate end-effector force. The crank has the moment of inertia I and the viscous friction at the joint connecting the crank and the base. If the arm end-effector position is defined as (x, y) in a Cartesian coordinate system whose origin is at the center of the crank denoted as (x_0, y_0), the coordinates x and y can be expressed as

$$\begin{aligned} \begin{pmatrix} x \\ y \end{pmatrix} &= \begin{pmatrix} -r\sin\varphi + x_0 \\ r\cos\varphi + y_0 \end{pmatrix} = \begin{pmatrix} l_1 c_1 + l_2 c_{12} \\ l_1 s_1 + l_2 s_{12} \end{pmatrix} \\ \begin{pmatrix} \ddot{x} \\ \ddot{y} \end{pmatrix} &= \begin{pmatrix} r\sin\varphi\dot{\varphi} - r\cos\varphi\ddot{\varphi} \\ -r\cos\varphi\dot{\varphi} - r\sin\varphi\ddot{\varphi} \end{pmatrix} = \frac{\partial J(\theta)^2}{\partial^2\theta}, \end{aligned} \quad (5)$$

where J is the Jacobian matrix of $[x, y]^T$. φ and θ_i are the crank angle and the i-th joint angle, respectively. l_i is the length of the i-th link. c_1, c_{12}, s_1 and s_{12} denote $\cos\theta_1$, $\cos(\theta_1+\theta_2)$, $\sin\theta_1$ and $\sin(\theta_1+\theta_2)$, respectively. r is the radius of the crank. Eq. (5) can be rearranged as follows:

$$J(\theta)\ddot{\theta} + \dot{J}(\theta,\dot{\theta})\dot{\theta} = r(u(\phi)\ddot{\phi} - v(\phi)\dot{\phi}^2), \quad (6)$$

where u is the tangential unit vector and v is the normal unit vector at the outline of the crank as shown in Fig. 5, respectively.

Now the dynamic equations of the crank and the arm are given in the following form.

$$I\ddot{\varphi} + C\dot{\varphi} = ru(\varphi)^T F \tag{7}$$

$$M(\theta)\ddot{\theta} + V(\theta,\dot{\theta}) + G(\theta) = \tau - J(\theta)^T F \tag{8}$$

$$\tau = k(\theta_v - \theta) - b\dot{\theta}, \tag{9}$$

where M is the inertia matrix, V is the Coriolis/centripetal vector, and G is the gravity vector, k and b denotes the joint stiffness and viscosity matrixes, respectively (Gomi & Osu, 1998), θ_v is the output of the neural oscillator (see Eq. (2)), F is the contact force vector interacting between the crank and the end-efector. By solving Eqs. (7) and (8) simultaneously using Eq. (6), F is obtained as

$$F = \{J(\theta)M(\theta)^{-1}J(\theta)^T + r^2 I^{-1}u(\phi)u(\phi)^T\}^{-1}\{J(\theta)M(\theta)^{-1}(\tau - V(\theta)) + \dot{J}(\theta,\dot{\theta})\dot{\theta} + r\dot{\phi}(v(\phi)\dot{\phi} + CI^{-1}u(\phi))\} \tag{10}$$

It is very hard to properly tune parameters of the neural oscillator for attaining the desired rotation task. Moreover, this dynamic model is tightly coupled to crank dynamics as described in Eq. (10). Thus, the proposed parameter tuning approach is divided into the following two steps:

1) Step 1: Find initial parameters of the neural oscillator corresponding to desired inputs of each joint using the cost function given by:

$$\phi = \left|\frac{T - T_G}{T_G}\right| + v \cdot \max\left(\frac{|A_d - C|}{B} - 1, 0\right) \tag{11}$$

subject to

$$i)\ A_{\min} \le A_d \le A_{\max}$$

$$ii)\ |A_d - C| \le B$$

where $C=(A_{max}+A_{min})/2$, $B=(A_{max}-A_{min})/2$; A_d is the desired amplitude of the neural oscillator for the rotation task, A_{max} and A_{min} are the maximum and minimum amplitude constraints, respectively; T and T_G denote the desired and measured natural frequencies of the output generated by the neural oscillator, respectively. v is the performance gain.

2) Step 2: Using the initial parameters obtained by Step1, run the proposed SA algorithm as illustrated in Fig. 4. The cost function for the crank rotation includes the velocity of the rotation, torque, and consumed energy.

Implementing Step 1 and Step 2 in sequence, we are able to acquire the appropriate initial and tuned parameters as seen in Table 1. Figure 6 (a) indicates a cooling state in terms of cooling schedule. Cooling or annealing gain K is set as 0.95. It can be observed in Fig. 6 (b) that the optimal process was well operated and a better solution at the lowest cost function was obtained iteratively. As expected, when the tuned parameters are employed to perform the given task, a stable motion could be accomplished as shown in Fig. 6. It is evident in Fig.

6 (c) that initial transient responses disappear due to the entrainment property of the neural oscillator. This property enables the arm to sustain the given task against changes in parameters of arm kinematics and dynamics as well as disturbances.

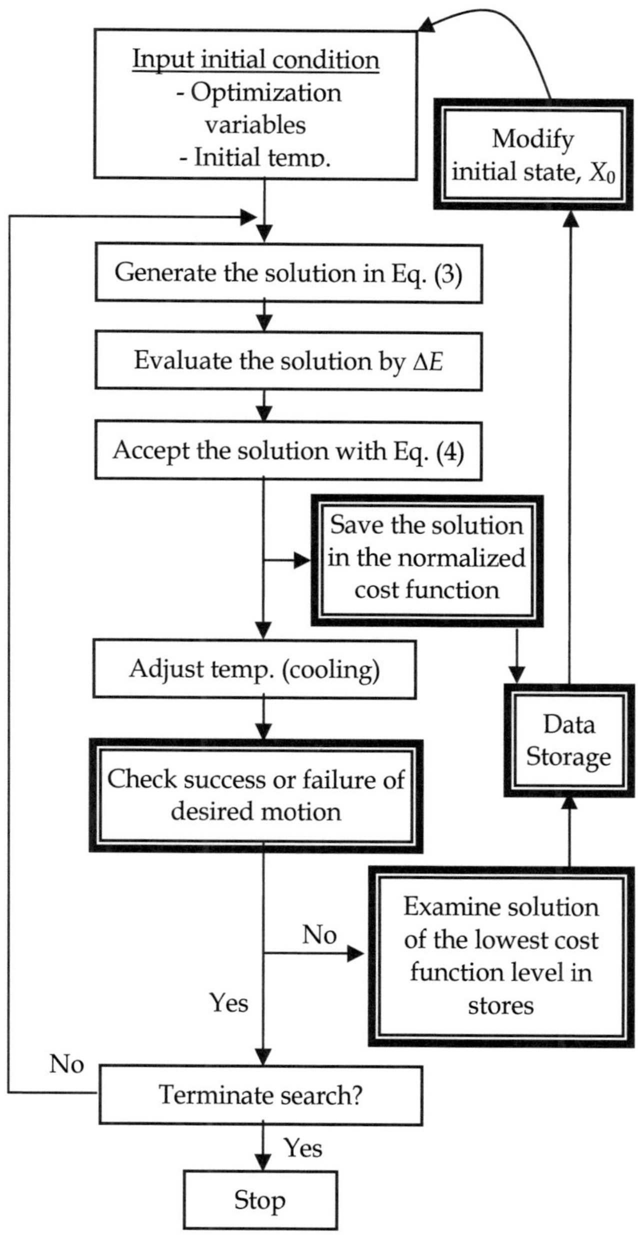

Fig. 4. Flowchart of the upgraded SA for task based parameter optimization.

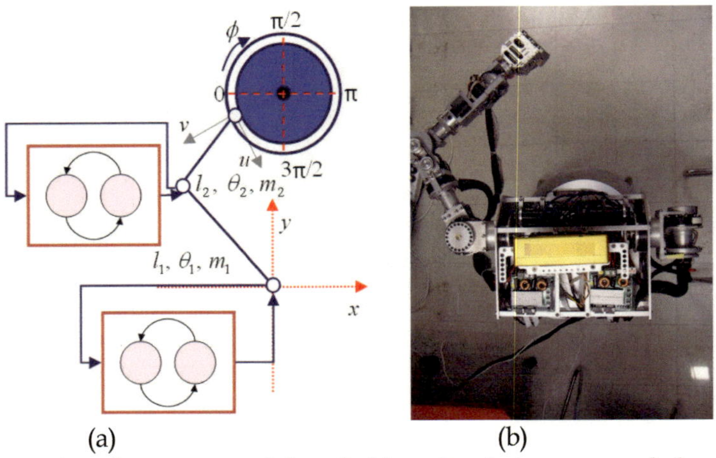

Fig. 5. (a) Schematic robot arm model and (b) real robot arm coupled with the neural oscillator for experimental test

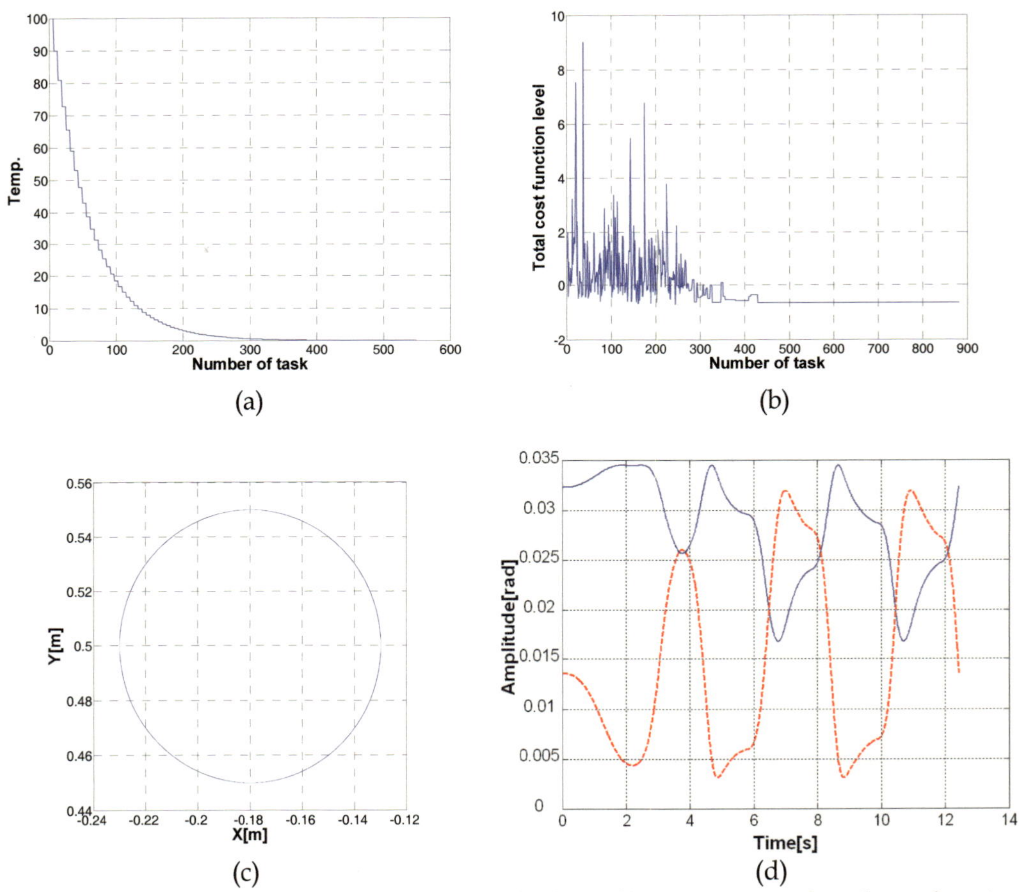

Fig. 6. (a) Temperature transition for cooling schedule, (b) A transition of total cost function level, (c) The end-effector trajectory of two-link arm (d) The output of joint angle. The red dash line is the first joint angle and the second joint angle is drawn by the blue thin line

Initial parameters		Optimized parameters	
Inhibitory weight (w)	2.0	Inhibitory weight (w)	4.012
Time constant (T_r)	0.25	Time constant (T_r)	1.601
(T_a)	0.5	(T_a)	3.210
Sensory gain (k)	1	Sensory gain (k)	10.010
Tonic input (s)	60	Tonic input (s)	57.358
Robot Arm Model			
Mass 1 (m_1),	Mass 2 (m_2)	2.347kg,	0.834kg
Inertia 1 (I_1),	Inertia 2 (I_2)	0.0098kgm^2,	0.0035kgm^2
Length 1 (l_1),	Length 2 (l_2)	0.224m,	0.225m

Table 1. Initial and tuned parameters of the neural oscillator with robot arm model

5. Experiments with a Real Robot Arm

To validate the proposed control scheme described in Section 4, we employed a real robot arm with 6 degrees of freedom (see Fig. 5 (b)) and constructed a real time control system. This arm controller runs at 200 Hz and is connected via IEEE 1394 for data transmission at 4 kHz. ATI industrial automation's Mini40 sensor was fitted to the wrist joint of the arm to detect external disturbances. The optimized parameters in Table 1 were used for the neural oscillator.

Figure 7 shows the arm kinematics. Since the crank motion is generated in the horizontal plane, $q_1$and q_3 are set to 90°. The initial values of q_5 and q_6 are set to 0°, respectively. q_2 and q_4, corresponding to θ_1 and θ_2 in Fig. 5 (a), respectively, are controlled by the neural oscillators and the constraint force given in Eq. (10). The constraint force enables the end-effector to trace the outline of the (virtual) crank. Hence, the end-effector can draw the circles as shown in Fig. 8 (see the overlapping circles in the center part of the figure).

Now, we will examine what happens in the arm motion if additive external disturbances exist. Arbitrary forces are applied to the end-effector at 15s, 28s, 44s, 57s, 73s and 89s sequentially as shown in Fig. 9. We first pushed the end-effector along the minus x direction. The force sensor value in the x and y direction are added to Eq. (10). Then, the joint angles change according to the direction of the applied force, which makes the neural oscillators entrain the joint angles as shown in Fig. 10. The solid line is the output of the neural oscillator connected to the first joint (q_2) and the dashed line indicates that of the neural oscillator connected to the second one (q_4). Hence a change in the output of the neural oscillator causes a change in the joint torque. Finally the joint angles are modified as shown in Fig. 11, where the bottom plot is the output of q_2 and the top one is the output of q_4. Fig. 12 shows the snap shots of the simulated crank motion by the robot arm, where we can observe that the end-effector traces the circle well, and adapts its motion when an external force is applied to it.

Table 2 compares the power consumption of the robot arm performing the above task with different parameters of the neural oscillator. The parameters were drawn arbitrary among the ones that guarantee a successful completion of the task. If the optimized parameters (set D) were employed, the most energy-efficient motion was realized.

	Parameter set A	Parameter set B	Parameter set C	Parameter set D (optimized)
Inhibitory weight (w)	2.0	2.503	4.012	**4.012**
Time constant	0.25	0.896	1.601	**1.601**
(T_r)	0.5	5.0	3.210	**3.210**
(T_a)	1.0	1.241	15.010	**10.010**
Sensory gain (k)	60.0	60.660	57.358	**57.358**
Tonic input (s)				
Measured current [A]	1.871	0.794	0.591	**0.572**
Power [W] consumption	89.808	38.112	28.368	**27.456**

Table 2. Power Consumption according to the selected parameter set of the neural oscillator

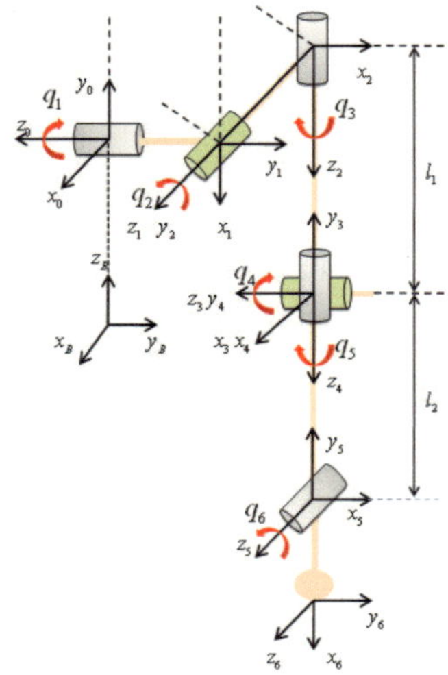

Fig. 7. Kinematic parameters of the robot arm

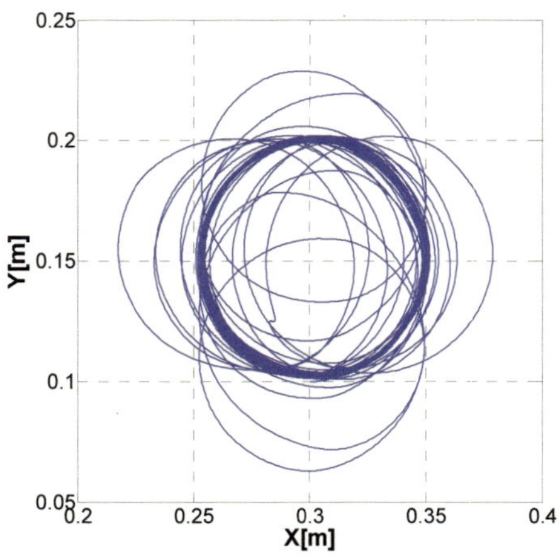

Fig. 8. The trajectory drawn by the end-effector of the arm

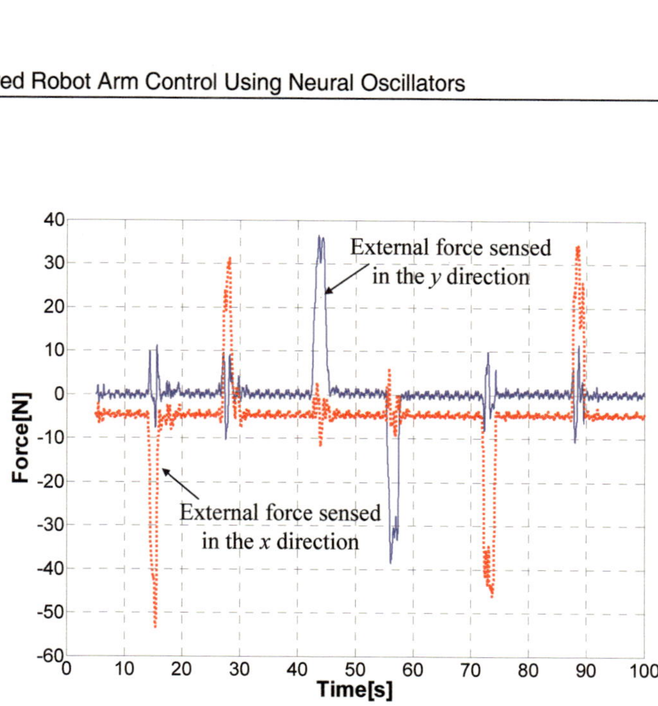

Fig. 9. The external forces measured by the force sensor in the x and y direction

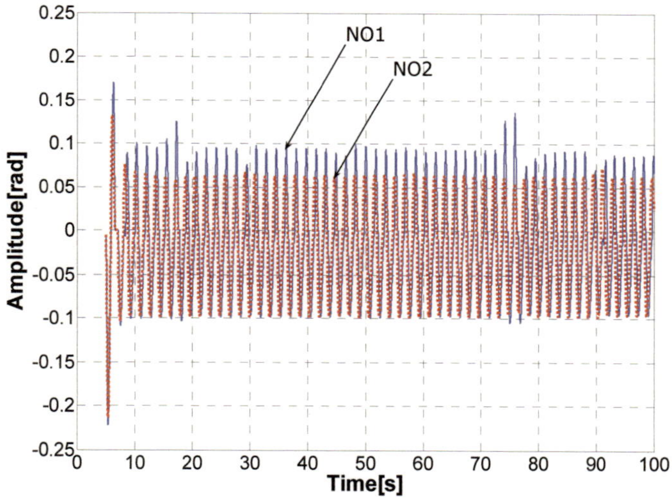

Fig. 10. The output of the neural oscillator coupled to the joints of the arm

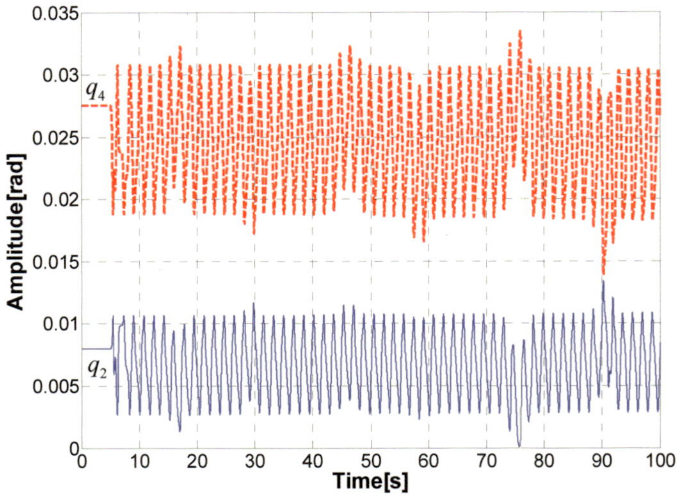

Fig. 11. The output of the first joint (q_2) and the second joint (q_4)

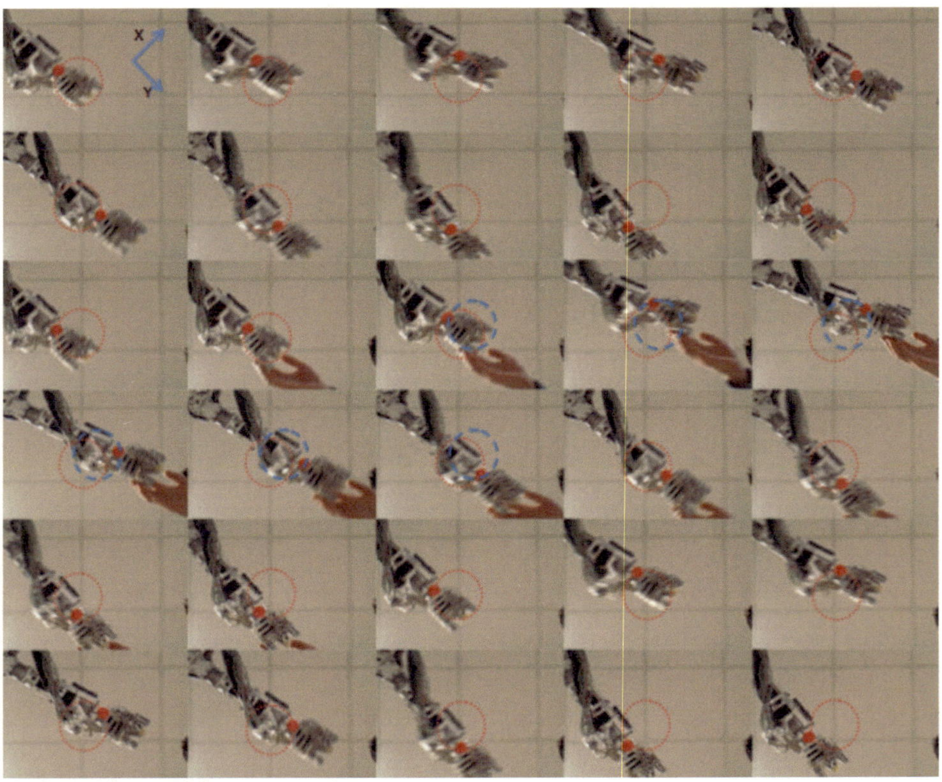

Fig. 12. Snap shots of the arm motion

6. Conclusion

This chapter presents an example of human-like behavior of a planar robot arm whose joints were coupled to neural oscillators. In contrast to existing works that were only capable of rhythmic pattern generation, the proposed approach allowed the robot arm to trace a trajectory correctly through entrainment. For successfully achieving this, we proposed an optimization approach for obtaining the parameters of the neural oscillator modifying the simulated annealing method. Simulation and experimental results showed the effectiveness of the proposed approach. Moreover, it was demonstrated that the robot arm could adaptively behave responding to external disturbances keeping the shape of the trajectory unchanged. This approach will be extended to a more complex behavior toward the realization of biologically inspired robot control architectures.

7. Acknowledgement

This work was supported in part by Korea MIC & IITA through IT Leading R&D Support Project. This research was also conducted as a program for the "Fostering Talent in Emergent Research Fields" in Special Coordination Funds for Promoting Science and Technology by Japan Ministry of Education, Culture, Sports, Science and Technology.

8. References

Matsuoka, K. (1985). Sustained Oscillations Generated by Mutually Inhibiting Neurons with Adaptation, *Biological Cybernetics*, Vol. 52, No. 6, pp. 367-376, ISSN 0340-1200, October 1985

Matsuoka, K. (1987). Mechanisms of Frequency and Pattern Control in the Neural Rhythm Generators, *Biological Cybernetics*, Vol. 56, No. 5-6, pp. 345-353, ISSN 0340-1200, July 1987

Taga, G.; Yamagushi, Y. & Shimizu, H. (1991). Self-organized Control of Bipedal Locomotion by Neural Oscillators in Unpredictable Environment, *Biological Cybernetics*, Vol. 65, No. 3, pp. 147-159, ISSN 0340-1200, July 1991

Taga, G. (1995). Model of the Neuro-musculo-skeletal System for Human Locomotion, *Biological Cybernetics*, Vol. 73, No. 2, pp. 91-111, ISSN 0340-1200, August 1995

Miyakoshi, S.; Taga, G.; Kuniyoshi, Y. & Nagakubo, A. (1998). Three-dimensional Bipedal Stepping Motion Using Neural Oscillators-Towards Humanoid Motion in the Real World, *Proceedings of IEEE/RSJ Int. Conf. on Intelligent Robots and Systems*, pp. 84-89, Canada, October 1998, Victoria B.C.

Fukuoka, Y.; Kimura, H. & Cohen, A. H. (2003). Adaptive Dynamic Walking of a Quadruped Robot on Irregular Terrain Based on Biological Concepts, *Int. Journal of Robotics Research*, Vol. 22, No. 3-4, pp. 84-89, ISSN, 0278-3649, March-April 2003

Endo, G.; Nakanishi, J.; Morimoto, J. & Cheng, G. (2005). Experimental Studies of a Neural Oscillator for Biped Locomotion with QRIO, *Proceedings of IEEE Int. Conf. on Robotics and Automation*, pp. 596-602, Spain, April 2005, Barcelona

Yang, W.; Chong, N. Y.; Ra, S.; Kim, C. & You, B. J. (2008). Self-stabilizing Bipedal Locomotion Employing Neural Oscillators, *Proceedings of IEEE-RAS Int. Conf. on Humanoid Robots*, pp. 8-5, Korea, December 2008, Daejeon

Williamson, M. M. (1996). Postural Primitives: Interactive Behavior for a Humanoid Robot Arm, *Proceedings of 4th Int. Conf. on Simulation of Adaptive Behavior*, pp. 124-131, ISBN-10:0-262-63178-4, USA, September 1996, MIT Press, Cape Cod

Williamson, M. M. (1998). Rhythmic Robot Arm Control Using Oscillators, *Proceedings of IEEE/RSJ Int. Conf. on Intelligent Robots and Systems*, pp. 77-83, Canada, October 1998, Victoria

A. M. Arsenio, A. M. (2000). Tuning of neural oscillators for the design of rhythmic motions, *Proceedings of IEEE Int. Conf. on Robotics and Automation*, pp. 1888-1893, USA, April 2000, California

Yang, W.; Chong, N. Y.; Kim, C. & You, B. J. (2007). Optimizing Neural Oscillator for Rhythmic Movement Control, *Proceedings of IEEE Int. Symp. on Robot and Human Interactive Communication*, pp. 807-814, Korea, September 2007, Jeju

Yang, W.; Chong, N. Y.; Kim, C. & You, B. J. (2008). Entrainment-enhanced Neural Oscillator for Rhythmic Motion Control, *Journal of Intelligent Service Robotics*, Vol. 1, No. 4, pp. 303-311, ISSN 1861-2776, October 2008

Slotine, J.-J. E. & Li, W. (1991). *Applied Nonlinear Control*, Prentice Hall., ISBN 0-13-040890, New Jersey

Kirkpatrick, S.; Gelatt, C. D. & Vecchi, M. P. (1983). Optimization by Simulated Annealing, *Science*, Vol. 220, No. 4598, pp. 671-680, ISSN 0036-8075, May 1983

Geman, S. & Geman, D. (1984). Stochastic Relaxation, Gibbs Distributions, and the Bayesian Restoration of images," *IEEE Transactions on Pattern Analysis and Machine Intelligence*, Vol. 6, No. 6, pp. 721-741, ISSN 0018-9340, November 1984

Gomi, H., & Osu, R. (1998). Task-dependent viscoelasticity of human multijoint arm and its spatial characteristics for interaction with environment," *Journal of Neuroscience*, pp. 8965-8978, ISSN 0270-6474, November 1998, Washington DC

9

Knotting a Flexible Rope using a High-speed Multifingered Hand System based on Synthesis of Knotting Manipulation Skills

Yuji Yamakawa [1], Akio Namiki [2],
Masatoshi Ishikawa [1] and Makoto Shimojo [3]
1 Graduate School of Information Science and Technology, Univ. of Tokyo
Japan
2 Graduate School of Engineering, Chiba University
Japan
3 Graduate School of Electro-Communications, Univ. of Electro-Communications
Japan

1. Introduction

Until now, the dexterous manipulation of a rigid body has been achieved. In our laboratory, the dynamic manipulations such as batting (Senoo et al., 2004), dribbling (Shiokata et al., 2005), regrasping (Furukawa et al, 2006), pen spinning (Ishihara et al., 2006), and throwing (Senoo et al., 2008) (http://www.k2.t.u-tokyo.ac.jp/fusion/) have been performed by using the high-speed robot system and the high-speed sensory feedback. In the future, the high-speed and dexterous manipulation of a flexible body should be carried out in order to improve the manipulation ability. Therefore, we aim at the high-speed and dexterous manipulation of a flexible body.

Manipulation of flexible linear objects is a significantly important problem in the robotics research, and the number of studies looking at this problem has been increasing recently.

We have developed strategies for the skills required by a multifingered hand in a knotting task (Yamakawa et al., 2007, Yamakawa et al., 2008). To improve the robustness of these strategies, we proposed high-speed visual and tactile sensory feedback control. We demonstrated successful completion of a so-called overhand knot by sequential application of the individual skills involved.

However, it has not been explained that many knots can be produced using only these proposed skills. Therefore, in this research we examine this point based on an original knot theory. The goal of this research is to analyze the knotting process and to define the relationship between the knotting process and the individual skills that can be performed by a robot hand (Yamakawa et al., 2008).

Finally, two kinds of knot (overhand knot and half hitch) were achieved using a high-speed multifingered hand with high-speed tactile sensors and a high-speed vision.

In this paper, "task" is defined as one manipulation carried out by the robot, and "skill" is defined as a minimum required element to achieve the task and a meaningful element that

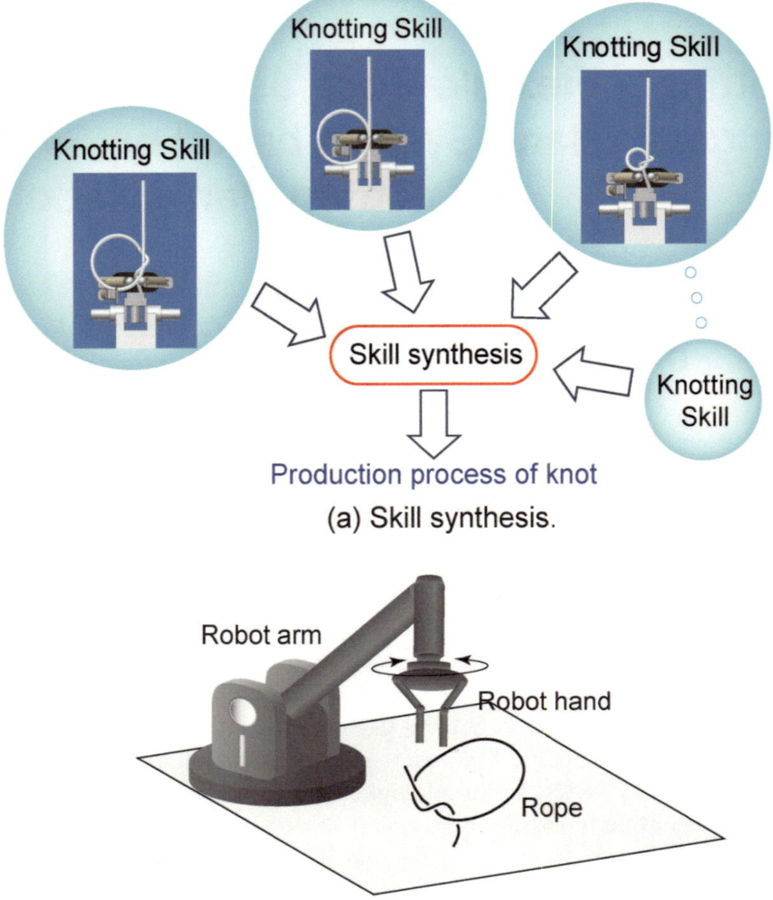

(a) Skill synthesis.

(b) Concept.

Fig. 1. Purpose of This Research

changes the rope state. Therefore, one "task" is achieved by the combination of a certain number of "skills". In order to further develop the manipulation theory, it is extremely important not only to achieve many tasks, but also to integrate the skills that are needed for task realization (Fig. 1(a)).

2. Related Work

Some researchers have focused on manipulation of flexible linear objects. For instance, Inoue & Inaba proposed a method of knotting a rope using visual feedback (Inoue & Inaba, 1984). Matsuno et al. used an imaging system to recognize the shape of a rope and demonstrated knotting using dual manipulators (Matsuno et al., 2001). Saha & Isto suggested a topological motion planner for manipulating flexible linear objects (Saha & Isto, 2006). Morita et al. suggested a knot planning technique by using a system called Knot Planning from

Observation (KPO) (Morita et al., 2003). Wakamatsu et al. formulated a description of knots and a process of knot manipulation (Wakamatsu et al., 2006).

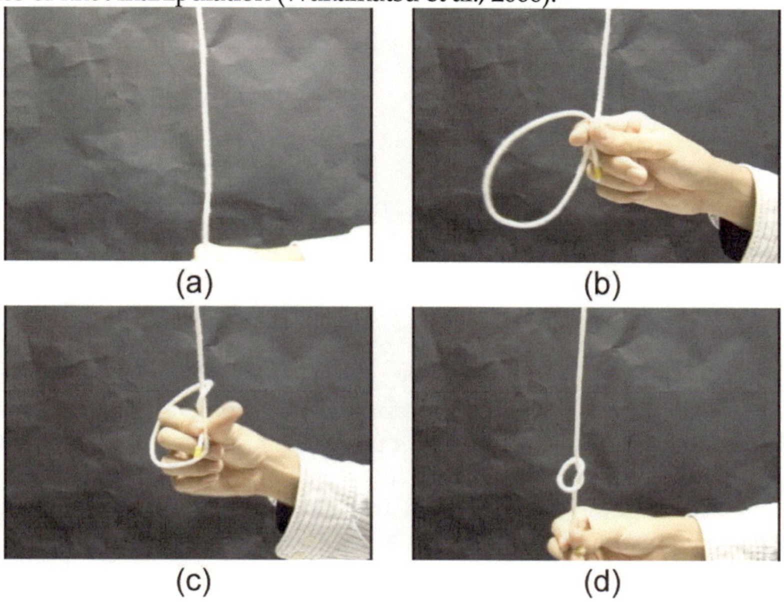

Fig. 2. Human Knotting

In this research, we focus on the relationship between the structure of a knot and the individual robot hand skills required to implement the knot, and we propose a new method of producing a knot with a robot hand. In addition, we also propose a new description that takes the grasp configuration of the intersections of a knot into account.
Knotting manipulation of a flexible rope has been demonstrated in numerous studies. These studies generally considered the geometry of the knot and proposed a knotting process; however, they did not match the actual type of manipulation performed by the robot hand. For example, even if the geometric relation of a rope is the same, the knotting process substantially changes depending on the grasp configuration (for example, loop production). Thus, to consider the actual manipulation by a robot hand, it is necessary to determine the relationship between the knotting process and the individual manipulation skills of the robot hand. In addition, a description of the knotting process that is appropriate to the robot hand should be considered.
The aim of this research is to identify the knot manipulation skills required by a general robot arm and robot hand to achieve a knotting process, as shown in Fig. 1(b).

3. Extraction of Knotting Skills

It is extremely important to identify the skills that are appropriate to the knotting task. There are two main reasons why the manipulation of a flexible linear object is difficult:
1) Rope deformation during manipulation, and
2) Difficulty of predicting the rope deformation.
To overcome these difficulties, the following strategies are considered:
1) Using sensory feedback control by real-time measurement of the rope deformation,

2) Identifying a manipulation skill that cancels rope deformation, and
3) Identifying a manipulation skill that exploits the rope deformation.

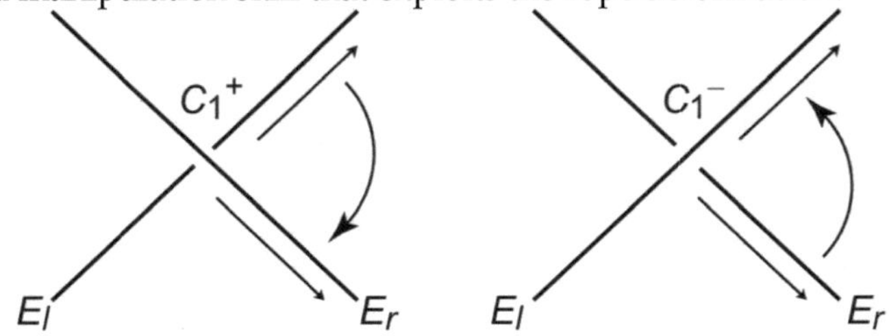

(a) Clockwise: + (b) Counter clockwise: –

Fig. 3. Description of Intersection (Wakamatsu et al., 2006)

Based on these strategies, in order to identify manipulation skills required for knotting, we analyzed a one-handed knotting process performed by a human hand, as shown in Fig. 2. Fig. 2(a) is the initial state, Fig. 2(b) shows loop production using rope deformation by arm motion, Fig. 2(c) shows rope permutation by finger motion, and Fig. 2(d) shows rope pulling by arm motion. As a result, we identified three skills required to achieve one-handed knotting: loop production, rope permutation, and rope pulling.

Loop production is categorized as a manipulation skill that utilizes the rope deformation. Rope permutation is categorized as a manipulation skill that cancels rope deformation. Rope pulling is a motion required to complete the knot.

In addition, in order to allow handling of the ends of the rope, one more manipulation skill called rope moving is added.

In summary, the essential manipulation skills for knotting are:

1) Loop production,
2) Rope permutation,
3) Rope pulling, and
4) Rope moving.

4. Description of Intersection and Hand Skill

First, we define the description of a rope intersection that constitutes the knot, by referring to the method proposed in the previous work (Wakamatsu et al., 2006).

The beginning of a rope is represented by E_l (or l_1 or r_1) as the initial location. In the same way, the end of the rope is represented by E_r (or l_2 or r_2) as the final location. Descriptions of the intersections $C_i^{\{+,-\}}$ ($i = 1, 2, 3 \cdots$) are assigned for all intersections from E_l to E_r, where i is an intersection number. The sign "+, –" shows the intersection sign. The sign "+" means clockwise, and the sign "–" means counterclockwise, as shown in Fig. 3. As a result, the description of the intersection is defined as shown in Fig. 3.

Moreover, we introduce a new description here. First, the two directions of the rope at "outlets" of the intersection are summed. Second, the line that connects the two fingers grasping the intersection is considered. Finally, the relationship between the summed rope direction and this connecting line is examined. If the relationship is parallel, the description of the intersection shown in Fig. 4(a) is defined. If the relationship is orthogonal, the description of the intersection shown in Fig. 4(b) is defined. This description method is used in loop production, in particular.

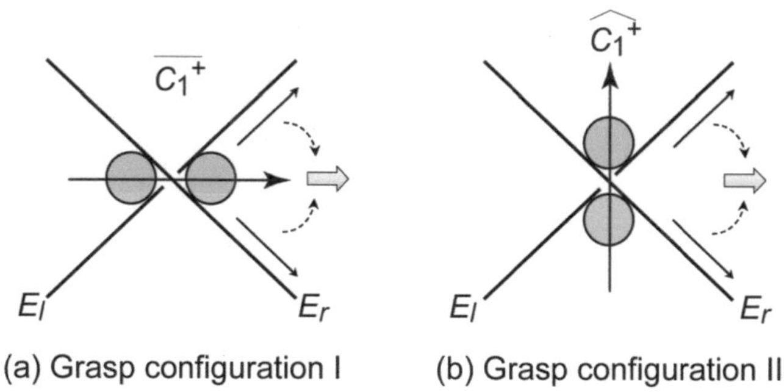

Fig. 4. Grasp Configurations

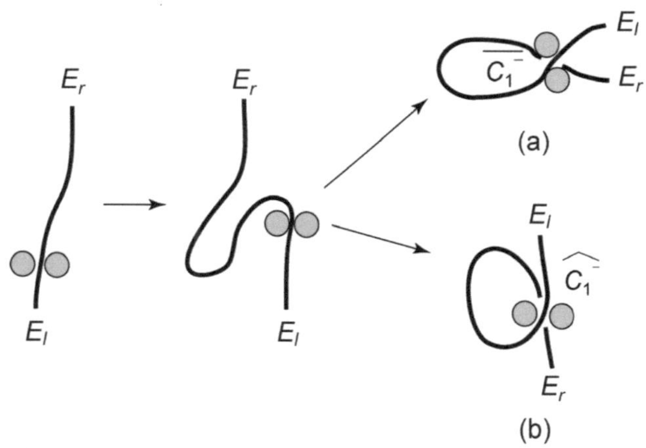

Fig. 5. Loop Production

Next, the characteristics of the identified hand skills are clarified based on the intersection description.

In this paper, a skill is defined as a minimum required element to achieve a task. In addition, the objective of this study is to clarify the relationship between the knot production process and manipulation skills. This relationship can be obtained from a description of the knot and analysis of the knot. Thus, recognition of the rope by a visual sensor is not taken into account in this paper.

4.1 Loop Production

Fig. 5 shows the loop production. Loop production is an operation that makes one intersection. Thus, the description of the intersection is given by the followings.

$$E_l - E_r \to E_l - \overline{C_1^{\{+,-\}}} - \overline{C_1^{\{+,-\}}} - E_r \quad \text{Fig. 5(a)}$$
$$E_l - E_r \to E_l - \hat{C}_1^{\{+,-\}} - \hat{C}_1^{\{+,-\}} - E_r \quad \text{Fig. 5(b)}$$

This skill is equivalent to Reidemeister move I in knot theory. The intersection sign depends on the direction of the loop.

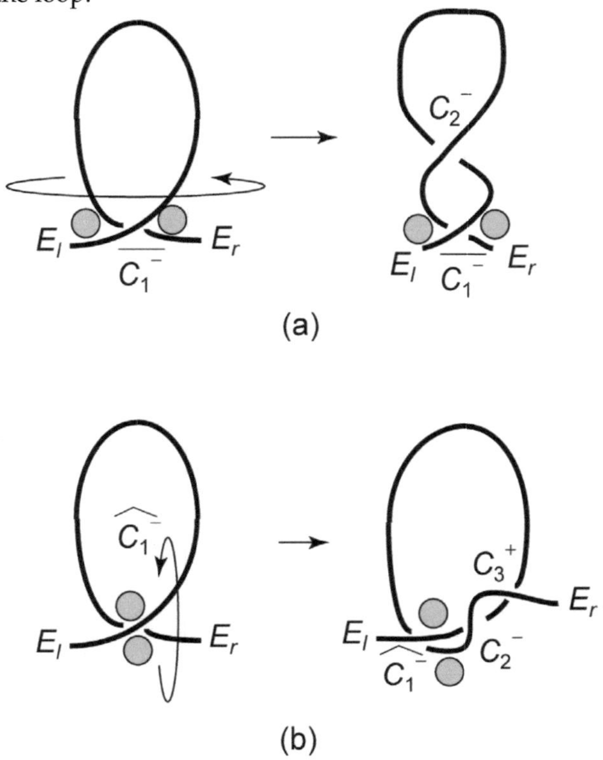

Fig. 6. Rope Permutation

<u>Loop production strategy</u>
In this skill, a loop that serves as the starting point of the knot is produced on the rope. First, the rope is grasped by two fingers. A loop is produced by the two dimensional translational motion of the hand. There are two different types of loop production depending on the grasp configuration of the two fingers, as shown in Fig. 5. This difference is important for performing the actual knotting process by robot fingers.
To achieve robust loop production, the hand motion should be controlled by real-time visual feedback (Yamakawa et al., 2008).

4.2 Rope Permutation

Fig. 6 shows the rope permutation. Rope permutation is divided into two types, depending on the grasp configuration. In the case shown in Fig. 6(a), the description of the intersection is the following:

$$E_l - \overline{C_1^{\{+,-\}}} - \overline{C_1^{\{+,-\}}} - E_r \rightarrow$$
$$E_l - \overline{C_1^{\{+,-\}}} - C_2^{\{+,-\}} - C_2^{\{+,-\}} - \overline{C_1^{\{+,-\}}} - E_r$$

In the case shown in Fig. 6(b), the description of the intersection is the following:

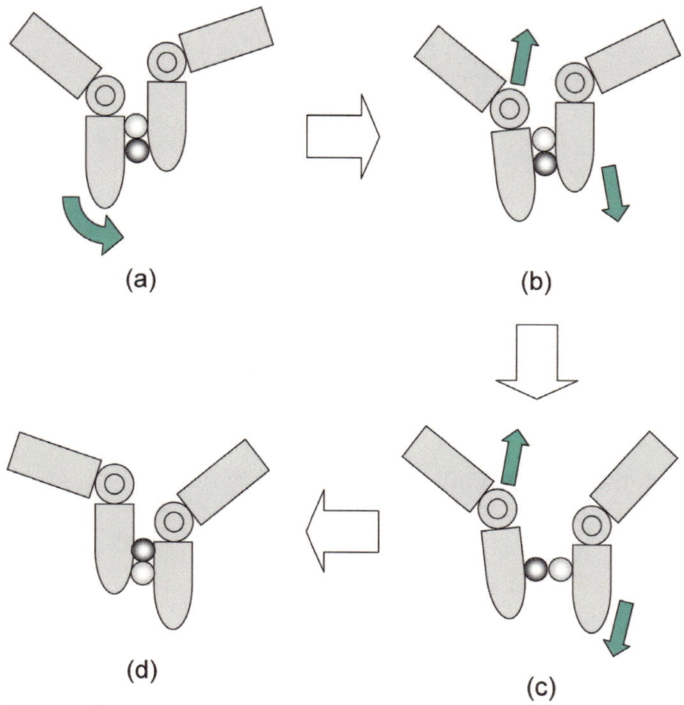

Fig. 7. Strategy of Rope Permutation

$$E_l - \hat{C}_1^{\{+,-\}} - \hat{C}_1^{\{+,-\}} - E_r \rightarrow$$
$$E_l - \hat{C}_1^{\{+,-\}} - C_2^{\{+,-\}} - C_3^{\{-,+\}} - \hat{C}_1^{\{+,-\}} - C_2^{\{+,-\}} - C_3^{\{-,+\}} - E_r$$

The difference between the two types of rope permutation depends on the grasp configuration used in the loop production.

Rope permutation strategy

The strategy of rope permutation is shown in Fig. 7. Both fingers are moved while remaining parallel. At some point, two sections of the rope engage each other by virtue of friction (Fig. 7(b)).

By continuing to move the two fingers in parallel, the two sections of the rope are permuted (exchange places) (Fig. 7(c)). At the moment of permutation, the grasp force is increased. In order to achieve smooth permutation, it is necessary to increase the distance between both fingers at this moment. Therefore, grasp force control based on tactile feedback is implemented (Yamakawa et al., 2007).

4.3 Rope Pulling

Rope pulling is an operation that basically removes the intersection, as shown in Fig. 8. This skill is effective when used together with rope permutation. Two operations can be realized

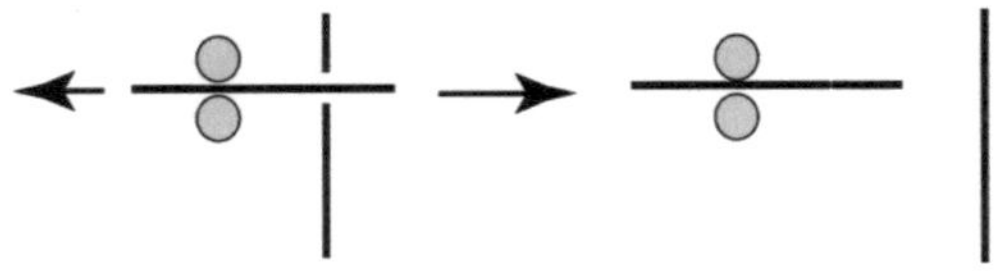

Fig. 8. Rope Pulling

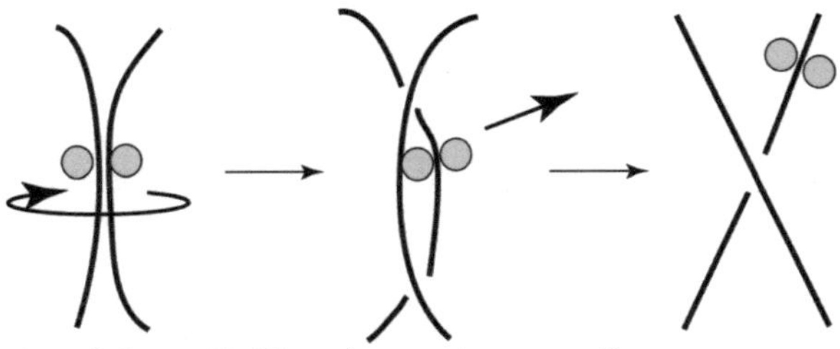

Fig. 9. Rope Permutation & Rope Pulling (operation type I)

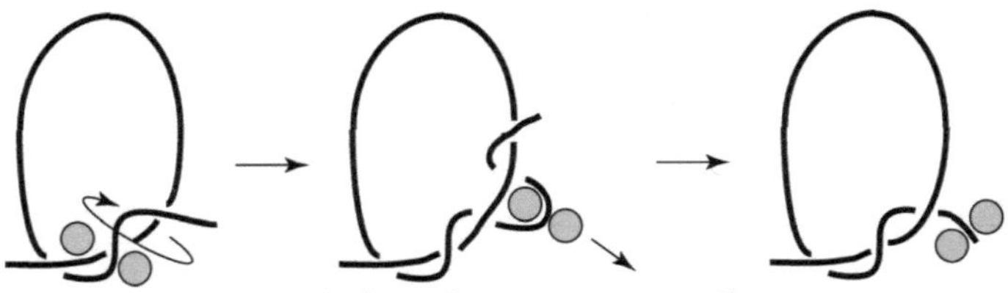

Fig. 10. Rope Permutation & Rope Pulling (operation type II)

by utilizing these two skills: One is the ability to create one intersection. Another is the ability to change the sign of any intersection. When one intersection is produced (Fig. 9), the description of the intersection is given by the following.

$$E_l - E_r \rightarrow E_l - C_1^{\{+,-\}} - E_r$$

When the sign of an intersection is changed (Fig. 10), the description of the intersection is given by the following.

$$E_l - C_1^{\{+,-\}} - \cdots - C_i^{\{+,-\}} - \cdots - C_1^{\{+,-\}} - E_r \rightarrow$$
$$E_l - C_1^{\{+,-\}} - \cdots - C_i^{\{-,+\}} - \cdots - C_1^{\{+,-\}} - E_r \rightarrow$$

Rope pulling strategy

In general, rope pulling can be achieved by the translational motion of the robot hand.

4.4 Rope Moving

Rope moving is defined as the three-dimensional free motion of the end of the rope without other parts of the rope acting as obstacles. Assuming that the rope is located in a plane, this skill is the free motion in the space above (or below) the plane. Rope moving is used to cross the two sections of the rope or to set one section close to another. When the two sections are

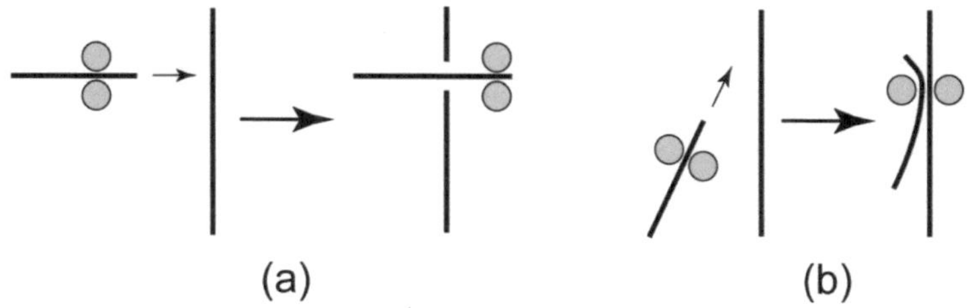

Fig. 11. Rope Moving

intersected as shown in Fig. 11(a), one intersection is made. The description of the intersection can be written as

$$E_l - E_r \rightarrow E_l - C_1^{\{+,-\}} - E_r.$$

When one section of the grasped rope approaches another, as shown in Fig. 11(b), the intersection is not created.

Rope moving strategy

Rope moving can be carried out by the three-dimensional motion of the robot hand.

5. Analysis of Knot and Skill Synthesis

In this section, we analyze the feasibility of some knots based on synthesis of the individual skills described above. Then, a knot production process can be obtained by using the analysis method.

Wakamatsu et al. have proposed a process of knot manipulation based on four skills that consist of Reidemeister moves and rope crossing (Wakamatsu et al., 2006). Although they described a systematic knot production process, the relationship between the knot production process and the required knot manipulation skills was not discussed.

In order to achieve actual knotting, it is necessary to consider not only the knot production process, but also the relationship between the production process and individual hand skills. For this reason, we propose the following analysis method.

Analysis Method

1) Represent a knot based on the description of the intersections that constitute the knot.
2) Unravel one intersection of the knot, starting from the intersection nearest the end of the rope.
3) Iterate step 2) until the intersections disappear. As a result, a sequence of operations to remove the intersections is obtained.

4) Apply appropriate skills to the sequence, while following the sequence obtained in step 3) in reverse.

By this analysis method, the location and sign of each intersection is identified. As a result, it can be determined how to generate the intersections. This method can be applied not only to a knot generated by one rope, but also to a knot generated by one rope and one object and a knot generated by two ropes. Although the knot production process obtained by the

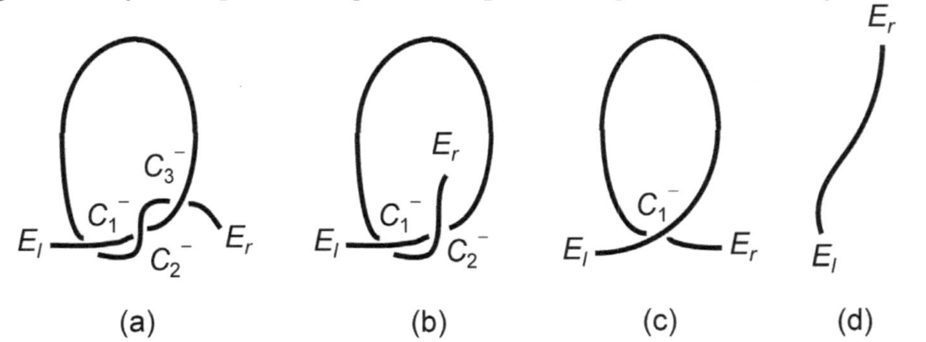

Fig. 12. Analysis of Overhand Knot

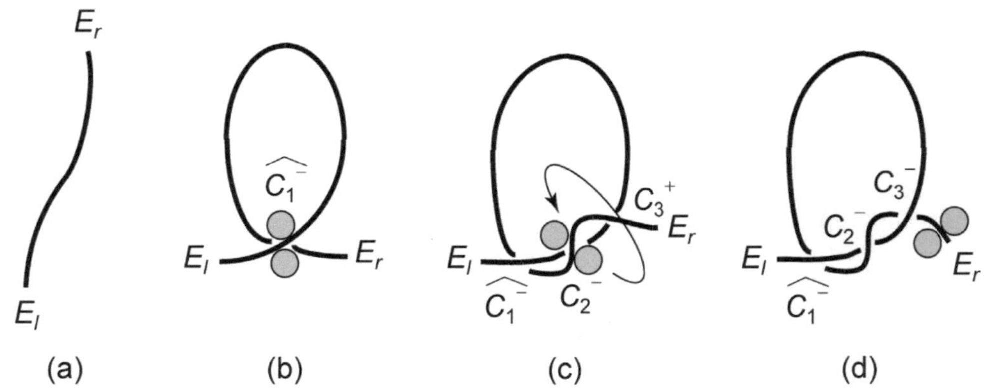

Fig. 13. Production Process of Overhand Knot

proposed analysis method may not be optimal, this method can always provide one solution of the knot production process.

5.1 Knot Generated by One Rope

Here, we consider the description and production method of a knot generated by one rope. The start of the rope is located at the beginning of the rope (represented by E_l). The end of the rope is represented by E_r. As an example, here we analyze an "overhand knot".

Overhand knot (Fig. 12(a))

An overhand knot is the simplest knot that is created on a rope. This knot prevents the rope unraveling.

Analysis of overhand knot

First, the description of the intersection in the overhand knot is

$$E_l - C_1^- - C_2^- - C_3^- - C_1^- - C_2^- - C_3^- - E_r \,. \text{ (Fig. 12(a))}.$$

Next, removing one intersection (C_3^-), starting from the intersection near the end (E_r) of the rope, gives the following description of the intersections of the overhand knot:

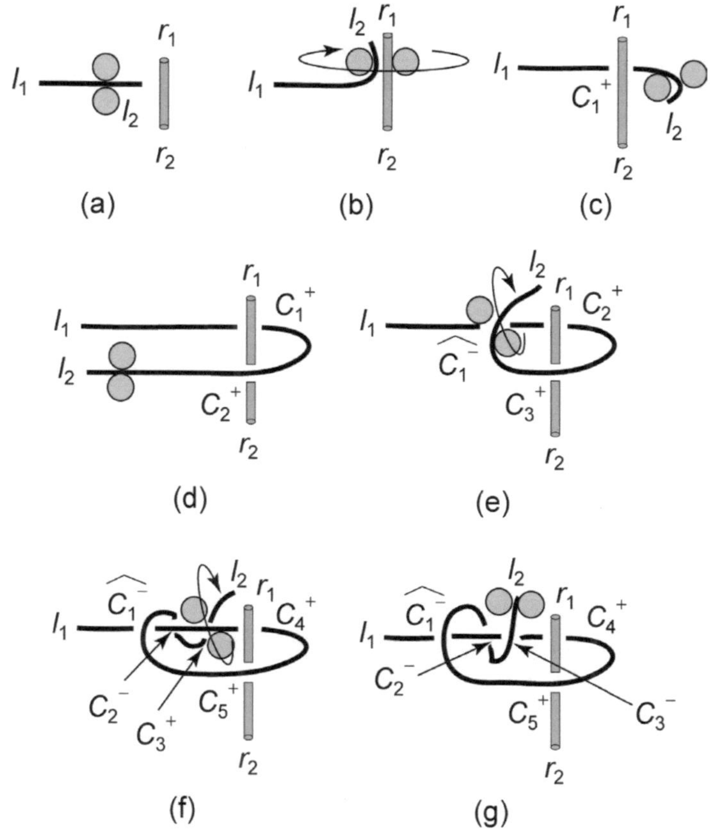

Fig. 14. Production Process of Half Hitch

$$E_l - C_1^- - C_2^- - C_1^- - C_2^- - E_r \text{ (Fig. 12(b))}$$

Iterating this operation until the intersections disappear yields the following description of the intersection

$$E_l - C_1^- - C_1^- - E_r \text{ (Fig. 12(c))}$$

$$E_l - E_r \text{ (Fig. 12(d))}.$$

The production process of the overhand knot can be obtained by following this process in reverse while considering the description of the intersections.

Production process of overhand knot

First, loop production is performed, and the intersection $\hat{C}_1^-$ is created (Fig. 13(a)). Second, the intersection C_2^- is produced. However, it is not effective to produce only the intersection C_2^-. By checking the final type of knot, it is found that two intersections should be made after the intersection $\hat{C}_1^-$. In addition, the final intersection C_3^- should pass under the rope. For these reasons, the state shown in Fig. 13(b) can be produced only by performing rope permutation (Fig. 6(b)). Finally, using rope permutation and rope pulling (Fig. 10), the overhand knot is achieved, as shown in Fig. 13(c). The description of the intersections for the overhand knot production process can be represented by the following:

$$E_l - E_r \text{ (Fig. 13(a))}$$

$$E_l - \hat{C}_1^- - \hat{C}_1^- - E_r \text{ (Fig. 13(b))}$$

$$E_l - \hat{C}_1^- - C_2^- - C_3^+ - \hat{C}_1^- - C_2^- - C_3^+ - E_r \text{ (Fig. 13(c))}$$

$$E_l - \hat{C}_1^- - C_2^- - C_3^- - \hat{C}_1^- - C_2^- - C_3^- - E_r \text{ (Fig. 13(d))}$$

The eight knot and the stevedore's knot can be produced based on skill synthesis in the same way.

5.2 Knot Generated by One Rope and One Object

In this section, we consider the knotting process of a knot generated by one rope and one object. As an example, we analyze a "half hitch".

Half hitch (Fig. 14)

The half hitch is one of the knots that make a connection between a rope and an object. Although it is an easy task to make this knot, the strength of the knot is very low. However, the strength can be increased by combining the half hitch with other knots.

Production process of half hitch

Here, we omit the intersection description of the half hitch. The left end and the right end of the rope are represented by l_1 and l_2, and the left end and the right end of the object are represented by r_1 and r_2. The description of intersections on the rope and the object is performed in the order of initial location.

First, the intersection C_1^+ is created by rope permutation (Fig. 14(b), (c)). Second, the rope is wrapped around the stick by rope moving to produce the intersection C_2^+ (Fig. 14(d)). Next, the intersection $\hat{C}_1^-$ is made by loop production (Fig. 14(e)). Finally, the half hitch is finished by performing rope permutation twice and rope pulling once (Fig. 14(f), (g)).

A knot generated by two ropes can be considered in the same way.

5.3 Knot Generated by Two Ropes

In this section, the production process of a knot generated by two ropes is explained. As an example, we analyze the "square knot" (Fig. 15). The description of the intersection is the same as that for the knot generated by one rope and one object.

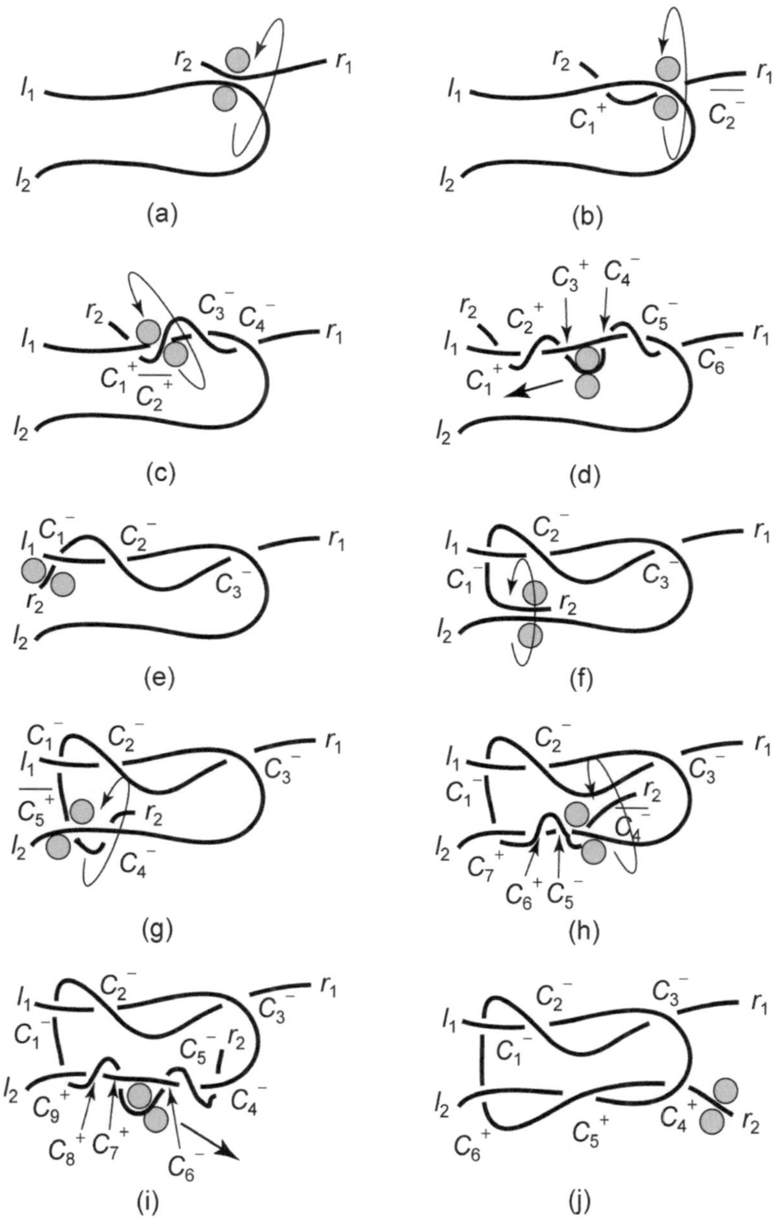

Fig. 15. Production Process of Square Knot

Square knot (Fig. 15)
Square knot is one of representative knots generated by two ropes. This knot is used when two ropes that have same diameter are connected. To make this knot is very easy, and the strength of this knot is very high.

Production process of square knot
First, the two ropes are set as shown in Fig. 15(a). Second, the intersections (C_1^-, C_2^- and C_3^-) are produced by three rope permutations and rope pulling, as shown in Fig. 15(b)-(e). Third, the two ropes are set as shown in Fig. 15(f) by rope moving. Lastly, the intersections (C_4^+, C_5^+ and C_6^+) are produced by three rope permutations and rope pulling, as shown in Fig. 15(g)-(j).
As a result, the production process of a knot can be obtained by skill synthesis. Namely, it is considered that the extracted skills are reconfigurable ones.

6. Experiment

6.1 Experimental System

The experimental system is composed of a high-speed multifingered hand (Namiki et al., 2003), high-speed tactile sensors (Ishikawa & Shimojo, 1982), and a high-speed visual sensor (Nakabo et al., 2000).
The high-speed multifingered hand has three fingers and two wrist joints. Moreover, each finger is divided into a top link and a root link. A newly developed compact harmonic drive gear and a high-power miniature actuator are installed in each finger link. The design of this actuator is based on the new concept that the maximum power output should be increased beyond the rated power output. The joints of the hand can be closed at a speed of 180 deg. /0.1 s.
The high-speed tactile sensor can measure the centre position of a two-dimensionally distributed load, and the total load can be measured within 1 ms. This sensor is installed to each top link in order to perform the grasp force control during the rope permutation (Yamakawa et al., 2007).
The high-speed vision can measure the centre position and the angle of the principal axis of inertia within 1 ms. This sensor is mounted on the 2-DOF (pan and tilt) active mechanism. This sensor is used for the wrist-joint control during loop production in order to achieve robust control against rope deformation.
The cycle time of sensor feedback and control processing is set at 1 ms. Thus, the robot can react quickly to target motion in unpredictable conditions.
In order to prevent the rope from slipping on the fingers of the robot hand, a fingerstall is attached to each top link. One end of the rope is grasped by the robot hand, and the other end is held in a pulley so as to allow free up and down motion.

6.2 Experimental Results

Fig. 17 and Fig. 19 show sequences of continuous photographs of the knotting tasks (overhand knot and half hitch). The knotting strategy used was the one proposed in the previous papers (Yamakawa et al., 2007, Yamakawa et al., 2008).

Overhand knot

The experimental system is shown in Fig. 16. Fig. 17(a)-(e) show loop production. In Fig. 17(f), the rope sections are pressed by the free finger to strengthen the contact state between the two sections. Fig. 17(g)-(i) show rope permutation. Fig. 17(j)-(l) show rope pulling.

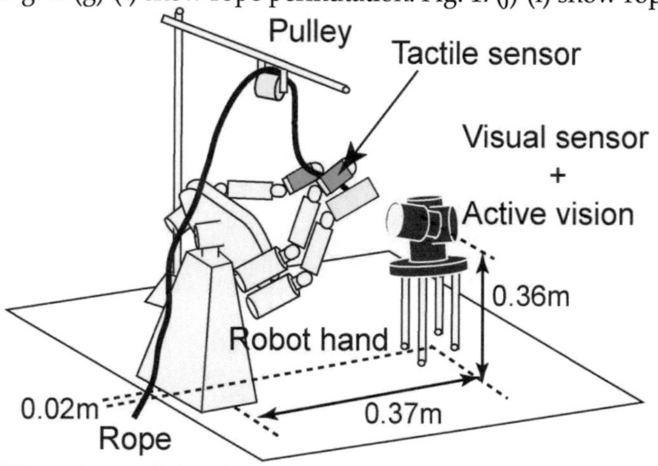

Fig. 16. Overall of Experimental Condition (Overhand Knot)

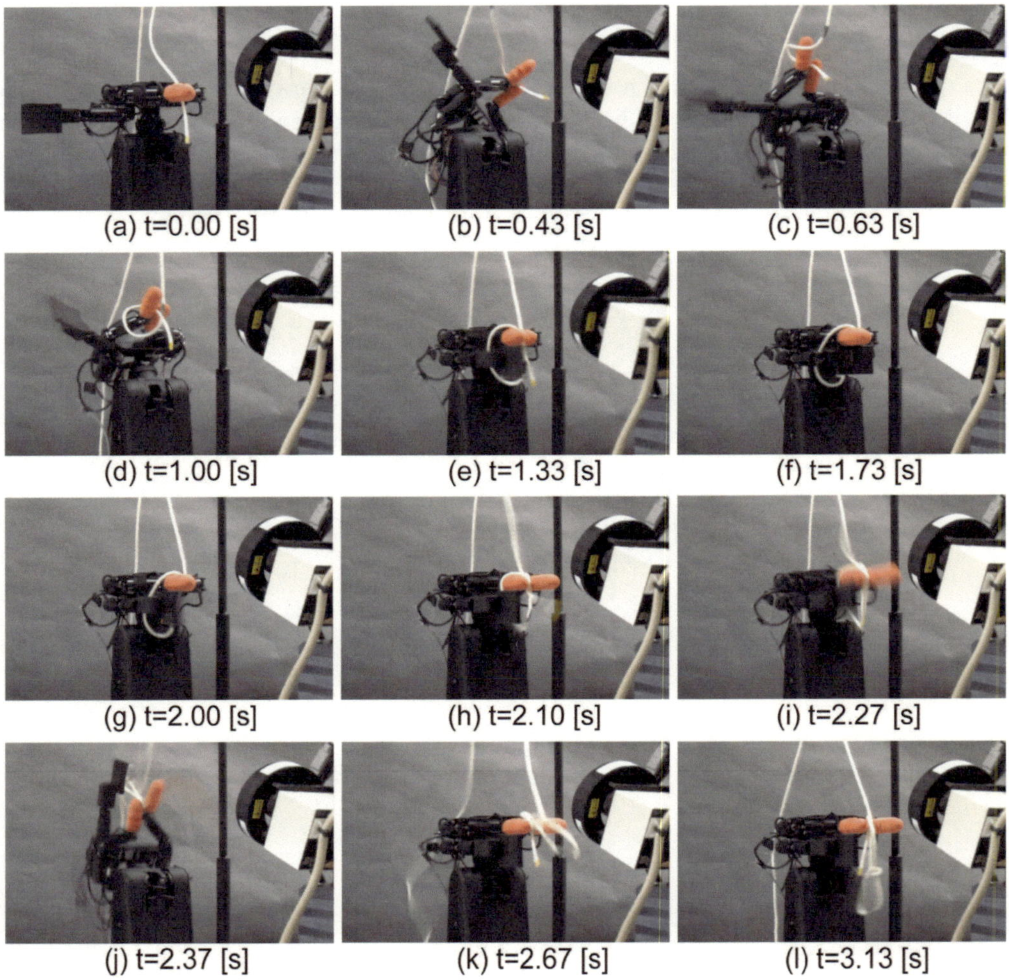

Fig. 17. Experimental Result of Overhand Knot

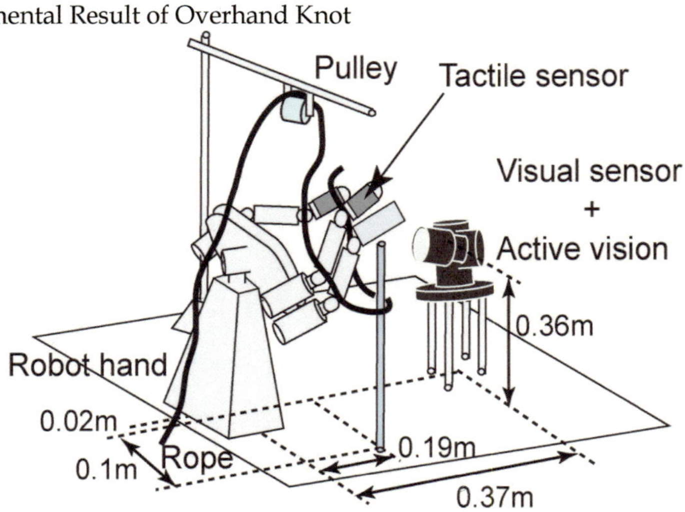

Fig. 18. Overall of Experimental Condition (Half Hitch)

Half hitch

The experimental system is shown in Fig. 18. In the initial state, the rope is wrapped around the object, as shown in Fig. 18. Fig. 19(a)-(c) show loop production. In Fig. 19(d), the rope sections are pressed by the free finger to strengthen the contact state between the two sections. Fig. 19(e)-(g) show rope permutation. Fig. 19(h) and (i) show rope pulling. Finally, Fig. 19(j)-(l) show additional rope pulling by a human hand to tighten up the knot.

These video sequences can be viewed on our web site (http://www.k2.t.u-tokyo.ac.jp /fusion/SkillSynthesis/).

7. Conclusion

The aim of this research is to obtain the production process of a knot and to clarify the relationship between the production process of the knot and the manipulation skills for knotting.

First, to identify the necessary skills for knotting, we analyzed a knotting action performed by a human subject. As a result, we identified four skills such as "loop production", "rope permutation", "rope pulling" and "rope moving".

And then, in order to analyze a knot, we suggested a new description method of the intersection that constitutes the knot.

Next, we proposed a method to produce a knot. The proposed method is based on a description of the intersections, and it is described by the sequence of operations achieved using the four identified skills. We analyzed three types of knot: a knot generated by one rope, a knot generated by one rope and one object, and a knot generated by two ropes. These knots could be produced by the synthesis of the four skills. In addition, we also determined the relationship between the knot production process and the individual skills required by the robot hand in knot manipulation.

Finally, we demonstrated productions of an overhand knot and a half hitch by using a high-speed multifingered hand with high-speed visual and tactile sensory feedback. In the future, we will attempt to apply our approach to other types of knots.

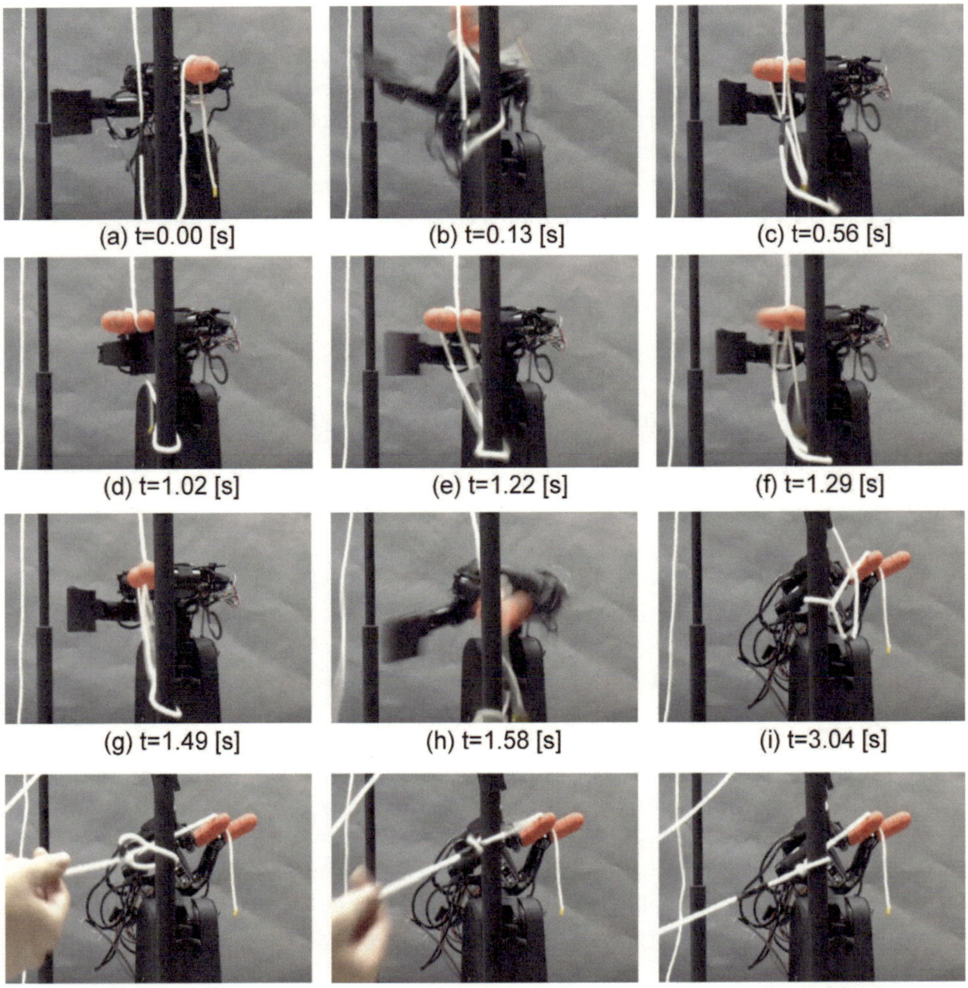

(a) t=0.00 [s] (b) t=0.13 [s] (c) t=0.56 [s]
(d) t=1.02 [s] (e) t=1.22 [s] (f) t=1.29 [s]
(g) t=1.49 [s] (h) t=1.58 [s] (i) t=3.04 [s]
(j) t=4.98 [s] (k) t=5.58 [s] (l) t=6.50 [s]

Fig. 19. Experimental Result of Half Hitch

8. References

Furukawa, N.; Namiki, A.; Senoo, T. & Ishikawa, M. (2006). Dynamic Regrasping Using a High-speed Multifingered Hand and a High-speed Vision System, *Proc. IEEE Int. Conf. on Robotics and Automation*, pp. 181-187

Inoue, H. & Inaba, M. (1984). Hand-eye Coordination in Rope Handling, *Robotics Research: The First International Symposium, MIT Press*, pp.163-174

Ishihara, T.; Namiki, A.; Ishikawa, M. & Shimojo, M. (2006). Dynamic Pen Spinning Using a High-speed Multifingered Hand with High-speed Tactile Sensor, *Proc. IEEE RAS Int. Conf. on Humanoid Robots*, pp. 258-263

Ishikawa, M. & Shimojo, M. (1982). A Method for Measuring the Center Position of a Two Dimensional Distributed Load Using Pressure-Conductive Rubber, *Trans. The Society of Instrument and Control Engineers*, Vol. 18, No. 7, pp. 730-735 (in Japanese)

10

Screw and cable actuators (SCS) and their applications to force feedback teleoperation, exoskeleton and anthropomorphic robotics

Philippe Garrec
CEA List
Interactive Robotics Unit
France

1. Introduction

Some years ago, the CEA developed a new actuator - the Screw and Cable System - to motorize a teleoperation force feedback master arm that would be more economical than previous machines such as the MA23 master arm, a pioneering machine originally designed in 1974 by Jean Vertut and his team also at CEA. The new master arm has been since industrialized and is now manufactured by Haption® under the name Virtuose™ 6D 40-40.
Shorly after, we also designed, upon the same SCS actuator, a new force feedback slave arm for radioactive waste disposal inside a well (STeP: Système de Téléopération en Puits). After these achievements, we recognized that SCS could be interestingly integrated inside manipulator's articulated structure instead of being concentrated at its base. Our laboratory then engaged in the successful design of the upper limb exoskeleton today named ABLE. This is indeed a new type of anthropomorphic, open robot that also offers true linear torque capability without force sensor. A low inertia of the structure and motors altogether lead to a high transparency.

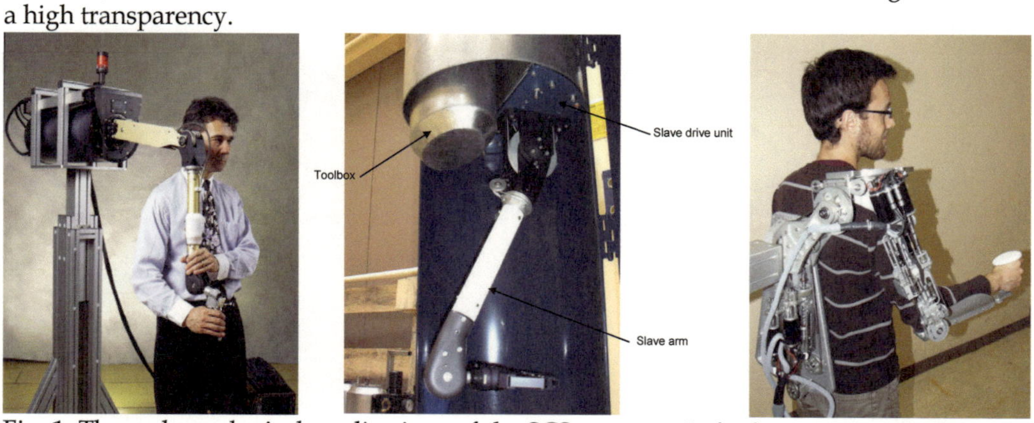

Fig. 1. Three chronological applications of the SCS actuator: Left, the master arm Virtuose™ 6D 40-40 (CEA/Haption) ; Center, the slave arm STeP ; Right, A 4 axis version of the ABLE, a upper limb exoskeleton (CEA)

Its mechanical architecture also features several dedicated innovations - shoulder articulation, adjustable segments, forearm-wrist articulated cage – which all work in tight synergy with the actuators. Evaluation of this device for rehabilitation purpose is undergoing and future applications of the SCS actuators to low-limb exoskeletons and anthropomorphic assistive arms are also planned.

2. Genesis of the SCS actuator

2.1 The problematic of linear torque amplification in Electrical Master Slave Manipulator (EMSM)

The SCS actuator is originally a new answer to the problem of electrical motor torque amplification, a domain pioneered by electrical master-slave manipulators in which our laboratory has been tightly associated: (Goertz et al., 1955) ; (Galbiati et al., 1964) ; (Flatau, 1965) ; (Flatau & Vertut, 1972) ; (Vertut et al., 1975) ; (Köhler, 1981) ; (Vertut & Coiffet, 1984). In these types of manipulators, force feedback is simply obtained through mechanical reversibility and a high linearity of force transmission. The absence of torque/force sensor and associated drift and calibration procedure contribute to a high reliability of the machine. For example, the Mascot EMSM system used by Oxford Technologies Ltd under the name DEXTER has performed over 7,500hrs of remote handling tasks inside the JET (Joint European Torus, UK) with a system availability above 95% in tough conditions. However industrially proven machines, built under strict quality requirements, are expensive and rather bulky. Fig. 2 shows important pioneering machines each of them associated with their torque amplification solutions.

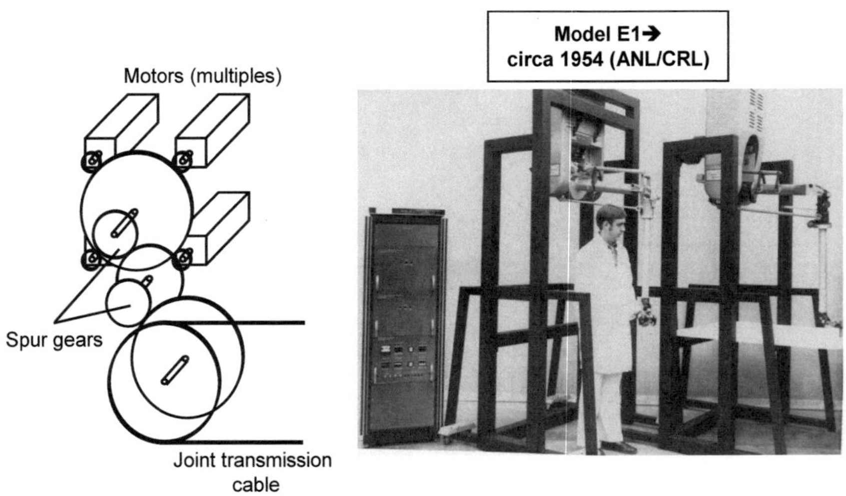

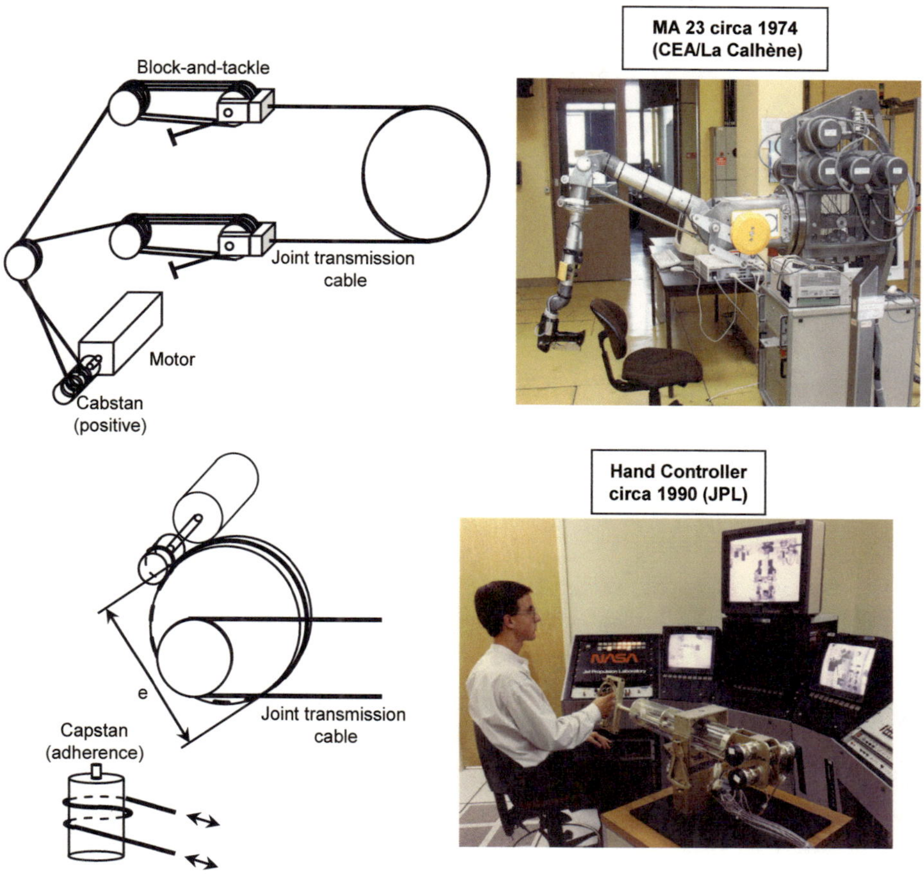

Fig. 2. Landmarks in torque amplification in electrical master-slave telemanipulator (EMSM)

The first principle has been used by R. Goertz on all his designs from the E1 model (the first servomanipulator) to the E4 and Model M. Motor torque is amplified using high-precision spur gears driving the joints either directly (translation joints) or, like the scheme shows, through transmission cable (for remote rotation joints). The second is due to J. Vertut and is team for the MA 23. Motor torque is amplified using block-and-tackle cable (or tape) arrangements which drives a transmission cable (or tape). The last one, the capstan has been used on the Hand Controller. The cable is wrapped around pulleys to increase the adherence, thus enabling the capstan to transmit more torque with very low tension in the cable resulting in a very low friction threshold. For this reason, this is today the more sensitive device for torque amplification and it is most commonly found on haptic devices.

2.2 Force reflection and force transmission in a mechanical linkage

Force reflection (or force feedback) can be defined as the force exerted by the operator on the master device to balance the force exerted by the load on the slave device. This force may be altered in intensity and sense depending on the properties (reversible/irreversible or self-locking) and performances of the mechanical transmission used (Fig. 3).

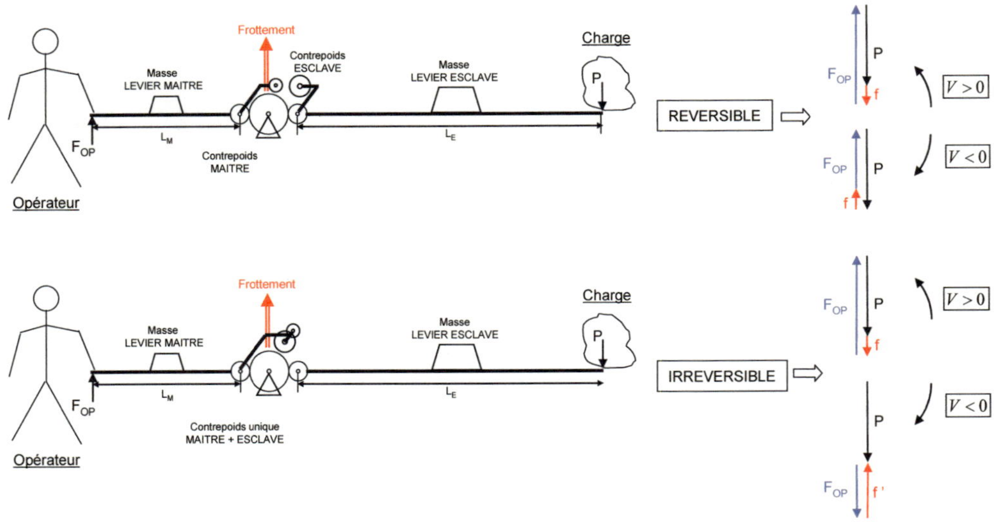

Fig. 3. The concept of force reflection and its alteration with mechanical transmission properties (Top: reversible; Bottom, irreversible)

We can see that irreversibility makes the force feedback incoherent and is thus always avoided in mechanical telemanipulation.

Reversible/Irreversible – Bilateral and Backdrivable

It should be noticed that a reversible transmission is always paired with a bilateral (or back-drivable) behaviour whereas an irreversible (self-locking) transmission may be given a bilateral behaviour through assistance in a closed loop mode with a force sensor. This is why it is useful to avoid confusion between the mechanical property of the transmission obtained by construction with its behaviour. Table 1 summarizes the various cases encountered.

Mechanical type (constructive property)	Behaviour	
	Non assisted (open loop)	Assisted (closed loop)
Irreversible	Unilateral - Self-locking	Bilateral - Backdrivable
Reversible	Bilateral - Backdrivable	

Table 1. Mechanical properties and behaviour of transmissions

Force transmission and force amplification diagram

It is possible to use a universal input-output force transmission diagram to represent the concept of force transmission and amplification for any kind of mechanism (Garrec, 2002). Fig. 4 is a simplified diagram of force transmission for a reversible transmission. Intersections (I, J) between characteristics are only fictive.

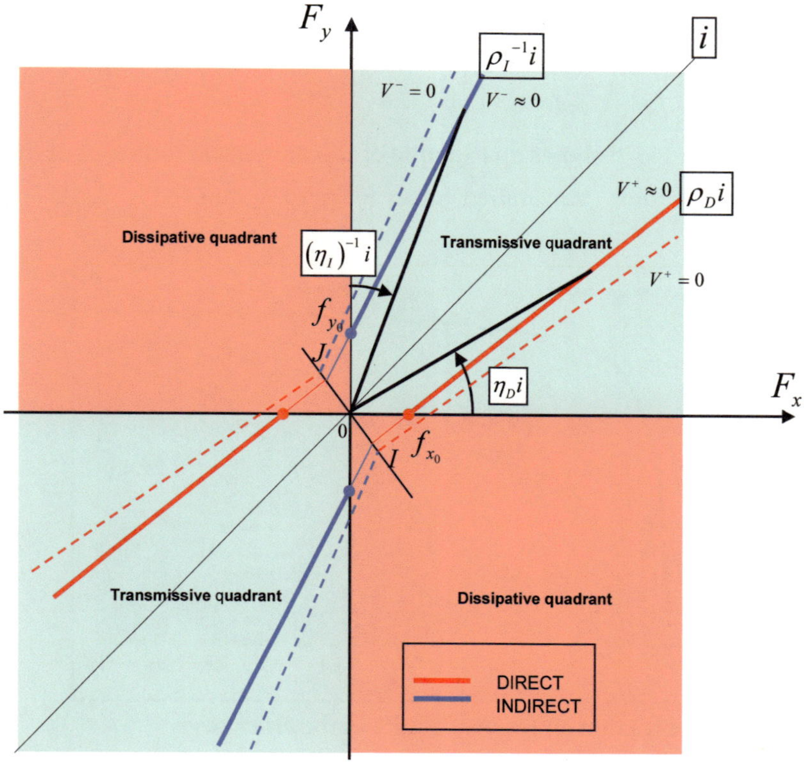

Fig. 4. Force transmission diagram for a reversible transmission

To discuss the basic performances of the transmission, it is sufficient to restrain the representation to the dry friction (Coulomb law). It can be shown that adding a viscous friction would only enlarge the bi-conical diagram. Since mechanical components may transform torque in force, input and output axis do not necessarily have the same unit, F_x, F_y must be considered as generalized efforts. The reference characteristic (i coefficient) corresponds to the kinematic ratio, so in reference to the chosen coordinates, it represents a strictly linear amplification/conversion of forces/torques without friction. Dotted lines correspond to the static dry friction (no speed) and plain lines correspond to the kinematic dry friction (low speed). Red (DIRECT) and blue (INDIRECT) characteristics have the respective coefficients ρ_D and ρ_I. For any mechanism comprising an incline (screw, worm gear, etc.), ρ values are potentially different producing an asymmetry.

The minimum friction in the mechanism created by internal constraints, leads to minimum input and output friction (sometimes called no-load input/output friction or hysteresis). The transmissive quadrant (in blue) corresponds to a real transmission of energy between input/output or vice versa. In the dissipative quadrant (in pink), the mechanism is dissipating the energy supplied by both the input and the output.

In the transmissive quadrant, the efficiency $\eta = F_y / iF_x$, can be defined and plotted as a function as the input force in relative scale $F_x / F_{x_{\max}}$. It represents the effective output

force available for a given input force, which can be interpreted as a default of transparency of the transmission. Efficiency is null for the minimum friction $f_{x_0}/F_{x_{max}}$ and tends to ρ for its maximum value $F_x/F_{x_{max}} = 1$.
The notional diagram Fig. 5 shows an example of the dramatical influence of the minimum friction on the output force (transmitted force) for ρ =0,95 and for $f_{x_0}/F_{x_{max}}$ respectively equal to 2% and 10%.

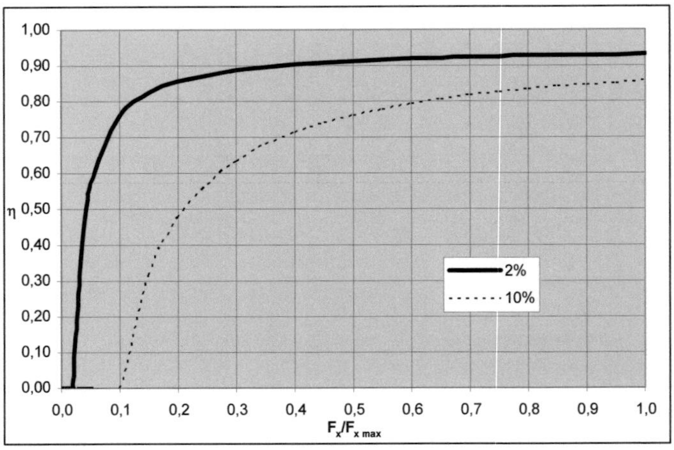

Fig. 5. Effect of relative friction on the availability of the efficiency

Note: For an irreversible mechanism ρ_I is negative and the corresponding characteristics are located in the dissipative quadrants (Fig. 6). In this case, ρ_I parameter is no longer an expression of an efficiency.

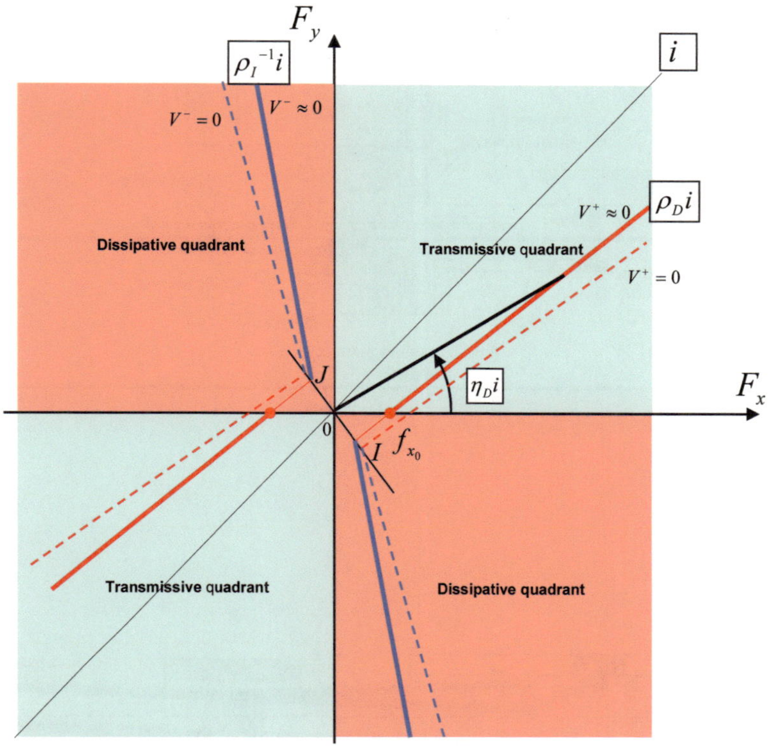

Fig. 6. Force transmission diagram of an irreversible transmission

Linear force transmission

We can now define the conditions to be fulfilled to obtain a linear force transmission:

- the mechanism must be reversible
- the minimum input-output friction must be minimized. A classical quantitative criteria has been proposed in the context of telemanipulator (Vertut & Coiffet, 1984). It can be defined as the ratio of the minimum friction on the maximum capacity of the transmission (sometimes called relative friction). Its is a fundamental performance criterium in force reflecting manipulator
- the divergence of the characteristics must be minimum (ρ maximum)
- ρ_D and ρ_I values should be ideally equal for symmetry purpose

2.3 The Screw and Cable mechanics and its application to the master arm Virtuose 6D

In the late nineties our laboratory was trying to design a new teleoperation, force-feedback, master arm that would be less costly than the MA23 (CEA-La Calhène), a machine that has been consistently used in French teleoperation systems since its creation around 1974. This work resulted in the creation of the screw-and-sable transmission or SCS (Garrec, 2000) as well as the construction of a prototype of the master arm Virtuose 6D (Garrec et al., 2004).

Mechanics of the Screw and Cable actuator

SCS basic principles are presented in Fig. 7.

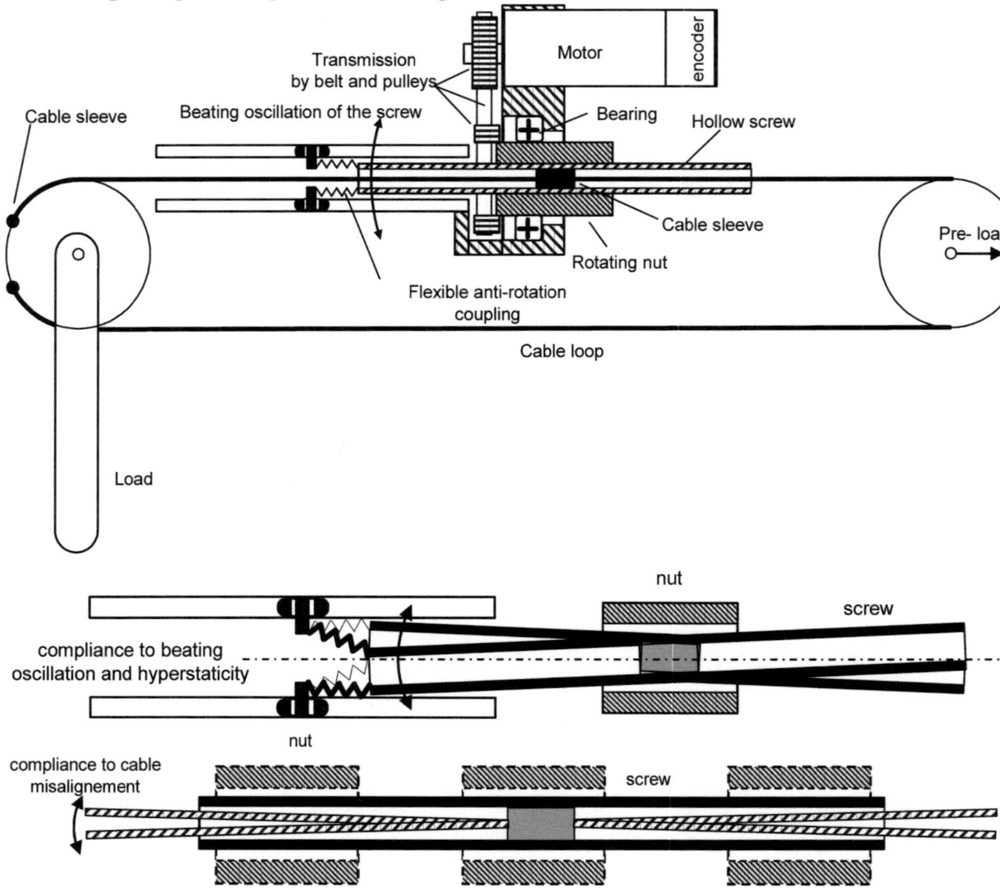

Fig. 7. SCS basic principles (patented)

A rotative joint is driven by a standard push-pull cable. On one side, the cable is driven by a ball-screw which translates directly in its nut without any linear guiding (the screw is locked in rotation thanks to rollers moving into slots). The nut rotates in a fixed bearing and is driven by the motor thanks to a belt transmission. Alternatively, pan-cake direct drive motors can be used (Fig. 13).

First of all, the ball-screw is free to oscillate thanks to a flexible coupling. These oscillations are known as beating oscillations and are amplified by the deliberate absence of centering device such as a linear bearing. Complementary the screw is bored and the cable passes inside with a radial play and is attached in its center. The scheme shows the various positions of the cable attachment relative to the nut when the screw is translated. This minimalist and compliant mounting almost completely isolates the screw from bending moments and thus guarantees a low and regular friction. The result is a highly linear force amplifier and transmitter which is also unusually compliant to manufacturing imperfections and structural deformations.

Regarding the performance in torque amplification linearity, the Fig. 8 presents a torque-force transmission diagram for a typical SCS using a THK BNK 1010 ball-screw (Diameter: 10mm ; Lead: 10 mm). Both DIRECT and INDIRECT (ρ) coefficient (maximum efficiency) are close to 0,94 and no-load friction represents approximately 1/1000 of the maximum load capacity of the screw. These values show that in terms of force transmission quality and symmetry, a SCS competes with the best existing transmissions, the capstan excepted.

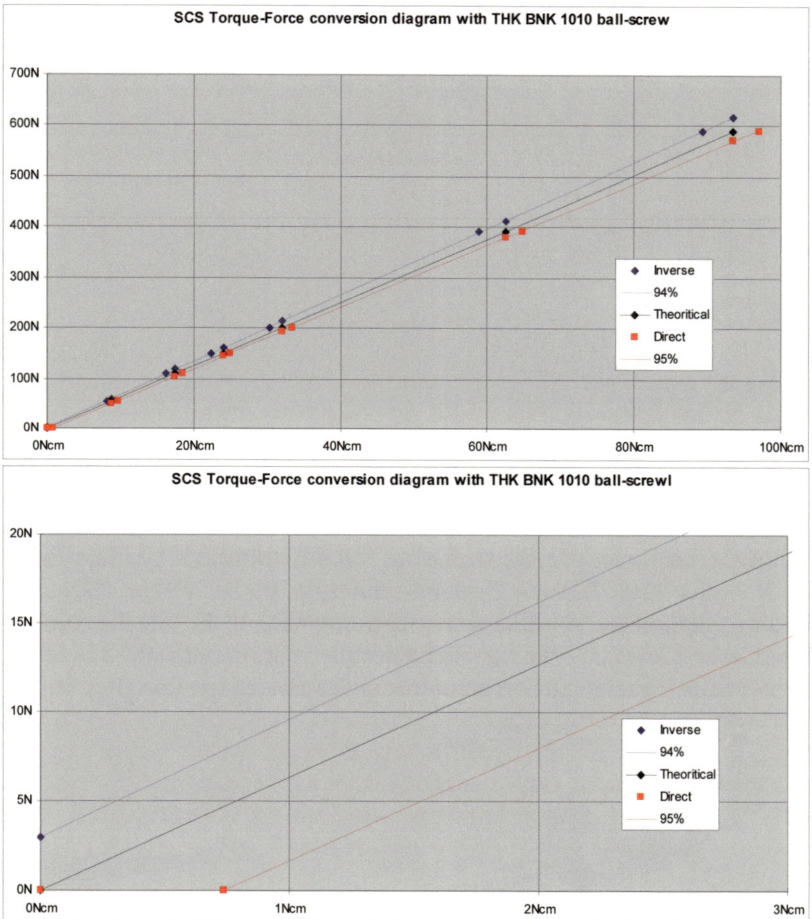

Fig. 8. A typical SCS force transmission diagram (in the transmissive quadrant): Top, default of linearity ; Bottom, a magnified view of the input/output friction thresholds

<u>Filtering effect of the centered attachment of the cable</u>

A simple modeling of the effect of cable tension on the efforts created between the screw and the nut demonstrate the efficiency of the centered attachment in comparison with a conventional attachment at the extremities of the screw. In the example Fig. 9, the travel of the screw is +/- 100 mm. Both the bending moment and transversal force created by the cable are reduced about 10 times (on this example) in comparison with a standard attachment at the extremities of the screw. This important result explains, in the existing realizations, the suppression of any detectable friction irregularities, even for strokes up to 200mm.

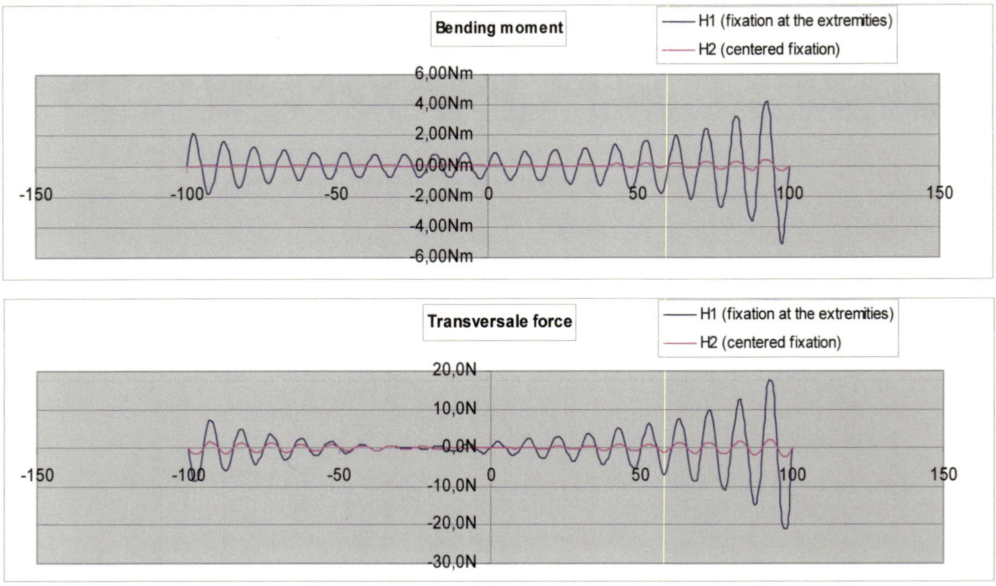

Fig. 9. Typical filtering effect of the centered attachment of the cable

The master arm Virtuose™ 6D 40-40

The first prototype has been presented during the 9th American Nuclear Society Topical Meeting on Robotics and Remote Systems congress in Seattle in 2001 and is today manufactured by Haption® under the name Virtuose™ 6D 40-40. It is the combination of an articulated arm issued from an existing mechanical telemanipulator, the MA 30 (La Calhène) and a motorization unit packing 6 SCS actuators at the base of the arm (Fig. 10).

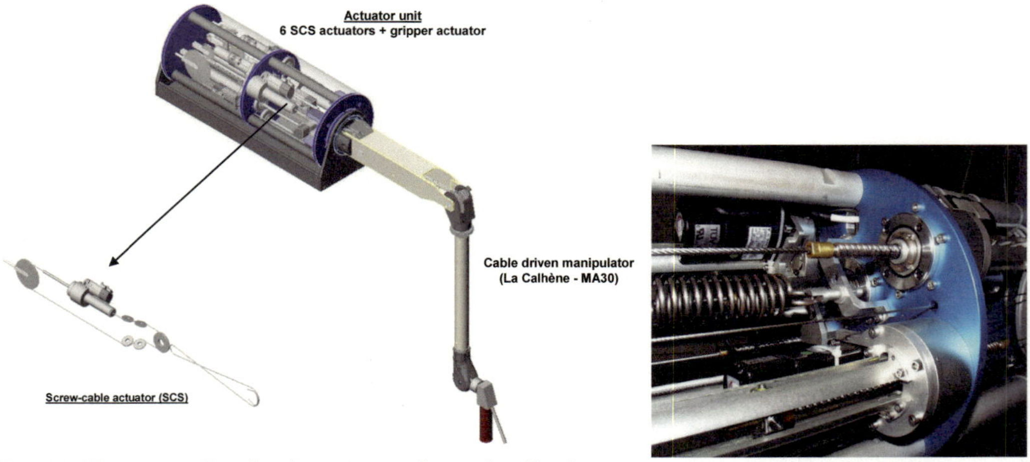

Fig. 10. Virtuose 6D: the first 6 axis force feedback master arm powered by 6 screws and cables

Gravity compensation is realized by computed torques provided by the motors, excepted on the first axis where a spring maintains the first limb around the horizontal at rest.
The following pictures shows the industrial version, Virtuose™ 6D 40-40, used in the thru-the-wall telescopic teleoperator MT 200 TAO developed for the needs of AREVA's reprocessing plant hot-cells (Garrec et al. 2007).

Fig. 11. The MT 200 TAO (CEA/AREVA) developed for AREVA La Hague hot-cells.

The following table summarizes the main specifications of the prototype equipped with brushless DC motors.

DoF	6
Cartesian working volume	400 mm ´ 400 mm ´ 400 mm
Continous force	40 N
Continuous moment	2 Nm
Theorical resolution in translation (12 bit)	0,02 mm
Theorical resolution in rotation (12 bit)	0,1 milliradian
Dry friction (translation)	2 to 3 N
Total mass of the arm	35 kg

Table 2. Virtuose 6D prototype main specifications

2.4 The STeP teleoperation system

Shortly after completing Virtuose 6D, a custom designed slave arm based on the same philosophy was design for the need of the teleoperation system STeP (STeP: Système de Téléopération en Puits) dedicated to retrieve radioactive material in a well (Fig. 12). The specifications of this system have been previously presented (Goubot & Garrec, 2003).

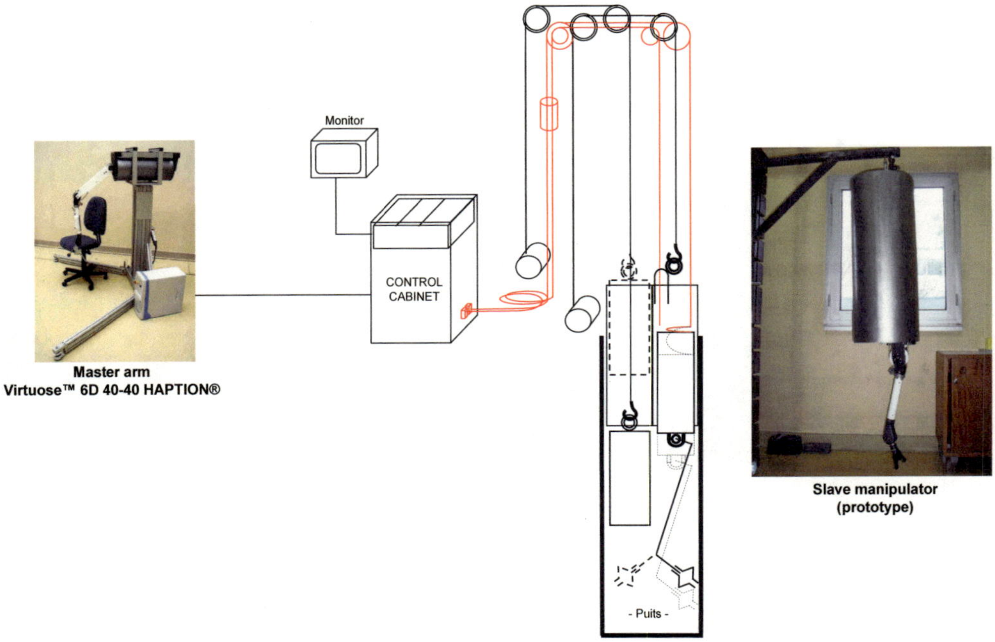

Fig. 12. A general view of the STeP teleoperation system for interventions in well

The slave arm is designed to work in a radioactive environment and it occupies only the half well's section. A tool box travels in the left space to bring tools and retrieve material. It is a simplified 5 axis arm equipped with a gripper, the first movement being a vertical translation. All joints are driven by cables and are provided with force feedback. We expanded upon the same SCS mechanics but this time we opted for direct-drive concentric pan-cake motors in order to pack its 6 actuators inside a half-cylinder housing which also integrates counterweights to compensate the actuator unit's weight on its vertical travel (Fig. 13).

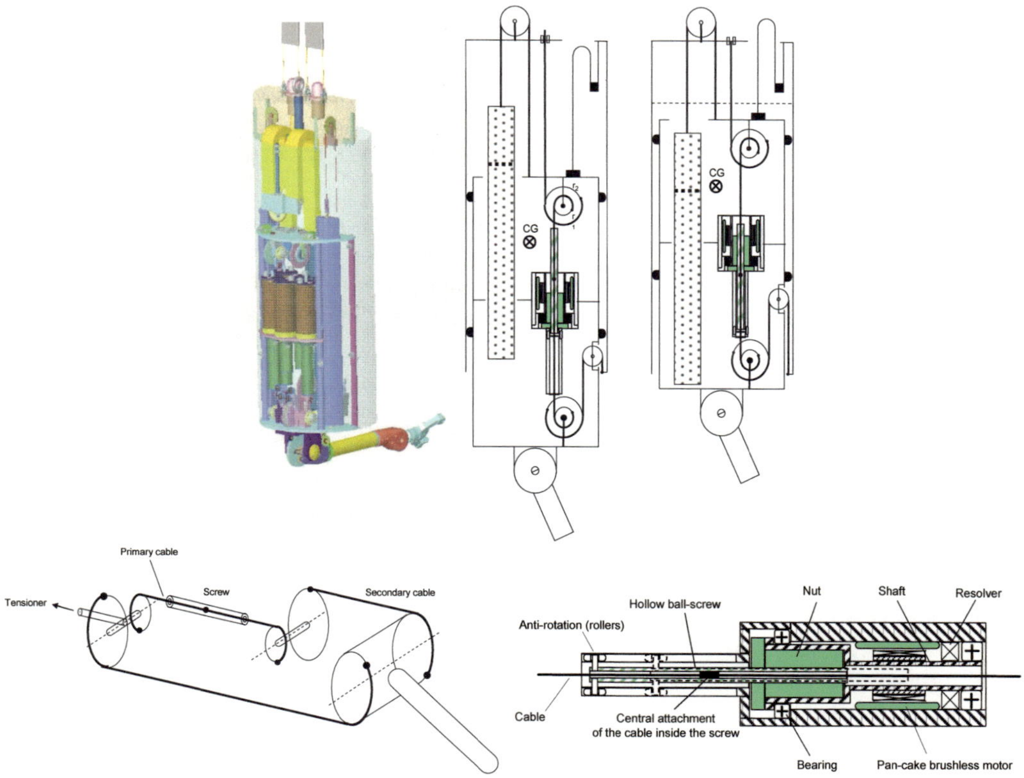

Fig. 13. Top, slave unit and a detail of its typical balanced translation motion ; Bottom, special type of loop used to increase linear travel and a SCS driven by concentric pan-cake motors

In comparison with master arms, slave arms have greater joint amplitude. This is why we had to slightly increase the complexity of cable loops to increase the linear displacement of the cable and thus stay within the allowed longitudinal dimension of the unit. Here again, the SCS actuators contribute to an extremely compact motor unit.

2.5 The new trade-off offered by SCS

All of the previous solutions - spur gears, block-and-tackle, capstan - have the common drawback of a transversal motor compared to the direction of cables. The SCS on the contrary is the only one where the motor is parallel to the cable and this enables a transversal joint to be driven without minimal losses, avoiding bevel gear. In addition to the well-known advantages of cable transmissions (shock absorption, smoothness, high efficiency, and design versatility for intricate routings through joints) the basic advantages of the SCS are:

- high force capacity (with ball-screws for instance)
- high linearity in force amplification allow force control without force sensor (reliability, absence of drift and calibration procedure, electromagnetic immunity, simplified wiring)

- motor aligned parallel to cable: compact arrangement to actuate transversal without beveled gearboxes
- low inertia (with appropriate lead)
- high linear stiffness
- highly tolerant to manufacturing incertitude and to structure flexibility (wide choice of structural material)
- cable endurance (large cable section and low speed)

Regarding drawbacks in comparison with other tendon driven mechanism, the SCS presents a potential asymmetry in terms of stiffness, as soon as one of the cable portion looses its tension.

3. The design of the upper limb exoskeleton ABLE 4 axis

In the first applications of the SCS the advantage of the alignment of the motor with the cable benefited to the compactness of the actuator based unit. We realized that it was possible to go further by integrating the SCS in the moving parts of the arm which would reduce the length of the cable and simplify its routing. Correlatively, in order to limit the detrimental effect of the increased moving mass (both in terms of gravity torque and inertia) we chose to reposition the dead mass of motor near the upstream articulation of the arm using lightweight shafts to transmit the torque. This is actually the application to the SCS of a known idea (Flatau & Vertut, 1972). Altogether this combination represented a new tradeoff (Fig. 14).

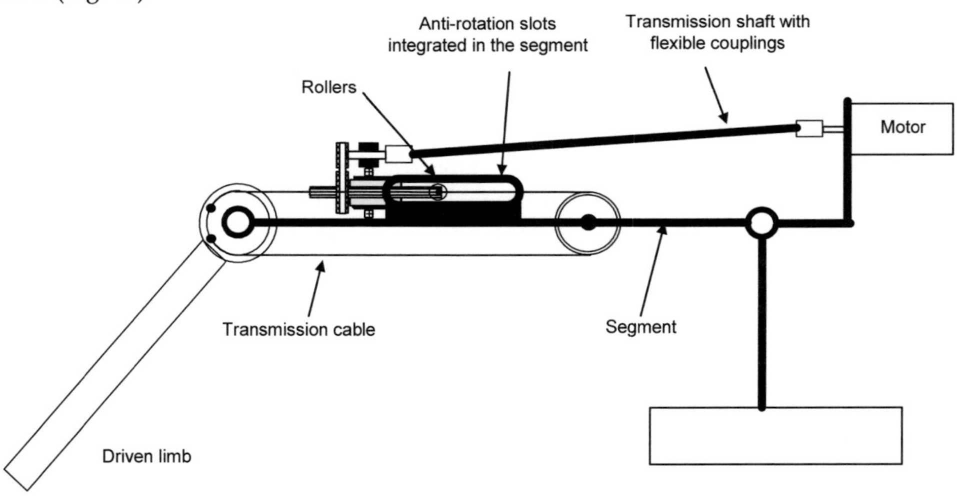

Fig. 14. Embedded SCS principle

The design of an upper limb exoskeleton appeared then as an appropriate application of this principle paired with an exciting design challenge.
The second option was to take advantage of the flexibility of the cable to pack two SCS's in the arm module, each of them actuating a transversal axis (shoulder and elbow joint). The overall result is a streamline arm module where the two SCS's perform alike artificial electrical muscles (Fig. 15).

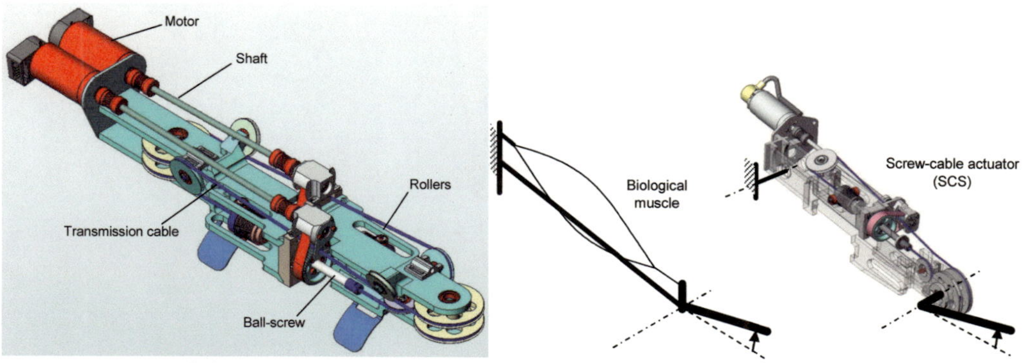

Fig. 15. Arm module twin actuators and its analogy with biological muscles

In a second phase we designed the shoulder joint and the back module. The scheme Fig. 16 shows the resulting kinematics of the 4 first joints. The shoulder articulation is a spherical articulation made of three orthogonal pivots whose common intersection approximately coincides with the center of the person's shoulder.

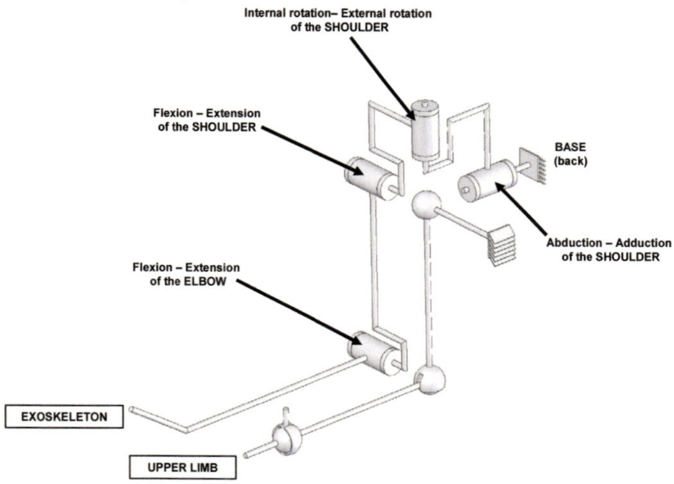

Fig. 16. ABLE - 4 axis kinematics

However, the major difference with previous designs is that the second joint is realized with a circular guide. Such an arrangement is both free of singularity and not invasive as shown on Fig. 17.

The back module incorporates two SCS which drive the first and second joints whereas the third joint is driven transversally by one of the two embedded SCS of the arm module (Fig. 18). The coupling effect between the two first joints is classically compensated by the control.

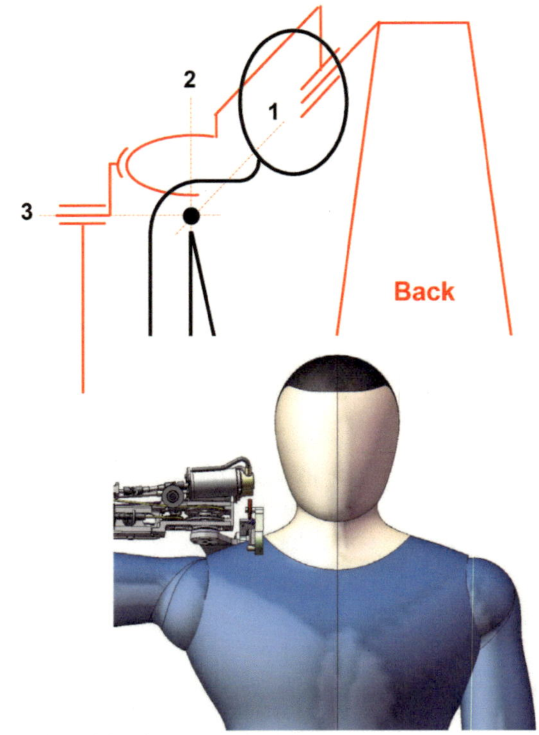

Fig. 17. ABLE - 4 axis shoulder kinematics

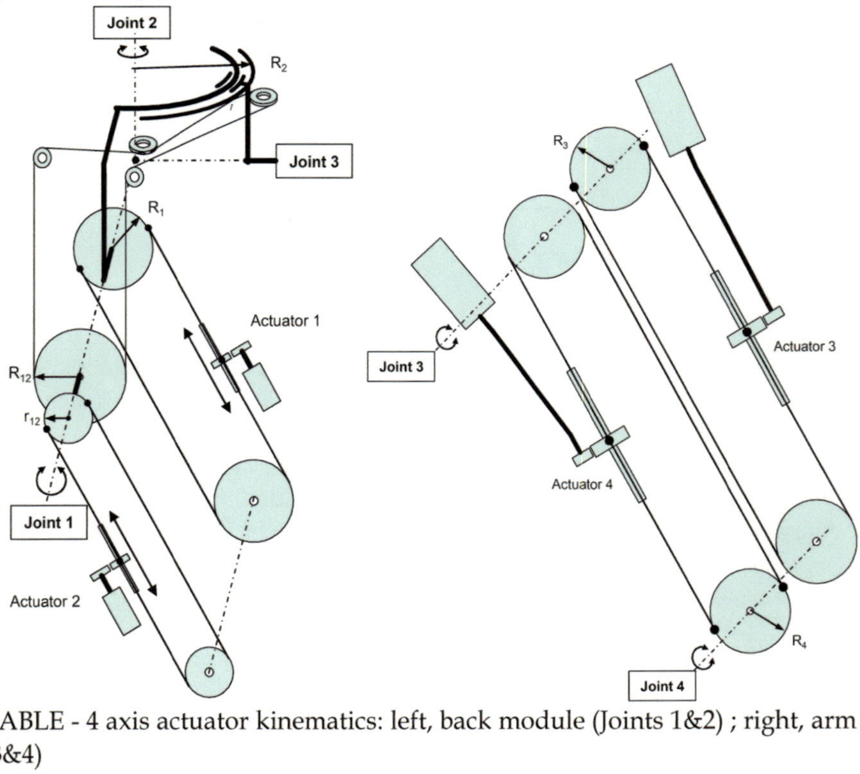

Fig. 18. ABLE - 4 axis actuator kinematics: left, back module (Joints 1&2) ; right, arm module (Joints 3&4)

The result is a simple, integrated and morphologically compatible design combined with a distributed actuator mass and volume along the structure (Fig. 19). The ABLE 4 axis general architecture and design was previously presented in (Garrec et al. 2004) ; (Garrec et al. 2006); Garrec et al. 2008)

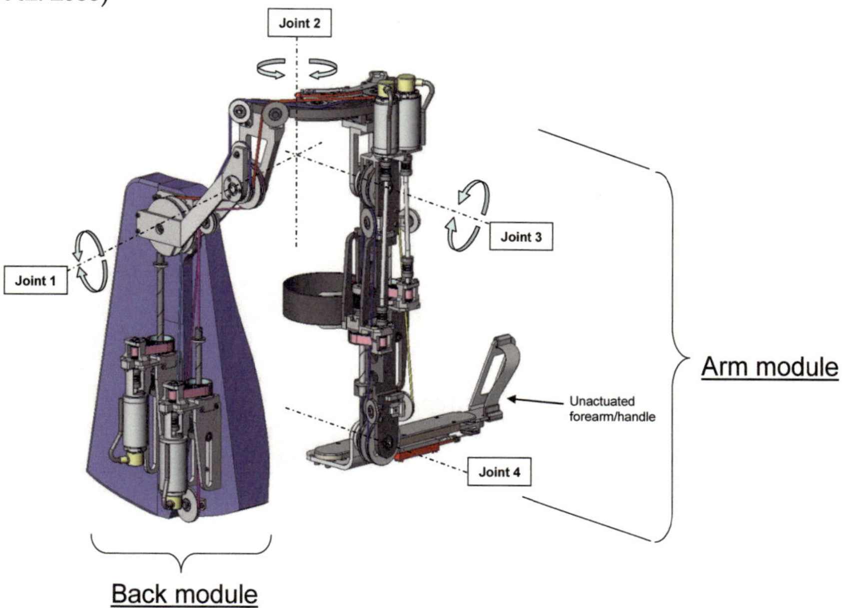

Fig. 19. ABLE - 4 axis architecture

It is important to note that since the two SCS only occupy the half of the back module, it is possible to integrate the motorization of a second exoskeleton without any change of its volume. The Table 3 summarizes the main specifications of ABLE - 4 axis.

MODULE	BACK		ARM	
JOINT	Axis 1	Axis 2	Axis 3	Axis 4
	Abduction / Adduction	Rotation Internal / External	Flexion / Extension	Flexion / Extension
ARTICULATION	SHOULDER			ELBOW
Amplitude	110 °			130 °
Motors	DC ironless Faulhaber type			
Transmission	Belt + Ball-screw and cable (SCS)			
Ratio	106	107	71	71
Max. velocity in hand (approx.)	1 m/s			
Joint torque (continuous)	18 Nm	18 Nm	13 Nm	13 Nm
Continuous effort in hand (approx.)	50 N	50 N	40 N	40 N
No-load friction in hand (approx.)	3 N		2 N	

Table 3. ABLE 4 axis main specifications

4. Toward ABLE 7 axis: the design of an innovative forearm-wrist module actuated by SCS

Looking for completing ABLE with a forearm and wrist that would preserve its general design options, we realized that existing structures described in previous literature

(Bergamasco et al., 1994) ; (Gupta & Malley, 2006) ; (Frisoli et al., 2007) ; (Perry et al., 2007) did not address simultaneously the following requirements:

- open structure (for safety reason and psychological acceptance)
- lightweight
- low friction
- low inertia
- streamline
- scalable

We decided then to focus on the forearm design only, looking for a structure that would allow a sufficient axial rotation (prono-supination) and address the listed requirements.

4.1 Forearm

Parallel articulated structure (prono-supination)

In its basic version, the forearm mechanism is presented in Fig. 20. It combines a parallel structure made of 3 rods, connecting a mobile arch to a fixed arch thanks to ball-joints, and a fixed cantilever mast attached to the fixed arch, which extremity supports a bearing made of 3 rollers which determine the mobile arch to rotate around its center. The mechanism is altogether acting as a forearm structure and a circular guiding for the axial rotation (prono-supination). Incidentally, this movement evokes those of radius-ulna bones. We can notice that more rods could be arranged in parallel without changing the proprieties of the structure, for example to increase the resistance or stiffness. In this case inner constraints would necessary appear (hyperstaticity) and should be limited by a suitable accuracy of the realization. Shear forces are balanced by the fixed cantilever mast whereas bending moments are balanced by traction/compression forces in the rods. This decomposition of the static torsor makes an optimal use of each element: miniature bearings can easily sustain the desired shear forces (about 50 N) whereas lightweight rods/ball-joints can easily transmit the necessary traction/compression forces (about 150 N). Compared to an existing industrial product such as THK HCR circular bearing or previous published designs, this mechanism presents a better compromise between volume, weight and friction and it can be easily scaled to different person sizes.

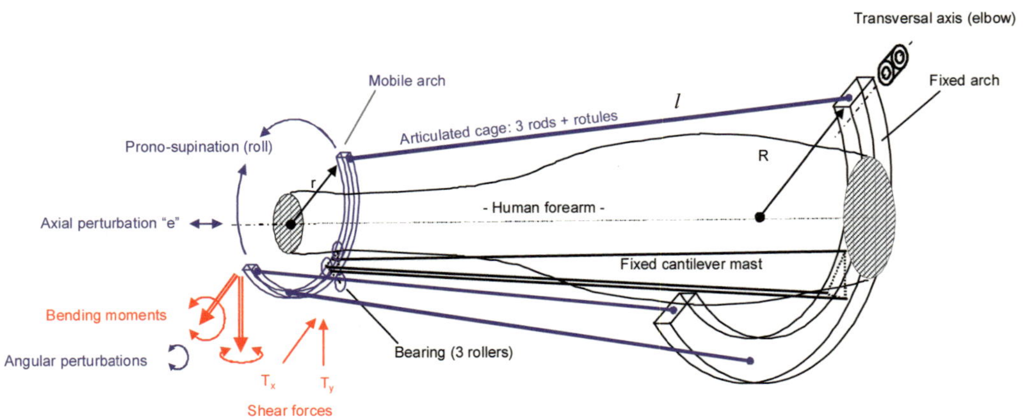

Fig. 20. Forearm parallel mechanism allowing an axial rotation concentrically to the human forearm (prono-supination)

However, the amplitude of the rotation is naturally limited by interferences between rods and fixed parts but also because an axial translation of the mobile arch (Fig. 21) occurs when it rotates (non linear coupled motion).

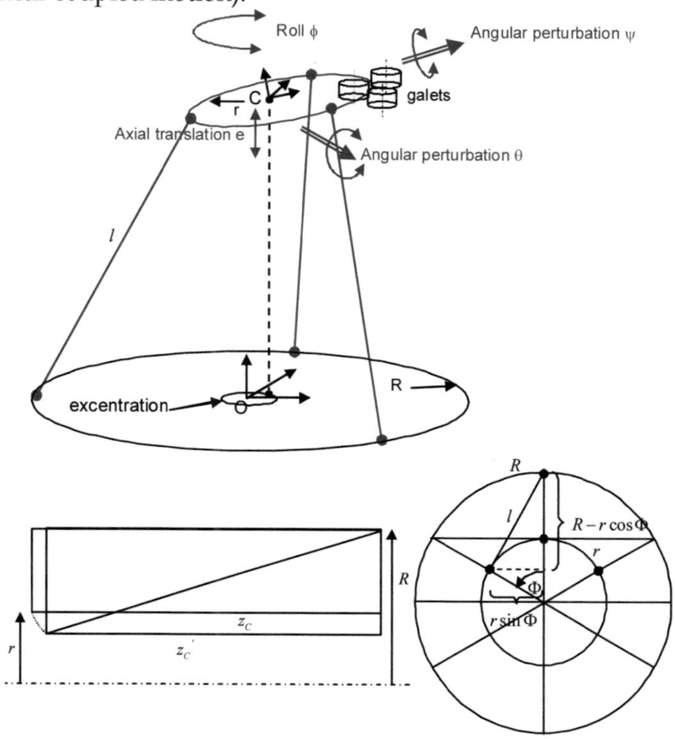

Fig. 21. Kinematic perturbations of the parallel mechanism

The expression of the perturbative translation is given by the equation (1).

$$e = z_c - z'_c = z_c\left(1-\sqrt{1+\frac{2Rr(\cos\Phi-1)}{z_c^{2}}}\right) \quad (1)$$

An order of magnitude of the perturbative translation is given for an average-size human forearm in Table 4.

Prono-supination amplitude	∠∠	120°
	∠	60°
Fixed arch radius	R	65mm
Mobile arch radius	r	45mm
Nominal distance between arches	z_c	250mm
Rod length	l	251mm
Minimum distance between arches	z'_c	244mm
Maximum translation	e	6mm

Table 4. Estimated translation for a human forearm

We notice that in comparison with the flexibility of the skin and muscle, this perturbation (≈ 2%) is relatively small and thus likely to not be felt by the user. A second cause of perturbation is the offset of the center C of the mobile arch (fabrication incertitude, play or elastic deformation under the load) which will induce an oscillation of its plane. In our current design (Fig. 23), its amplitude is estimated to be kept below 1° which should be again undetectable.

Axial rotation drive

The perturbative translation of the arch is obviously a major constraint for the design of the drive. Here again the SCS actuator brings an appreciable tolerance and the torque amplification offered by the screw and the belt is sufficient to avoid any complementary gear. The transmission cable has a particular routing through the fixed mast (Fig. 22).

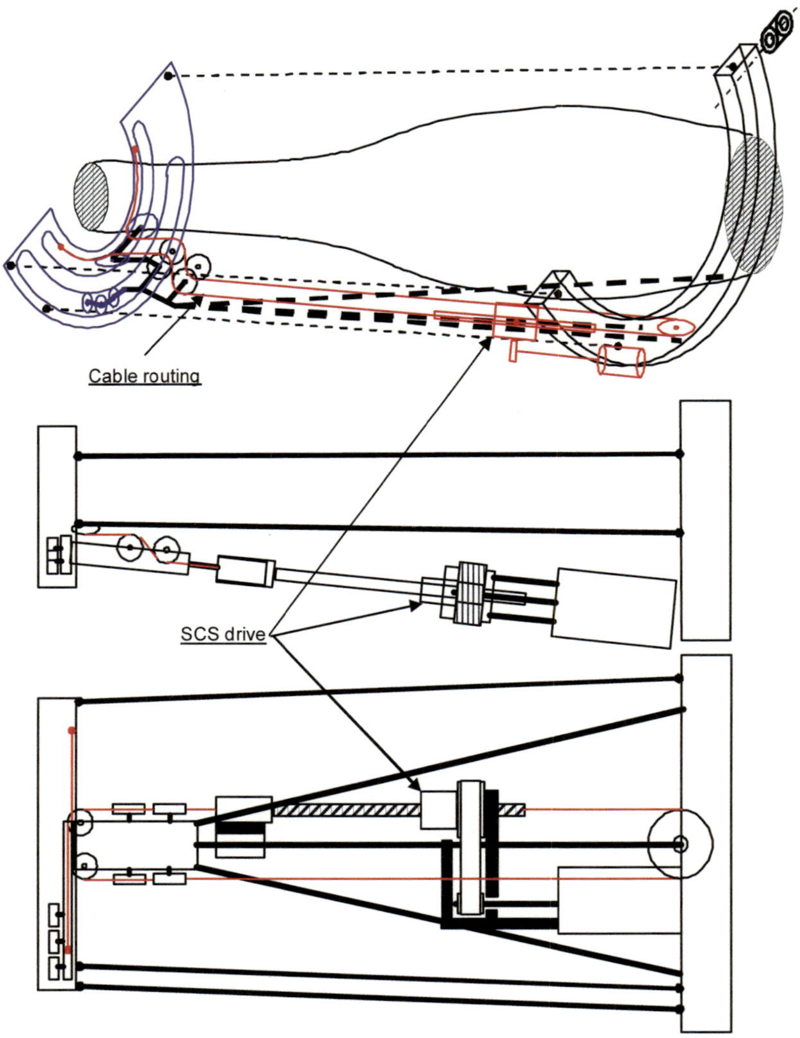

Fig. 22. Axial rotation drive by an SCS showing the cable transmission routing

The specific design of the SCS enables its integration within the mast overall volume. The combined rotation-translation movement of the arch creates a non conventional trajectory that requires a careful design of the grooves to limit cable length variations. Adjustments are also provided both in length and laterally. Fig. 23 shows a view of a current project.

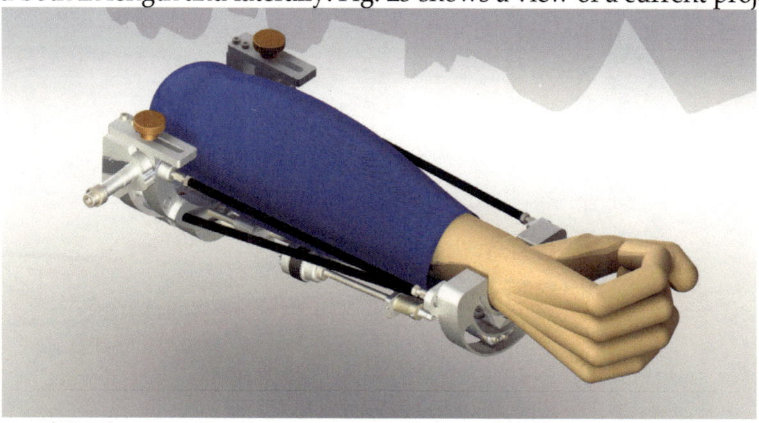

Fig. 23. A view of the forearm 1 axis structure and drive (SCS)

4.2 Wrist articulation and drive

The wrist is classically formed of two transversal perpendicular axis (equivalent of a U-joint) attached to the mobile arch. They are driven by a pair of SCS mounted on a structure that replaces one of the 3 rods of the parallel mechanism (Fig. 24).

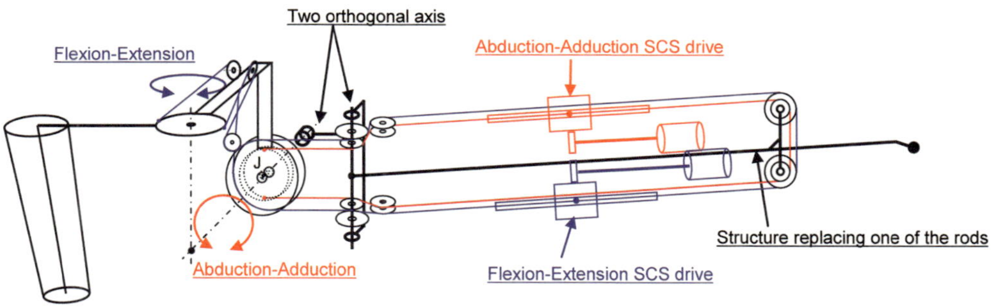

Fig. 24. Wrist 2 axis articulation and SCS drives with their respective cable routing

The examination of the routing of the cables indicates that there is a main linear coupling (red on blue) and also other non linear couplings of a low magnitude due to the deflexion of the cables through the two orthogonal axis (influence of the kinematics of the parallel linkage). The synergy of SCS actuators with the parallel linkage is potent because actuators can be harmoniously aligned with the rods and additionnally, the reflected inertia of the actuators is reduced because rod velocity decreases towards the fixed arch.

4.2 Forearm and wrist structure

Finally the structure is obtained by assembling the two precedent structures (Fig. 25).

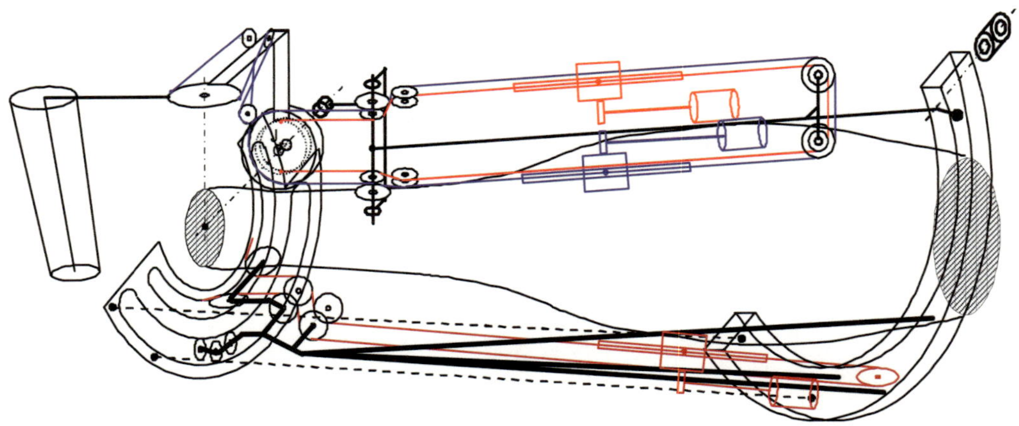

Fig. 25. Forearm and wrist 3 axis structure with its SCS drives

The heterogeneous linking devices on the mobile arch (two ball-joints on one hand and two orthogonal axis on the other hand) generate some other angular perturbations but fortunately of a low magnitude (below 1° for our design) and again the cable transmission is able to absorb this tiny default. Because the chosen interface with the person is a handle, the human forearm will translate of about 6 mm relatively to the fixed arch. The absence of enclosure and the flexibility of the human skin should make it insensitive however. Fig. 26 shows the current project of a forearm-wrist unit planned for the 7 axis ABLE version.

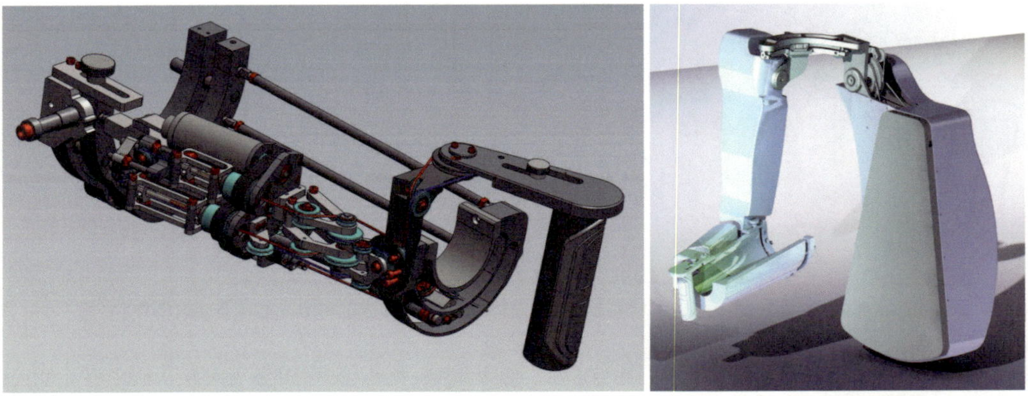

Fig. 26. Top, a view of the 3 axis forearm-wrist ; Bottom, a possible 7 axis ABLE equipped with the forearm-wrist

Finally, this new forearm-wrist mechanical structure is a lightweight open and truly anthropomorphic device, that provides a satisfactory amplitude for the critical prono-supination movement. The SCS integrates well in such a complex structure and guarantees a high linearity of force transmission paired with pretty high values of torque.

MODULE	FOREARM-WRIST		
JOINT/ARTICULATION	Axis 5	Axis 6	Axis 7
	Prono-supination	Abduction-Adduction	Flexion / Extension
ARTICULATION	FOREARM		
Amplitude	110°		
Motors	DC ironless Faulhaber type		
Transmission	Belt + Ball-screw and cable (SCS)		
Ratio	117	49	47
Max. velocity in hand (notional)	1 m/s		
Joint torque (continuous)	2 Nm	2 Nm	2 Nm
Continuous effort in hand (approx.)	19 N	22 N	20 N
No-load friction in hand (approx.)	0 N		0 N

Table 5. Forearm-wrist main specifications

5. Further development of ABLE and expansion of SCS applications to anthropomorphic robotics

Furthermore, we suggest that derived versions of the above forearm mechanism could be used for low limb exoskeletons as well (Fig. 27).

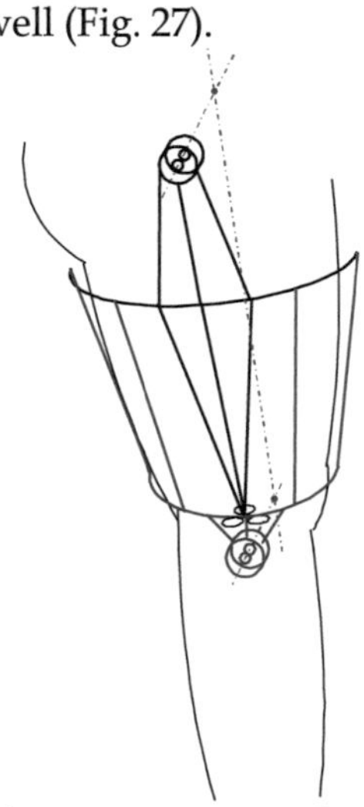

Fig. 27. Example of adaptation of a forearm-wrist mechanism to a low limb exoskeleton (thy-knee articulation)

6. Conclusion

Less than a decade ago the SCS actuator technology was introduced in several prototypes and 2 industrial master arms have been since in used in various teleoperation systems at AREVA and CEA. Similar actuators have been successfully integrated in articulated architecture of the upper limb exoskeleton ABLE leading to an open and anthropomorphic machine. Evaluations are undergoing for its use as an orthosis for rehabilitation (Jarrassé et al. 2008). A new complete 7 axis version is also planned in 2010, to be used in a

teleoperation/haptics context. These research projects will contribute to specify a multi-role exoskeleton with a higher potential for industrialization.
SCS actuators are also well adapted to anthropomorphic robots such as lightweight, assistive slave arms embeddable on a mobile platform, biped robots and low limb exoskeletons. Other industrial applications may envisioned, for example in the field of assistance devices to reduce musculoskeletal disorders.
However some more developmental work must be done to secure industrial applications. For example the control of cable tension should be ideally continuous along the life of the machine to allow a predictive maintenance. A software tool would be useful to facilitate the design of transmission cable in complex routings and the integration of all mechanical parameters of the cable to automatically compute the performances of the machine.

7. References

Bergamasco M., Allotta B., Bosio L., Ferretti L., Parrini G., Prisco G. M., Salsedo F., and Sartini G.," An arm exoskeleton system for teleoperation and virtual environments applications", in Proc. IEEE Int. Conf. Robot. Autom., vol. 2, 1994, pp. 1449–1454

Flatau C. R., "Development of Servo Manipulators for High Energy Accelerator Requirements' Proc. 13th Conf.Remote Sist. Technolol., 29, 1965

Flatau C. R., Vertut J."MA22-A compact bilateral servo master-slave manipulator", Proc. of 20th Conf. on Remote Systems Tech., 1972

Frisoli L., Borelli A., Montagner, et al,"Arm rehabilitation with a robotic exoskeleleton in Virtual Reality", Proc. of IEEE ICORR 2007, Intern. Conf. on Rehabilitation Robotics

Galbiati et al., "A Compact Flexible Servo System for Master Slave Electric Manipulator", 1964

Garrec P., "Systèmes mécaniques", in: Coiffet. P et Kheddar A., Téléopération et télérobotique, Ch 2., Hermes, Paris, France, 2002

Garrec P., French Patent: « Transmission à vis, écrou et câble attaché à la vis » for screw and nut transmission and cable - FR0101630, 2000 (EUR 01938347.0-2421 and US 10/296,740

Garrec P., Friconneau J.P., Louveau F.," Virtuose 6D: A new force-control master arm using innovative ball-screw actuators", in Proceedings of ISIR 35th International Symposium in Robotics, Paris, March 2004

Garrec P., Martins J.P., Friconneau J.P., "A new Technology for Portable Exoskeletons", AMSE 2004 - Vol. 65 n° 7/8, pp 13-22.

Garrec P., Martins J.P., Gravez, Perrot Y., Measson Y., "A New Force-Feedback, Morphologically Inspired Portable Exoskeleton" Proceedings of 15th IEEE International Symposium on Robot and Human Interactive Communication (RO-MAN - 2006), September 2006, Hatfield, UK.

Goertz R. C. et al., "Master-Slave Servo-Manipulator" Model 2", Proc.4th Ann. Conf. Hot Lab. And Equip. 1, 1955

Goubot J.M., Garrec P., "STeP: an innovative teleoperation system for decommissioning operations", Workshop "Decommissioning challenges: An industrial reality ?", French Nuclear Energy Society, 23-28 November 2003, Avignon, France

Gupta A., O'Malley M.K.," Design of a haptic arm exoskeleton for training and rehabilitation", Mechatronics, IEEE/ASME Transactions on Volume 11, Issue 3, June 2006 Page(s):280 - 289

Köhler G.-W.,« Manipulator Type Book », Verlag Karl Thiemig, München, 1981

Perry J.C., Rosen J., Burns S.,"Upper-Limb Powered Exoskeleton Design", IEEE/ASME Transactions on Mechatronics, Volume 12, Issue 4, Aug. 2007 Page(s):408 - 417

Vertut J., and al."Bilateral Servo Manipulator MA23 in Direct Mode and Via Optimized Computer Control" 2nd Remotely Manned Syst. Technol.Conf., 1975

Vertut J., Coiffet P., Téléopération « Evolution des techniques », vol. 3A, Hermes, Paris France, 1984

A novel verticalized reeducation device for spinal cord injuries: the WalkTrainer, from design to clinical trials

Stauffer Yves[1], Mohamed Bouri[1], Reymond Clavel[1],
Yves Allemand[2] and Roland Brodard[2]
[1]*EPFL, LSRO Station 9, 1015 Lausanne, Switzerland* [2]*FSC, Rue du Commerce 2,1870 Monthey 2, Switzerland*

1. Introduction

In industrialized countries, 2 persons out of 100'000 undergo a spinal cord injury (SCI) every year. Which brings the total population in these countries to roughly 70 persons for 100'000 inhabitants. A lesion to the spinal cord results in a partial or complete loss of motor and/or sensory control below injury. Fortunately in over 50% of the cases some functions are spared (Jackson, 2008). The loss of motor control is only one of the consequence of such a lesion. Indeed cardiovascular weakness, muscular atrophy, osteoporosis appear rapidly after the injury and may severely impede the reeducation of the patient (Freeman-Somers, 2001).

Treadmill training assisted by therapists, standing or swimming are part of the currently used reeducation techniques (Figure 2b). For walking relearning the manually provided assistance can require up to three therapists (Grundy, 2002). Their work is of course strenuous, especially in the case of a spastic patient. In addition no quantification of their effort is possible and the applied mobilization may differ from one session to another.

In that context robotic assistive devices have been developed and are now being introduced in clinical environments (Colombo et al., 2000) (Schmidt et al., 2007) (Metrailler et al., 2006) (Figure 2a & Figure 2c). Such robots can advantageously provide long training, with quantifiable mobilization (force and position). But in the event of a complete lesion the patient's muscles remain completely passive. To overcome that issue electrical muscle stimulation can be employed (Dohring & Daly, 2008). The complex nature of the muscles associated to the dynamics of walking, require subtle stimulation schemes which rely on position and force information provided by the robot.

The combination of muscle stimulation (preferably in a closed loop manner to avoid rapid exhaustion and guarantee better control) and robotic mobilization optimizes the afferent feedback provided to the subject and allows an effective muscle training. The WalkTrainer, that is one of these reeducation devices, will be extensively presented in this chapter (Stauffer & al., 2009).

2. Generalities

2.1. Medullar lesions

The spinal cord serves as a bidirectional communication link between the brain and the rest of the body (Germann, 2001). In the event of a lesion, orders will be prevented from being sent to the muscles (efferent) and sensory information will of course also not be able to reach the brain (afferent). Depending on the severity of the lesion some information can still be transmitted, in that case the lesion is referred to as incomplete. In addition the level (height) of the lesion also matters. Lesion at the cervical level result into tetraplegia (the four limbs are affected), whereas lower lesions lead to paraplegia (only the legs are affected) (Freeman-Somers, 2001).

The absence of voluntary muscle control is only the most well known consequence of SCI (Freeman-Somers, 2001) (Aito, 2003). The more recurrent issues are summarized below (Figure 1):

- Cardiovascular: autonomous pathway can lose the supervision from the brain, alteration of heart control. Absence of muscular pumping: bad venous return, can cause blood clots and other complications (pressure sores on the skin for instance).
- Respiratory: weakened inspiration, results in less effective oxygenation of the blood. Weakened expiration can cause the building up of secretion in the lungs, which results in infections.
- Osteopenial (osteoporosis): bone density decreases fast after SCI, which leads to fracture prone conditions.
- Muscle fatigue: the muscle masse diminishes and muscle fibers become less fatigue resistant.
- Bladder and bowel issues: sphincter muscle control can be lost. As a consequence retention can occur, which can lead to incontinence.

2.2. Mobilization and electrical muscle stimulation

Current reeducation of paraplegic subjects mostly relies on mobilization of the lower limbs. The motion can be applied either by physiotherapists or by a robotic device (Colombo et al., 2000) (Schmidt et al., 2007) (Figs. 1a-1c). With the development of smaller and more effective sensors and the increase of computational power, the second option, namely robotic reeducation devices, is more and more present. Not only can the device move the subjects' limbs in a predefined way for a longer time than therapists. But these robots can also adapt to the effort provided by the patient, as a therapist would (Jezernik, 2003). Such treatment is beneficial for:

- maintenance of the articulation mobility,
- maintenance of muscle compliance,
- bone loading,
- stimulation of intestinal regulation by the vertical posture.

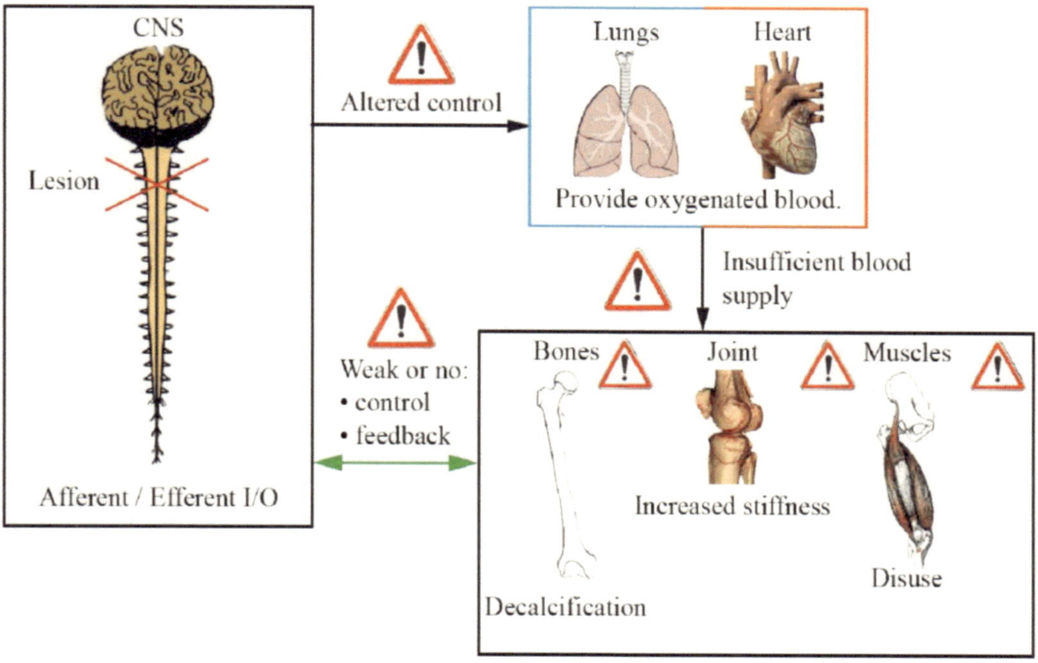

Fig. 1. Consequences of SCI greatly affect the walking capabilities.

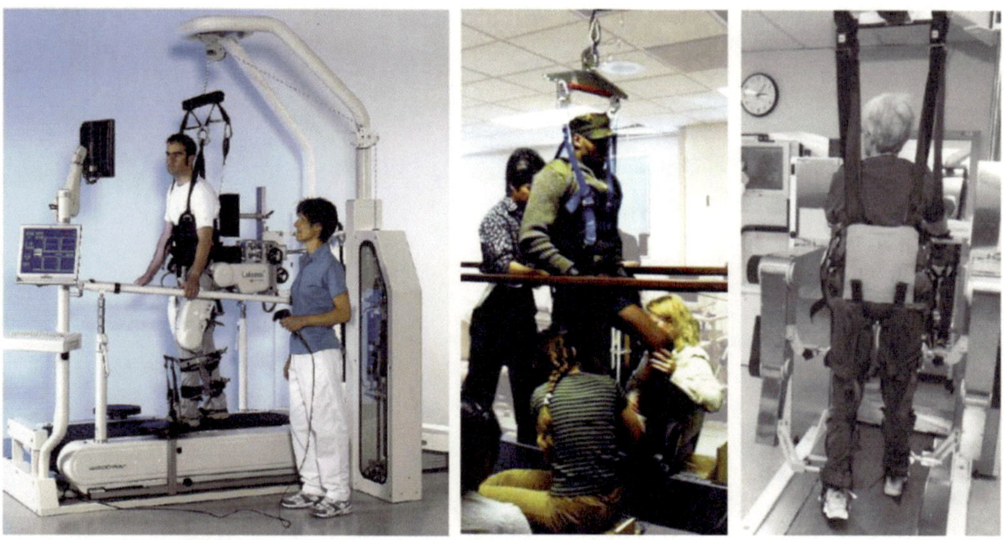

Fig. 2. Walking training: (a) Lokomat from Hocoma. (b) Manual training with three therapists. (c) AutoAmbulator.

However, if the patient is unable to initiate the motion by himself or if his voluntary control is too weak, the training is completely passive. In the sense that the robot provides the whole effort. In addition the muscle spindle stretch reflex can actually cause the muscles to contract, but in opposition to the desired motion. For instance if the robot is applying a knee flexion motion. This can trigger the muscles spindles of the quadriceps muscles (involved in knee extension). Their information is sent to the spinal cord where two actions are performed. First, the quadriceps muscles are ordered to contract. Second, the hamstrings (involved in knee flexion) are ordered to relax. As a consequence not only are the wrong muscles contracting, but the muscle that should contract is relaxing (Figure 3a).

In order to overcome these issues functional electrical stimulation (FES) can be applied. By using surface, intramuscular or implanted electrodes, electrical pulses are generated to elicit an action potential on the motor neuron of the concerned muscle. In other words, the electrodes are replacing the lacking commands in order to generate a muscular contraction. In addition to stimulating the motor neuron, the Ia fibers from the muscle spindles (i.e. fibers that send the position/velocity information back to the spinal cord) are also stimulated. This has two important implications. First it causes the motor neuron of the stimulated muscle to become more receptive to voluntary control. Second, it inhibits the antagonistic muscle (Frischknecht & Chantraine, 1989) (Figure 3b).

Unfortunately controlling muscles precisely with FES is not trivial. Also muscles of paraplegic subjects are usually weak (atrophied) and FES tends to induce fast fatigue (high stimulation frequencies have to be applied to obtain a fused contraction and fatigable muscle fibers are recruited first).

Up to day using purely FES based strategies for autonomy enhancement of paraplegic subjects have failed, or allow only very limited walking distances (Graupe, 2003). But they provide advantages in terms of muscular contractions and enhanced proprioceptive feedback. On the other hand robotized reeducation devices can now be trained to assist the patient only as much as needed. The combination of FES and robotics is thus a logical choice for generating new reeducation devices.

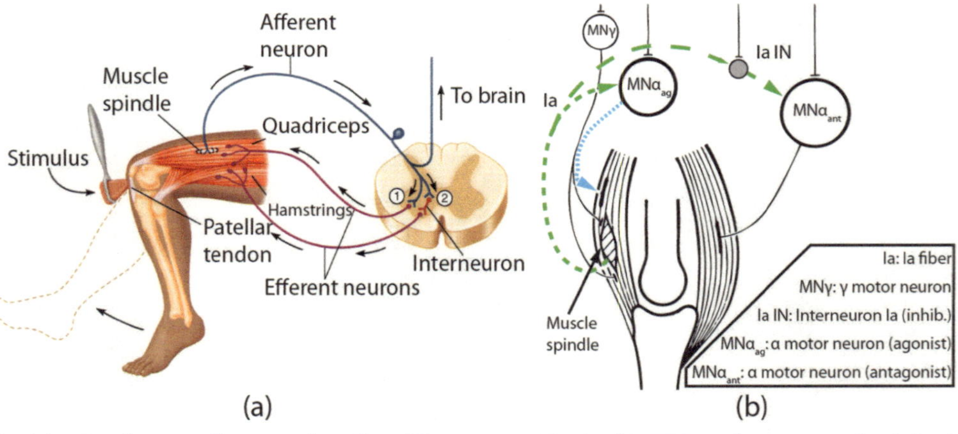

Fig. 3. (a) Muscle spindle stretch reflex (Germann, 2001). (b) FES and reciprocal inhibition. Small dashed arrow (blue): effect of FES on agonist muscle contraction. Average arrow (green): positive feedback on agonist motor neuron. Large dashed arrow (green): antagonist inhibition. (Frischknecht & Chantraine, 1989).

2.3. Plasticity and central pattern generators

Plasticity refers to the ability of a system to adapt to a changing environment. In the case of the nervous system, such plasticity is of course very active during childhood, but remains active during the whole life of a subject. Plasticity can occur at different levels, in the brain, the spinal cord or even at the neuromuscular junctions. If a motor neuron is injured another motor neuron can reinervate the muscle fibers. In that case one would speak of axon sprouting. In other occasions the strength of a synaptic connection is adapted. Such processes can be dependent on feedback from the effector organs in order to calibrate the strength of the input output connection. This is of great importance for incomplete SCI. In that case plasticity can cause the connections to adapt in a way that enables the person to have a better control of her limbs.

It has been shown that involving the patient actively in the therapy (i.e. asking him to do the motion) leads to greater outcomes (Kaelin-Lane, 2005). The same has been observed if training modalities are combined. For instance, treadmill, with body weight support and FES (Barbeau, 1998). Which means that plasticity can be stimulated by summing various stimulation modalities. As a summary plasticity is "improved" if the patient is participating to the training *(supervision)* and if the training stimulates many senses coherently *(summation).* These two considerations have of course to be taken into account when developing reeducation devices.

Central pattern generators (CPG) are present in the lower regions of the voluntary control hierarchy (brainstem and spinal cord) of animals. These structures are able to generate cyclic signals autonomously. Their activity is modulated by the brain and by feedback from the limbs. Spinalized cats can be placed on a treadmill and walking motions can be generated. If the treadmill speed is increased the gait shifts to trot and eventually gallop. For humans the possible existence of a CPG was first observed by (Calancie et al., 1994). Later walking activity during treadmill walking was also reported (Dietz, 1995)(Dobkin, 1995). In addition rhythmic motions seem to have positive effects on neural plasticity (Pearson, 2000). Such motions are thus employed on rehabilitation devices such as the Lokomat (Jezernik et al., 2003) or the WalkTrainer (Stauffer et al, 2009).

2.4. Conclusion: implication on the development of the WalkTrainer

The foundations for the development of the WalkTrainer reeducation device have been given in this section. As a summary these are:

- Active muscular participation of the subject: coherent muscular activity, muscle build up, high afferent feedback.
- Robotic mobilization: correct motion and longer training sessions.
- Verticalized posture: coherent feet loading, control of balance.
- Overground deambulation: correct visual feedback, natural walking.

3. Robotic design and realization

The development of the WalkTrainer was done with the idea of optimizing the feedback to the subject. The following sections will briefly highlight the key features of the five main components of the WalkTrainer.

3.1 Mobile frame or deambulator

"Should we go for a walk? Yes, sure." The mobile frame was developed with that

perspective (Figure 4a). Most rehabilitation devices are treadmill based; up to day the WalkTrainer is the only rehabilitation device that is rolling over ground with leg and pelvis orthoses. The deambulator is equipped with four wheels. Two castor wheels at the front for stabilization and two motorized wheels mounted in differential at the back. The WalkTrainer can thus roll along straight or curved paths at velocities up to 1.6 m/s. The training can take place indoor, in large corridors or outside on training tracks. This kind of real walking is not only more motivating for the subject. But visual information (i.e. moving scenery) is naturally provided to the user.
In addition the mobile frame serves as support to the other orthotic devices. As well as for the on board PC (control and wireless communication) and the 2 kWh LiPo battery (2 hours autonomy)

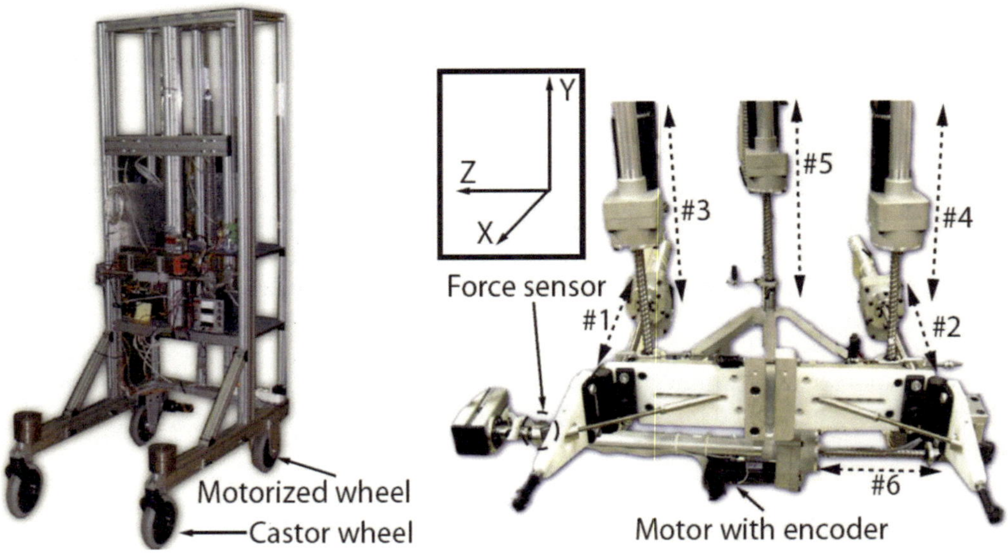

(a) (b)

Fig. 4. (a) Mobile frame. (b) Pelvic orthosis. The motorized axes are numbered.

3.2. Pelvic orthosis (PO)

Trying to walk without getting the pelvis to move is almost impossible. During natural walking motions along the six degrees of freedom (DOF) occur. The combination of these motions allows the overall reduction of motion of the center of gravity of the body, which leads to a more energy efficient walking. In addition stability is also taken care of by pelvic motions. For instance when only one foot is in contact with the ground (stance phase) the center of gravity is shifted laterally (Z axis (Figure 4b)) towards that leg.

Naturally during reeducation therapists try to mobilize the pelvis in order to facilitate the walking process (Trueblood, 1989). Of course the WalkTrainer is expected to be able to do the same. For that reason a six DOF pelvis orthosis was developed (Figure 4b). Each motorized axis is composed of a backdriveable ball screw transmission with a redundant position measurement (absolute and incremental). In addition force sensors allow to precisely quantify what force the orthosis is applying to the subject, that information can be used for diagnostic or security issues (Stauffer et al., 2007).

3.3. Leg orthosis (LO)

As for the pelvis, the legs are mobilized and interactions monitored by using force sensors in series with every actuator (Allemand et al., 2009) (Figure 5a-b). Motorized mobilization is provided for the three main articulations (hip, knee and ankle) in the sagittal plane. Motorization is provided by backdriveable ball screw transmissions. As for the pelvic orthosis redundant position measurement is performed. The leg orthosis is also equipped with force sensors. Those are used mainly for the control of the closed loop electrical muscle stimulation. Using the force sensors for monitoring or diagnostic purposes is also feasible.

The kinematics of the leg orthosis allow a high lateral stiffness at the ankle (i.e. prevents the scissoring of the legs) and allows at the same time the hip to move in all directions. As a consequence the upper part of the leg orthosis is "following" the motions of the pelvic orthosis. In addition the actuation and linkages are located at the rear of the user. Which not only allows the arms of the subject to swing freely, but also reduces the apparent size of the device.

The motorized orthosis is connected to a lightweight exoskeleton that is to be put on the user first. Doing so allows to have a greater surface area to apply the force. Which of course reduces the applied pressure and prevents the apparition of skin problems. The hip actuation is provided by a linkage at the thigh. The knee motorization is connected to the shank and custom built shoes are used to transmit the actuation to the ankle articulation .In addition there is a link between the leg and pelvic orthosis (Figure 5c). That serves to distribute the unloading force (vertical) between the pelvic and leg orthosis.

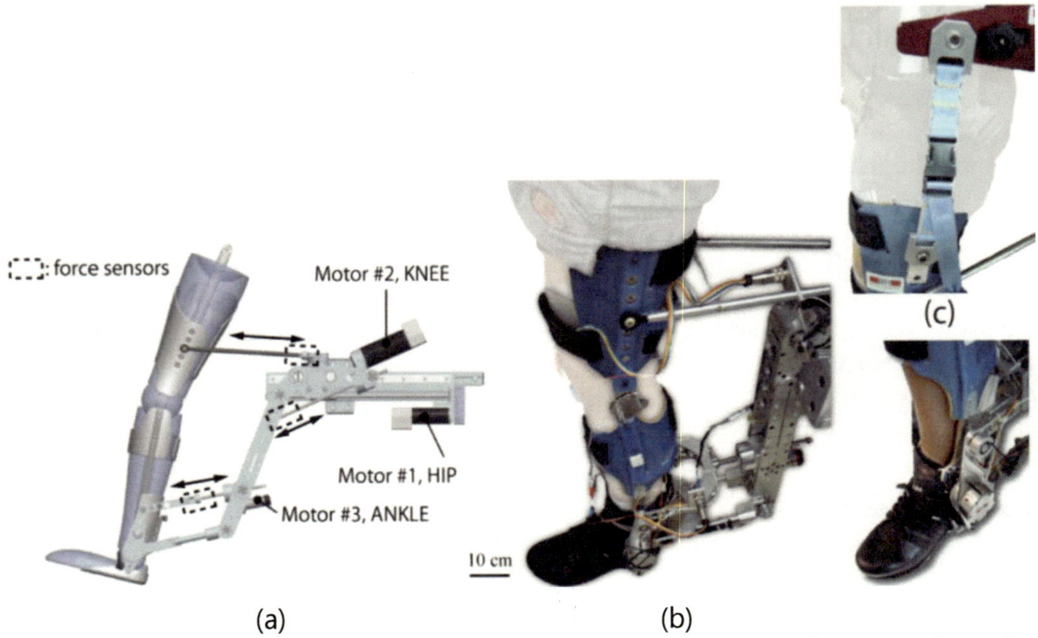

Fig. 5. (a) Leg orthosis schematic. Hybrid kinematics (combination of series and parallel kinematics). (b) Leg orthosis realization. (c) Link between leg and pelvic orthosis.

3.4 Active body weight support (BWS)

The weak muscular condition of paraplegic subjects restrains them from bearing all their weight on the legs. For that reason a certain amount of unloading is required. Furthermore in the case of high lesions, control of abdominal or trunk muscle can also be lost. The stability of the patient must then be guaranteed by the assistive device. These two conditions could be fulfilled with a simple unloading mechanism (a spring for instance). But it has been shown that the correctness of the weight bearing pattern felt under the foot is also important during training. For these reasons a more subtle motorization of the system is preferred.

In order to be energy efficient a spring in series with a motor allows to unload the subject by a fixed percentage of his body weight. A second motor (also in series) is then employed for the regulation of the unloading as a function of the gait cycle (i.e. during training). Again both motors are equipped with incremental and absolute position sensors. A force sensor mounted on the system also allows to control precisely the unloading.

The interface with the user consists of a specially developed harness (Figure 6b) that is fixed at two locations: at the pelvis (behind the user, Figure 6c) and above the subject's head. The second fixation point is a passive 3 DOF mechanism (translation, rotation, rotation), prevents the user from feeling like a pendulum (Figure 6a).

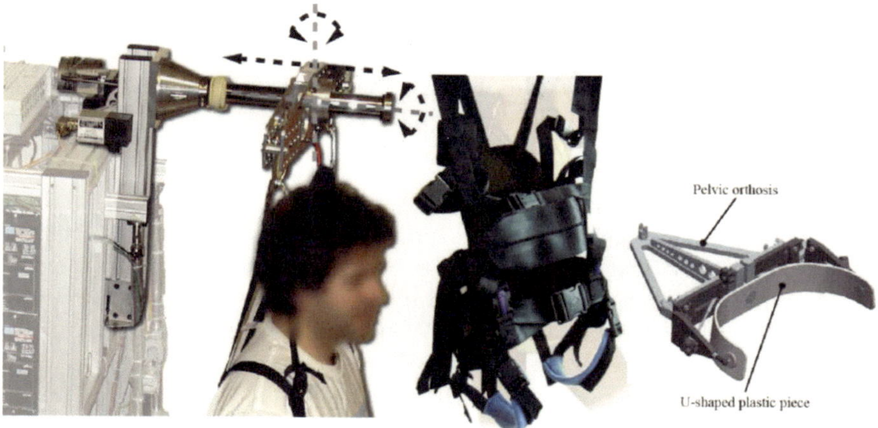

(a) (b) (c)

Fig. 6. (a) Active body weight support. The DOF of the upper fixation point are highlighted (b) Developed harness. (c) Fixation between the pelvic orthosis and the harness. Only the plastic piece that is fixed to the harness is shown.

3.5 Electrical muscle stimulator (StimMaker)

In order to replace the lacking voluntary muscle control, electrical muscle stimulation is provided in real-time. We developed a 20 channel stimulator that is to be used with standard surface electrodes (Figure 7a). On the WalkTrainer the following muscles are stimulated: gluteus maximus (GM), rectus femoris (RF), vastus medialis and lateralis (VM, VL), hamstrings (HA), tibialis anterior (TA) and gastrocnemius (GA). Each electrode pair (or channel) can be provided with currents up to 150 mA. By default the admissible stimulation intensity is lowered to values between 80 and 100 mA. The stimulator communicates with a PC by using a standard RS232 protocol.

Electrodes can either be individually connected to the device (frontal connection) or two times 10 channels connectors are provided at the back. The second solution is of course more practical for clinical applications. Especially since the WalkTrainer has been equipped with connectors that regroup the electrodes (one connector per thigh and one per shank (Figure 7b)).

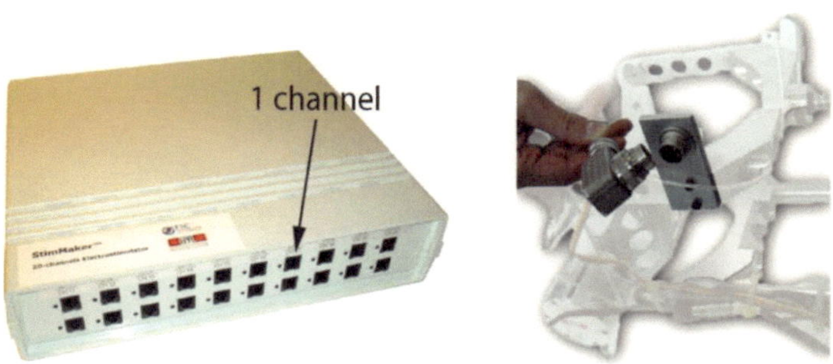

(a) (b)

Fig. 7. (a) Electrical muscle stimulator (front view). (b) Regrouped electrode connection for the shank muscles (TA, GA) on the ankle part of the leg orthosis.

4. Rehabilitation strategy on the WalkTrainer

As introduced in before rehabilitation with the WalkTrainer relies on the combination of robotic mobilization and muscle stimulation. This section describes how these were developed and implemented.

4.1 Leg and pelvis mobilization

During manual reeducation physiotherapists aim at mobilizing the patients' legs and pelvis with the correct trajectory. The WalkTrainer is excepted to do so, but during longer training sessions and with a greater accuracy. The therapists knows how to adapt to a tall, thin or elder subjects. The trajectory generator of the WalkTrainer is expected to be able to do the same.

For the pelvis a four step procedure was undertaken:

1. Motion measurement on valid subjects
2. Motion model identification (offline)
3. Implementation in the trajectory generator of the WalkTrainer
4. Validation on valid and paraplegic subjects

1. A measurement campaign was undertaken on twenty healthy subjects. Each person was requested to walk at four different speeds, every time the 6 DOF of the pelvis were saved for later analysis (Stauffer et al., 2008).
2. The average curve of the twenty subject was then computed. Several models (linear with/without parameters) were then proposed to compute the peak to peak amplitude of each DOF as a function of various parameters (size, gender, walking speed, ...) (Figure 8a).
3. One model was then selected and implemented in the trajectory generator of the WalkTrainer. Since the pelvic orthosis was employed alone, the phase and frequency synchronization for the motion control was provided by a footswitch placed under the subject's right heel (Stauffer et al., 2007).
4. Finally the effect of using the model to adapt the reference pelvic trajectories was successfully assessed. As expected a reduction of the measured force applied by the pelvic orthosis is observed when the model is used to adapt the trajectory (Stauffer et al., 2007) (Figure 8b).

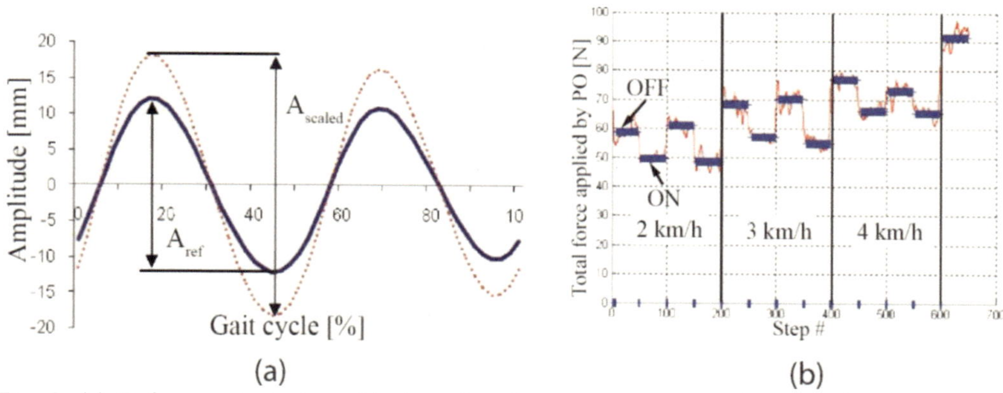

Fig. 8. (a) Reference trajectory (A_{ref}) and scaled trajectory by using the developed model (A_{scaled}). (b) Total force measured by the pelvic orthosis as a function of the step number at different walking speeds. The model is switched from OFF to On every 50 steps.

Speed dependent leg trajectory measurements have also been undertaken but not implemented on the WalkTrainer yet. In addition preliminary pelvic motion measurements along curved paths have also been performed.
As a conclusion, the motorized orthoses are providing precise mobilization and the force sensors serve as input to the muscle stimulation.

4.2 Electrical muscle stimulation

An off line EMG based stimulation scheme was implemented. Indeed EMG represent well muscular activity and also muscular force (to a certain extent). For every of the 7 muscles the main working point (as a function of the gait cycle) was found (Figure 9c). This working point then served as identification position of the force-stimulation intensity relation (force-intensity (Figs. 9a-b)). These EMG measurements are also used to create a so called enable windows, that allow the muscle to be stimulated only at certain periods of the gait cycle (in order to be coherent with the motion and antagonistic inhibition).
Then for every articulation a target torque has to be generated as a function of the gait cycle. One can chose either to exaggerate some features such as: push-off or foot clearance. Or use the force sensors of the orthosis in order to minimize the interactions between the leg orthosis and the subject's leg by tuning the muscle stimulation.
Finally the so obtained target torque (hip, knee and ankle) has to be converted into a muscular force. By using the identified force-intensity relation of each muscle on the 3 articulations, one can find a set of rules to transform articular torques into muscular force.
Finally all these elements are combined and a coherent muscle stimulation strategy is obtained (Stauffer, 2009) (Figure 10). The torque generator, generates target torques for the hip, knee and ankle articulations as a function of the gait cycle by using predefined algorithms. That can either increase the physiologically needed torque, apply a constant torque or minimize the interaction between the user and the exoskeleton. This articular torque is then transformed into a muscular target torque, this operation is critical since some muscles span over several joints and joints are crossed by more than one muscle. By using the muscle model the required stimulation intensity is then computed. The so obtained intensity is checked for coherence by comparison to the timing obtained by offline EMG

measurements. Finally the muscle stimulator performs its task by stimulating the muscles and the interactions between the user and exoskeleton are measured and sent back to the torque generator.

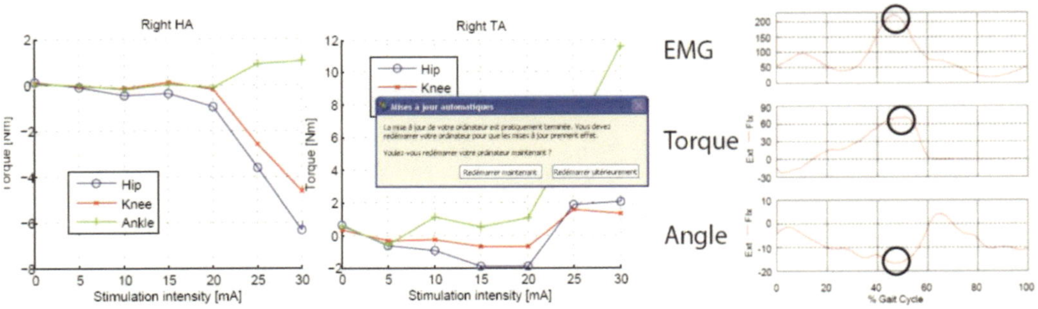

(a) (b) (c)

Fig. 9. (a) Hamstring muscle identification result (biarticular: hip and knee). (b) Tibialis anterior muscle identification result (monoarticular: ankle). (c) EMG activity of the triceps surae, ankle torque and ankle angle as a function of the gait cycle. Adapted from (Kirtley, 2006)

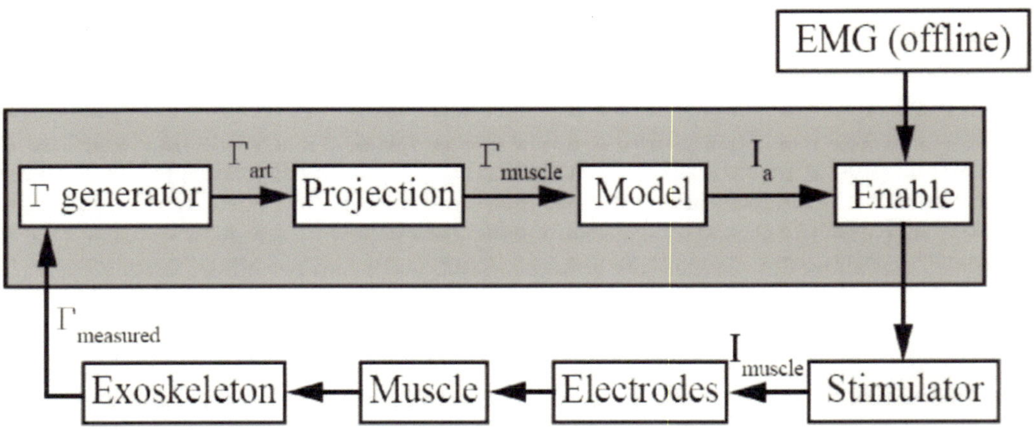

Fig. 10. Closed loop muscle stimulation strategy on the WalkTrainer.

4.3 The complete rehabilitation strategy

In addition to the orthotic mobilization and closed loop muscle stimulation the remaining components of the WalkTrainer are also utilized. The active body weight support motion is synchronized with the pelvic orthosis. This guarantees a coherent unloading and thus provides a good foot loading. At the same time the forward velocity of the mobile frame is matched with the walking speed programmed on the leg orthosis.

As a result coherent position information are provided to the legs and pelvis of the subject. The same is true for muscular loading, since it is either provided by the user himself (partial lesion) or by the FES only or by a mix of both. The joint work of the pelvic orthosis and body weight support creates a coherent loading of the users' feet. Finally the rolling of the

WalkTrainer provides natural visual feedback and further motivates the subject (Stauffer, 2009) (Figure 11a).

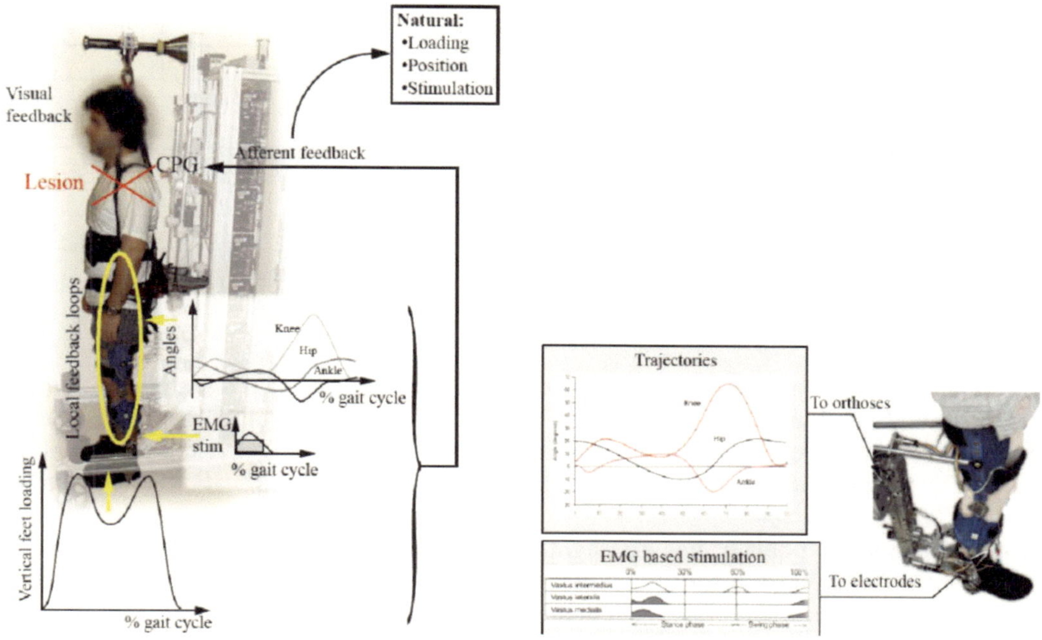

(a) (b)

Fig. 11. (a) Main feedback pathways targeted by the rehabilitation strategy. (b) Integration of robotic mobilization and EMG based stimulation.

5. Clinical trials

At the beginning of 2008 a three month clinical trial was held at the Clinique romande de réadaptation (CRR). The objectives were of two kinds. First, practical questions had to be answered such as: how easily can paraplegic subjects be installed, do they have apprehension, do they appreciate the training. Second, therapeutic questions had also to be addressed. These were for instance: muscular increase, better coordination, reduction of spasticity (Allemand & Stauffer, 2009).

5.1 Phases and patients

In order not to provide the patient with too many new sensations at one time. And also to have a progressive testing of the device, a four phase clinical protocol was developed.

Phase 1: the WalkTrainer is mounted on a treadmill, no muscle stimulation is applied to the patient.

Phase 2: the WalkTrainer is mounted on a treadmill, muscle stimulation is applied to the patient.

Phase 3: the WalkTrainer is rolling over ground, no muscle stimulation is applied to the patient.

Phase 4: the WalkTrainer is rolling over ground, muscle stimulation is applied to the patient (Figure 12a-b).

Due to clinical constraints the patient population was quite heterogeneous (Table 1). Also most of the patients couldn't stay during the whole trials, but merely 2 to 3 weeks. Which corresponds to their annual stay at the CRR.

Patient	Gender	Age	Lesion (motor/sensory)	# training Sessions	Phases
1	M	25	T12/T12	12	1-4
2	F	39	C7/C7	6	1-4
3	M	31	C4/C2	2	1
4	F	21	C8/C3	2	1
5	M	52	L2/L3	3	1-2
6	M	26	D5/D5	2	1,4

Table 1. Summary of the patients that participated to the clinical trial.

5.2 Results

The robotic mobilization performed as expected. The patients were successfully mobilized by the leg and pelvic orthoses. The electrical muscle stimulation algorithms also worked. For instance the stimulation intensities were higher on the more affected leg and the stimulation adapted to the remaining voluntary participation (i.e. decrease of stimulation) (Allemand & Stauffer, 2009).

From a therapeutic point of view a slight decrease of spasticity could also be observed (measured manually). Unfortunately no muscular force increase could be measured. This is not a surprise with a single weekly training session.

The WalkTrainer also turned out to be an effective diagnostic device. A comparison between the measured articular torques (with and without voluntary participation of the patient) confirmed the observations of the physiotherapists regarding the patient's remaining capabilities.

5.2 Conclusions for the clinical trials

The main objectives were reached. Transfer and installation of patients (even ASIA A) could easily be done with the adequate material, for instance a cigogne (Figure 12c). Also patients had absolutely no apprehension and even liked the training.

From a robotics point of view the WalkTrainer also succeeded his first trials, indeed no major incident or damage occurred to the device. From a control point of view everything turned out as expected. The motion generation as well as the muscle stimulation functioned perfectly.

The clinical outcomes are more contrasted, indeed some results have been obtained (reduction of spasticity, mobilization and stimulation, use of the WalkTrainer as a diagnostic device). However longer clinical trials with a higher training load should now be conducted.

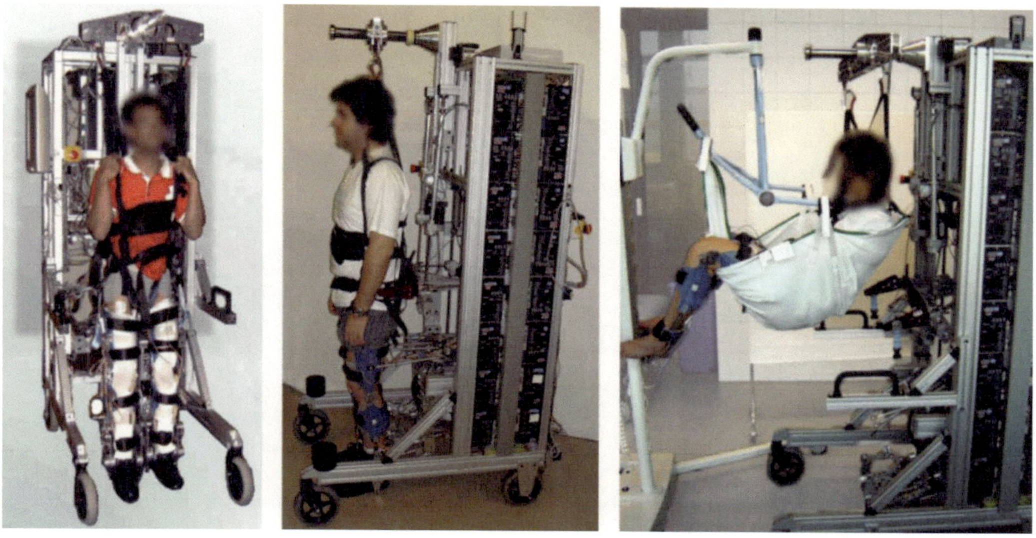

(a) (b) (c)
Fig. 12. (a) Subject ready for training (front view), (b) side view. (c) Transfer using a cigogne.

6. Conclusion

The WalkTrainer is a novel reeducation device that combines optimal mobilization (personalized trajectories) by the means of a 6 DOF pelvic orthosis and a 2 times 3 DOF leg orthosis, coupled to closed loop electrical muscle stimulation. Its development was carried out at EPFL (LSRO) in collaboration with the FSC (Swiss Foundation for Cyberthosis) as a CTI project (7485.2 LSPP-LS) with SWORTEC SA as main industrial partner.

It is up to day the only reeducation device that combines leg and pelvis mobilization to electrical muscle stimulation, while walking over ground. In addition the control strategy is also new. First generation robots or stimulation strategies merely played back predefined patterns (motion or FES). Second generation robots or stimulation strategies adapted their effort (motion or FES) to the patient's participation. Third generation reeducation devices are the combination of second generation robotic control and second generation FES. The WalkTrainer can thus be viewed as the first third generation reeducation robot (Figure 13).

The clinical trials carried out at the Clinique romande de réadaptation (CRR) have proven that the WalkTrainer is functional and that its reeducation capabilities are real. The software and hardware have been tested thoroughly and are now ready for more intensive clinical trials.

Future development of the WalkTrainer will involve walking along curved paths. In order to do so a new leg orthosis will have to be designed. Indeed the current version cannot perform internal/external rotation. Then after a new set of clinical testing, the industrialization will begin.

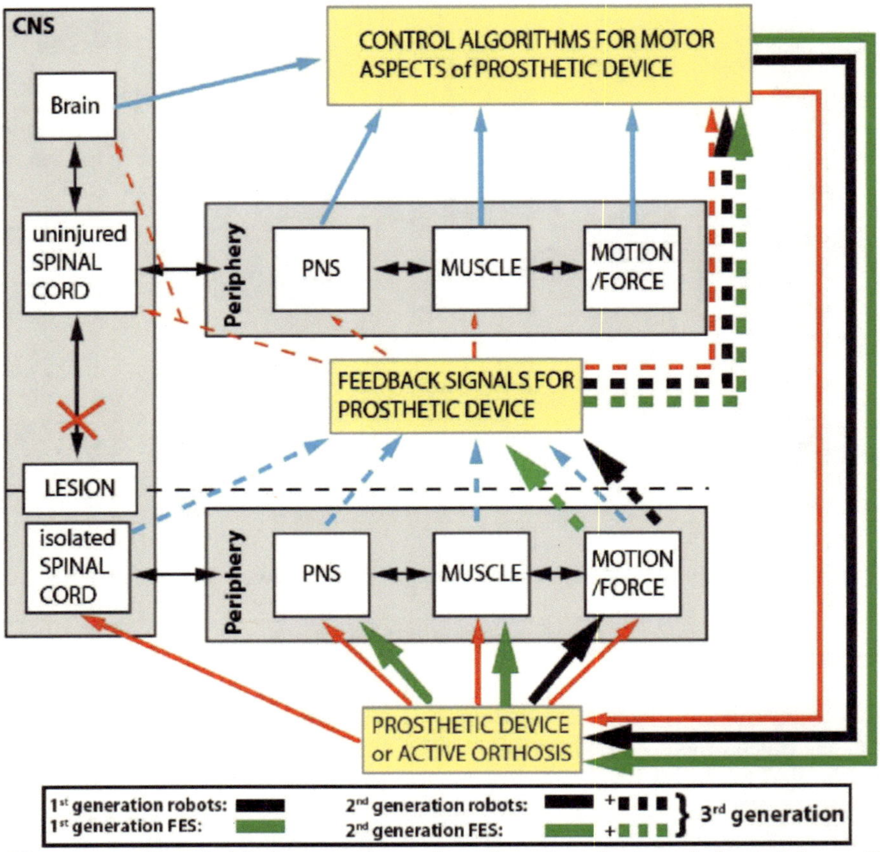

Fig. 13. First, second and third generation of reeducation devices (adapted from (Giszter, 2008)).

7. Acknowledgments

The Cyberthosis team would like to thank: the CTI/KTI (project: 7485.2 LSPP-LS) and the Loterie Romande for their financial support, as well as the Clinique romande de réadaptation (CRR) for the precious help before and during the clinical trials that took place in their facilities.

8. References

Allemand, Y and Stauffer, Y. (2009). Overground Gait Rehabilitation: First Clinical Investigation with the WalkTrainer, *Technically Assisted Rehabilitation,* March 2009

Allemand, Y., Stauffer, Y., Clavel, R. & Brodard, R. (2009). Design of a new lower extremity orthosis for overground gait training with the Walktrainer. *In ICORR,* Japan, June 2009

Aito, S. (2003). Complications during the acute phase of traumatic spinal cord lesions. *Spinal Cord,* 41(11):629-635

Barbeau, H., Norman, K., Fung, J., Visintin, M. & Ladouceur, M. (1998). Does neurorehabilitation play a role in the recovery of walking in neurological populations? *Annals of the New York Academy of Sciences,* 860:377-392

Calancie, B., Needham-Shropshire, B., Jacobs, P., Willer, K., Zych, G., & B. A. Green. (1994),
Involuntary stepping after chronic spinal cord injury. evidence for a central rhythm generator for locomotion in man. *Brain,* 117(5):1143-1159

Colombo, G. Jorg, M., & Dietz, V. (2000). Driven gait orthosis to do locomotor training of paraplegic patients. *In Annual International Conference of the IEEE Engineering in Medicine and Biology,* - Proceedings, pages 3159-3163, USA, July 2000.

Dietz, V., Colombo, G., Jensen, L., &. Baumgartner, L. (1995). Locomotor capacity of spinal cord in paraplegic patients. *Annals of Neurology,* 37(5):574-582

Dobkin, B. H., Harkema, S., Requejo, P. & Edgerton, V. R. (1995). Modulation of locomotor-like emg activity in subjects with complete and incomplete spinal cord injury. *Journal of neurologic rehabilitation,* 9(4):183-190

Dohring, M. E. & Daly, J. J.(2008). Automatic synchronization of functional electrical stimulation and robotic assisted treadmill training. *IEEE Transactions on Neural Systems and Rehabilitation Engineering,* 16(3)

Freeman-Somers, M. (2001). *Spinal cord injury: Functional rehabilitation,* Prentice Hall, 0-0-8385-8616-3, 2nd edition

Frischknecht, R. & Chantraine, A. (1989). The treatment of spasticity by physical methods [traitement de la spasticity par les agents physiques]. *Annales de Reeadaptation et de Meedecine Physique,* 32(4):475-493

Germann, W. (2001). *Principles of human physiology,* Benjamin Cummings, 0-8053-5622-2 , 4th edition

Giszter, S. F. (2008). Spinal cord injury: Present and future therapeutic devices and prostheses. *Neurotherapeutics,* 5(1):147-162

Graupe, D. (2002) An overview of the state of the art of noninvasive FES for independent ambulation by thoracic level paraplegics. *Surgical Neurology,* 24(5):431-442

Grundy, G. (2002). *ABC of spinal cord injury,* BMJ Books, 0-7279-1518-5, 4th edition

Jackson, A. & Mott P. (2008). Spinal cord injury information network, http://www.spinalcord.uab.edu

Jezernik, S., Schärer, R., Colombo, G. & Morari, M. (2003). Adaptive robotic rehabilitation of locomotion: A clinical study in spinally injured individuals. *Spinal Cord,* 41(12):657-666

Kaelin-Lane, A., Sawaki, L. & Cohen, L. G. (2005). Role of voluntary drive in encoding an elementary motor memory. *Journal of Neurophysiology,* 93(2):1099-1103

Kirtley, C. (2006). Clinical gait analysis theory and practice. Churchill Livingstone, 1st edition

Metrailler, P., Blanchard, V., Perrin, I., Brodard, R., Frischknecht, R., Schmitt, C., Fournier, J., & Clavel, R. (2006) Improvement of rehabilitation possibilities with the MotionMaker™, *Proceedings of the First IEEE/RAS-EMBS International Conference on Biomedical Robotics and Biomechatronics: BioRob 2006,* pp. 359-364

Pearson, K. G. (2000). Neural adaptation in the generation of rhythmic behavior. *Annual Review of Physiology,* 62:723-753

12

Human Machine Interface in Assistive Robotics: Application to a Force Controlled Upper-Limb Powered Exoskeleton

Malek Baklouti[1,2], Jamil AbouSaleh[3], Eric Monacelli[2] and Serge Couvet[1]
(1) THALES Training and Simulation, France (2) LISV Lab, University of Versailles, France (3) PAMI Lab, University of Waterloo, Canada

1. Introduction

The current work deals with Human-Machine Interfaces for assistive robotic systems. Such systems emerge to assist humans in their daily life tasks, such as in personal hygiene, eating, educational, and independent movement tasks. The goal is to improve the mobility of the disabled person while making the system transparent and non invasive to the user's comfort and mobility.

ESTA is a project that aims to develop an assistive exoskeleton for the upper limb. The work on this project took off in 2007. The exoskeleton has been designed to compensate for the loss of mobility in the upper limb, especially at the shoulder and elbow joints. Designing a proper Human-Machine Interface has been a primary concern in developing this system. Due to their many limitations and restricted operability, old usual interfaces were not convenient to use, and hence, the development of new adaptive interfaces that focus on optimizing the interaction between the user, the system and the environment is a main necessity.

In this chapter, a brief overview of some of the existing exoskeleton systems and the recent assistive robotic technologies for upper limb will be presented. Most of such control architectures use force sensors to control exoskeleton, and are based on motor torque calculation through dynamic laws, which require the knowledge of the mass and inertia matrix of each part of the orthosis, along with the presence of a dynamic model for the arm. However, in our proposed approach, motors are controlled with speed input instead. A description of our proposed system, along with the approach pursued to control the exoskeleton using pressure sensors will be presented in details through the flow of the chapter.

2. Designing a Rehabilitation System

The aim of assistive robotics is to reduce the human deficiency by enabling disabled people to be independent in carrying their daily life tasks. A key word to be underlined is that *capability* and *disability* are not opposites. As a matter of fact, disability should be seen as a reduced range of capabilities in a certain context, and this pattern of capabilities varies widely between individuals. This makes the design of a generic system a fundamental concern.

2.1 Requirement and challenges

When designing a rehabilitation system, two main notions should be considered: *Operability* and *Effectiveness*. While effectiveness measures the extent to which the deployment of the device improves the consumer's living situation, *Operability*, on the other hand, measures the ease of manipulating the device, and the response accuracy to the consumer's operative commands.

Simplicity and *Portability* are also major concerns when designing such systems. These notions require a rehabilitation robotic system to be adaptable, flexible, cost effective, and able to fit with the cognitive, perceptual and motor skills of the human operator. To fulfil these requirements, designing a proper system interface becomes a primordial issue since it represents the bridge connecting the user's intention and his/her action.

Usability is also an important part of user acceptance. According to Nielsen system acceptability includes a social and a practical part (Nielsen 1993, Birch & Cameron 1990). Usability includes easy learn ability, efficiency in use, remember ability, lack of errors in operation as well as subjective pleasure.

Hence, assistive systems should be endowed with interfaces that are specifically designed for disabled people in order to enable them to control the system with the most natural and less tiring mean. This is the primary concern of many researchers working in the field of Human-Machine Interfaces.

In the following, some of the challenges that emerge when designing a human-machine interface for assistive robotic systems are discussed.

2.2 The need of an adequate Human-Machine Interface

The major requirements when designing a HMI for disabled people can be summarised as follow (Hillman and Jepson 1992):

- **The need for simple and intuitive interface:** Both the robot and the user must be able to communicate and cooperate in a straightforward manner. This will require that the system is capable of indicating the intentions and/or the internal state of the robot to the user, and that the user can send commands back to the robots using intuitive and natural verbal and gesture tools.
- **The need for perception modelling:** In a natural human interaction, a person uses different ways for communication. For example, a person will look, point and simultaneously ask for a specific object to be brought to him. Meaning that intention is expressed through the gesture, gaze point, as well as the voice. Thus, sensing and predicting the intention of the user is a major step when designing natural human-machine interfaces.
- **Flexible and genuine interface accessibility:** Human factors and the ability of the operator should also be considered especially when dealing with assistive systems. In specific scenario, as in the case of a physically disabled person, the operator might be endowed with restricted margin of movement. It is in fact a considerable challenge to make interfaces that are most generic and accessible for all, especially when dealing with people with cognitive, sensory and mobility impairments.

- **The need for an adequate feedback:** The importance of feedback to assist the user has long been recognized as a fundamental principle in the design of good interfaces. Offering adequate feedback for user's actions is thus essential in designing HMI.
- **Keep the user in the control loop of the system:** Using the "user in the loop" paradigm is essential when designing assistive systems. Though interfaces dedicated to disabled people should be intelligent enough to sense the intention of the user, they should not control the system autonomously and exclude the user from the loop. The user should always feel control of the system.
- **Cost:** Designing sophisticated interfaces requires sophisticated tools and technologies which might be of very high cost. Therefore, a proper compromise between cost and operability of the system should be addressed.

Many efforts have been pursued to develop assistive interfaces. In the following, we will only focus on interfaces that fit the need of physically impaired persons. Such interfaces can be classified into the following classes:
The first class of interfaces is represented by the modified "usual" devices, such as keyboard, trackballs and joystick. These devices were transformed to be more ergonomic to fit the person's disability (Hobday 1996). However, disabled people may not conform to a standard pattern; in fact, while some classes of disability are more common than others, there cannot be a single pattern which fits everyone. Thus considering individual solutions is exceedingly expensive in practice. Moreover, this kind of devices is only adapted to people who retain a good measure of manual dexterity. Therefore, having interfaces that are intelligent enough to react differently and appropriately to different stimuli are of big importance when tackling the problem of assistive devices.
The second class of interfaces represents those that sense signals from the user to predict and synthesise an adequate control command. The major examples of such interfaces are the ones that sense not only the user's eye movements (Ohno and Mukawa 2003), head gestures, and body movements but also EEG or EMG signals (Ferreira et al. 2008, Rani & Sarkar 2005, Artemiadis & Kyriakopoulos 2006). These interfaces have the advantage that their input device (sensor) is kept unmodified, and the whole intelligence and adaptability module consists of processing the signals in an adequate manner to extract/synthesise the control command. Thus, by the mean of intelligent algorithms and processing tools, and without remodelling the hardware, the essential requirement for matching the system's and the user's need can be fulfilled without incurring a prohibitive cost of constructing each system individually.

3. Force Controlled Exoskeleton: "ESTA" Project

In this chapter, we address assistive systems that are dedicated to help impaired people to interact with their external world by providing them with the ability of taking physical actions, as well as manipulating other objects. Two classes of systems can be distinguished: the "robotic manipulator" and "assistive exoskeleton".
The robotic manipulators have been first extensively used for industrial applications for repetitive tasks; then recently, they were used as a manipulation aid for disabled people. Such systems mainly consist of actuated joints and an end-effector for grasping objects.

On the other hand, assistive exoskeletons are systems that are worn by the operator as an orthotic device. Their joints and links correspond then to those of the human body. Assistive exoskeletons are used to amplify the user's fragile power, thus keeping the grasping task under the full control of the operator. The operator provides control signals, while the exoskeleton provides most of the mechanical power required to carry out the task.

3.1 Exoskeleton for Rehabilitation

3.1.1The History of Assistive Exoskeleton

In this section, a brief overview of the most assistive robotic systems introduced in the last half century will be presented.

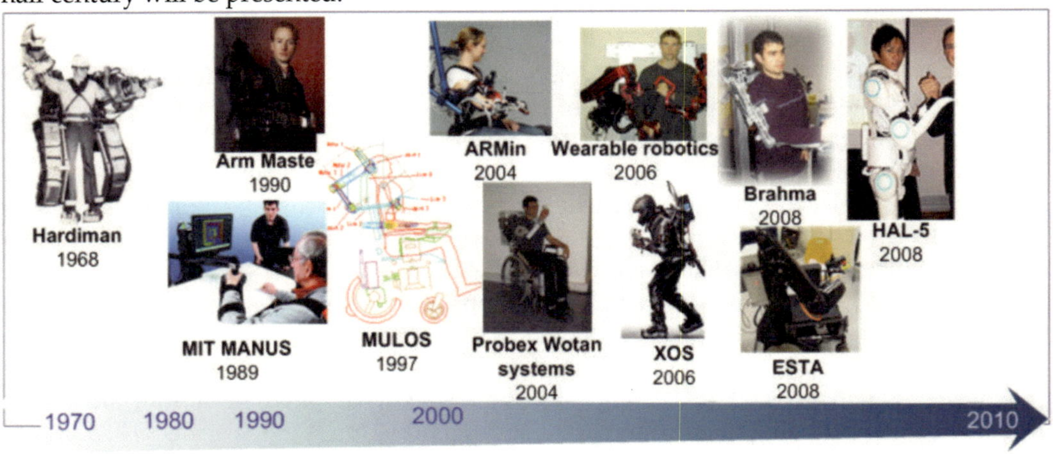

Fig. 1. The History of assistive exoskeleton

The first attempt to build a powered exoskeleton appears in 1968 with the exoskeleton Hardiman presented in Fig. 1. Hardiman is a 30-DoF exoskeleton composed of two legs and two arms. This system was initially designed for military purposes.

The first articulated upper limb orthosis dedicated for assistive purposes was the 'Balanced Fore Arm Orthosis' (BFO) (Alexander and al; 1992). Developed in 1965, this passive exoskeleton mounted on a wheel chair is intended to help disabled persons to accomplish movements on a horizontal plane. A motorized version of the BFO was developed by the Burke rehabilitation centre in 1975, but no significant results were achieved.

The first illustrative model for assistive exoskeleton systems are the ArmMaster (1990) and MANUS. The "Exoskeleton Force ArmMaster", developed by Exos corporation, is a five motorized DOF for shoulder and elbow joints. The system is completely back mounted and can exert a torque that able to go from 13Nm on the forearm to 40Nm on the arm (Burke 1992).

Another important system is the MIT MANUS (Kwee et al. 1988, Hogan et al. 1992), which is an articulated arm designed for the physical therapy of stroke victims. Although the first version of the MIT MANUS permitted to enable movement on a horizontal plane, the robotic arm was not parallel to the human one, and the orthosis was not mobile. MIT kept improving their system and upgraded it to possess an exoskeleton with 6 degrees of freedom.

The EU Telematics program started in 1997 the MULOS project (Motorized Upper Limb Orthosis System) MULOS is a 5 DoF powered upper limb orthosis, which allows the

movement of the shoulder (3 DoF), the elbow, and the forearm. The orthosis was designed to work in three different modalities: assistive, continuous passive motion, and exercise (Johnson & Buckley 1997). However, despite the good potential of this orthosis, the development was discontinued in 1997.
The French company Wotan Systems has developed since 2000 a prototype of force controlled arm exoskeleton (Probex). The system is equipped with force sensors that sense the effort exerted by the operator on the system. The exerted force is then analyzed by a PC located in the wheelchair, and then amplified by pneumatic artificial muscles.
A more recent developed system is the exoskeleton ARMin. It supports spatial movements of the shoulder and elbow joint. Its control strategy takes into consideration the patient effort.
The most current exoskeleton systems are XOS, Brahma, and HAL. However, both XOS and HAL gained most of the interest and fame. Although such systems were dedicated for military purposes, we chose to briefly present them because of their high level of impact within the research community. XOS, a project initiated by DARPA, was developed to help doing some military work in the battlefield. The first prototype of the exoskeleton was designed in 2008 by Sarcos company.
HAL system is designed by a spin-off company from the Tsukuba University and will be commercialised under the name of "Robot Suit HAL". Two versions of this system exist: HAL-3 for the lower limbs, and HAL-5 for the full body.

3.1.2 How are they controlled?

Clearly, designing a proper interface is crucial in rehabilitation robotics. But the question remains, how do we define "proper", and what kind of interface should be used?

Exoskeleton system	Human Machine interface used
HAL system Kawamoto & Sankai 2002)	operator's intention estimated using EMG and impedance adjustment
ARMin (Nef & Riener 2005)	EMG signals : "patient-cooperative" control that recognizes the patient's movement intention in terms of muscular efforts
Werable Robotics	EMG signals
MIT MANUS (Hogan et al. 1992)	A six degrees-of-freedom force sensor is mounted on the robot end-effector. Motors torques were computed using dynamic laws applied on the robotic arm.
Golden Arm, (Allen et al. 1972) Rancho Los amigos Hopital	(1) Joint per Joint control using seven tongue operated switches ; (2) eye trackers[Moe1973].
Case Institute of Technology (LeBlanc & Leifer 1982)	(1)Pre recorded tasks;(2) a head-mounted light source that triggers light sensors in the environment.

Table. 1. Summury of the control strategy

Therefore, many challenges should be taken into consideration when designing a human machine interface. Table 1 shows some of the interfaces used to control the different assistive systems.
The skin surface electromyogram (EMG) is often used to detect the user's motion intention in the latest systems. These biological signals directly reflect the user's motion intention.
HAL exoskeleton operates by sensing weak electrical impulses from muscles via electrodes on the skin of the operator. An onboard computer analyzes them and activates the servos of the suit, mimicking the motions of the wearer.
Other systems use force sensor to synthesise the control law. MANUS was the first one controlled by force sensors. From the force applied on the extremity of the arm, motors torques were calculated by using dynamic laws applied on the robotic arm.
Less conventional interfaces have been used, such as tongue switches, light sensors, as well as image based control (Madentec 2008, Vaidyanathan et al. 2007, Betke et al. 2002).

3.2 ESTA system

"ESTA" project emerged from different reflections and specifications of the user. The goal was to design a low-cost exoskeleton controlled by force sensors that allows disabled population to achieve arm movements.
Myopathy is a neuromuscular disease in which the muscle fibres do not function, resulting in muscular weakness. This project is an industrial project conducted between different partners and consists in developing an innovative 4DoF assistive exoskeletal orthosis. It aims to develop an assistive technology for people suffering from myopathy, by assisting their upper limb movements.
The exoskeleton has been designed to compensate for the loss of mobility in the upper limb, especially for joints at the shoulder and elbow. The specifications were developed by the AFM, and they involve mainly the architecture, the control command, as well as the security and energy consumption of the orthosis.

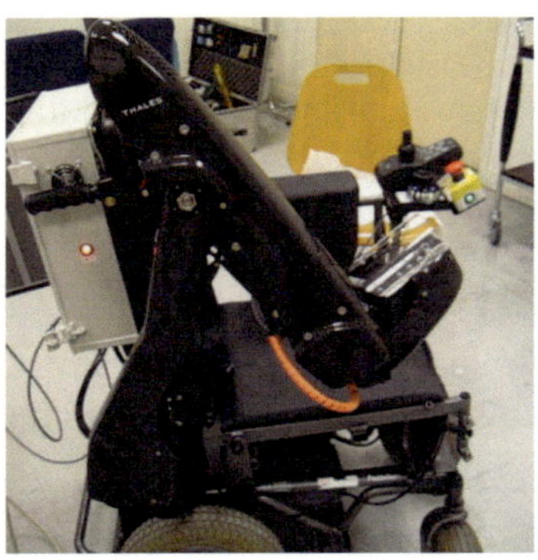

Fig. 2 . the prototype of the exoskeleton

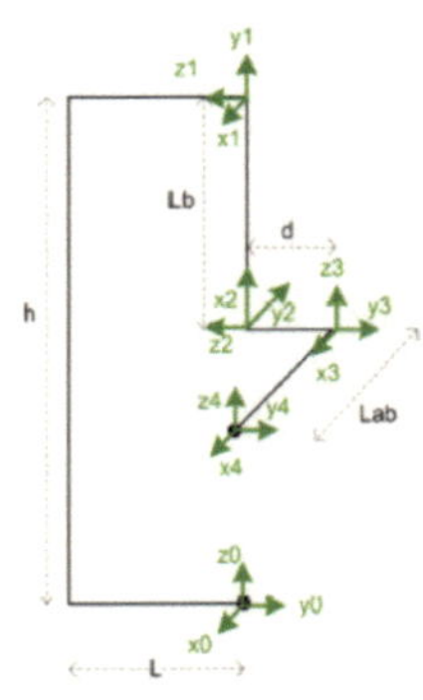

Fig. 3. Coordinate systems for D-H parameters

Each joint of the orthosis is actuated with a 50 Watt DC brushless motor that is powerful enough to allow the exoskeleton to lift a 3Kg weight. We use digital servo drive supplied by Elmo to control these motors. The low-level speed control of the motors is done by the Elmo whistle.

3.2.1 Force measurement

FSR sensors (Force Sensing Resistors), placed into two bracelets (Fig. 4), are used to measure the forces exerted by the patient on the orthosis. The bracelets are attached around both the arm and the forearm of the patient as shown in Fig. 5.

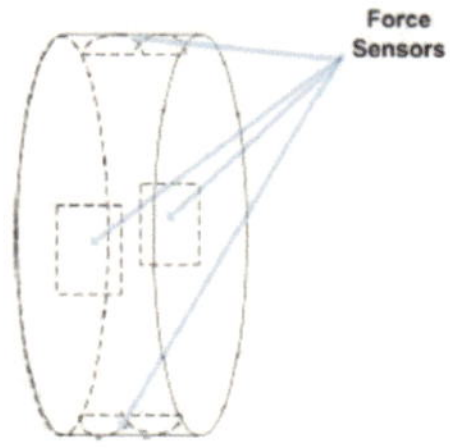

Fig. 4. Bracelet equipped with force sensors

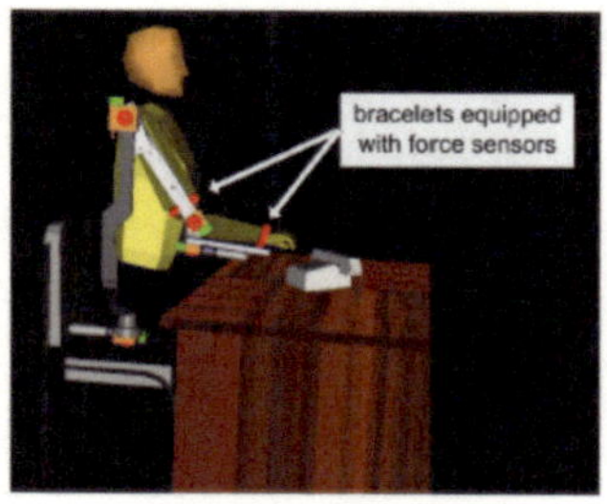

Fig. 5. Model of the system

When analysing the signals coming from the sensors, it was noticed that (Fig. 8) they represent the superposition of the following components:

1) Very low-frequency components:
   - an offset due to the sensor intrinsic properties and the variation of temperature.
   - an offset due to the pressure exerted by the bracelet on the arm (this pressure is not constant, due to the relaxation of the bracelet materials).
2) Low-frequencies components:
   - a low frequency signal due to the mass of the arm (the mass perceived by the sensors depends on the position of the exoskeleton).
   - the inertia of the arm.
3) The command signal frequency that we want to extract.

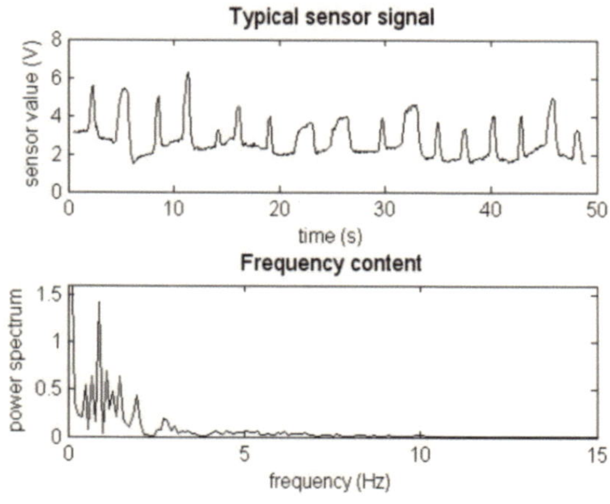

Fig. 6. A typical command signal and its spectrum

Figure 6 presents a typical sensor signal and its Fast Fourier Transform (FFT). The computation of the FFT shows that all the energy is concentrated in the low frequencies, ranging from 0 to 5 Hz.

In the following, we propose a new approach to control the exoskeleton using pressure sensors. Most of the existing control architectures that use force sensors to control exoskeleton are based on motor torque calculation through dynamic laws. Such control systems require the knowledge of the mass and inertia matrices of each part of the orthosis, as well as a dynamic model of the arm. In our approach, we propose the design an algorithm that control the motors with speed consigns instead of torque consigns. Therefore, evaluating the mass and inertia matrices of the orthosis become unnecessary.
A free-delay digital filter is first used. This filter removes non-interesting information coming from the sensors, such as offsets, and forces due to the mass and inertia of the arm of the patient. The filtered signals from each sensor are then interpreted as speed vectors. Each of these vectors corresponds to the desired speed of the orthosis at the point where the sensor is mounted.

3.2.2 Filter design

The analysis of the signal should be done in real time. Signal processing tools would introduce an unacceptable delay such as wavelet analysis, and hence cannot be used to extract the command signal. In the following section, we will describe a filtering method that is used to extract the information without introducing any delay.
As we have seen earlier in power spectrum of the signal, the frequencies corresponding to the command signal (1.5 to 5 Hz) are very close to the undesirable frequencies that we would like to remove (0 to 1.5 Hz). The signal component that is due to mass and inertia of the arm can be removed by filtering the frequencies below 1.4 Hz. We draw the window of the filter in Fig. 9. To have a significant attenuation of the offset, the arm weight, and inertia, we choose:

$$\begin{cases} a = -3 ; b = -20 \ \% \\ \dfrac{f_0}{f_1} = \dfrac{2}{1.5} \end{cases} \tag{1}$$

We compute the FIR (finite Impulse Response) filter coefficients that minimize the weighted, and integrated squared error between the ideal piecewise linear function and the magnitude response of the filter. To match with the designed window, the order n of the filter should be at least 30. The delay introduced by such a filter can be then calculated as:

$$\frac{n}{f_{sampling}} = \frac{40}{100} = 0.4 \, s \tag{2}$$

3.2.3 Motor speed calculation

Finally, we use an inverse kinematics module to compute each motor speed.
The high-level control loop of the system is presented in the Fig. 7. After filtering the signal from the sensors, the kinematic model of the orthosis is used to compute the appropriate motors speed. The control loop is finally feedback closed by the mean of a PID (Proportional Integral Derivative) controller.

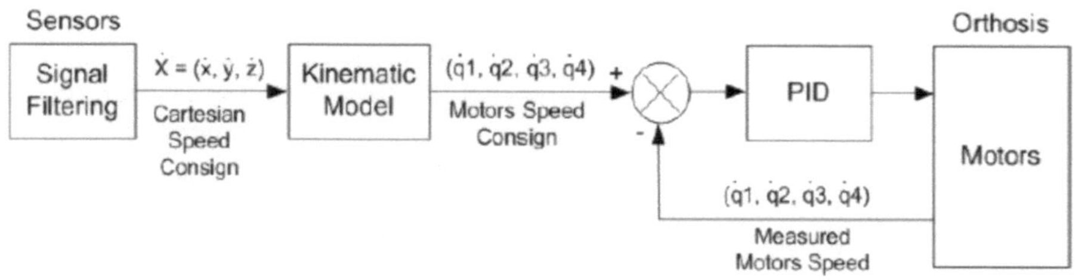

Fig. 7. High level control loop

The force applied by the user on one sensor is interpreted as a desired speed of the exoskeleton at the point X_{sensor} where the sensor is mounted. $\dot{X}$ represents a vector that is orthogonal to the plane of the sensor. We can then calculate the corresponding speed of each motor by using the Jacobian in the following equation:

$$\dot{X}_{sensor} = J\,\dot{q} \tag{3}$$

Where:

- $\dot{X}_{sensor}$ represents the velocity
- J represents the Jacobian of the orthosis
- $\dot{q} = \left\{\dot{q}_1, \dot{q}_2, \dot{q}_3, \dot{q}_4\right\}$ represents the joints speed

Instead of using the pseudo inverse of the Jacobian matrix to solve the system, we compute each joint speed by minimizing the following function :

$$f : (\dot{q}, \dot{X}_{sensor}) \rightarrow f(\dot{q}, \dot{X}_{sensor}) = \dot{X}_{sensor} - J \cdot \dot{q} \tag{4}$$

For a system composed of **n** sensors, it is first necessary to calculate the **n** Jacobian matrix J_i that corresponds to each sensor position X_{sensor}. Then, the motors speed consign vector $\dot{q}_i$ is calculated for each sensor as shown in equation 3. Finally, the vector $\dot{q}$ is obtained by taking the average of all $\dot{q}_i$.

4. Conclusion

The work in this chapter deals with Human-Machine Interfaces for assistive robotic systems that aim to improve the mobility of the disabled people. Many requirements and challenges rise when designing such systems, such as operability, effectiveness, simplicity, and portability. These notions require the rehabilitation robotic system to be adaptable, flexible, cost effective, and able to fit with the cognitive, perceptual and motor skills of the human operator.

This chapter presented ESTA as an example of assistive systems. It is a project that aims to develop an assistive exoskeleton for the upper limb. The work on this project took off in 2007. The exoskeleton has been designed to compensate for the loss of mobility in the upper limb, especially at the shoulder and elbow joints.

A novel HMI interface has been designed for ESTA orthosis based on pressure sensors. While most of the existing control architectures that use force sensors to control exoskeleton are based on motor torque calculation through dynamic laws, which requires the knowledge of the mass and inertia matrices of each part of the orthosis, as well as a dynamic model of the arm, our new approach to control the exoskeleton uses speed consigns instead of torque consigns; Therefore, evaluating the mass and inertia matrices of the orthosis become unnecessary.

5. References

Nielsen, J. ; (1993). Usability Engineering. *Academic Press Inc.* pp. 362

Hillman, M. ; Jepson, J. (1992) Evaluation of a robotic workstation for the disabled. *Journal of Biomedical Engineering* ,Vol. 14, No.3 , (187-192)

Beitler, M.T. ; Stanger, C.A. and Howell, R.D. (1994) The design of an integrated interface to an educational robotic system, *Proceedings of the RESNA 94 Annual Conference*, Nashville, pp. 448-450, USA

Birch, G.E. and Cameron, W. (1990), User Acceptability in Robotics Assistive Devices, *Proceedings of the International Conference of Rehabilitation Robotics*, pp. 83-94, (1990).

Hobday, S. W. (1996) The Maltron Keyboards, *Proceedings of the IEE Colloquium on Interfaces*, Dundee, UK, pp. 1-8.

Rani P. and Sarkar M. (2005) EMG-based high level humanrobot interaction system for people with disability, *Proceedings of the IEEE International Workshop on Robot and Human Interactive Communication*, Nashville, Tennessee, pp. 280-285

Artemiadis P. and Kyriakopoulos K. (2006), EMG-based teleoperation of a robot arm in planar catching movements using ARMAX model and trajectory monitoring techniques, *Proceedings of the IEEE International Conference on Robotics and Automation*, Orlando, Florida, pp. 3244-3249

Ferreira, A.; Celeste, W, C; Cheein, F. A.; Bastos-Filho, T. F.; Sarcinelli-Filho, M. and Carelli, R. (2008) Human-machine interfaces based on EMG and EEG applied to robotic systems, *Journal of NeuroEngineering and Rehabilitation* 03 pp. 5-10

Ohno, T. and Mukawa, N. (2003). Gaze-Based Interaction for Anyone, Anytime. *In Proceedings of HCI International 2003*, Crete, Greece, June 22-27, Vol. 4, pp. 1452-1456.

Alexander, M.A.; Nelson, M.R. and Shah, A. (1992) Orthotics, adapted seating and assistive devices. *In Pediatric Rehabilitation*, 2nd ed. Baltimore, MD: Williams and Wilkins, pp. 186–187.

H. Kwee; J. Duimel; J. Smit; A. T. de Moed; J. van Woerden and L. v. d. Kolk (1988) The Manus Wheelchair-Mounted Manipulator: Developments towards a Production

Model, *In Proceedings of the 3rd International Conference of the Association for the Advancement of Rehabilitation Technology,* pp. 460-462, (1988).

Hogan, N.; Krebs, H.I.; Charnnarong, J.; Srikrishna, P.,;and Sharon, A. (1992). MIT-MANUS: A workstation for manual therapy and training. *In Proceedings of the IEEE Workshop on Robot and Human Communication,* Tokyo, Japan, pp. 161–165.

Burke, J.; Exoskeleton (1992) Master Arm, Wrist and Effector Controller with Force Reflecting Telepresence, *Technical Report AL/CF-TR-1994-0146,* Odentics Inc., Anaheim, CA, December 1992, pp. 104.

Johnson, GR; Buckley, MA (1997) Development of a New Motorized Upper Limb Orthotic System (MULOS). *Proceedings of the Rehabilitation Engineering Society of North America.* Pittsburgh PA; June 1997 . p . 399-401.

Allen JR; Karchak A Jr.; Bontrager EL (1972) Final project report, design and fabricate a pair of Rancho anthropomorphic manipulator. *Technical Report, The Rancho Los Amigos Hospital Inc* ., 12826, Hawthorm Street, Downey CA 90242, 1972.

LeBlanc M; Leifer L . (1982) Environmental control and robotic manipulation aids . *Engineering in Medicine and Biology Magazine* 1982 ; December : 16-22.

Kawamoto H and Sankai Y (2002) Power assist system HAL-3 for gait disorder person. *In Computer Helping People with Special Needs: 8th International Conference,* ICCHP 2002. Lecture Notes in Computer Science. Edited by: Miesenberger K, Klaus J, Zagler W. Berlin: Springer Verlag; 2002:196-203.

Nef T and Riener R. (2005) ARMin—Design of a novel arm rehabilitation robot. *In: Proceedings of the 9th International Conference on Rehabilitation Robotics;* 2005 Jun 28–Jul 1; Chicago, Illinois. Madison (WI): Omnipress; 2005. p. 57–60.

Madentec Inc. (2008), assistive technologies. *http://www.madentec.com.* Accessed January 2008.

Vaidyanathan, B.; Chung, L.; Gupta, H. ;Kook, S.; Kota and J. D. West; (2007) Tongue-Movement Communication and Control Concept for Hands-Free Human-Machine Interfaces. *IEEE Transactions on Systems, Man, and Cybernetics - Part A: Systems and Humans,* 37(4):533-546

Betke M.; Gips J.; and Fleming P. (2002) The Camera Mouse: Visual tracking of body features to provide computer access for people with severe disabilities. *IEEE T Neur Sys Reh* 10(1):1-10

Development of a New Vehicle Based on Human Walking Movement with a Turning System

Misato Nihei, Takeshi Ando, Yuzo Kaneshige,
Takenobu Inoue and Masakatsu G. Fujie
The University of Tokyo, Waseda University,
Research Institute of National Rehabilitation Center for Persons with Disabilities
Japan

1. Introduction

In engineering research and technology, a considerable amount of attention has recently been focused on developing devices to assist the elderly. For elderly persons, walking is one way of improving the functions of the mind and body, and is therefore effective in increasing the quality of life (J. Jutai et al., 1999). Among the techniques that can be used to assist mobility among the elderly are improvement in physical ability by training or by using power-assisted devices (Y. Sankai et al., 2000, C. J. Walsh et al., 2006), reducing loads on the lower limbs by sustaining the weight of the individual (R. Riener et al., 2000, T. Tani et al., 1996), and expanding the range of environments that can be accessed by users of mobility devices, for example, by the development of wheelchairs that are capable of climbing stairs (R. Morales et al., 2006, E. S. Boy et al., 2002). Furthermore, many intelligent robots can anticipate the intensions of humans (R. C. Simpson et al., 2005, Q. Zeng et al., 2006), but it is difficult to use this ability in meeting the needs of the elderly in terms of facilitating movement and mobility.

Fig. 1. Tread-Walk 1

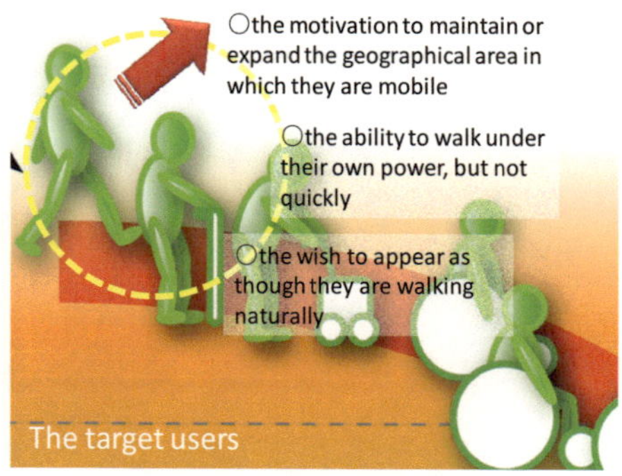

Fig. 2. The target user of Tread-Walk1 and Tread-Walk2

We have designed and developed two new mobility-aid systems for the elderly that improve their walking speeds. These devices are called Tread-Walk1 (TW-1), which is shown in Fig. 1, and Tread-Walk2 (TW-2). They are mobile treadmill-like systems that are designed to meet the following three requirements: the maintenance of bodily function, the extension of mobility (based on a trend survey of elderly persons (M. Nihei et al., 2008)), and the maintenance of the appearance a natural gait (M. Nihei et al., 2007). The targeted users, as shown in Fig. 2, have the following characteristics: the ability to walk under their own power, but not quickly; the motivation to maintain or expand the geographical area in which they are mobile; and the wish to appear as though they are walking naturally.
The control of the system's belt velocity is based on the natural walking velocity of the user. The user's intended walking speed is detected by the kicking and braking forces applied by the user's feet as he or she walks on the treadmill belt, which in turn is connected to a dc motor. These forces are counteracted by the propulsion force of the belt.
To control this system more accurately, signals for rotational movement should also be derived directly from the walking motion. In this chapter, we focus on human rotational movement, and we describe a new system that uses two belts to derive separate signals from each foot during walking (M. Nihei, 2008).
There are types of two-belt treadmill system. One has force plates built into it (A. Belli et al., 2001) and is mainly used for the analysis of walking movement, because it can measure the reaction force from the floor during walking. The other has an impedance control system using an ac servomotor (T. Tani et al., 1995); in this system, the burden on the lower limbs during walking is controlled by the rotational velocity of the belt. The applications of this kind of system in rehabilitation and training to prevent falling have already been the subject of research (K. Hase et al., 2002).
In this chapter, we explain our developments and we describe our evaluation of a new system called Tread-Walk2 (TW-2), which effects changes in direction by using as an input signal the differences between the patterns of kicking and braking forces applied to the treadmill belt by the soles of the right foot and the left foot. To identify the advantages and disadvantages of the TW-2, we compared its functions with those of the TW-1.

2. Development Concept

2.1 The Functions of Tread-Walk1

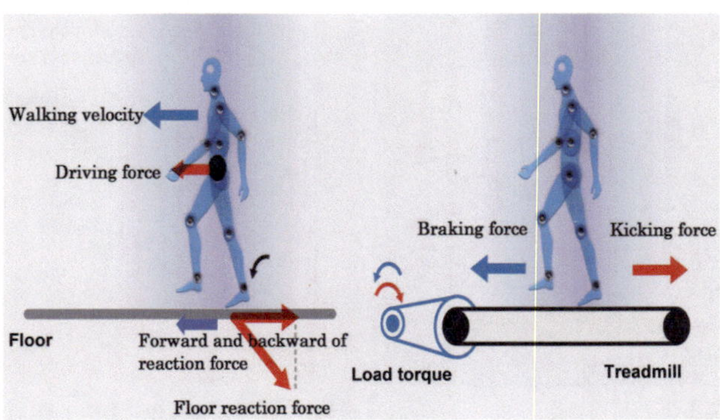

(a) Natural walking (b) Walking on a treadmill system

Fig. 3. The system for detecting the forward and backward components of the reaction force of the ground

TW-2 is basically a functional enhancement of TW-1. TW-1, which is illustrated in Fig. 3, amplifies the velocity of the treadmill, based on the input signal, to drive a wheel that, in turn, moves the device along the ground (M. Nihei, 2008). In this way, the slow gait of the user is transformed into a more-rapid movement of the device in the same direction. In TW-1, the sum of the forward and backward components of the reaction force of the ground applied to both feet during the walking movement is used as an input signal to accelerate and decelerate the system, as shown in Fig. 3 (Y. Kaneshige et al., 2006). In addition, the angle of rotation of the handle is used a second input signal to allow the system to be steered to the right or left. Therefore, the rotational movement that controls the steering of TW-1 is performed by a movement of the user's upper limbs. Other functions of TW-1 are discussed below.

1) The Basic Concept

The prototype of the TW-1 system is shown in Fig. 4. The prototype consists of a treadmill on which the user walks, a drive wheel, and a set of handlebars. Fig. 5 shows a diagram of the system, along with the walking and driving components. The control of the speed of the treadmill is activated by a dc motor that is attached to the treadmill belt. The driving system operates according to the following protocol. When the user starts to walk on the treadmill, the belt is rotated by the kicking force of the user. The resulting rotational torque is transferred to the shaft of the motor, where it is detected as a load current. A microprocessor control unit calculates the load current from the kicking/braking force and determines the walking speed; the treadmill then moves at the same speed so that the user stays in a position that is stationary relative to the treadmill base. At the same time, the walking speed is amplified by the driving wheels, so that the vehicle moves forward at a greater speed than that at which the user walks. The relationship between the walking speed and the driving speed is determined by the optional gain. In addition, the steering system controls the relative speeds of the right and left wheels.

Walking movement on TW-1 was analyzed kinematically by comparing it with the corresponding walking movement on flat ground (T. Ando et al., 2008). The angles of the joints of the lower limbs were found to be almost the same in these two walking movements; however, as shown in Fig. 6, the foot kicked backward more strongly, and the trajectory of the foot after toe-off became higher and larger in the direction opposite to the travel direction because of the belt-velocity control algorithm of TW-1. In addition, the clearance before heel contact was higher when walking on the TW-1.

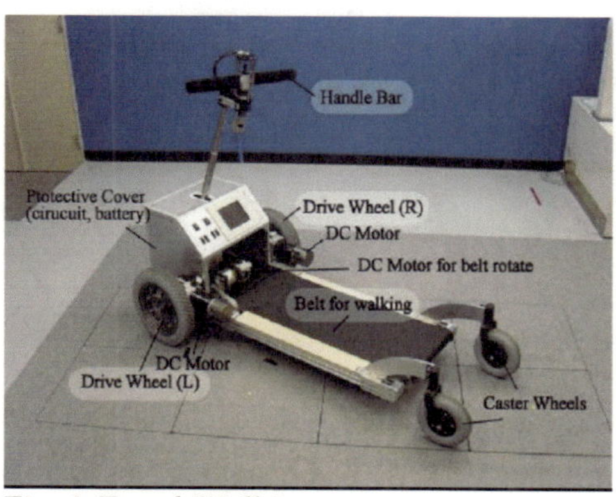

Fig. 4. Tread-Walk1

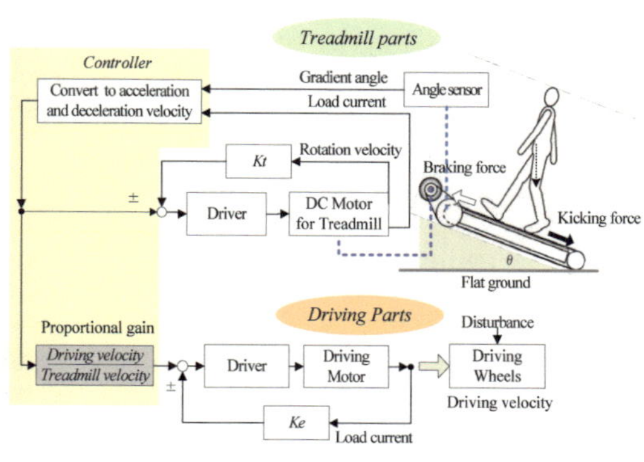

Fig. 5. A diagram of TW-1

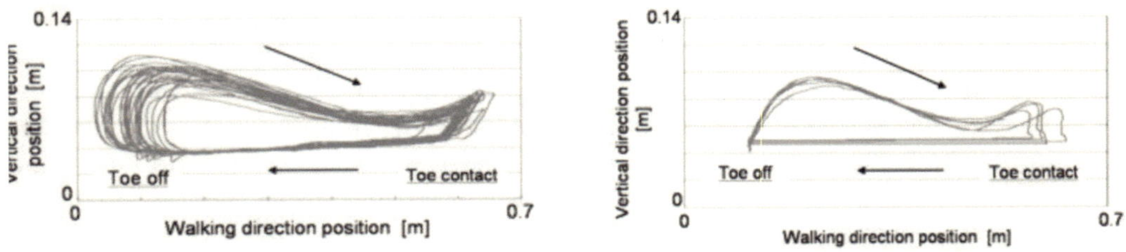

(a) Tread-Walk1 (b) Flat walking

Fig. 6. Toe trajectory in different walking movements. The driving wheels of TW-1 were fixed, each walking movement was conducted at a velocity of 2.0 km/h and a cadence 58 or 96 steps/min. Measuring system: 3D motion-capture system [Vicon 612 (Vicon, Los Angeles, CA, USA; 100 Hz; accuracy 1 mm].

1) Steering system

We proposed a method for steering control that enables the user to turn the device safely and in a stable manner. The system is designed to allow the user to maintain his or her balance in a standing position during a turning operation; to achieve this, the rotational velocity at which a comfortable balance can be maintained in a standing position when the user is subjected to a centrifugal force was established, as shown in Fig. 7(T. Ando et al., 2008).

2) Slope

We developed a new system that facilitates walking up and down slopes, as well as walking on flat ground. Fig. 8 shows a trial of walking down a slope. To realize this, we detected the forward and backward components of the floor reaction force using a newly developed system for measuring the torque load on the motor. In addition, we developed a model that can express the relationship between the torque load during walking and the gradient angle of the slope. By considering the gradient angle as a feedback element, we demonstrated that it was possible for the user to walk straight on slopes as well as on flat ground. The results of a number of driving experiments using the prototype machine on slopes showed that a user can walk smoothly up or down a slope of up to ±5.5° in an interior environment and one of ±7° in an exterior environment without stopping (M. Nihei, et al., 2006).

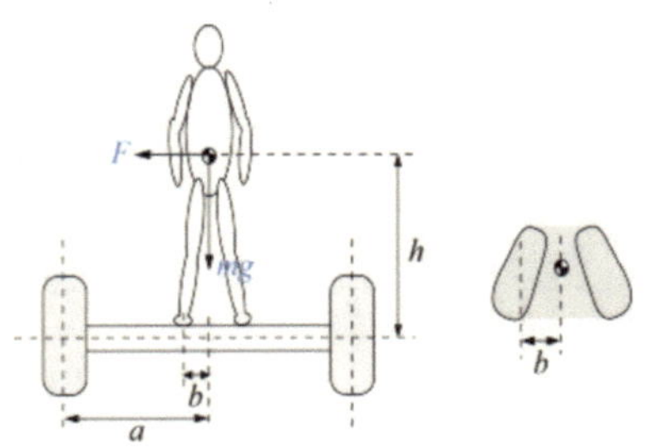

Frontal plane Support basal area

Fig. 7. The relationship between the centrifugal force and the center of gravity: a is the distance from the wheel to the center of the vehicle, b is the distance between the center of the sole of either foot and the projection of the center of gravity onto the platform, h the height of the centre of gravity of the user above the platform, and F is the centrifugal force.

Fig. 8. Photograph showing the device being driven down a slope of 7°

However, to control this system more accurately, signals for rotational movement should be derived directly from the walking motion. Below, we focus on human rotational movement and describe a new system that uses two belts to derive a separate signal for each foot during walking.

2.2 Directional change by rotational movements

In the human gait, the swing phase occurs when one leg is brought forward, and the stance phase occurs when an external force is applied to the ground alternately by each foot. The rotational movement needed to adjust walking speed and to change direction is made by adjusting the force and moment between the foot and the ground.
Rotational movements can divided into the spin turn, in which the movement that is required to change direction is made without moving forward, and the pivot turn, in which the movement required to change direction is made while the individual is moving forward. Such rotational movements are realized by adjusting the kicking and braking forces of the system.
As shown in Figs. 9 and 10, the rotational center of the human body is present on the front-horizontal axis. In the spin turn (Fig. 9), there is a rotation about the front-sagittal axis, corresponding to the center of the trunk, and about the spinning foot. In the case shown, the left leg is on the rotational axis, and the turn is realized when the left ankle and hip joint are adducted or the left leg kicks the ground strongly. On the pivot turn (Fig 10), on the other hand, the rotational center is outside the body of the user and changes because of the rotational angle. This turn is performed when the kicking forces of the right foot and the left foot are different.

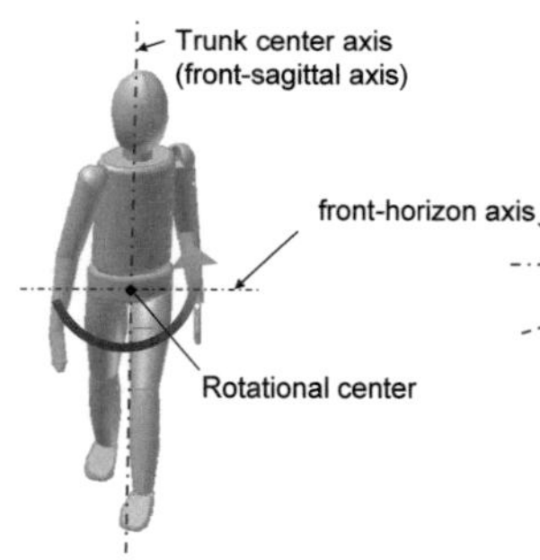

Fig. 9. Spin turn

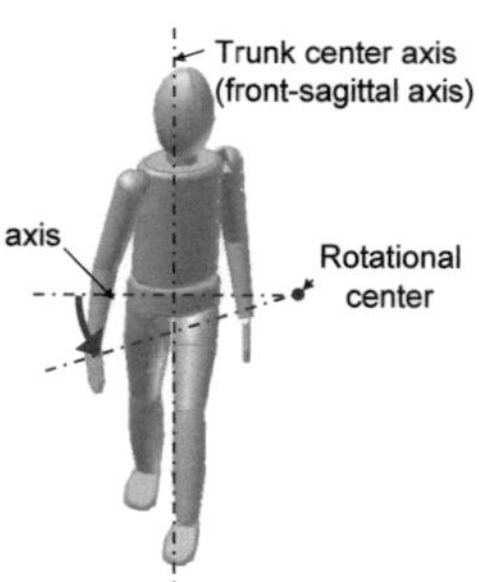

Fig. 10. Pivot turn

2.2 Design and Required Specifications

The spin turn and pivot turn of TW-2 are realized by using the movements involved in changes in direction as explained above. The system detects the kicking force of each foot of the user, and a mechanism turns the vehicle in response to the detected spin turn or pivot turn movements made by the user.

1) The system detects the kicking force of the user's feet

As with TW-1, TW-2 detects the kicking force of both feet by detecting the forward and backward component of the ground reaction force applied to each foot by the load current of the dc motors connected to the walking belts. The treadmill of the TW-2, however, consists of two walking belts to detect the kicking force of each foot independently.

2) Vehicle turns based on the user's movements

To conduct the turn movement with a similar sensationnatural walking, the center of rotation of the TW-2 should be close to that of the human user. Therefore, as shown in Fig. 9, the two driving wheels are arranged under the center line of the twin-belt treadmill. In addition, two passive casters are positioned to the front and two more are positioned to the rear to improve stability. So that the TW-2 can be used on sidewalks, its size must conform to the JIS (Japan Industrial Standardization) standards for motored wheelchairs (JIS T9203, 2003), that is, a maximum length of 1200 mm, a maximum width of 700 mm, a maximum height of 1090 mm, a maximum speed of 6.0 km/hr, and sufficient rotational performance that the TW-2 can make a right-angle turn in a 1.2-m-wide intersection. Moreover, the weight must be less than 100 kg, the same as that of a standard electric cart.

3. Development of the TW-2

3.1 Mechanism

1) The overall design

TW-2 is composed of three subsystems: *the walking (treadmill) subsystem*, which extracts the walking pattern, controls the velocity of the treadmill belt, and extracts the rotational movement; *the driving subsystem,* which moves the vehicle along the ground; and *the operation-lever subsystem,* which supports the trunk of the user and recognizes the kind of rotational movement that is being made. The prototype is shown in Fig. 11, and Table 1 list the specifications for the TW-2.

2) The walking (treadmill) subsystem

As shown in Fig. 12, the walking subsystem consists mainly of the walking belts and a walking board; the latter supports the user's weight. The size of the walking board is 460 × 720 mm, which takes into account the typical walking stride length of older people. In addition, a tape is used to decrease the friction on the upper surface of the walking board.

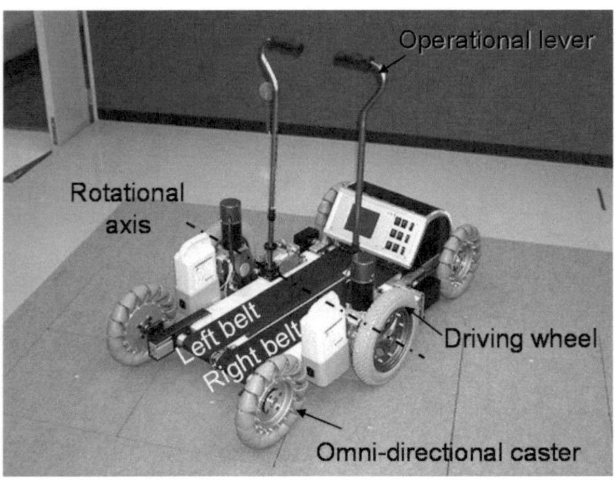

Fig. 11. Tread-Walk2

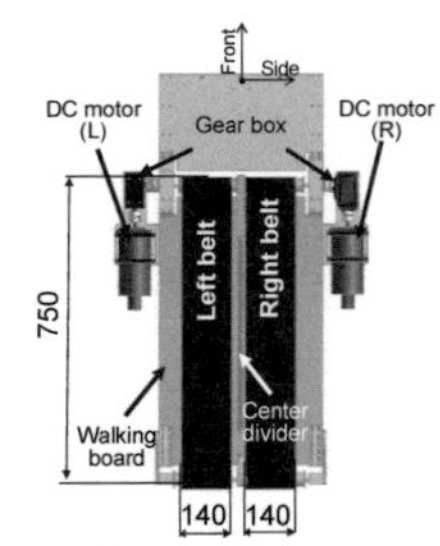

Fig. 12. Walking subsystem

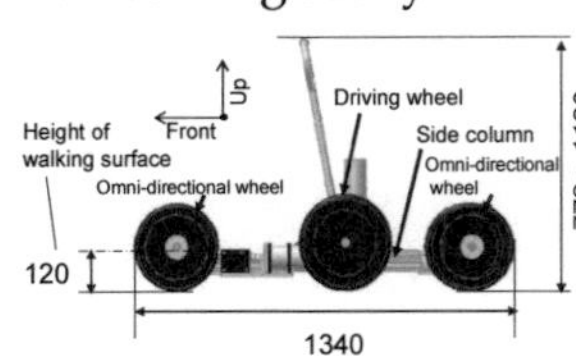

Fig. 13. Driving subsystem

The width of each belt is 140 mm, which is wider than the average foot width of 100 mm (NEDO database). The height of the walking subsystem (140 mm) is as low as possible to minimize the chances of falling off and to reduce the burden of mounting and getting down from the TW-2.

		TW-2	TW-1
Size (mm)		1340 × 840 × 1120	1520 × 760 × 1120
Minimum rotational radius (mm)	Spin turn	840	
	Pivot turn	1120	
Weight (kg) (including battery)		90	75
Max. vehicle velocity (km/h)		6	
Max. walking belt velocity (km/h)		3	
Continuous use time (h)		~3	~6
System motors		2 wheel-driving motors 2 treadmill-driving motors	2 wheel-driving motors 1 treadmill-driving motor

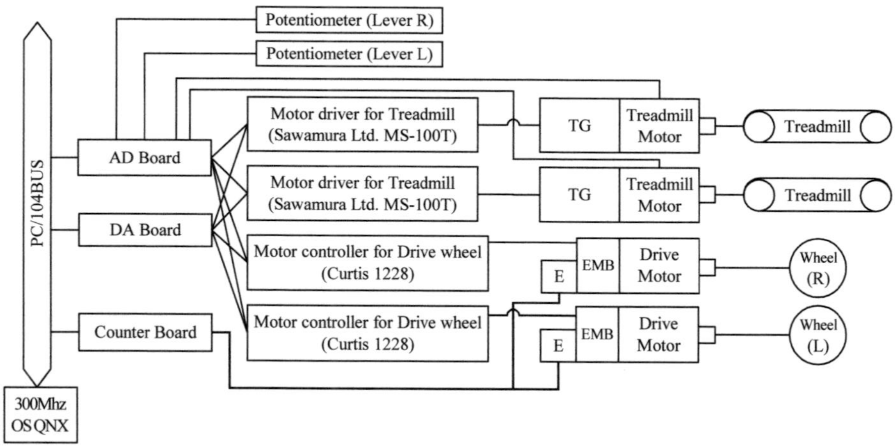

Fig. 14. System structure

The forces applied to each walking belt are transferred to a dc motor that connects with the corresponding belt through a gearbox (gear ratio: 5:1). The dc motor, motor drive, and tacho-generator were selected for use in the walking subsystem on the assumptions that the maximum load would be 80 kg and the maximum velocity is 4.0 km/h.

3) Driving subsystem

As shown in Fig. 13, the size of TW-2 is 1340 × 840 mm, which takes into account the typical width of a sidewalk in Japan. The driving subsystem includes six wheels: two central or driving wheels, and four front and rear wheels that are omnidirectional to improve the performance of translatory and rotational movements. There are two dc-geared driving motors that are controlled by a velocity control driver (Curtis Inc.). The right and left driving wheels are aligned with the center of the treadmill belt to match the centre of rotation of the user with that of the TW-2. The front wheels are set 630 mm ahead of the driving wheels, and the rear wheels are placed 480 mm behind the driving wheels.

4) Operational lever subsystem

The operational lever subsystem consists of a supporting column, a gas spring, and a potentiometer. The height of the lever is adjustable from 770 to 1120 mm to accommodate users of various heights without affecting their walking movements. In addition, the lever is placed about 100 mm in front of the center of the walking belt to maximize the likelihood that the user will walk on the center of the belt.

3.2 System structure

Figure 14 shows the structure of the TW-2 electronics system. Power is supplied from a 24-V Ni-H battery that actuates two dc motors that rotate the two treadmill belts, and another two dc motors that turn the driving wheels. The control microprocessor is powered by a 5-V dc source provided from the power supply via a dc–dc converter. Encoders are used to measure the number of rotations made by the two driving wheels. The continuous operating time of the TW-2 is about two hours.

4. Rotational Movements of Humans and the System

4.1 Rotational Method

The parameters for determining the direction of travel from rotational movements made by the user are the point of application of the floor reaction force, the magnitude of the force, and its moment. In TW-2, however, the only detected signal that is used to conduct rotational movement is the forward and backward component of the ground reaction force. This can cause problems. First, it is possible that the translatory performance will be decreased because differences between the walking patterns of the right and left feet may occur at the start of the walking movement and/or when the walking velocity is low. Secondly, it is difficult to recognize the pattern when a rotational movement is conducted at the first step of a walking motion.

Therefore, in the TW-2, there is a component in the lever subsystem that solves this problem by allowing the user to go straight or to turn.

4.2 Sensing the Walking Pattern of Each Leg

An examination was conducted to measure the pattern of the forward and backward components of the ground reaction force. In this examination, only the treadmill subsystem was actuated, and the walking pattern was detected by measuring the load current values of the right and left belt-driving motors during the walking movement. In particular, it was important to detect the value of the current when the treadmill motor was controlled at a constant velocity. The relation between the load current value of a motor and the forward and backward reaction force applied to its belt is shown in Equation 1.

$$f = A \times K_t \times i / r \qquad (1)$$

where f is the forward or backward component of the ground reaction force, A is the measured current value, K_t is the torque constant, i is the reduction ratio, and r is the shaft radius.

The kicking and braking force of each leg during walking movement on the treadmill belt is transferred to a dc motor connected to the belt for the appropriate leg, and the load torque that is generated by the kicking and braking force is detected as a load current value.
Fig. 15 shows the load current values for a period of 20 seconds of constant walking at a velocity of 2 km/h. As shown in the Fig. 15, the right foot and the left foot currents alternate at constant frequencies. It should be added that the frequency of the vibrations corresponds to a walking cadence of 58 steps/min, which we set as a typical value of the walking cadence for an elderly person.

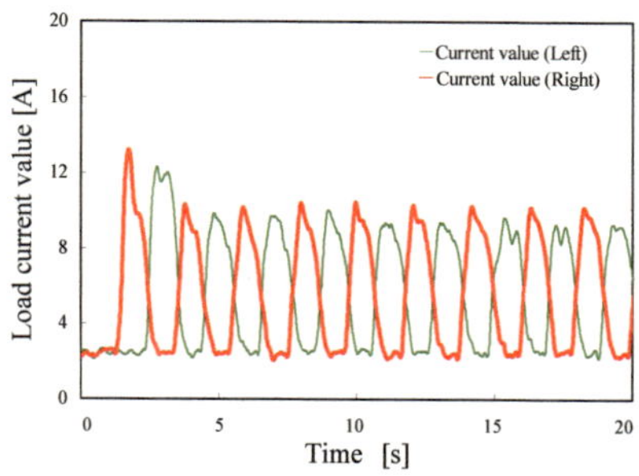

Fig. 15. Load current values for the left and right feet during walking movements

Therefore, this figure confirms that the right and left walking patterns can be independently estimated from the frequency and the peak value. Moreover, the figure also confirms that the braking force applied to the walking belt on contact of the heel raises the load current value, because the resulting braking force is in the direction of deceleration of the rotational velocity of the motor. On the other hand, the kicking force applied as the toe lifts off decreases the current value, because the kicking force accelerates the rotational velocity of the motor. Thus, it is certain that the peaks shown in Fig. 15 correspond to braking forces, whereas the valleys correspond to kicking forces.

4.3 Operating Movement

Movements of the TW-2 system can be divided into a spin-turn mode, a pivot-turn mode, and a straight mode. Table 2 shows the relation between each of these modes and the lever manipulation and walking movements.

Operation of lever switches P1 & P2	Walking movement	TW-2 movement
ON and ON		 Straight

<table>
<tr><td>ON and OFF
or
OFF and ON</td><td>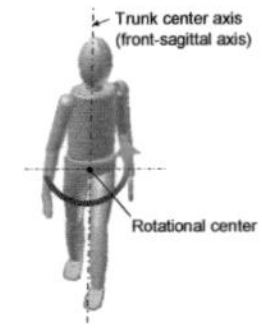</td><td>
Pivot turn</td></tr>
<tr><td>OFF and OFF</td><td>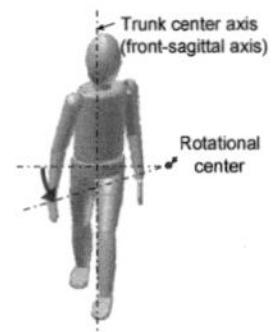</td><td>
Spin turn</td></tr>
</table>

Table 2. TW-2 Operations

1) Straight Movement

In straight movement, the belt velocity is controlled in the same manner as it is controlled during the walking movement, even if the forward and backward components of ground reaction force of the right and left legs are different. The increasing and decreasing values of the right and left treadmill velocities, $Tr_{Straight}$ and $Tl_{Straight}$, are calculated by multiplying a constant by the sum of the integral value of the forward and backward components of the ground reaction force of the right and left legs, $r(t)$ and $l(t)$, from a time t_0 to the current time t.

$$Tr_{Straight} = Tl_{Straight} = \alpha\left\{\int_{t_0}^{t} r(t)dt + \int_{t_0}^{t} l(t)dt\right\} \tag{2}$$

Acceleration or deceleration is performed when this value reaches a threshold value for acceleration or deceleration. The value of t_0 is reset every time the vehicle accelerates or decelerates.

2) Turning movement

The forward and backward components of the ground reaction forces of the right and left legs are used independently as input signals for altering the velocity of the control belt. Therefore, the values for recognizing the increasing and decreasing velocities of the right and left belts Tr_{Turn}, Tl_{Turn} are defined as follows:

$$Tr_{Turn} = \beta\int_{t_{r0}}^{t} r(t)dt, Tl_{Turn} = \beta\int_{t_{l0}}^{t} l(t)dt \tag{3}$$

where β is a constant.

4.4 Rotational Movements and Rotational Angles

1) Spin-turn mode

(a) Spin turn

The relation between each kicking force and the rotational angle is as follows: TW-2 turns to the right when the left foot applies the kicking force and turns to the left when the right foot applies the kicking force. On the basis of the values of kicking forces for the right and left feet, the turn direction is decided by means of threshold recognition. For example, when the kicking force is applied by the right foot, a braking force is simultaneously generated by the left foot. When the generated kicking and braking forces are above arbitrary threshold values, the rotational angle is calculated by using an angular acceleration function. On the basis of the relation between the impulse of the rotational movement and the momentum during the walking motion, the angular acceleration function is selected to make the relation between the kicking force during rotational movement and the rotational angular acceleration linear. The angular acceleration function is derived as follows:

$$\ddot{\theta} = k_s \cdot l / I \cdot f(t) \qquad f(t) = \begin{cases} f_R \\ -f_L \end{cases} \tag{4}$$

where $\ddot{\theta}$ is the angular acceleration of the person and the TW-2, k_s is a correction coefficient, l is the distance between the right and left feet of the user, I is hypothetical inertia moment, and f is the kicking force of the right or left foot. In this case, the values of l, I, k_s are constant and K_s (the correction coefficient between the angular acceleration and kicking force) is defined as follows.

$$Ks = ks \cdot l / I \tag{5}$$

Therefore, the following equations are derived.

$$\ddot{\theta} = K_s \cdot f(t) \tag{6}$$

$$\theta = \iint K_s \cdot f(t) \cdot dt \cdot dt \tag{7}$$

That is, the angular acceleration is generated from the impulse produced by the kicking force. Hence, it is possible to adjust the rotational angle depending on the kicking force in a way that is similar to adjusting the rotational movement of a rigid body. In other words, it is easy for the user to control the rotational operation of the TW-2.

Furthermore, when K_s, is large, a small kicking force can result in a large angle of rotation, whereas when K_s is small, a large kicking force can result in a very small angle of rotation. Additionally, K_s is adjustable for each user, because K_s is set to an arbitrary value that is matched to the user's physical condition. In TW-2, the corrective coefficient K_s is set by trial and error until the user feels it is easy to operate the rotational movement of the system.

In the spin-turn mode, the thresholds for the kicking and braking force of each leg are set as f_{thk} and f_{thb}, to recognize the spin turn and pivot turn, and the spin turn is conducted when the forces simultaneously exceed f_{thk} and f_{thb}. These thresholds can be adjusted by trial and error to recognize even small kicking and braking forces, so the user feels it is easy to control the rotational movement of the vehicle. Additionally, the maximum angular velocity $\dot{\theta}$ is set at a comfortable level for the user.

(b) Spin turn operation in the TW-2

As shown in Fig. 16, the angular velocity of TW-2 is determined by the rotational velocities of the right and left driving motors and the distances between the right and left driving wheels.

$$\dot{\theta} = (v_R - v_L) / 2d \tag{8}$$

where $v_R = -v_L$, $\dot{\theta}$ is the angular velocity of TW-2, v_R is the rotational velocity of the right motor, v_L is the rotational velocity of the left motor, and $2d$ is the distance between the right and left driving wheels.

Fig. 17 shows the measured rotational angles and kicking forces during a spin turn (f_{thk} = 0.42 A, f_{thb} = 3.0 A). It was found that TW-2 rotated 92.9° to the left when the kicking force of the left foot was large (less than 0.42 A). Similarly, it was confirmed that right and left rotations of the TW-2 are controlled by the right and left kicking and braking forces.

Fig. 18 shows the trajectories of the right and left driving wheels and the center point between them, which is equal to the rotational center of the human body under the same experimental condition, as shown in Fig. 10. The right and left driving wheels are rotated around the center point of the TW-2. When TW-2 turns 92.9° in the spin-turn mode, the displacements of the midpoints of the right and left driving wheels are about 36 mm along the x-axis and 38 mm along the y-axis from the original points, that is, about 4.3% of the overall width and 2.8% of the overall length. Thus, it was confirmed that the amount of vibration of the rotational axis is small.

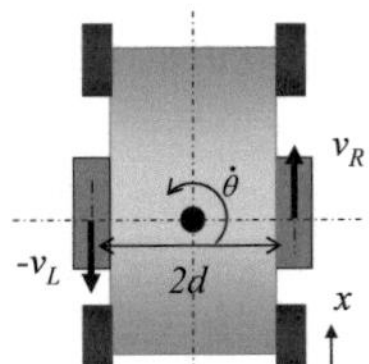

Fig. 16. Spin turn model of TW-2

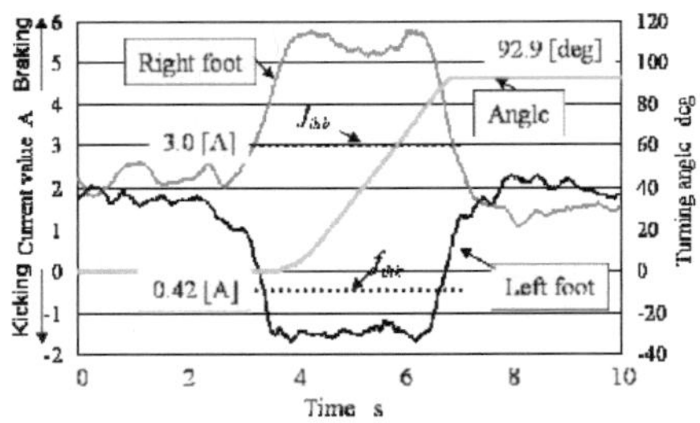

Fig. 17 Spin turn and walking pattern

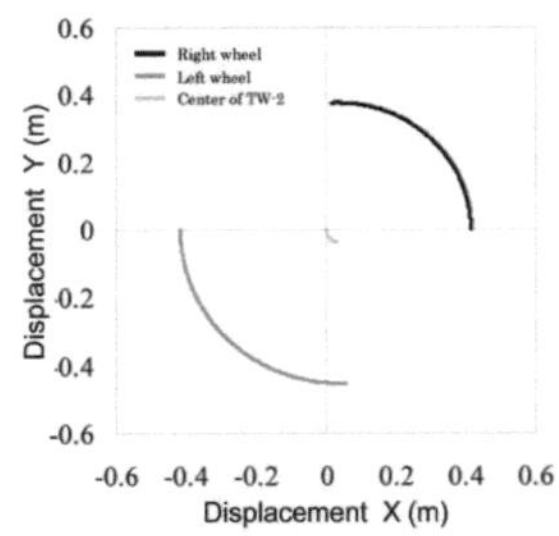

Fig. 18. Vehicle turn (spin turn mode)

2) Pivot-turn mode

(a) Rotational movement in the pivot turn

As in the case of the spin turn, the relation between the kicking force and rotational angular velocity is expressed as:

$$\ddot{\theta} = k_p \cdot l / I \cdot f(t) \qquad f(t) = \begin{cases} f_R \\ -f_L \end{cases} \tag{9}$$

where $\ddot{\theta}$ is the angular acceleration of the person, and TW-2, k_p is a corrective coefficient, l is the distance between the right and left feet or belts, I is the hypothetical inertia moment of a human body, f_R is the kicking force of the right foot, f_L is the kicking force of the left foot. In this case, l, I, and k_p are constant, and K_p (the correction coefficient for the angular acceleration and the kicking force) is defined as:

$$K_p = kp \cdot l / I \tag{10}$$

Therefore, the following values are derived:

$$\ddot{\theta} = K_p \cdot f(t) \tag{11}$$

$$\theta = \iint K_p \cdot f(t) \cdot dt \cdot dt \tag{12}$$

As is the case of the spin turn, an angular acceleration is generated from the impulse of the kicking force. It is therefore possible to adjust the rotational angle depending on the kicking force. In other words, pivot-turn operation of TW-2 is also easy.

Furthermore, the larger the value of K_p, which is the corrective coefficient between the angular acceleration and the kicking force, the larger will be the angle of rotation, even if the kicking force is small. On the other hand, when K_p is small, it becomes possible to control a minute rotational angle by a large kicking force. Therefore, because K_p is set to an arbitrary value matched to the user's physical condition, control of rotational movement is easily adjustable. In TW-2, the corrective coefficient K_p is set by trial and error until the user feels that it is easy to control the rotational movement of the system.

In the pivot-turn mode, the threshold of the kicking force is set to f_{th}, so that the pivot turn is executed only when the arbitrary kicking force exceeds f_{tk}. This threshold is adjustable by trial and error, so that rotational movement can be executed even by a small kicking force that can be easily applied by the user. Additionally, the maximum angular velocity $\dot{\theta}$ is set at a level that is comfortable for the user.

In our system, the rotational velocities of the right and left motors are adjusted to correspond to the rotational centers of the wheels.

(b) Pivot-turn operation

As shown in Fig. 19, the angular velocity of TW-2 (v is the translational velocity of TW-2) is determined by the rotational velocities of the right and left driving motors and the distances between the right and left driving wheels.

$$\dot{\theta} = (v_R - v_L)/2d \tag{13}$$

$$\rho = 2d \cdot (v_R + v_L)/(v_R - v_L) \tag{14}$$

where $\dot{\theta}$ is angular velocity of TW-2, v_R is the rotational velocity of the right motor, v_L is the rotational velocity of the left motor, $2d$ is the distance between the right and left driving wheels, and ρ is the distance from the rotational center to the center of TW-2.

Fig. 20 shows the rotational angles and load current values corresponding to the right and left kicking forces when the TW-2 makes a right-angle turn at an arbitrary angular velocity. The load current values for the right and left motors are alternately detected during straight movement. As shown in the figure, the kicking force of the left foot increases at around 3 seconds. The threshold of the kicking force was set to be 0.2 A, after user trial and error indicated that it was easy to operate TW-2 with this force. Therefore, it was confirmed that TW-2 starts to turn by 94.6° when the kicking force falls below a threshold of 0.2 A. (Fig. 13 includes data for the right foot as a reference.)

Fig. 21 shows the trajectories of the right and left wheels of TW-2 and the midpoints of the right and left wheels. It was found that TW-2 is rotated by making the left driving wheel the center of rotation.

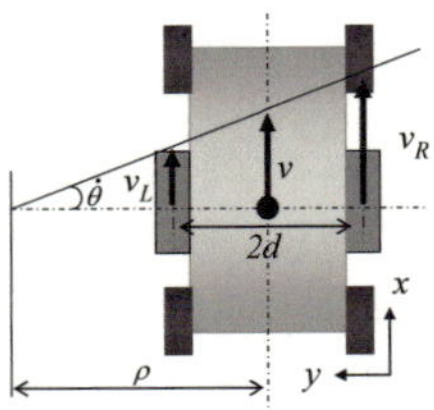

Fig. 19. Pivot turn model

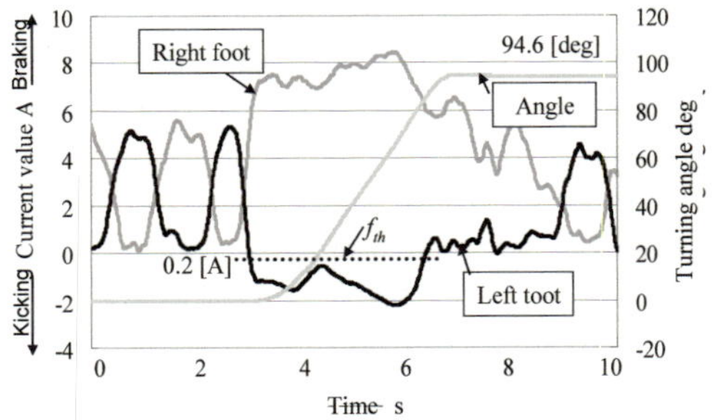

Fig. 20. Walking pattern and pivot-turn mode

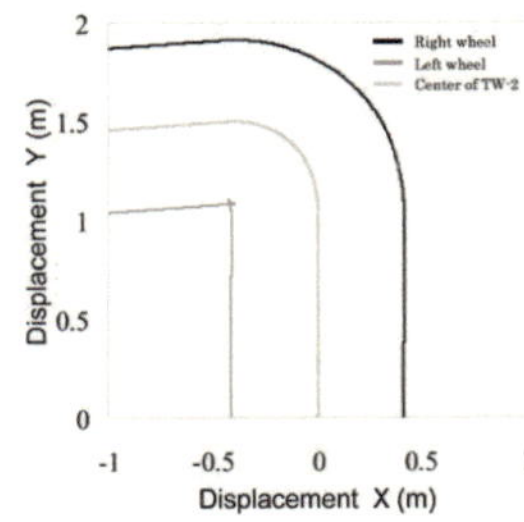

Fig. 21. Trajectory of TW-2 in pivot turn

5. Driving Evaluation for a Right-Angle Intersection

5.1 Method and Conditions

In this experiment, which is based on the JIS T9023 Rotational Performance Test, the subjects each drove the TW-2 into a 1.2-m-wide virtual right-angle intersection (defined by plastic tape). The subjects were 10 young people and 5 elderly people, all of whom were healthy. The subjects practiced controlling the TW-2 in advance. After that, they repeatedly drove the TW-2 until they were able to control it without contacting the intersection plastic tape. The test drives were conducted in spin-turn and pivot-turn modes.

As shown in Fig. 22, markers were placed at the outside front tip of the right front wheel [A: front wheel (R)], at the outside center of the left driving wheel [B: middle wheel (L)], at the outside rear tip of the right rear wheel [C: rear wheel (R)], and at the midpoint of the right and left driving wheels (O: TW-2 center). The trajectory of TW-2 while being driven through the right-angle intersection was measured by using a three-dimensional position sensor (Vicon 612).

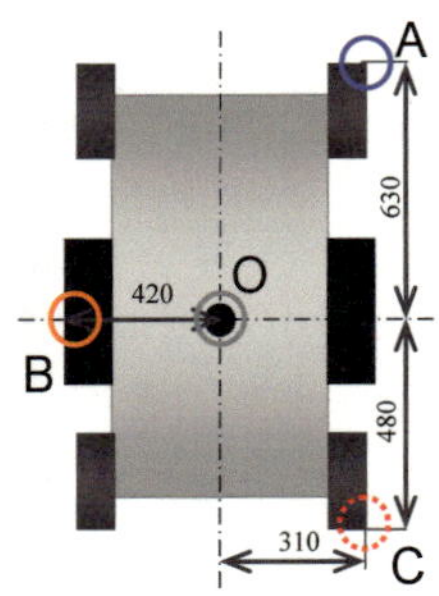

Fig. 22. Marker positions

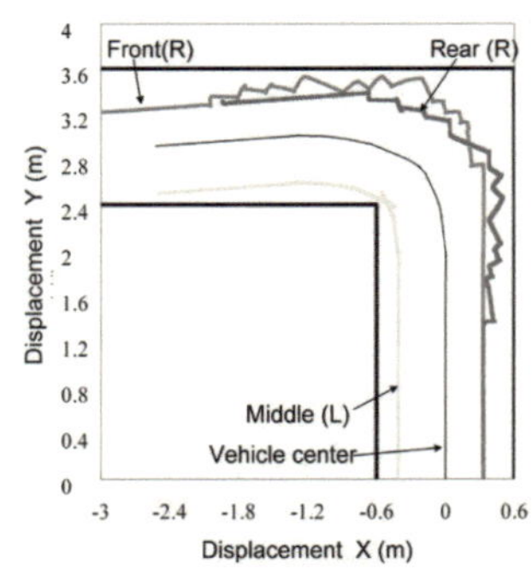

Fig. 23 The trajectory of spin turn

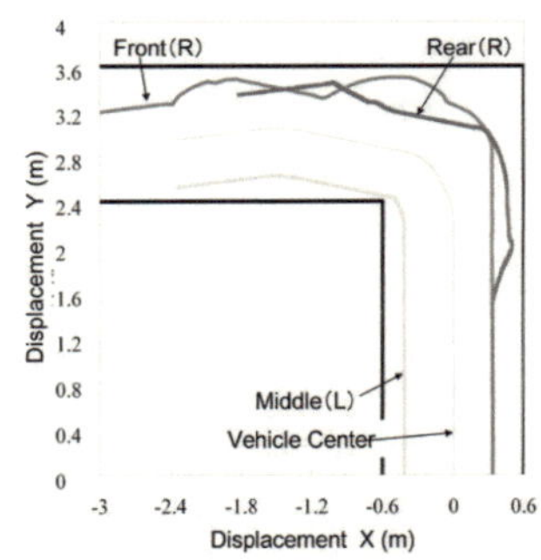

Fig. 24 The trajectory of pivot turn

5.2 Experimental Results

Table 3 shows the numbers of trial times needed before the subjects could successfully control the TW-2 at the right-angle intersection without contacting the virtual edges of the intersection. In the spin-turn mode, all the subjects succeeded within three trials. On the other hand, in the pivot-turn mode, all the young subjects succeeded within three attempts, whereas it took one elderly subject five attempts to achieve successful control.

Some examples of the results of the experiments are shown in Fig. 23 and Fig. 24. The left front and rear wheels rotated in a ragged fashion around the intersection in the spin turn mode, as shown in Fig. 23. On the other hand, it was confirmed that the right front and rear wheels rotated smoothly around the intersection in the pivot-turn mode, as shown is Fig. 24. In both modes, it was found that TW-2 did not contact the edges of the virtual intersection.

Attempt Number		1	2	3	4	5
Spin turn	young person	7	1	2		
	elderly person	2	2	1		
Pivot turn	young person	4	2	4		
	elderly person	0	2	1	1	1

Table 3. Experimental Results Showing Numbers of Successful Navigations at Each Attempt

5.3 Discussion

It was found that both young and elderly people were able to steer the TW-2 successfully using the rotational operation method developed in this research. This experimental result shows that the vehicle motion in the pivot turn is smoother than that in the spin turn. The spin-turn trajectory shows that when the vehicle turns a narrow corner, the user has to operate two patterns of motion: straight driving and a spin turn. On the other hand, it was also found that more trials were needed to control the TW-2 at the right-angle intersection in the pivot turn mode than in the spin turn mode. Therefore, the spin turn mode is more suitable than the pivot turn mode for narrow sidewalks or roads where subtle operation is needed. Moreover, it was found that it is more difficult for elderly people than for young people to manipulate TW-2 in the pivot turn. This is because the pivot turn is more difficult to execute as it requires both subtle movements and complex lever operations (as shown in Table 2). In addition, it is assumed that the motion of the pivot turn is easier to understand because the center of rotation is located near the rotational center of the human body.

To sum up, we found that the rotational movement of TW-2 can be realized by using the forward and backward components of the ground reaction force during the rotational walking movement. In addition, we found that the system requires a new operating method for easier operation than is provided by the current system.

6. Functional Comparison of the TW-1 and the TW-2

We developed a treadmill system with one or two belts (one for each foot) that are used to detect the forward and backward component of the ground reaction force, which is the kicking force during walking movement. Additionally, we developed a new vehicle, called TW-2, that could conduct rotational movements based on the pattern of the kicking forces applied by the right and left feet of the user as he or she walks on a two-belt treadmill. As a result, the operations for both straight movement and rotational movement are realized directly by using the human gait.

Because the driving movement can easily be matched to the user's physical condition, and an arbitrary driving velocity can be set in the system, it is expected that elderly people could use TW-2 to maintain their walking function while improving their overall mobility.

	TW-1	TW-2
Overall	Weight: lighter Driving time (lead battery); long but heavy Center of rotation: similar to that of a bicycle	Weight: heavier Driving time (Ni-H); light but short Center of rotation: similar to a human turning
Steering subsystem	Handle: user-friendly; similar to that of a bicycle	Switching lever: Safe, but not user-friendly
Walking subsystem	One motor: fewer parts	Two belts: permits measurement of the force for each leg and enables the walking phase to be analyzed.
Driving subsystem	Front drive: simple	Center drive and omni-wheel: similar to human turning, but requires many parts
Target person and purpose	Young to healthy elderly with a functional ability to walk; prevention of need for care; amusement and sport	Young to healthy elderly with functional ability to walk; prevention of need for care; amusement and sports. Possibly adaptable for paraplegic patients (if safety measure are adopted)

Table 4. Comparison of Characteristics

It became clear from the functional evaluation of the two prototypes that such systems show merits and demerits. Table 4 shows a comparison of characteristics for various function performed by the TW-1 and TW-2.

Three functions of TW-1 were modified in TW-2. The steering subsystem has a two-switch lever for selection of a spin turn or a pivot turn; however, this switching system is a little complex and requires some skill on the behalf of the operator. The walking subsystem has two belts and two motors; this system realized the requirement for turning the vehicle by measuring the force differential between each leg during walking. However, because this system requires one more motor than is used in the TW-1, it adds weight to the system and requires additional power. The driving subsystem adopted a center drive and an omni-wheel system; this system realized the requirement to be centered on the center of the human operator. However, the design requires two more casters than is the case for the TW-1. Overall, the TW-2 was adequate for its required function, but some problems, such as weight and usability, emerged. Therefore, to improve the TW-2 system, we suggest that the handle of the steering system requires adaptation to be more user friendly, and the structure requires modifications to achieve weight savings. For the walking subsystem, we suggest that there needs to be improvements in the system for evaluating the walking ability of users in clinical use.

7. Conclusion

In this chapter, we describe that the development of Tread-Walk2 (TW-2), which effects changes in direction by using the difference between the patterns of kicking and braking

forces applied to the treadmill belt by the soles of the right and left feet of the user as an input signal. TW-2 can conduct rotational movements that are based on the pattern of the right and left kicking forces applied by the feet of the user as he or she walks on the two-belt treadmill. As a result, the operations for both straight movement and rotational movement are realized by using the human walking movement. Furthermore, on the basis of a comparison of the functions of the three subsystems of TW-1 and TW-2, the advantages and disadvantages of each were clarified. From the result of this comparison, TW-2 was found to be adequate for its required function; however, we recognize that some problems remain, and it is necessary to make improvements in the weight of the vehicle and in the usability of the steering system.
In future, more work is needed to improve the TW-2 so that it is more useful and easier to operate than the other vehicles that elderly people use daily. Specifically, the walking subsystem of TW-2 needs to be developed so that it accommodates people whose right-leg and left-leg walking patterns and performances differ, e.g. as in the case of hemiplegia; in such case, it will be necessary to consider and improve the assistance parameters for each leg individually.

8. Acknowledgements

This work was supported in part by the 21st Century Center of Excellence Program "The innovative research on symbiosis technologies for human and robots in an elderly dominated society", Waseda University, Tokyo, Japan, the "Establishment of a Consolidated Research Institute for Advanced Science and Medical Care", Encouraging Development Strategic Research Canters Program, the Special Coordination Funds for Promoting Science and Technology, Ministry of Education, Culture, Sports, Science and Technology, Japan, the robotic medical technology cluster in Gifu prefecture," Knowledge Cluster Initiative, Ministry of Education, Culture, Sports, Science and Technology, Japan and KAKENHI 207004600.

9. References

Ando, T., Nihei, M., Kaneshige, Y., Inoue, T., Fujie, M.G. (2008). A Steering System of a New Mobility-Aid Vehicle with walking: Tread-Walk, *The second IEEE RAS/EMBS International Conference on Biomedical Robotics and Biomechatronics.*

Ando, T., Nihei, M., Ohki, E., Kobayashi Y., Fujie, M. G. (in preparation). Kinematic Walking Analysis on a New Vehicle Tread-Walk with Active Velocity Control of Treadmill Belt.

Belli, A., Bui, P., Berger, A., Geyssant, A., Lacour, J-R. (2001). A Treadmill Ergometer for Three- dimensional Ground Reaction Forces Measurement during Walking. *Journal of Biomechanics*, 34, 105-112.

Boy, E. S. Teo, C. L. Burdet, E. (2002). Collaborative Wheelchair Assistant. *Proceedings of the IEEE/RSJ CIRS*, Vol.2, 1511-1516.

Hase, K., Obuchi, S., Horie, T., (2002). Rehabilitation System to Prevent Falls during Walking in Older Adults (Planning of Design Concept and Primary Experiments). *Transactions of the Japan Society of Mechanical Engineers*, C 68 (668), 1245-1250.

Japanese Industrial Standarization, Motored wheel, JIS T9203, 2003.

J. Jutai, et al. (1999). Quality of life impact of assistive technology. *Rehabilitation Engineering, RESJA (in Japanese)*, 14, 2-7.

Kaneshige, Y., Nihei, M., Fujie, M. G. (2006). Development of New Mobility Assistive Robot for Elderly People with Body Functional Control -Estimation walking speed from floor reaction and treadmill-. *Proceedings of the IEEE RAS-EMBS*, 79.

Morales, R. Feliu, V., Gonzallez, A., Pintado, P. (2006), Coordinated Motion of a New Staircase Climbing Wheelchair with Increased Passenger Comfort. *Proceedings of the 2006 IEEE ICRA*, 3995-4001.

NEDO, Physical function database of elderly people, walking width in free walking, (Available 2007.2) : http://www.hql.jp/project/funcdb1993/.

Nihei, M., Kaneshige, Y., Fujie, M. G., Inoue, T. (2006). Development of a New Mobility System Tread-Walk -Design of a Control Algorithm for Slope Movement-, *Proceedings of the 2006 IEEE International Conference on Robotics and Biomimetics*, 1006-1011.

Nihei, M. Kaneshige, Y. Inoue T. Fujie M. G. (2007). Proposition of a New Mobility Aid for Older Persons -Reducing psychological conflict associated with the use of Assistive Technology, *Assistive Technology Research Series*, AAATE 07, Vol.20, 80-84.

Nihei, M. Inoue T., Fujie M. G. (2008).Psychological Influence of Wheelchairs on the Elderly Persons from Qualitative Research of Daily Living", *J. of Robotics and Mechatronics Vol.20 No.4*, 641-649.

Nihei, M., Ando, T., Kaneshige, Y., Fujie, M. G., Inoue, T.(2008). A New Mobility-Aid Vehicle with a Unique Turning System. *Proceedings of the 2008 IEEE/RSJ IROS*, 293-300.

Walsh, C. J. Pasch, K. Herr, H. (2006). An autonomous, under actuated exoskeleton for load-carrying augmentation. *Proceedings of the 2006 IEEE/RSJ IROS*, 1410-1415.

Riener, R., Lunenburger L. Jezernik Saso, Anderschitz M., Colombo, G. Dietz V. (2000). Patient-Cooperative Strategies for Robot-Aided Treadmill Training First Experimental Results. *IEEE Trans. NEURAL SYSTEM AND REHABILITATION ENGINEERING*, Vol.13, No.3, 380-394.

Sankai, Y., Kawamura, Y., Okamura, J., Woong, L. S. (2000). Study on hybrid power assist system HAL-1 for walking aid using EMG, *Proceeding of the JME on Ibaraki Symposium*, 269.

Simpson, R. C. (2005). Smart wheelchairs: A literature review, *J. of Rehabilitation Research & Development*, Vol. 42, No. 4, 423-436.

Tani, T., Koseki, A., Sakai A., Hattori, S. (1996), System Design and Field-testing of the Walk Training System. *Proceedings of the IEEE/RSJ IROS96*, 340-344.

Tani, T., Koseki, A., Sakai, A., Hattori, S., Control Methods of Walk Training System, *Transactions of the Japan Society of Mechanical Enginerrs*. C 62(597), 1996, 1928-1934.

Walsh, C. J. Pasch, K. Herr, H. (2006). An autonomous, under actuated exoskeleton for load-carrying augmentation. *Proceedings of the 2006 IEEE/RSJ IROS*, 1410-1415.

Zeng, Q., Teo, C. L., Rebsamen, B., Burdet, E. (2006). Design of a Collaborative Wheelchair with Path Guidance Assistance. *Proceedings of the 2006 IEEE ICRA*, 877-882.

Flexible Mono-tread Mobile Track (FMT) - A New Mobile Mechanism Using One Track and Vertebral Structure -

Tetsuya Kinugasa*, Yuta Otani**, Takafumi Haji*, Koji Yoshida*, Koichi Osuka*** and Hisanori Amano****
**Okayama University of Science, **Pacific Software Development Ltd., ***Osaka University, ****National Research Institute of Fire and Disaster*
Japan

1. Introduction

Lots of disaster such as huge earthquakes, the 1995 Kobe earthquake in Japan (as shown in Fig.1), followed by the 2004 Indian Ocean earthquake and the 2008 Sichuan earthquake and so on, in addition, 11th September 2001 attack on the World Trade Center, have led us to recognize the necessity to utilize robots in search and rescue at disaster areas.

Fig. 1. Collapsed Japanese wooden houses in Kobe (1995, the left figure) and Noto (2007, the right figure) earthquakes

Research and development activities for the utilization of robot technology to assist humans in rescue operations have resulted in a challenging field of robotics: Rescue Robotics. The rescue robots must function in extremely hazardous and very complex disaster environments, moreover, the composition of rubble in the disaster area varies due to regional circumstances and types of disasters, etc. Hence, it is very important to develop mobile mechanisms that enable robots to travel across such the rubble and access to the interior of the rubble pile. The importance of development of powerful mobile mechanisms has been also indicated in ``the special project for earthquake disaster mitigation in urban areas'' by The Ministry of the Education, Culture, Sports, Science and Technology in Japan (DDT-report, 2006). Then, serpentine robots (Takayama & Hirose, 2003, Arai et al., 2004 & 2008, Osuka & Kitajima, 2003, Kamegawa & Matsuno, 2007, Miyanaka et al., 2007) have been developed as search robots to travel across the rubble of collapsed buildings for the purpose of finding victims trapped in the rubble with expectation of powerful tools for the purpose. They have surely high mobility. However, as indicated in (Miyanaka et al., 2007) these robots have such problems as follows: if the bottom sides of the robots, which are not covered by tracks, get on an edge of obstacles, the robots become hung in the air. See the left figure of Fig.2. And if the connecting joint parts (The serpentine robots usually consists of some segments which are connected by joints) get on an edge of obstacles, the robots get stuck. See the right figure of Fig.2.

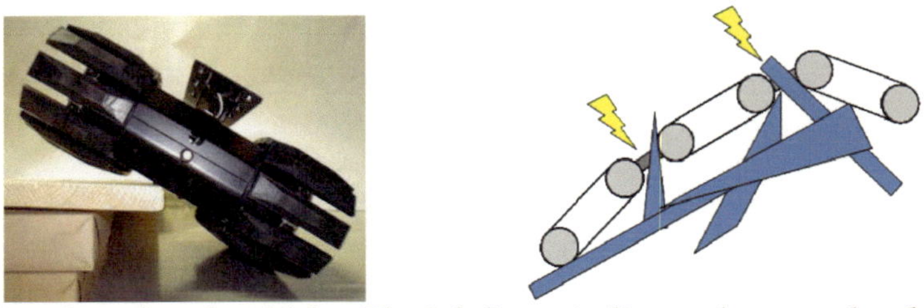

Fig. 2. Stuck situation of track vehicles. The left figure indicates that a track vehicle becomes hung in the air. The right figure shows that connecting joint parts get on an edge

Hence, in this chapter, to get over the problems, we propose a new mobile mechanism: Flexible Mono-tread Mobile Track (FMT). FMT has only *one* track which wraps around the vehicle body, and the body flexes in three dimensions when the vehicle turns, climbs up and down stairs, and so on (see the left and right figures of Fig.3).

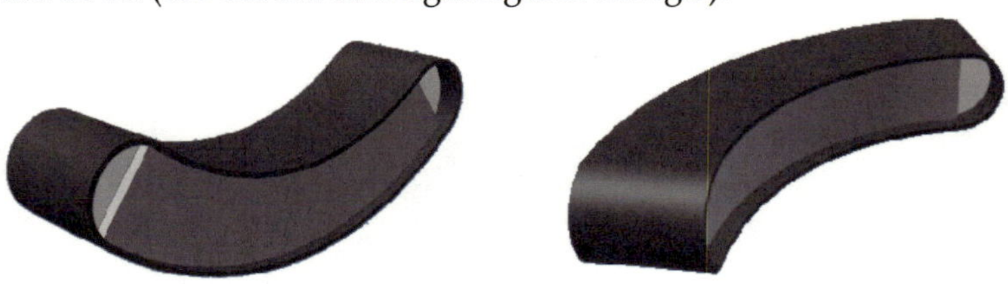

Fig. 3. Flexible mono-tread mobile track (FMT). The left and the right figures indicate retro-flexion and lateral flexion postures of FMT.

2. Track Vehicle

Though the search robots would need various functions, among them, travelling across rough terrain such as the rubble of collapsed buildings would be one of the most important. Tracks are known as very effective for vehicles to travel on extremely rough terrain. And the mobility of the vehicles using tracks would be higher as the overall length of the tracks gets longer. Hence, for getting longer tracks, serpentine robots have the same structure in common; the robots consist of several segments each of which has tracks and they are connected by joints. However as indicated in previous section, these robots have such problems as described above. In addition, this architecture would need more actuating mechanisms as the number of joints increase, then problems would also arise that increase of overall weight, more complicated control system, less reliability. Then in this section we consider the conventional mobile mechanisms with tracks and make clear the problems of them.

2.1 Differential Tracks

This type of mobile mechanism employs a pair of tracks as shown in Fig.4. It is typical type, and adopted in tanks, construction equipments, and so on. The difference of velocities between left track and right one enables the vehicle to turn to left or right. Also, the vehicle can turn in its own radius by driving the tracks in opposite direction. This type needs only one actuating mechanism and operation can be done intuitively. However since the units of the tracks move on the ground with slip, it is difficult for the vehicle to go straight or to trace curved lines without any help of some kinds of control subsystems. The height of obstacles that the vehicle can climb over is determined by mainly the radius of sprocket for track belt, hence improving the height would need to enlarge the vehicle. Moreover, if the bottom side which is not covered by tracks get on an edge of obstacles, the vehicle becomes hung in the air. Using wider tracks would solve the problem, however, it results in increasing running and turning resistance. The vehicle cannot turn at worst caused by large resistance of the wider and longer tracks.

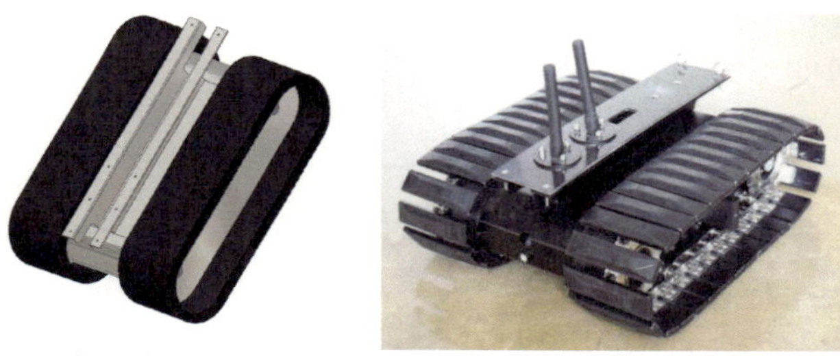

Fig. 4. Differential tracks, the right figure is 'Frigo-D' developed by National Research Institute of Fire and Disaster

2.2 Articulated Forward Tracks

When a rescue robot becomes stuck for a certain reason, articulated forward tracks (``Flipper tracks'') would be very effective to get out of the situation, e.g., Hibiscus (Koyanagi, 2006) has such tracks. The mobile robots with flipper tracks have been introduced to collect useful data in places inaccessible to humans at the area of WTC disaster, and the effectiveness of the flipper tracks has been proven. However, the flipper tracks and body section of vehicle may get object between them, and the joints that connect them may entangle something soft or fibers, etc. Those would force the flipper tracks to be useless, thereby the vehicle to be out of control. Moreover, problems might arise due to adding the flipper tracks: less reliability, difficult operation, increase of overall weight. To solve one of the problems, there have been studies (Ohno et al., 2007) to control flipper tracks automatically to help the vehicles to climb stairs or traverse gaps.

2.3 Serially connected Tracks

Not only for traveling across the rubble, but also accessing to the interior of the rubble pile through voids or opening, many search robots have adopted an architecture: some track-equipped segments are connected through joints as shown in Fig.5. For example, Souryu-I, II, III, IV, V (Takayama & Hirose, 2003, Arai et al., 2004 & 2008), MOIRA (Osuka & Kitajima, 2003), KOHGA (Kamegawa & Matsuno, 2007, Miyanaka et al., 2007). This type of vehicles can turn by differential movement of a pair of tracks on each segment (Arai et al., 2008, Kamegawa & Matsuno, 2007, Miyanaka et al., 2007) and the movement of joints between segments. Vertical joint actuators give higher ability for climbing over steps than that of individual tracked segments. On the other hand, running resistance would be more. The demerit can be removed by using shorter tracks on each segment, however, that would increase the number of segments, hence result to increase the number of actuating mechanisms of tracks and joints.

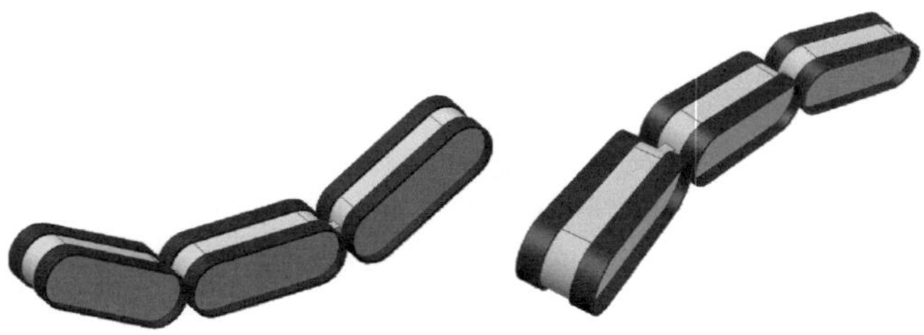

Fig. 5. Images of serially connected track

2.4 Mono-tread tracks

2.4.1 Serially connected mono-tread tracks

Souryu-V (Arai et al., 2008) has been the first vehicle which has the architecture that mono-tread track-equipped segments are connected through joints. The bottom side and top side of each segment are completely covered by track belts, hence the vehicle is hard to get stuck. Also that enables to avoid accommodation of debris between tracks and drive sprockets. On the other hand, if the vehicle turns with short turning radius, running resistance increases due to large slip friction between tracks and contacting surface. And stuck situation shown in the right figure in Fig.2 still might occur. Increasing the number of segments would increase the overall weight of the vehicle and complexity of the mechanism, as an inherent of the architecture.

2.4.2 Mono-tread mobile track

The serially connected track vehicles, which have extended bodies in longitudinal direction for access to the interior of the rubble pile, have the problems mentioned above due to the architecture. An idea to solve the problems might be to wrap around flexible body of a vehicle by only *one* track belt. Based on the idea, Fukuda (Fukuda, et al., 1994) have developed a robot which consists of only one mono-tread tracked segment, hence, mono-tread mobile track for cleaning the walls of buildings. The robot turns left or right by bending in shape around its yaw axis and climbs over steps by bending around its pitch axis. However, the bending around two axes rotate such as pivots, and the track belt is made of rubber, hence bending of the body is limited to be mild. Schempf (Schempf, 2003) have developed "AURORA" that can bend more laterally (net plus and minus 60 deg). But lateral bending has been realized by two rotational joints, thus its mechanism has been complicated. In addition, retro-flexion has not been enough to climb relatively high obstacles (over half of its height). Tanaka (Tanaka, 2006) has studied the realization of mono-tread mobile track that can bend more around each axis by using "flexible chain" (which is described in detail below). The effort by Tanaka, et al. has been done almost at the same time as ours independently. But they have not yet implemented.

3. Flexible mono-tread mobile track

Then we propose a new mobile mechanism: Flexible mono-tread mobile track (FMT) to get over the problems as mentioned above and describe on FMT in detail in this section. FMT has only one track which wraps around the vehicle body. By employing ``flexible chain'' (see Fig.6) and vertebral structure (see Fig.7), the body flexes in 3D (see Fig. 8), and could flex much enough to change the direction of its head part. It is called ``WORMY''. Table I shows the basic specifications of WORMY and Table II the devices used for WORMY.

Fig. 6. Flexible belt: Configuration of belt segments (the left), Flexible belt (the middle) and belt with grousers (the right)

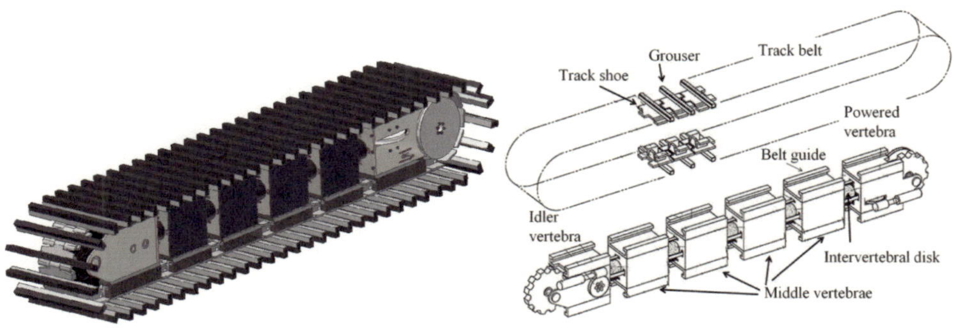

Fig. 7. Vertebral structure of FMT: 3D CAD image (the left) and its anatomy chart (the right)

Fig. 8. Flexible Mono-tread Mobile Track (FMT). Lateral flexion (the left), retro-flexion (the middle) and twisting (the right)

Weight	Length	Hight	Width
12.4kg	1.2m	0.2m	0.2m
Vertebra {Ends (2 cells), Middle (4 cells)}			
Composite monocoque (CFRP,Aluminum,Plastic)			
	Intervertebral disk (5 cells)		
Rubber	ϕ40mm	length 40mm	
	Turning	Retroflexing	Twisting
Radius	640mm	750mm	-
Angle	$\pm$90deg	$\pm$70deg	$\pm$20deg

Table 1. Specification of the prototype WORMY

Track set	Tsubakimoto Chain Co.
Track belt	TPU-862-T (R_{min}=500)
Sprocket	MB200-58641-SPR
Idler wheel	MB144-6633-IW
Motors	Maxon Motor
Track	RE30,GP32(679:49),MR
Flexion	RE30,GP32(24:5),MR
Gear and belt	
Track	Belt&pulley (6:1)
Retroflexion	Worm,Belt&pulley(24:1)
Lateral flexion	Worm,Belt&pulley(25:8)

Table 2. Devices of prototype WORMY

3.1 Mechanism for smooth flexing of the body

We adopted the vertebral structure for the body of WORMY. It consists of six segments as vertebrae and cylindrically shaped flexible materials such as rubber or springs are put between the segments as inter-vertebral discs. The flexible material allows a segment to rotate in small extent relative to adjacent segment around each of roll, pitch and yaw axes. Thereby, the body, as a whole, flexes in shape symmetrically around yaw axis (which is ``lateral flexion'') and around pitch axis (which is ``retro-flexion''), to make a smooth circular arc. Also, twisting around roll axis, the body conforms to rough terrain compliantly. Fig.8 shows the lateral flexion (the left), retro-flexion (the middle), and twist (the right). Actuators are located in both terminal segments to enable the body to have the lateral flexion and retro-flexion, as shown in Fig. 9. The first and the second figures from the left of Fig.9 illustrate the drive sprocket for the track belt and actuating mechanism for the lateral flexion, the first and the second figures from the right illustrate the idler sprocket and actuating mechanism for the retro-flexion. The actuating mechanisms for lateral flexion and retro-flexion employ toothed urethane belts. As shown in the left figure, the belt for lateral flexion is driven by a pulley and the both ends of the belt are fixed in another terminal segment (: Idler vertebra). Each segment has two holes for the belt which have been drilled in positions symmetry about the centre of the segment. The belt starts from the fixed point in Idler vertebra and goes through each hole made on a position of each middle segment, then goes around the pulley and returns to another fixed point through each hole made on another position of each middle segment. The tension caused by pulley deforms the rubber materials to result in uniform flexion of the body. Like-wise, WORMY can also retro-flex.

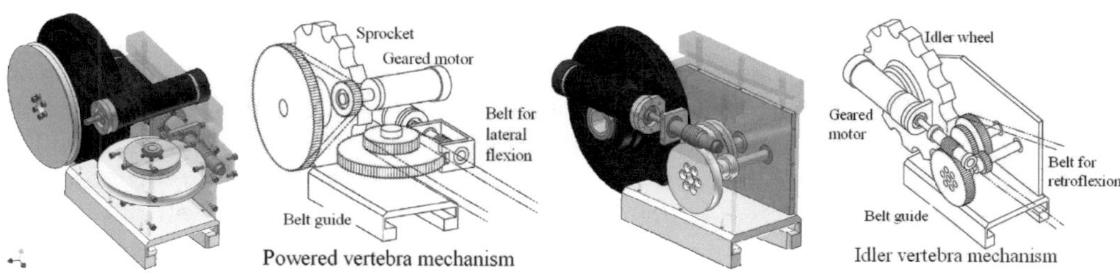

Fig. 9. Mechanism for track belt drive and flexion: Powered vertebra (the first and the second figures from the left) and Idler vertebra (the first and the second figures from the right)

3.2 Maneuverability

When FMT turns left or right, the turning radius is determined by the flexion of the body. Then, we can change the turning radius while the vehicle moves forward or backward, hence it is not difficult to let the vehicle to trace winding lines, moreover the slip would not arise between the grouser and the ground in contact area. These characteristics allow us to operate FMT in the same manner as car-like vehicles; we are familiar with how to drive (operate) them. Twisting motion of the body around roll axis is passive.

4. Geometry on the length of track belt for FMT

Since FMT employs the vertebral structure, when it retro-flexes, the track belt moves along a straight line between the vertebrae, i.e., in the part of inter-vertebral disc. Hence, as a whole, the track belt moves along a polygonal path to wrap around the body. Then the overall length of the track belt would be different between when FMT retro-flexes and when it is in straight line. The same situation would occur if FMT has other architecture: several segments connected by active joints. In the followings we consider on geometry on the length of track belt.

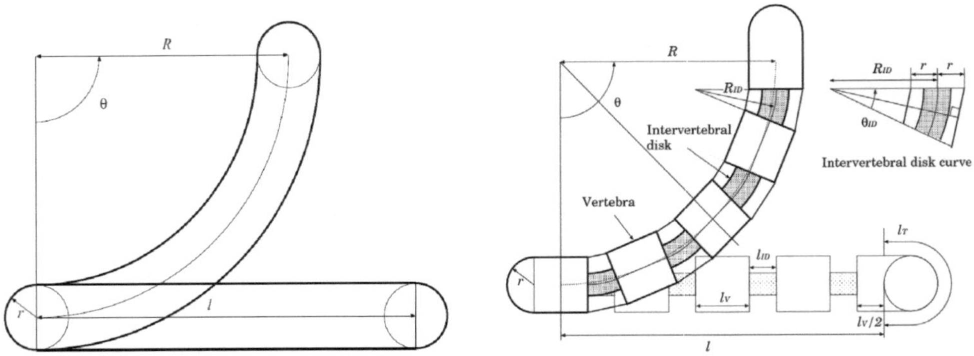

Fig. 10. Geometry on the length of track belt: Continuous flexibility (the left) and vertebral structure (the right)

4.1 In case of continuous flexibility

The left figure of Fig.10 shows schematic picture of side view when FMT retro-flexes and when it is in straight line. As shown in the figure, let l denote the length of centerline which connects the centers of drive sprocket and idler sprocket, r the radius of the sprockets. The centerline of the vehicle runs through the geometrical center of each cross section of the vehicle. As mentioned above, when the vehicle retro-flexes, it flexes symmetrically, hence the centerline comes to be circular arc. Then let R denote the gyration radius length and θ the central angle for the arc and we call the central angle as ``flex angle''. The left figure of Fig.10 shows the retro-flexion in case that θ is equal to $\pi/2$. The overall length of the track belt L_s when the vehicle is in straight line is described as

$$L_s = 2\,l + 2\,\pi\,r = 2\,(l + \pi) \tag{1}$$

Here, we assume that the body of the vehicle is continuously flexible; the arcs formed by the bottom side and top side of the body have the same center as that of centerline. And we also assume: $l = R\,\theta$. Then, overall length of the track belt L_r when retro-flexes is

$$L_r = 2\,\pi\,r + (R-r)\,\theta + (R-r)\,\theta \;= 2\,(R\,\theta + \pi\,r) \tag{2}$$

Using the assumption, we obtain $L_s = L_r$.

4.2 In case of vertebral structure

Next we treat FMT. Both terminal vertebrae have sprockets, then picking up the half length of the middle vertebra, we assign it to a part of the overall length of terminal vertebrae; the partial length of terminal vertebrae are taken to be involved in flexing. The number of vertebrae involving terminal ones is k+1, that of inter-vertebral discs is k. The right figure of Fig.10 shows the FMT with five vertebrae and four inter-vertebral discs. We assume the length of center line is conserved when it retro-flexes. Let l_V denote the vertebra length and l_{ID} the inter-vertebral disc, let R_{ID} denote the flex radius of the inter-vertebral disc and θ_{ID} the flexed angle of them as shown in Fig.10. Then,

$$R_{ID}\ \theta_{ID} = l_{ID} \tag{3}$$

holds. The length of track belt in the part of as inter-vertebral disc is

$$2\ (R_{ID} + r)\ \sin(\theta/2k), \tag{4}$$

at the bottom side of the body,

$$2\ (R_{ID} - r)\ \sin(\theta/2k), \tag{5}$$

at the top side of the body. Here, l_T denote the length shown in Fig.10. We can obtain the overall length of track belt Lr when retro-flexing:

$$L_r = 2\ l_T + 2k\ l_V\ + 2\ R_{ID} \sin(\theta/2k) \tag{6}$$

If k is large enough, $\theta/2k$ is small enough, hence $L_r = 2\ l_T\ + 2\ k\ (l_V + l_{ID})$. It means $L_r = L_s$ (:the track belt length when straight).

4.3 In case of multi segment mechanism

When a vehicle which employs the multi-segment mechanisms with joints bends its body, the shape should be polygonal: some straight line segments and corners. Universal joints are often employed to connect the segments for enabling the body to bend in 3D. If two active joints of one degree-of-freedom are combined to be one joint for connecting the segments, the body can bend to form shapes which consists of some different circular arcs, e.g., s-shaped line (see the left figure of Fig.11). We call them ``multi-arc-up-down flexion". The overall length of track belt when retro-flexing gets near to that when straight if the number of joints increases and the length between segments decreases. The conclusion is the same as that in previous subsection.

4.4 In case of multi arc flexion

Next we give considerations on track belt length when FMT bends to form multi-arc flexion. The right figure of Fig.11 shows the side view when the vehicle retro-flexes and thereafter bends to be multi-arc-up-down flexion which consists of three circular arcs. We assume here that tangential line of each end of each arc coincides to that of adjacent arc and the length of centre line is conserved. Then the overall length of track belt when it flexes is expressed as

$$L_r = 2(\pi\ r + \Sigma_{i=1}\ l_i\ ) \tag{7}$$

where, n denotes the number of arcs. Using (7), we obtain

$$L_r = 2\ l_T + 2\ \Sigma\ \ l_i\ (l_T + l_{ID}) \tag{8}$$

where, super-script i means that the symbol with i is those for i-th arc. Equation (8) gives us a result that the track belt might loosen a little when FMT bends to form multi-arc flexion.

However, we can cope with that by adapting the tension of the belt or increasing the number of segments.

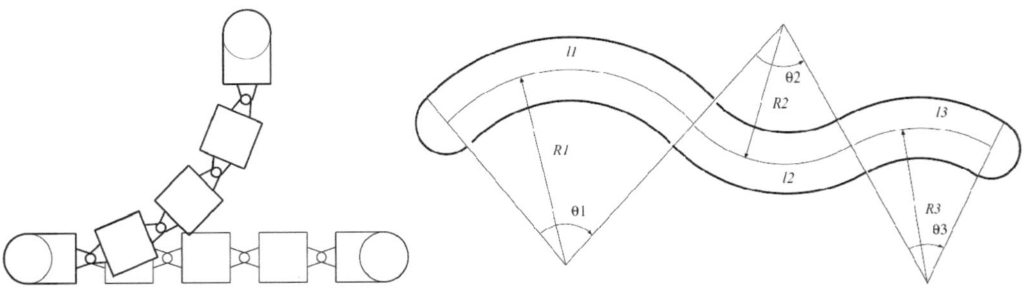

Fig. 11. Geometry on the length: Multi segment mechanism (the left) and multi arc flexion (the right)

5. Roll resistance of track belt in flexion

Roll resistance of the track belt of FMT varies depending on flexed angles when the vehicle retro-flexes and/or lateral flexion. It is worthwhile to make clear the characteristics of roll resistance for the purpose of speed control of the vehicle or autonomous mobility. Using WORMY, we have examined the relationship between flexed angles and roll resistance when WORMY lateral flexion. Travelling velocity of the vehicle has been proportional controlled to desired value. In experiments, the desired value was planned by using a step function which rises up to 1.25 m/s at the time 1s and get down again to 0 m/s at 8 s. P controller generates the control signal in voltage within the range of 5 V and outputs it to motor driver module. The module gives current to DC-motor which is proportional to the input. The left figure of Fig.12 shows the graph of velocity of WORMY versus time when it was in straight line, and the right figure the control signal (input to motor driver module) versus time. The graphs include waves of high frequency due to simple difference of angle data obtained by encoder. As is seen from the left figure of Fig.12, WORMY got to the velocity of about 1 m/s at about 1 s and maintained the velocity. As is seen from the right figure, the control signal (input) was about 4 V while WORMY was in the steady state. When the desired velocity rose up at 1 s, the control signal got up to saturate, hence, if we enlarge the range of control signal, the vehicle would get to steady state faster. Next, we have examined the relationship between the velocities in steady state and the control signals when the vehicle was turning on a flat plane. In the experiments, flexed angles were 0 deg, 30 deg, 40 deg, and 50 deg. The left figure of Fig.13 shows the graph of steady state velocity versus flexed angle. The right figure of Fig. 13 shows that the steady state velocity decreases and control signal increases as flexed angle increases and that the relationship is almost linear.

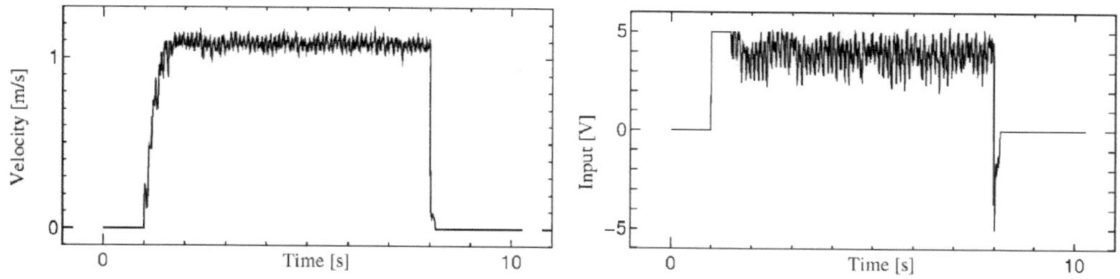

Fig. 12. Velocity and input voltage of prototype WORMY in case of straight posture: Velocity w.r.t. time (the left) and input voltage w.r.t. time

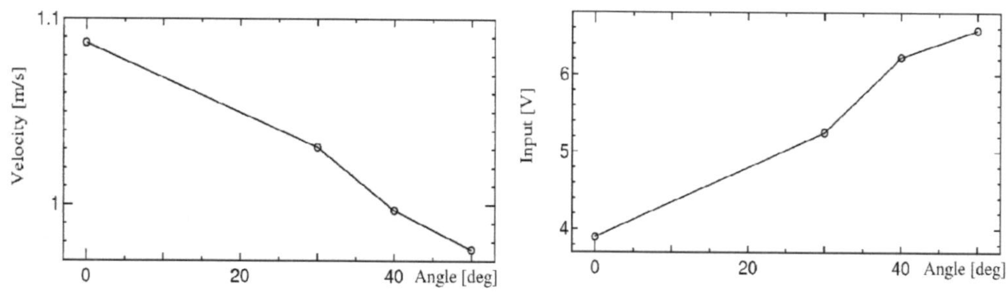

Fig. 13. Relationship between lateral flexion and track belt driving: Input voltage w.r.t. lateral flexion angle (the left) and velocity w.r.t. lateral flexion angle (the right)

6. Mobility

In order to assess the performance of WORMY, it has been tested. The test were basic ones; climbing over high steps, clearing wide gaps, climbing up and down stairs, and climbing slopes. Table 3 shows the performance of WORMY.

Wall climb ability		Gap clearability	
Maximum height		Maximum width	
Without retroflexion	Active	Without retroflexion	Active
185mm	450mm	450mm	550mm
Slope climb ability		Step climb ability	
45deg		Height	Going
		175mm	295mm

Table 3. Mobility of the prototype WORMY using battery (14.4 V)

6.1 Climbing over walls (steps)

Figure 14 shows the steps WORMY should take for climbing over a wall. First, the vehicle retro-flexes in front of the wall (0 s). Next, keeping retro-flexing it goes upward on the side of the wall (1 s). Then, leaning against the edge of the wall (2 s), it gets back to be straight (5 s) and goes forward (8 s). Finally, bending the body such that the head part turns downward, it goes forward (9 s). When the centre of gravity of the vehicle gets over the wall, it falls down on the ground to complete the task. As shown in Table 3, the maximum height is 450 mm by using retro-flex.

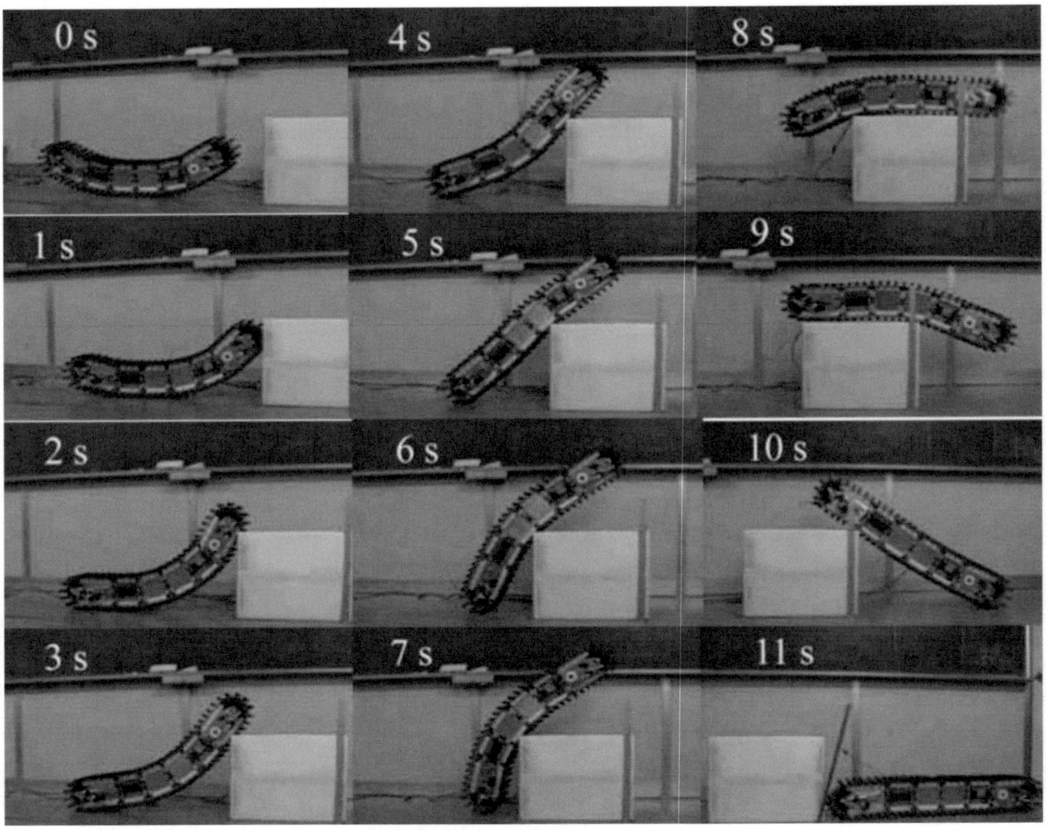

Fig. 14. Climb over walls (steps, height 450 mm)

6.2 Crossing ditches

Figure 15 shows the figures that WORMY traverses a ditch. In general, any robot can clear the gap whose length is shorter than the length between the front end of the robot and its center of gravity. Since WORMY has vertebral structure, the head part turns downward after it leaves the edge of gap if it goes forward without retroflex. Then, using retroflex it could clear the gap of 550 mm which is almost the same as the half length of WORMY.

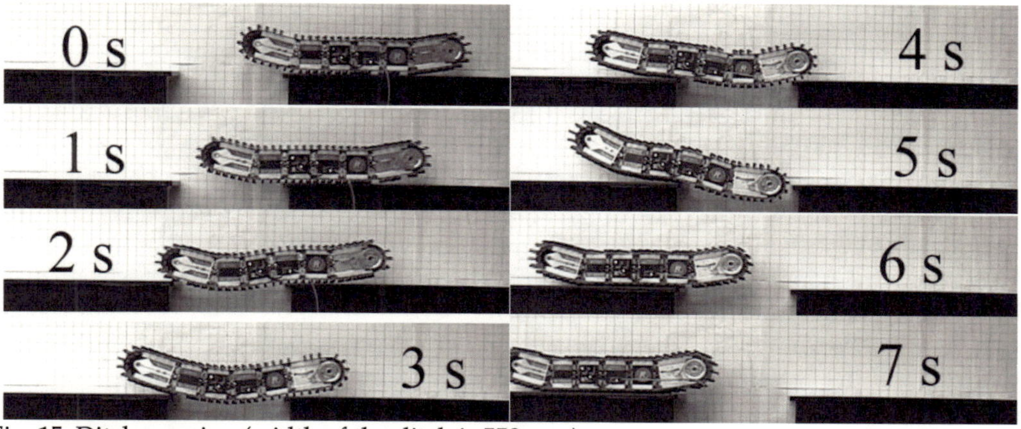

Fig. 15. Ditch crossing (width of the ditch is 550 mm)

6.3 Climbing slopes

Figure 16 shows the sequential photographs that WORMY climbs up a slope. We made slopes of inclined plywood whose length is 2 m. WORMY has been checked that it has enough power to drive up a 45deg inclined slope. We can ensure that it would climb up any slope of 45deg inclined if friction force is large enough.

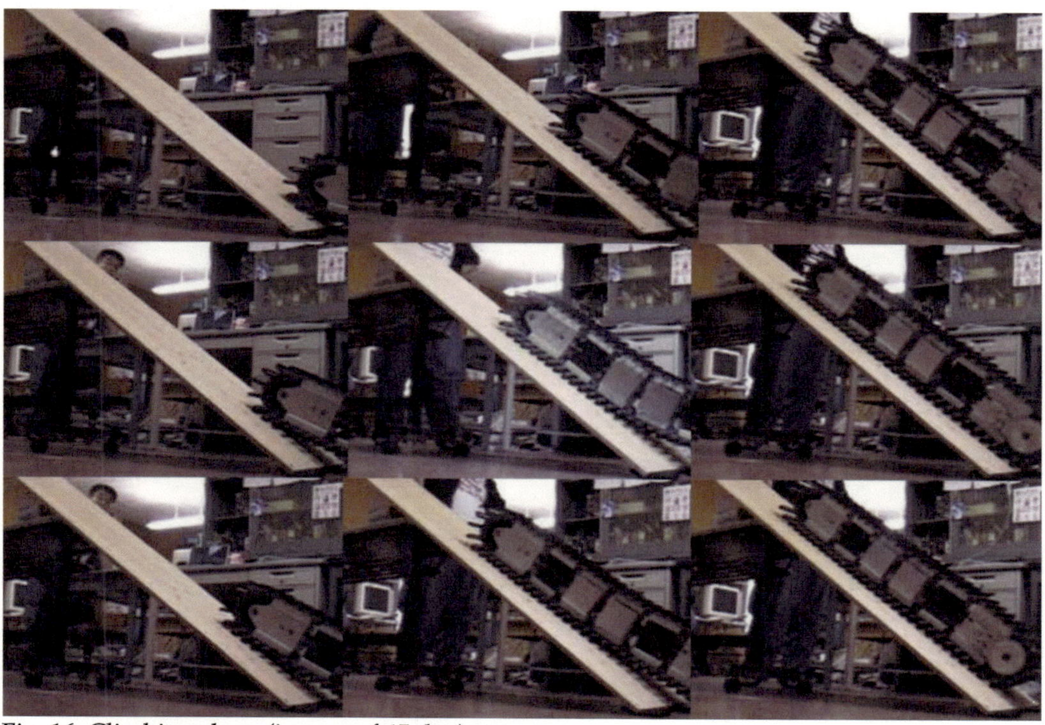
Fig. 16. Climbing slope (in case of 45 deg)

6.4 Climbing up and down stairs

We have tested WORMY using the existing stairs. Figure 17 shows the sequential figures that WORMY climbs up stairs. Figure 18 shows the schematic figure of the stairs that the vehicle could climb up. According to a standard for the design of stairs, we understood that the stairs in Fig.18 is close to those which are common in ordinary buildings. The stairs to subway station would be easily strewn with rubble to be hazardous and hard-to-reach after tragic events. The stairs to subway station are not so steep as that WORMY can climb up; the characteristics of a subway stairs are given in upper right corner of Fig. 18 as an example. We have found that when the vehicle goes down stairs, its head part passively turns downward, thereby impact force decreases which occurs when the body gets into contact with the tread of stairs. A problem has arisen that if the length of pitch line of the stairs coincides with the length that is equal to an integer multiplied by the length between two teeth, then the edges of treads of the stairs always get into contact on the part between teeth, thereby the vehicle might get stuck or lose grouser. WORMY would have a potential to climb up the stairs whose slope is 45 deg.

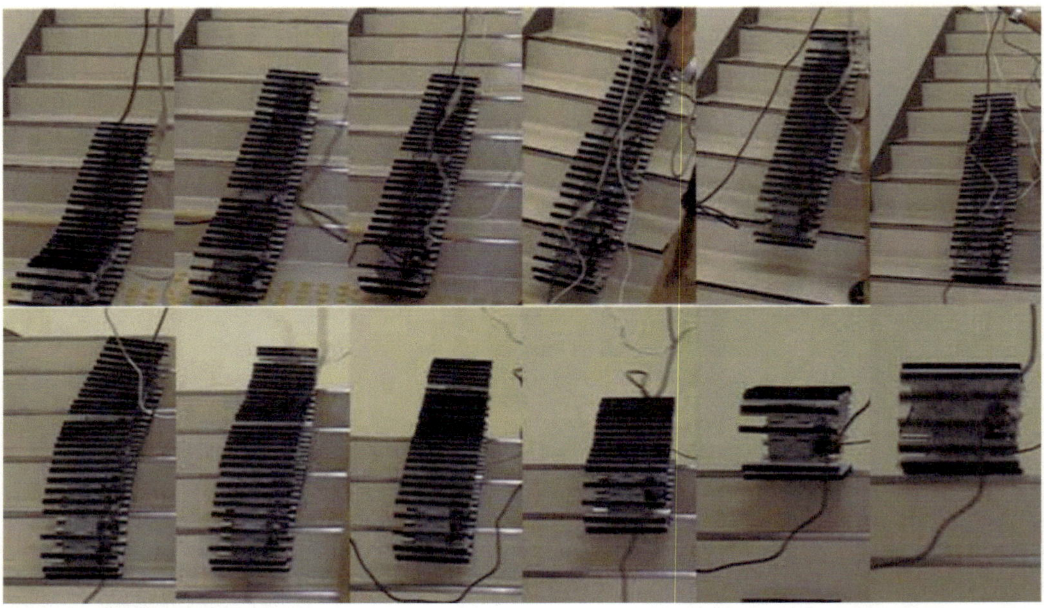

Fig. 17. Climbing up stairs

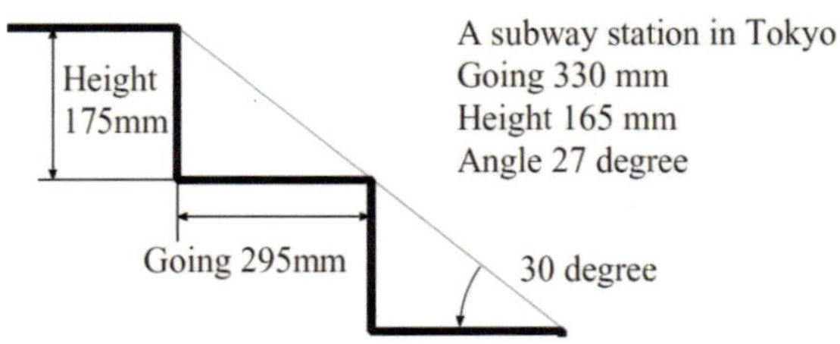

Fig. 18. Climbed Steps in experiments and an example of steps in public space

6.5 Plate Climbing

Figure 19 shows the thin plate climbing such that serially connected crawler may have some troubles. From the figure, FMT can climb over the thin plate without considering intrusion to joints by its slenderness. The serially connected crawler would be able to be recovered by bending or some other complicated mechanism such as puddles from stuck position. FMT has an advantage in the sense that it has no open joints and relatively large space.

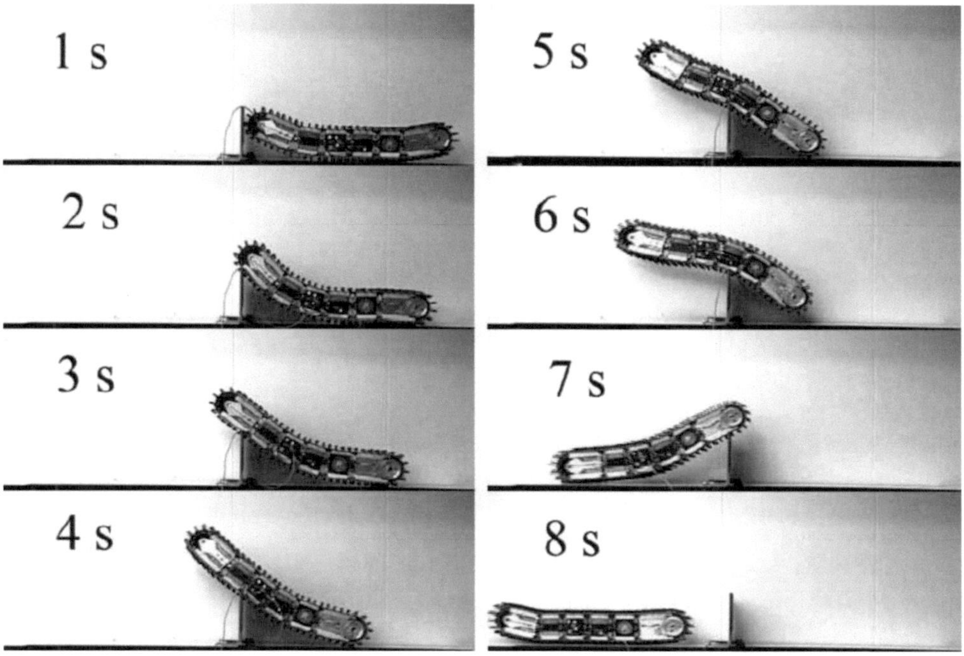

Fig. 19. Plate climbing (thickness 21 mm, height 300 mm and width 550 mm)

6.6 Recover-ability from lying position on its side

FMT has been checked that it can recover from lying position on its side. Figure 20 shows the sequential photographs that WORMY recovers from lying position. The test process is as follows: FMT is lying on its side (1s), retro-flexes upward (in direction of right hand side in the case, 2-4s), gets up by lateral flexion which makes center of FMT elevate (5-7s), and recovers to straight configuration.

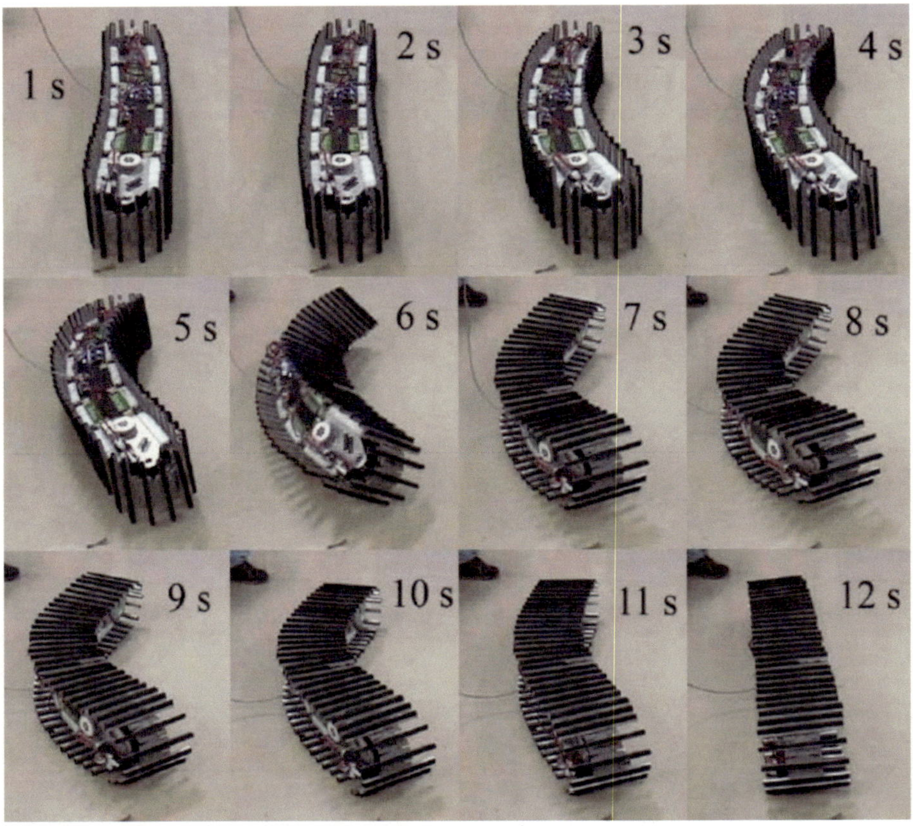

Fig. 20. Recover-ability from lying position on its side

6.7 Side winding

We have examined side winding through experiments using retro and lateral flexion skillfully. Figure 21 shows straight configuration (1s), upward retro-flexion which makes the both end vertebrae of FMT elevate (2s), lateral flexion (3-6s), downward retro-flexion which makes center of FMT elevate (8s), and recovering the straight configuration by lateral flexion (9-11s). As the result of the side winding, FMT could moves approximately 150 mm in lateral direction. Using the side winding, FMT can move lateral direction without track belt movement, and can avoid cutting back such as a non-horonomic system.

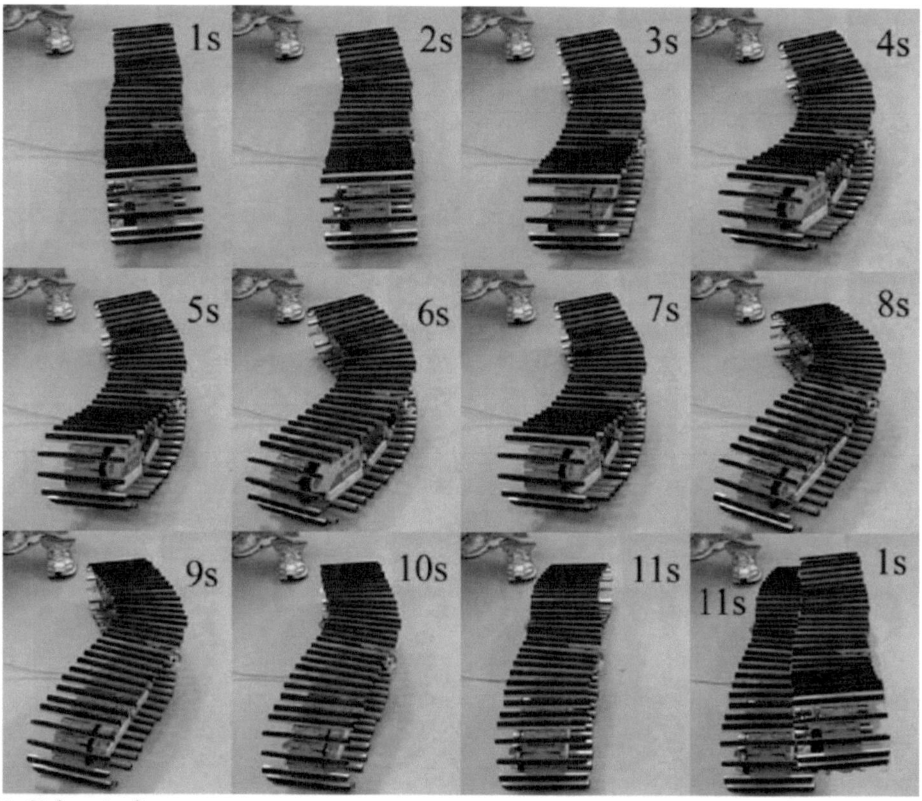

Fig. 21. Side winding

7. Conclusion

Serpentine robots, most of which have been developed as search robot, have high mobility on extremely rough terrain. However they have also some problems as described above. In this chapter we have proposed a new mobile mechanism: flexible mono-tread mobile track. It has the merits of serpentine robots and solved some of their problems. Key advantages of FMT are simple and light weight mechanism, covered by one track belt (FMT can recover even if FMT falls sideway) and 3D flexibility. Moreover we have developed a prototype ``WORMY'' and examined the performance on mobility. The followings are the important advantage of FMT:

1) Since the body is wrapped around by one flexible track, the vehicle has non-propulsion surface only on the sides of the body. Hence it can get around the problem of getting stuck on the edge of obstacles (FMT can recover by two way flexion even if FMT falls sideway).
2) Since the vehicle adopts vertebra-like structure and employs one actuating mechanism for flexing around each of pitch and yaw axes, the number of actuators would not change regardless of the overall length or the number of vertebrae (Light weight and simple mechanism can be achieved).
3) Since the body flexes smoothly in three dimensions, maneuverability is high, running resistance (slip) is less. The flexibility is effective against impact.

Dis-advantages are relatively large gyration radius, difficulty of smooth mechanism design for flexion such as belt, guide, tension and the number of segments, etc.

Future researches would be to equip various sensors on FMT for effective search of victims in search activity and to design a control scheme for autonomous movement.

8. References

DDT-report (2006). http://www.rescuesystem.org/ddt/H17-report/

Haji, T.; Otani, Y., Kinugasa, T., Yoshida, K., Osuka, K. & Amano, H.(2007). Development of Flexible Mono-Tread Mobile Track, *Proceedings of The 25th Annual Conference of the Robotics Society of Japan*, 3k28, Chiba Inst. of Tech., Sep. 2007, Narashino

Otani, Y.; Kinugasa, T. & Yoshida, K. (2006). Trial Manufacture of Robot for Inspection inside Rubble, *Proc. of the 15th Annual Conference of SICE Chugoku Branch*, pp 28-29, Hiroshima Univ. Oct., 2006, Higashi-Hiroshima

Kinugasa, T.; Otani, Y., Haji, T., Yoshida, K., Osuka, K. & Amano, H. (2008). A Proposal of Flexble Mono-tread Mobile Track - A New Mobile Mechanism Using Tracks -, *IEEE/RSJ International Conference on Interlligent Robots and Systems*, pp 1642-1647, Nice Acropolis Center, Sept., 2008, Nice

Kinugasa, T.; Otani, Y., Haji, T., Yoshida, K., Osuka, K. & Amano, H.; (2009). Flexible Mono-tread Mobile Track (FMT) - A New Mobile Mechanism Using Endless Track and Vertebrae Structure -, *Journal of the Robotics Society of Japan*, vol. 27, no. 2, pp 107-114, The Robotics Society of Japan, Jan., 2009

Kinugasa, T.; Haji, T., Yoshida, K., Osuka, K. & Amano, H. (2008) Mobility and Operability of Flexble Mono-tread Mobile Track (FMT), *Proceedings of IEEE International Workshop on Safety, Security and Rescue Robotics*, pp 65-70, Tohoku University, Oct. 2008, Sendai.

Takayama, T. & Hirose, S. (2003). Development of ``Soryu I & II": -Connected Crawler Vehicle for Inspection of Narrow and Winding Space-, *Journal of Robotics and Mechatoronics*, Vol. 15, No. 1, pp 61-69

Arai, M.; Takayama, T. & Hirose, S. (2004). Development of ``Souryu-III": Connected Crawler Vehicle for Inspection inside Narrow and Winding Spaces, *Proc. of 2004 IEEE/RSJ International Conference on Intelligent Robots and Systems*, pp 52-57, Ballys, Oct., 2004, Las Vegas

Arai, M.; Tanaka, Y., Hirose, S., Kuwahara, H. & Tsukui, S. (2008). Development of ``Souryu-IV" and ``Souryu-V:" Serially Connected Crawler Vehicles for In-rubble Searching Operations, *Journal of Field Robotics*, Vol.25, Issue 1, 2008, pp 31-65.

Osuka, K. & Kitajima, H. (2003). Development of Mobile Inspection Robot for Rescue Activities: MOIRA, *Proc. of IEEE/RSJ International Conference on Intelligent Robots and systems*, 2003, pp 3373-3377.

Kamegawa, T. & Matsuno, F. (2007). Development of a Remote-controlled Double Headed Serpentine Rescue Robot KOHGA, *Journal of the Robotics Society of Japan*, Vol.25, No.7, 2007, pp 52-59.

Miyanaka, H.; Wada, N., Kamegawa, T., Sato, N., Tsukui, S., Igarashi, H., & Matsuno, F. (2007). Development of an Unit Type Robot ``KOHGA2" with Stuck Avoidance

Ability, *Proc. of IEEE International Conference on Robotics and Automation*, 2007, FrB12.2.

Amano, H.; (2008). http://www.melos.co.jp/jigyou/product/index13.html

Koyanagi, E.; (2006). http://www.furo.org/robot/Hibiscus/

Ohno, K.; Morimura, S., Tadokoro, S., Koyanagi, E. & Yoshida, T. (2007). Semi-autonomous Control System of Rescue Crawler Robot Having Flippers for Getting Over Unknown-Steps, *Proceedings. of IEEE/RSJ International Conference on Intelligent Robots and Systems*, 2007, pp 3012-3018.

Casper, J. & Murphy, R., R; (2003). Human-Robot Interactions during the Robot-Assisted Urban Search and Rescue Response at the World Trade Center, *IEEE Transactions on Systems, Man and Cybernetics Part B*, Vol. 33, No. 3, 2003, pp 367—385.

Fukuda, T.; Nishibori, K., Matsuura, H., Arai, F. Sakai, S. & Kanasige, M. (1994). A Study on Wall Surface Mobile Robots : 2nd Report, Mechanism of Model with Variable Structural Crawler and Running Experimental Results, *Transactions of the Japan Society of Mechanical Engineers. C*, Vol.60, No.569, 1994, pp 211-217.

Schempf, H.; (2003). Less is More: AURORA - an example of minimalist design for tracked locomotion, Robotics Research, R.A. Jarvis and A. Zemlinsky (EDs.), Springer Verlag Berlin, pp.453--465, 2003

Yoshida; (2006). *A Study on 2 D. O. F. Mono-tread Mobile Track 'KYKLOS'*, Master Thesis of Depertment of Mechanical Engineering, Kobe University

Tadakuma, K.; (2008). Connected Two Units Crawlers to Realize Automatic Multiple Configuration as Search and Rescue Robot, Journal of Robotics and Mechatronics, Vol.20, No.1, 2008, pp 106-115

Hackney, L. S. & Froelich, J.; (1917). Traction Vehicle, United States Patent No.1219637, 1917.

Best, H. E.; (1970). Single Track Crawler Vehicle, United States Patent No.3548962, 1970.

Ames, V. H.; (1971). Steering, Driving and Single Track Support Systems For Vehicles, United States Patent No.3565198, 1971.

Stancy, Jr. J. C.; (1980). Terrain Vehicle Having a Single, Laterally Bendable Track, United States Patent No.4453611, 1980.

Bekker, M. G.; (1969). *Introduction to Terrain Vehicle Systems*, University of Michigan Press, 978-0472041442, Ann Arbor

Kimura, H.; Shimizu, K. & Hirose, S. (2004). Development of 'Genbu': Active-Wheel Passive-Joint Articulated Serpentine Robot, Design Engineering, Vol. 39, No. 5, 2004, pp 264-273.

Hirose, S.; (1987). *Bio-Mechanical Engineering*, Kogyo Chosa Kai, 978-4769320685

15

Design, Development, Dynamic Analysis, and Control of a Pipe Crawling Robot

Amir H. Heidari[1], Mehran Mehrandezh[1], Homayoun Najjaran[2] and Raman Paranjape[1]
[1]*University of Regina, Faculty of Engineering and Applied Science*
[2]*University of British Columbia, School of Engineering*

1. Introduction

Well functioning water networks are essential to the sustainability of a community. Large transmission and distribution water mains are often the most sensitive components of these networks since their failure can be catastrophic. Furthermore, due to the high cost of these pipes, the system does not usually provide redundancy to enable decommission for maintenance and rehabilitation. Hence, failure of such water mains often carries severe consequences including loss of service, severe damages and water contamination. Aging water mains often suffer from corrosion, tuberculation or excessive leakage. These problems can affect water quality and decrease hydraulic capacity of the mains contributing to water loss. In some cases, the main may be structurally weak and prone to breakage.

Prevention and/or early detection of such catastrophic failures need a comprehensive assessment of pipe condition. A proactive inspection approach is critical to the condition assessment as well as cost-effective repair and renewal of water mains. Regular cyclic inspections can provide information on the physical conditions of the pipes and on the rates of material deterioration. Nondestructive/non-intrusive technologies for evaluating pipe condition are essential tools for the early detection. However, more research is required to adapt existing technologies to the unique circumstances of large water mains that cannot be taken off service.

In this context, a robotic pipe crawler as an example of underwater robotic vehicles is designed to carry pipe inspection instruments including Nondestructive Testing (NDT) sensors used for inspection of in-service water mains of different materials. The robot can also provide real-time visual information about the interior surface of the pipe. The visual information and NDT data are synergistically used to make a more reliable decision about the condition of the pipe.

The on-board sensors would serve two purposes, namely (1) provide information for navigation and control of the robot, and (2) collect inspection data that can be post-processed. The proposed system has the following features:

- It remains operational with pipeline in service.
- It has a very simple structure (i.e., the minimum number of moving parts/actuators).
- It is stable enough, throughout its motion, to maximize the performance of the inspection sensors.

- It can suit pipes with inside diameters ranging from 6 to 10 inches.

It also allows for active condition assessment utilizing a variety of NDT methods to monitor defects such as mechanical damage, tuberculation, general wall loss, corrosion pitting, graphitization, cracks, reduced thickness of internal lining, and faulty joints. This can replace the traditional condition assessment methods, namely passive condition assessment, where only historical data are used to estimate the remaining service life of a pipe.

Precise control of the robot motion plays an important role in conducting effective assessment of the pipe condition. Nonlinear friction, backlash in mechanical components and hydrodynamic forces exerted on the robot would require a nonlinear control system design. However, nonlinear system theory is both limited and intricate, so the nonlinear system has to be linearized to take full advantage of linear system theory, which usually requires adjustments once the system departs from the design operating region.

To alleviate this problem, researchers have been recently examining the problem of designing systems that emulate functions of the human cognitive process (Chaudhuri et al., 1996). The challenge of research in this area is to design control systems that are autonomous (self-reliant) and intelligent in the sense that they satisfy the *Turing test* as follows: *if a man and a machine perform the same task and one cannot distinguish between the machine and the human by examining the nature of their performances then the machine is said to be intelligent, otherwise not* (Turing, 1950). Following this criterion some methods based on Artificial Neural Netwrok (ANN) (Hunt et al., 1992), Genetic Algorithms (GA) (Dimeo & Lee, 1995), and Fuzzy Logic (FL) (Lee, 1990) have been proposed in pursuit of modeling and control of nonlinear systems. Among these, FL has achieved increasing attention between control engineers and in industrial systems. The main idea of FL was introduced by Zadeh (Zadeh, 1973), and first applied by Mamdani (Mamdani et al., 1974) in an attempt to control structurally ill-modeled systems. An adaptive fuzzy system is a FL-system equipped with a training algorithm. Conceptually, it is constructed so that the linguistic information from experts can be directly incorporated through fuzzy IF-THEN rules, and numerical information from sensors is incorporated by training the FL-system to match the input-output (I/O) data and reduce the modeling error. However, the perfect match via an adaptive FL-system is generally impossible. Although the stability of an adaptive FL-system has been guaranteed in (Wang, 1994), (Wang, 1993), and (Wang & Mendel, 1992), the modeling error may deteriorate the tracking performance.

In order to improve the performance of the Fuzzy Logic Controller (FLC) and to meet the very basic requirements including stability and robustness, further tuning of the Membership Functions (MFs) and consequent parameters of the rules in Takagi-Sugeno-Kang (TSK) fuzzy systems (Takagi & Sugeno, 1985) is needed which demands optimization techniques and for that matter, incorporating evolutionary algorithms such as ANN and GA. This has led the researchers to introduce novel techniques like ANFIS, NEFCON, NEFCLASS and NEFPROX for this task (Nauck et al., 1997).

The performance of fuzzy controllers depends on two significant issues, namely the soundness of knowledge acquisition techniques and availability of human experts. These two severely restrict the application domains of FLCs. ANFIS bypasses the latter through tuning the FLC directly from a desired I/O data set.

In this context, an Adaptive Neuro-Fuzzy Inference System (ANFIS) (Jang, 1993) was adopted for the velocity servoing of a pipe crawling robot, where the parameters of the ANFIS were optimized based on experts' data obtained via a Human-In-The-Loop (HITL) real-time simulator.

This chapter is organized as follows: in the next section previous research in the field of inspection robots and more specifically pipeline inspection robots is reviewed. Next, in section III, the proposed design of the robotic pipe crawler studied in this research is detailed, which is followed by the investigation of kinematics and dynamics of the robot in section IV. Section V elaborates on the controller design for the robotic pipe crawler, including the structure of the controller, details of the human-in-the-loop system exploited in this research, tuning procedure of the controller and also some theoretical background on ANFIS. Simulation and experimental results are depicted and discussed in section VI. Finally, section VII encompasses the conclusion of the accomplished research and suggested future works.

2. Review of Previous Work

2.1 Conventional Inspection Methods

Statistical methods based on the number of pipe breaks per kilometer and reactive inspection techniques such as leak detection have been mainly used in the past for evaluation of water pipe condition. New testing technologies make it possible to develop more efficient and accurate approaches to maintain pipeline integrity through direct inspection. These techniques provide a variety of information about the condition of the pipes depending on their materials. Examples are the number of wires broken in a single section of the Pre-stressed Concrete Cylinder Pipe (PCCP), the depth of corrosion pitting in a ductile iron pipe, the extent of graphitization in a cast-iron pipe, or more generally the presence of leaking water (Grigg, 2006), (Eiswirth et al., 2001) and (Gummow & Eng, 2000).

2.2 Pipeline Inspection Vehicles

Remotely operated or autonomous vehicles moving inside pipes that can deploy NDT equipments have been studied extensively for the past two decades. An exhaustive review of the literature is impossible due to the limited space available. However, various locomotion systems developed and cited in literature for in-pipe operations can be categorized into three main groups as follows:

2.2.1 Pipe Inspection Gauges (PIG)

They are passive devices widely used for inspection of oil pipes and are designed so that sealing elements provide a positive interference with the pipe wall. Once inserted into a line, PIGs are driven through the line by applying pressure in the direction of required movement. A pressure differential is created across the PIG, resulting in movement in the direction of the pressure drop. Upon removal, the information logged using the PIGs onboard data storage unit is played back and analyzed. PIGs are normally employed for the inspection of pipelines with large diameters. Their inspection operations are limited to relatively straight and uninterrupted pipe lines operating in the high-pressure range. Short inspection runs are costly. Besides, the pipeline must be relatively clean for precise inspection.(Shiho et al., 2004),(Nguyen et al., 2001).

2.2.2 Floating Systems/Robots

Autonomous Underwater Vehicles (AUV) and underwater Remotely Operated Vehicles (ROV) are oceanographic locomotion interfaces used for data acquisition in subsea and deep-water missions. The applicability of existing floating robots in the confined environments such as pipes will be very limited. Further modifications will be needed to make them suitable for inspection of pressurized pipelines.(Griffiths, 2003),(Nickols et al., 1997)

2.2.3 Mobile Robots

Significant effort has been put into devising an effective mechanism to drive a robotic system carrying on-board sensors/testing devices through different pipe configurations. The sensors on these robots must be small in physical size, lightweight, and low in power consumption as compared to the other systems mentioned above. Academic researchers and industrial corporations have investigated many variations of drive mechanisms such as wheels, crawlers, wall press, walking, inchworm, screw and pushrods. Some systems have complex mechanisms and linkages, which in turn require complicated actuation and control. Wheeled systems claimed the edge over the majority due to their relative simplicity and ease of navigation and control. Comparatively, they are able to travel relatively fast and far. However, most of the mobile robots developed for this purpose have been residential in research labs because of their lack of ability to move inside pressurized pipes, e.g. (Koji, 1999), (Roh & Choi, 2005), and (Miwa et al., 2002). Some popular variants of mobile robots for pipe inspection are briefly described below.

- **Wheeled/tractor carriers:** These are the simplest drive mechanisms that are targeted for inspecting empty pipes. These remotely controlled vehicles are designed to serve as platforms to carry cameras and navigate through pipes and conduits.
- **Pipe Crawlers:** These are locomotion platforms that crawl slowly inside a pipeline. They can move down the pipeline independent of the product flow and maneuver past the physical barriers that limit inspection. They can even stop for detailed defect assessment. These robots are reconfigurable and can fit pipes with a variety of sizes. (Bradbeer et al., 2000)
- **Helical Pipe Rovers:** The robots developed at the University Libre de Bruxelles are considered as an example of a helical pipe rover (they are called HELI-PIPES). HELI-PIPE family consists of four different types of robots for in-pipe inspection. The robots have two parts articulated with a universal joint. One part (the stator) is guided along the pipe by a set of wheels moving parallel to the axis of the pipe, while the other part (the rotor) is forced to follow a helical motion thanks to tilted wheels rotating about the axis of the pipe. A single motor (with built-in gear reducer) is placed between the two parts (i.e., rotor and stator) to generate the forward motion (no directly actuated wheels needed). All the wheels are mounted on a suspension to accommodate slight changes in pipe diameter and also the curved segments of the pipe. These robots are autonomous and carry their own batteries and radio links. Their performance is, however, limited to very smooth and clean pipes. (Horodinca et al., 2002)
- **Walking Robots:** Wall-climbing robots with pneumatic suction cups and/or electromagnets have been used for inspection of vertical pipes, conduits, and steel structures (Glass et al., 1999). Walking robots are particularly useful for inspection of irregular and rough surfaces.

  Pipe inspection robots can be configured as tethered or wireless. They can be controlled remotely, or being totally autonomous. To the best of our knowledge, all existing pipe rovers are for inspection purposes only. In general, current mobile robotic systems are not yet adequate for on-the-fly repairs in a complex pipe environment.

  Development of the locomotion unit of a robot capable of inspecting in-service pressurized pipes remains a very challenging and novel research topic. Moreover, precise control of such a pipe inspection robot when subjected to flow disturbances necessitates development of nonlinear control strategies. This study addresses the mechanical design of a pipe crawling robot capable of moving inside pressurized pipes and a fuzzy-logic based control strategy to maintain a constant speed for the robot when moving inside live pipes.

3. The Proposed Design

3.1 Design Factors

Major factors considered in the design of the proposed pipe inspection robot are reviewed in this section. The principle objective put into practice in our design is to build a vehicle to serve as a highly stable platform capable of conducting precise sensing/scanning tasks. The stability of the platform in terms of having smooth motion with regulated cruise speed is necessary for accommodating sensor readings at a high bandwidth. Precise positioning of the vehicle is particularly important for using precision probes to inspect and evaluate the condition of the inner surface of the pipes. The main design requirements of the robot are as follows:

1. The vehicle should be capable of completing inspection without decommissioning the pipeline.
2. The vehicle has to be pressure tolerant up to 20 atmospheres. Freshwater transmission lines are operated at pressures of up to 16 atmospheres, therefore with a reasonable margin of safety we require the vehicle to be able to operate at 20 atmospheres, which corresponds to the hydrostatic pressure experienced at 200 meters of depth in open water.
3. The sensor payload of the vehicle has to be flexible and user interchangeable. The primary use of this vehicle is to carry a number of NDT sensors that are in various states of development. It is therefore necessary for the user to be able to swap and replace sensors within hours.
4. Autonomy of the inspection process:
    a. The length of the survey (several kilometers) makes a tethered vehicle impractical.
    b. Very detailed inspection should be done autonomously.
5. The robot should be designed in a way that it will not deteriorate the sanitation of the drinkable water when used in distribution water pipes.
6. The vehicle should be capable of traveling with any inclined pipe angle. The vehicle shall have the ability to travel vertically, negotiate multiple elbows, and potential obstacles protruding into the pipe up to 1/3 of the pipe diameter.
7. Travel speeds should be a minimum of 3 centimeters per second, with 30 centimeters per second as the desirable speed.
8. Finally, the vehicle should be able to stop and position itself at a specific location within the pipe using its onboard internal sensors, such as optical encoders.

3.2 The Proposed Vehicle Configuration

In our proposed system, we use a low drag cylindrical shape hull as a platform for carrying inspection/navigation sensors and NDT devices. The symmetrical shape of the hull can maintain a laminar boundary layer around the hulls outer surface. The low-drag property of the

main body enables the system to show superior stability against current in the pipe without loosing too much energy which is necessary in minimizing the size of the on-board battery pack required to travel long distances.
The hull consists of the following modules:

- *Nose Module* : This module accommodates a viewport for a digital still or a video camera.
- *Rechargeable Battery Module* : It provides power for propulsion, system hardware, and sensors during mission. The module contains Lithium-Ion rechargeable batteries with a total capacity of 1 *kWh*. The battery module has a built-in charger and can be charged separately from the vehicle as well as in the vehicle.
- *Actuator, Control and Communication Module* : it accommodates the vehicles actuator along with the control and communication electronics. Control instrumentation includes a *3 - axis* magneto-inductive compass, inclinometers, a temperature sensor, and an optical encoder. Communication is done via Bluetooth wireless module for short distances. For distances longer that 30 meters, the controller switches to autonomous operation. The actuator consists of a geared DC motor.

The main hull houses the actuator and the battery pack. The electronics responsible for power conversion, communication to the wireless transceiver, sensor integration, and various electric motor controls is housed in the second module connected to the main hull via a universal joint (see Fig. 1). Further details on the design of the proposed robot can be found in (Ratanasawanya et al., 2006). There is one set of driving wheels located at one end of the hull, pushing

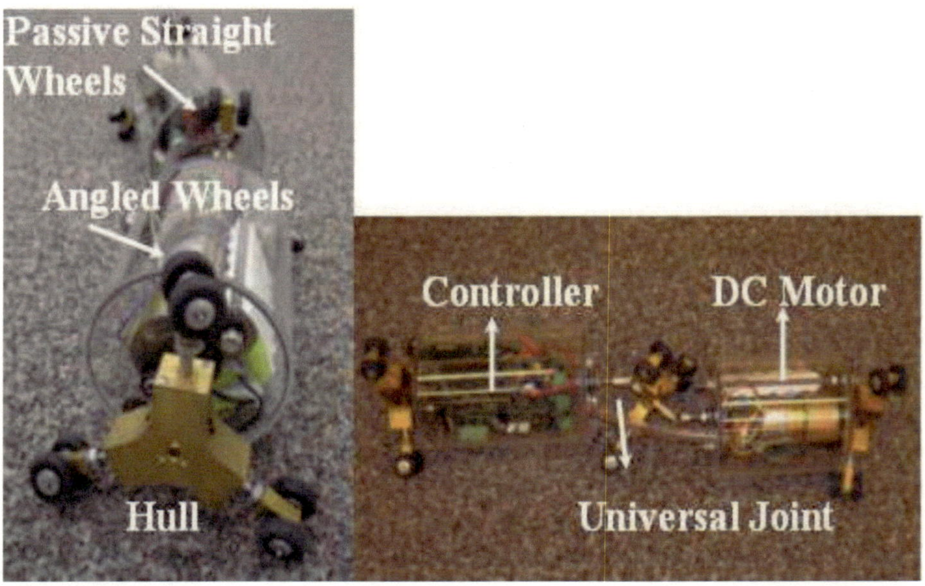

Fig. 1. The pipe inspection robot: (a) active and passive wheels. (b) side view of the robot.

against the pipe inner wall. These wheels are spring-loaded as depicted in Fig. 1. The driving wheels are approximately 4 centimeters in diameter with aluminum hubs and rubber tires. The tires have treads to provide additional traction. Larger compliant tires are appropriate for bumps and uneven internal surfaces. The driving wheels are actuated by a central geared DC motor which provides forward propulsion for the robot. The on-board electronics will be responsible for producing, filtering and controlling the power delivered to the motor for safe operation. Friction between the passive straight wheels attached to the hulls back end and the pipes wall, prevents the hull from spinning while the main actuator is providing smooth forward motion in the pipe.

Fig. 2 shows a simplified representation of the robots driving mechanism. One should note that,(1) only one pair of driving wheels are considered, and (2) the passive straight back wheels are not shown in this figure for simplicity. As can be seen from Figs. 1 and 2, the driving wheels are positioned at a small angle with respect to the vertical plane of the hull. The wheels are pushed against the inside wall of the pipe and driven along the circumference of the pipe. In this way, they generate a screw-type motion and move along the pipe. This mechanism, as schematically illustrated in Fig. 2, is analogous to a large screw being turned inside the pipe and consequently moving forward. When a reverse driving torque is applied to the wheels, the robot runs backward in the pipe. This design provides simplicity and compactness with minimal blockage of live pipes. Our proposed robot can negotiate pipes composed of straight and curved segments.

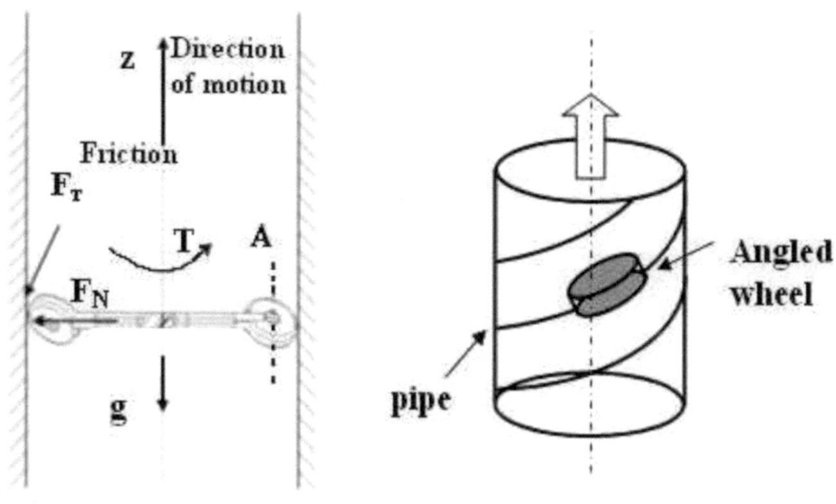

Fig. 2. The drive mechanism of the robot based on the principle of screw.

3.3 On-board Sensors

Three different types of sensors are incorporated into the design, namely (1) navigation, (2) communication, and (3) inspection sensors. However, some sensors potentially can be employed for both navigation and inspection. An optical encoder reading motors shaft displacement was used for localizing the robot inside the pipe. A vision sensor (i.e., a pinhole camera) along with an Omni-directional Stereo Laser Scanner (OSLS) were employed for navigation/inspection purposes. Unbounded position errors due to slippage in wheels is inevitable, therefore the OSLS can be superior over optical encoders to precisely measure lateral translational motion of the robot, namely, sway and two rotational motions, namely pitch and heave,

(Kulpate, 2006). A sensor fusion strategy would be required to integrate orthogonal information coming from different sensing units as the robot moves. It is also noteworthy that some temperature sensors were used in each module to continuously monitor the temperature build up in each water-tight unit.

4. Motion Analysis

In this section the kinematics and dynamics of the proposed robot moving inside a vertical straight pipe are investigated. For simplicity, the dynamic equations are derived based on the following assumptions:

1. The angle of the driving wheels cannot change on the fly;
2. The wheels apply a fixed amount of normal force to the pipe wall preventing the slippage (i.e., no on the fly extension in arms is allowed).

The vehicle model and coordinate systems used in this study are shown in Fig. 3. It is assumed that one DC motor drives the hub and also the wheels attached to the hull (or main body), as the prime actuator. From Fig. 3, frames *i, B,* and *W* represent the inertial fixed frame, the body frame attached to the main body of the robot, and the wheel frame attached to the wheel center of rotation, respectively. Physical parameters of the system in the presented dynamic model of the robot and their definition are given in Table 1.

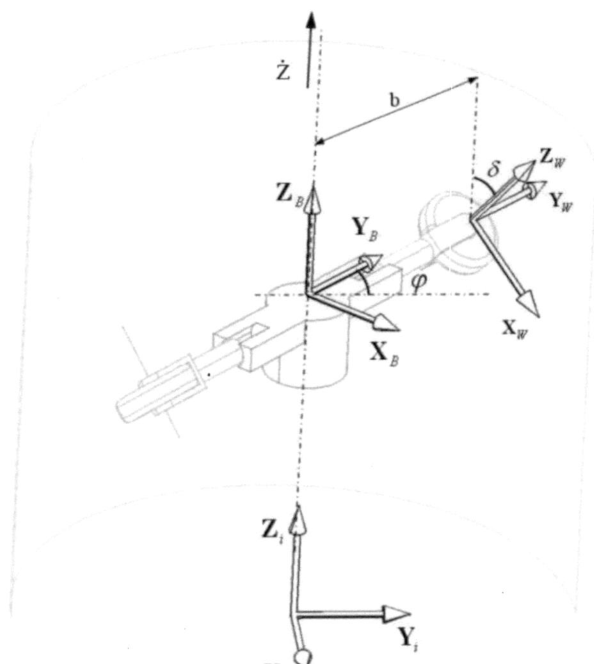

Fig. 3. The simplified model of the robot, with one pair of driving wheels, showing three reference frames.Passive wheels are not shown in this picture.

Physical Properties of the System		
Symbol	Definition	Unit
m	Wheel Mass	kg
M_h	Hull Mass	kg
M_m	Motor Mass	kg
r	Wheel Radius	m
A	Robot's Effective Cross-Sectional Area	m^2
C_d	Drag Coefficient	$-$
μ	Fluid Dynamic Viscosity	$\frac{kg}{m.s}$
v	Velocity of the Fluid	$\frac{m}{s}$
ρ	Fluid Density	$\frac{kg}{m^3}$
K_f	Damping Constant	$N.m.s$
K_m	Toque Constant	$\frac{N.m}{A}$
K_b	Back EMF Constant	$\frac{N.m}{A}$
R	Motor Resistance	Ω
L	Motor Inductance	H
I_B	Hull Polar Moment of Inertia	$kg.m^2$
I_{WZ},I_{WX}	Wheel Moment of Inertia	$kg.m^2$
I_m	Motor Moment of Inertia	$kg.m^2$
g	Gravitational Acceleration	$\frac{m}{s^2}$

Table 1. Physical Parameters of the Pipe Crawler System

4.1 Robot Kinematics

The infinitesimal translational displacement of the hull COG, dz and the angular displacement of the wheel $d\theta$ can be expressed in terms of the infinitesimal angular displacement of the hull $d\phi$ by:

$$dz = (b+r)d\phi \tan(\delta) \tag{1}$$

$$d\theta = \left(\frac{b+r}{r\cos\delta}\right)d\phi; \quad \delta \neq \frac{\pi}{2} \tag{2}$$

where δ is the wheel's inclination angle and b denotes the distance between the wheel's center of rotation and that for the hull.

4.2 Robot Dynamics

The dynamic equations of motion of the robotic vehicle can be derived using the standard Lagrangian approach. First we define Lagrangian as:

$$L = T - V \tag{3}$$

where T and V denote the *kinetic energy* and the *potential energy* due to the gravitational forces, respectively. The total kinetic energy of the robotic vehicle can be represented by:

$$T = T_{Motor} + T_{Hull} + \Gamma T_{AW} \tag{4}$$

where T_{Motor}, T_{Hull} and T_{AW} denote kinetic energies of the motor, hull and the angled wheels, respectively, and Γ denotes the number of angled (active) wheels. In (4), the kinetic energy of the passive straight wheels is disregarded. T_{Motor}, T_{Hull} and T_{AW} can be calculated as:

$$\begin{aligned} T_{Motor} &= \tfrac{1}{2} M_m \dot{z}^2 \\ T_{Hull} &= \tfrac{1}{2} M_h \dot{z}^2 + \tfrac{1}{2} I_B \dot{\phi}^2 \\ T_{AW} &= \left\{ (mr^2 + I_{WZ}) \left(\tfrac{bC_\delta}{b+r} \right)^2 + (mr^2 + I_{WX}) S_\delta^2 \right\} \tfrac{\dot{\theta}}{2} \end{aligned} \tag{5}$$

In (5), S_δ and C_δ represent the short form of $\sin(\delta)$ and $\cos(\delta)$, respectively. Considering (1) and (5) the total kinetic energy of the system can be written as:

$$T = \frac{1}{2} \left\{ \left((b+r) \frac{S_\delta}{C_\delta} \right)^2 \alpha_M + \Gamma b^2 \alpha_m + I_B \right\} \dot{\phi}^2 \tag{6}$$

where:

$$\begin{cases} \alpha_M = (M_m + M_h + \Gamma m + \Gamma \frac{I_{WX}}{r^2}) \\ \alpha_m = (m + \frac{I_{WZ}}{r^2}) \end{cases} \tag{7}$$

An infinitesimal change in the potential energy of the robot due to the gravity when moving in a vertical pipe can be calculated as:

$$dV = (M_m + N_h + \Gamma m) g dz \tag{8}$$

After substituting eqn. (1) in (8) one gets:

$$dV = (M_m + M_h + \Gamma m)(b + r) g d\phi \tan(\delta) \tag{9}$$

Considering the angle of rotation of the hull ϕ as the only generalized coordinate in the Lagrange formulation, one can write:

$$\frac{d}{dt} \left(\frac{\partial L}{\partial \dot{\phi}_i} \right) - \frac{\partial L}{\partial \phi} = Q \tag{10}$$

The generalized force Q applied on the robot moving inside the pipe is given by:

$$Q = T_m - T_f - T_D \tag{11}$$

where the right hand side of the above equation represents the non-potential generalized torques such as the electromechanical torque generated by the motor, T_m, the resisting torques due to the friction between the wheels and their axles T_f, and the resisting torque due to hydrodynamic drag force posed on the system T_D all projected onto the generalized coordinate, ϕ.

Friction plays a significant role in creating the motion of the robot. Insufficient friction at the point-of-contact between the wheels and the pipe wall leads to wheel slippage. The slippage constraint of a wheel is expressed as (using Coulomb friction law):

$$F_T \leq \mu F_N \tag{12}$$

where μ denotes the friction coefficient, and F_N denotes the normal force applied on the internal surface of the pipe by the robot's wheels. Therefore, the resisting torque due to the internal friction can be obtained from the following equation:

$$T_f = \Gamma \mu b F_N + K_{f_1}\dot{\phi} + K_{f_2}\dot{\theta} \tag{13}$$

One should note that in (13) :

1. $\Gamma \mu b F_N$ models the *Coulomb friction* applied to the hub acting on the wheels.
2. $K_{f_1}\dot{\phi}$ and $K_{f_2}\dot{\theta}$ model the *viscous friction* on the hub and the wheels, respectively.

From (2), the angular velocities of the hub and the wheels, namely $\dot{\phi}$ and $\dot{\theta}$ are related. Therefore one can write;

$$T_f = \Gamma \mu b F_N + K_f\dot{\phi} \tag{14}$$

where :

$$K_f = K_{f_1} + \frac{b+r}{rC_\delta}K_{f_2} \tag{15}$$

The hydrodynamic drag force induced by the flow on the robot, projected onto the generalized coordinate ϕ, can be expressed as follows:

$$T_D = bS_\delta \frac{\rho C_d A}{2}\left((b+r)\dot{\phi}S_\delta + v\right)^2 \tag{16}$$

where ρ, A, v and C_d are as listed in Table 1. One should note that in (16):

1. The effect of the rotational motion of the robot on the drag coefficient is not considered, therefore, the drag coefficient is assumed to remain constant as the robot moves.
2. Drag force on the wheels is negligible.

By substituting (14) and (16) in (11), the generalized force Q will be computed as:

$$Q = T_m - \Gamma \mu b F_N - K_f\dot{\phi} - bS_\delta \frac{\rho C_d A}{2}\left((b+r)\dot{\theta}S_\delta + v\right)^2 \tag{17}$$

Using (17) and substituting T and V from (6) and (9) into (10), the following closed form solution in form of a nonlinear 2nd-order differential equation for the wheels motion (and correspondingly the robot motion) can be obtained:

$$\ddot{\phi} = \frac{T_m - f(\dot{\phi}, v) - a_1}{a_2 + a_3 + I_B} \tag{18}$$

where:

$$\begin{cases} f(\dot{\phi}, v) = K_f\dot{\phi} + bS_\delta \frac{\rho C_d A}{2}\left((b+r)\dot{\phi}S_\delta + v\right)^2 \\ a_1 = \Gamma \mu b F_N + (M_m + M_h + \Gamma m)(b+r)g\tan(\delta) \\ a_2 = (M_m + M_h + \Gamma m + \Gamma \frac{I_{WX}}{r^2})\left((b+r)\tan(\delta)\right)^2 \\ a_3 = (m + \frac{I_{WZ}}{r^2})\Gamma b^2 \end{cases} \tag{19}$$

From (18), one can realize that the motion of the robot can be controlled by changing parameters such as the wheel inclination, δ the normal force exerted on the pipe wall via the wheels, F_N, and the torque applied to the wheels actuators, T_m. The only control input that can vary on the fly in our design is the motor torque, namely T_m. How to manipulate this torque in order to maintain a constant speed of motion when the robot is subjected to flow disturbances (i.e., variation in the flow speed, v) will be discussed in section 5.

4.3 Motor Dynamics

The dynamics of a permanent magnet DC motor is represented by :

$$\begin{aligned} T_m &= K_m i_a \\ \frac{di_a}{dt} &= -\frac{R}{L} i_a(t) - \frac{e_b}{L} + \frac{1}{L} v_{app}(t) \\ e_b(t) &= K_b \dot{\phi}(t) \end{aligned} \qquad (20)$$

where T_m is the mechanical torque generated by the motor, e_b is the back EMF of the motor and i_a is the armature current. Here v_{app} is the input voltage (i.e., the control variable) and i_a denotes the armature current. In (20) it is assumed that the DC motor is not geared (i.e., direct drive).

5. Controller Design

The primary objective of a controller is to provide appropriate inputs to a plant to obtain some desired output. In this research, the controller strives to balance hydrodynamic forces exerted on the robot due to the flow disturbances while maintaining a constant speed for the robot. Two sets of disturbance models in the form of step and also sinusoidal changes in flow velocity were generated randomly in a simulated environment. The controller tracks the response of the system to its user defined velocity set-point $\dot{Z}_{set}$ and sends a correction command in terms of the input voltage provided to the DC motor actuators.

We compare the behavior of two controllers in this research: a conventional *PID* controller and a fuzzy logic controller (FLC) trained using adaptive network-based fuzzy inference system (ANFIS) algorithm.

ANFIS generates a fuzzy inference system (FIS) that is in essence a complete fuzzy model based on data obtained from an operator through real-time HITL virtual reality simulator to tune the parameters of the FLC. More specifically parameters that define the membership functions on the inputs to the system and those that define the output of our system.

5.1 Servomechanism Problem

The servomechanism problem is one the most elementary problems in the field of automatic control, where it is desired to design a controller for the plant which satisfies the following two criteria for the system while maintaining closed-loop stability:

1.**Regulation** : The outputs are independent of the disturbances affecting the system.

2.**Tracking** :The outputs asymptotically track a referenced input signal applied to the system.

The controller's objective is to maintain a constant linear speed in robot's motion in the presence of disturbances. In general, robot's motion can be regulated by either changing the normal force F_N exerted on the pipe's wall via robot's wheels, changing active wheels' inclination angle δ offline, or by changing the input voltage provided to the DC motor on fly. The latter is adopted as the control variable.

5.2 Fuzzy Logic Control : An Overview

Recently, researchers have been exploiting *Artificial Intelligence* (AI) techniques to address the following two major issues where conventional control techniques still require improvement:

- Accuracy of nonlinear system modeling;

- The accommodation of plant dynamics;

The AI applications in the design and implementation of automatic control systems have been broadly described as "intelligent control". Such decision-making is inevitably autonomous and should result in improved overall performance over time. In this context, a neural-network-based fuzzy logic control strategy has been adopted in our system. The rational for this selection is that a precise linear dynamic model of our pipe crawler cannot be obtained. FLC's incorporate heuristic control knowledge in the form of *"IF-THEN"* rules and are a convenient choice when a precise linear dynamic model of the system to be controlled cannot be easily obtained.
Furthermore, FLC's have also shown a good degree of robustness in face of large variability and uncertainty in the system parameters (Wang, 1994),(Dimeo & Lee, 1995). An ANN can learn fuzzy rules from I/O data, incorporate prior knowledge of fuzzy rules, fine tune the membership functions and act as a self learning fuzzy controller by automatically generating the fuzzy rules needed (Jang, 1993). This capability of the NN was utilized to form an FL-based controller based on data obtained via Human-In-The-Loop (HITL) simulator.

5.2.1 Structure of the FLC

The rule-base of the proposed FLC contains rules of first order TSK type (Takagi & Sugeno, 1985). In our proposed FLC the two inputs to the controller are *error in linear velocity of the robot* $e(t)$ and the *rate of change in the error* $\dot{e}(t)$ as follows:

$$\begin{cases} e(t) &= \dot{Z}_{set} - \dot{z}(t); \\ \dot{e}(t) &= -\ddot{z}(t); \end{cases} \tag{21}$$

where "$\dot{Z}_{set}$" is the set-point in velocity. The controller output is the voltage applied to the DC motor of the hub, namely $v(t)$. The rationale for this selection of the input variables is that, intuitively speaking, human makes a decision about the value of $v(t)$ based on a visual feedback (detailed under human-in-the-loop simulator) of the change of the velocity of the robot (i.e. $e(t)$) and the rate of this change (i.e. $\dot{e}$(t)). This FLC adjusts the control variable, namely the input voltage provided to the hub's actuator in order to maintain a constant speed in the robot when subjected to flow disturbances.
The structure of ANFIS model implemented is based on :

- A first order TSK fuzzy model where the consequent part of the fuzzy *IF-THEN* rules is first order in terms of the premise parameters;
- To performs fuzzy "AND", algebraic "minimum" is manipulated as the *T-norm* ;
- To performs fuzzy "OR", algebraic "maximum" is manipulated as the *T-norm* ;
- Three sets of *product-of-two-sigmoidal* MF's on each input were implemented.

These MF's are depicted in Fig. 4 and are represented by :

$$f(x;\mathbf{q}) = \frac{1}{1+e^{-a_1(x-c_1)}} \times \frac{1}{1+e^{-a_2(x-c_2)}} \tag{22}$$

where $\mathbf{q} = [a_1, a_2, c_1, c_2]$.

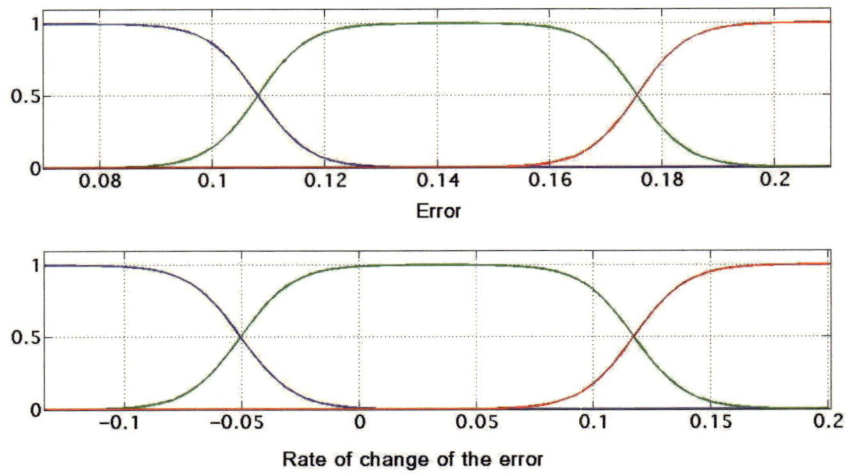

Fig. 4. Membership functions on the two inputs of the system : error and the rate of change in error before tuning.

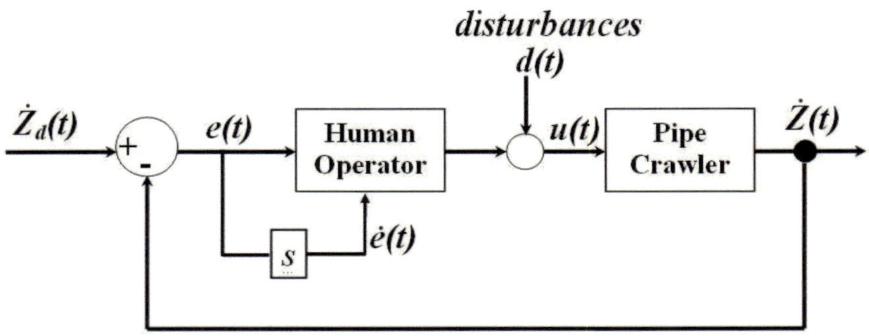

Fig. 5. Closed-loop system of the HITL simulator.

5.2.2 Human-In-the-Loop Simulator (HITL)

A real-time virtual reality HITL simulator was designed. Data acquired via this simulator was employed for training the ANFIS. The operator learns to control the velocity of the pipe crawler when subjected to flow disturbances, in the Human-Machine Interface (HMI) designed for this purpose. Fig. 5 shows the closed-loop system modeled in the HITL simulator. In this research we replace the "human operator" of the closed-loop with a stand-alone FLC whose parameters are tuned using the data acquired from the human operator, as depicted in Fig. 6

The disturbance on the system is simulated in the form of step changes in the flow velocity in the pipe. A snapshot of the HMI is given in Fig. 7. In this figure, $\dot{z}(t)$ and $\dot{Z}_{set}$ are depicted on top with a solid and a dashed line, respectively. The randomly generated flow disturbance (used for training) is also shown at the bottom of the figure. We will show through simulation that the controller tuned based on this type of disturbance is capable of rejecting different disturbances such as sinusoidal as well.

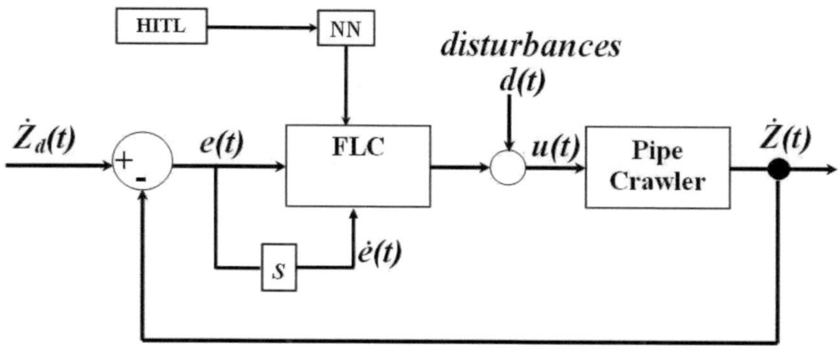

Fig. 6. FLC-based closed-loop system.

5.2.3 Acquiring Real-Time Data

The simulink model used for this purpose is depicted in Fig. 8. The disturbance in form of flow velocity and also the open-loop control signal in form of voltage (controlled by the trainee subject as explained below) are applied to the simulated system and the required data for training ANFIS (i.e. applied voltage $v(t)$, error $e(t)$ and the rate of change of error $\dot{e}(t)$) are captured and saved for manipulation in ANFIS. Also, the scope is the aforementioned HMI as in Fig. 7. A joystick was used as the *haptic device* to control the voltage applied to the DC motor actuator in the simulation environment and also experiment. The operator can continuously monitor the robot motion in real-time to correct its course of motion by varying the voltage provided to the motor.The objective is to make $\dot{z}(t)$ follow $\dot{Z}_{set}$ closely and consequently minimize the error.

Following the above procedure, we asked our trainee to accomplish the control task in the presence of step flow disturbance. The trainees go through a few trials in order to become an expert and the data provided by them can be used for training our ANFIS. The data acquisition time was set at 40s for the trainee to have enough time, between each of the four jumps in the flow velocity, to bring the system back to its set-point.

5.3 ANFIS Architecture

Here we elaborate on the ANFIS structure adopted in the proposed servomechanism control problem.

As explained previously (see section 5.2.1) there are three MF's on each input which yield a rule base with nine fuzzy if-then rules of first order TSK type (Turing, 1950).

$$\textit{Rule \#i}: \text{IF } e(t_h) \text{ is } A_{j1} \text{ and } \dot{e}(t_h) \text{ is } A_{j2} \text{ THEN } v_i = p_i e(t_h) + q_i \dot{e}(t_h) + r_i$$

where $i = \{1,\ldots,9\}$ is the rule number, $\{e(t_h),\dot{e}(t_h)\}$ are the numerical values of the error inputs at sampling time t_h and A_{jk}'s are linguistic variables (i.e. { *NEGATIVE* , *ZERO* , *POSITIVE* }). Also $j = \{1,2,3\}$ is the node number and $k = 1,2$ is the indicator of the input ("1" referring to a linguistic variable on "e" and "2" referring to a linguistic variable on "$\dot{e}$") .

The corresponding equivalent ANFIS structure is shown in Fig. 9. The node functions in each layer are of the same family.

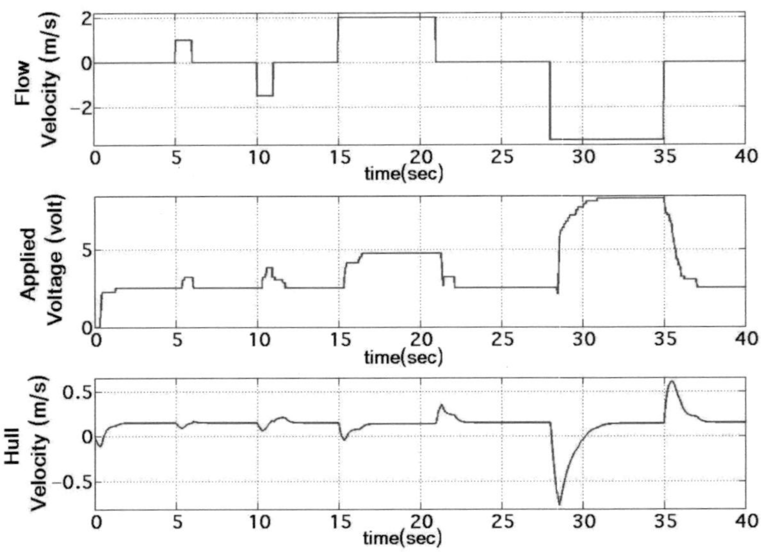

Fig. 7. A snapshot of the HMI used in this paper.

5.3.1 Hybrid Learning Rule

The architecture of ANFIS shows that the output can be expressed as: (Ghafari et al., 2006):

$$output = F(\vec{I}, S) \tag{23}$$

where $\vec{I}$ is the set of input variables S in the set of parameters. There will exist an identity function H such that the composite of $H \circ F$ is linear in some of the elements of consequent parameters S, then these elements can be identified by the Least Squared Estimation (LSE). More formally, if the parameter set S can be decomposed into two sets as:

$$S = S_1 \oplus S_2 \tag{24}$$

where $\oplus$ represents direct sum, such that $H \circ F$ is linear in the elements of S_2, then upon applying H to (23), we have:

$$H(output) = H \circ F(\vec{I}, S) \tag{25}$$

which is linear in the elements of S_2. Hence, given values of premise parameters S_1, we can plug P training data into (25) and obtain a matrix equation :

$$\mathbf{AX} = \mathbf{B} \tag{26}$$

where $\mathbf{X}$ is a vector of unknown parameters in S_2, and $\mathbf{A}$ and $\mathbf{B}$ are the set of inputs and outputs, respectively. Let $|S_2|$=M, then the dimensions of $\mathbf{A}$, $\mathbf{X}$ and $\mathbf{B}$ are $P \times M$, $M \times 1$ and $P \times 1$, respectively. As the number of training data P is usually greater than the number of linear parameters M, a least squared estimate is used to seek $\mathbf{X}$. On the other hand, the error measure for the *p-th* ($1 \leq p \leq P$) training data can be defined as the sum of squared errors:

$$E_p = \sum_{m=1}^{\#(L)} (T_{m,p} - O_{m,p}^L)^2 \tag{27}$$

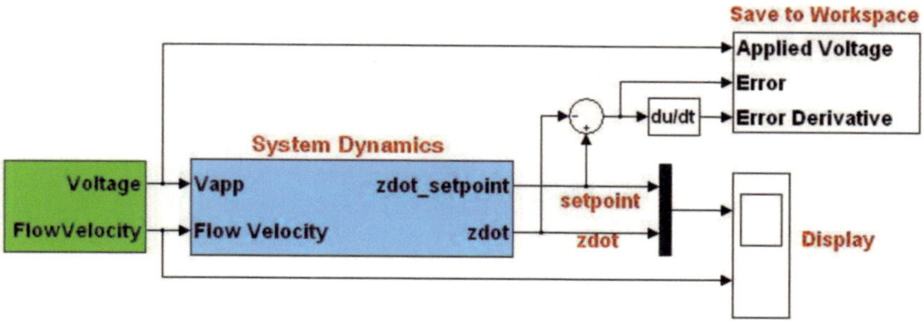

Fig. 8. Simulink model used for data acquisition.

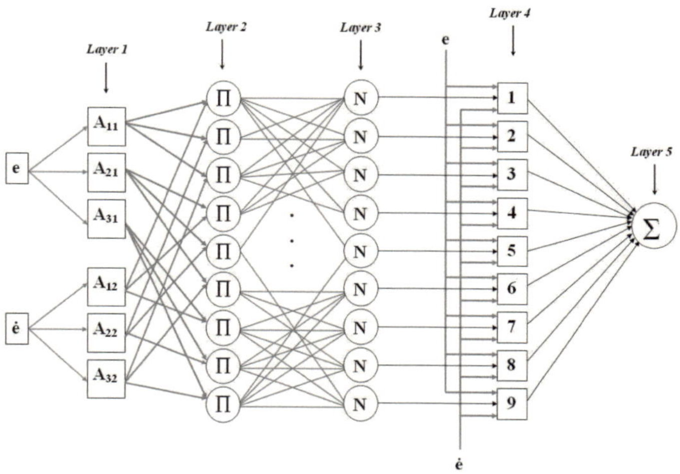

Fig. 9. The ANFIS structure adopted in this work.

where $T_{m,p}$ is the *m-th* component of the *p-th* target output vector, and $O^L_{m,p}$ is the *m-th* component of actual output vector produced by the presentation of the *p-th* input vector. Therefore, the overall error measure is equal to $E = \Sigma E_p$ and the derivative of the overall error measure E with respect to the premise parametes α is:

$$\frac{\partial E}{\partial \alpha} = \sum_{p=1}^{P} \frac{\partial E_p}{\partial \alpha} \tag{28}$$

The updated formula for the premise parameters α is :

$$\Delta\alpha = -\eta \frac{\partial E}{\partial \alpha} \tag{29}$$

where:

$$\eta = \frac{k}{\sqrt{\sum_{\alpha} (\frac{\partial E}{\partial \alpha})^2}} \tag{30}$$

is the learning rate for α and k is the step size and can be varied to change the speed of convergence.

5.3.2 Hybrid Learning Algorithm

Given the values of the premise parameters, the overall output of the proposed type-3 ANFIS structure can be expressed as a linear combination of the consequent parameters, i.e. the output v can be expressed as :

$$\begin{aligned} v &= \sum_{i=1}^{9}\left(\frac{w_i v_i}{\sum_{i=1}^{9} w_i}\right) \\ &= \sum_{i=1}^{9}((\bar{w}_i e)p_i) + \sum_{i=1}^{9}((\bar{w}_i \dot{e})q_i) + \sum_{i=1}^{9}((\bar{w}_i)r_i) \end{aligned} \tag{31}$$

which is linear in terms of the consequent parameters $\{p_i, q_i, r_i\}$.

a) Forward Pass : In the forward pass of the hybrid learning algorithm, the node outputs go forward till layer 4 where the consequent parameters are identified by the *Least Square Estimate (LSE)* from (26).

a) Backward Pass : In the backward pass, the error rates of each node output propagate from the output end toward the first layer, where now the premise parameters are updated by the gradient descent using (29).

Table 2 summarizes the activities in each path. This hybrid learning algorithm is shown to efficiently obtain the optimal premise and consequent parameters during the learning process.

	Forward Pass	**Backward Pass**
Premise Parameters	Fixed	Gradient Descent
Consequent Parameters	LSE	Fixed
Signals	Node Outputs	Error Rates

Table 2. The hybrid learning procedure for ANFIS in two passes (Jang, 1993)

5.3.3 Tuning the FLC using ANFIS

In order to tune parameters of both the linguistic variables' membership functions $\mu_{A_{jk}}(x_k)$ (i.e. the set $\{a_1, a_2, c_1, c_2\}$ as in (22)) and the parameters of the rules' consequents (i.e. $\{p_i, q_i, r_i\}$ for each rule i) we used the acquired data (see section 5.2.3) based on $\{e(t), \dot{e}(t)\}$ as inputs to the controller and the DC motor voltage $v(t)$ as the output of the system. In other words, using ANFIS, the objective is to find a relationship between the inputs and output of the controller of the form $v(t) = k_1 e(t) + k_2 \dot{e}(t) + k_3$ for each rule i. One can readily conclude by referring to (31) that :

$$k_1 = \sum_{i=1}^{9} w_i p_i \quad ; \quad k_2 = \sum_{i=1}^{9} w_i q_i \quad ; \quad k_3 = \sum_{i=1}^{9} w_i r_i$$

For this purpose, each trainee accomplishes the control task for 4000 time steps or 40 seconds in each trial (sampling time was set at $\delta t = 0.01 sec$). From each training run, 2000 data points were randomly selected to tune the FLC using ANFIS. After having been trained, ANFIS was tested with the remaining 2000 sampled data for verification.

6. Simulation and Experimental Results

6.1 Simulation Results

MATLAB VR2008a together with *SIMULINK*, the *Fuzzy Logic Toolbox* and *WinCon V5.0* from Quanser (Quanser, 2009) were used for real-time simulation of our proposed system. The control objective was to maintain a pre-set constant linear speed $\dot{Z}_{set}$ while moving the robot inside a vertical pipe in the presence of hydrodynamic forces due to flow. The SIMULINK model of the feedback-loop with the proposed FLC is shown in Fig. 10.

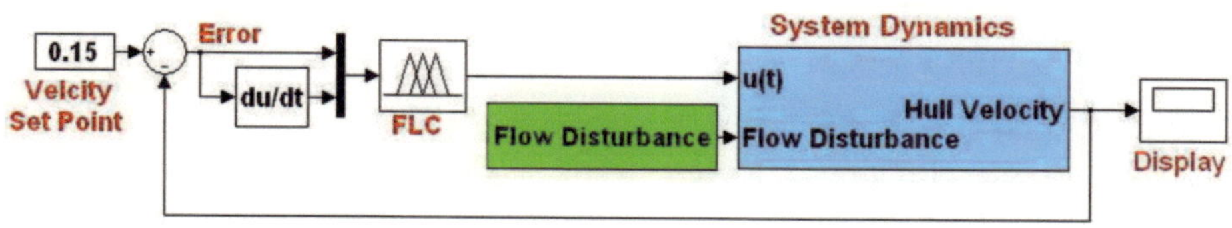

Fig. 10. Closed-loop system using stand-alone FLC used in simulation.

6.1.1 External Disturbance Models

Two flow disturbance models were used in the simulation environment : (1) *step* changes and (2) *sinusoidal* changes in flow velocity as depicted on top of Fig. 11.

A variety of simulations were conducted based on the classical PID and also the stand-alone

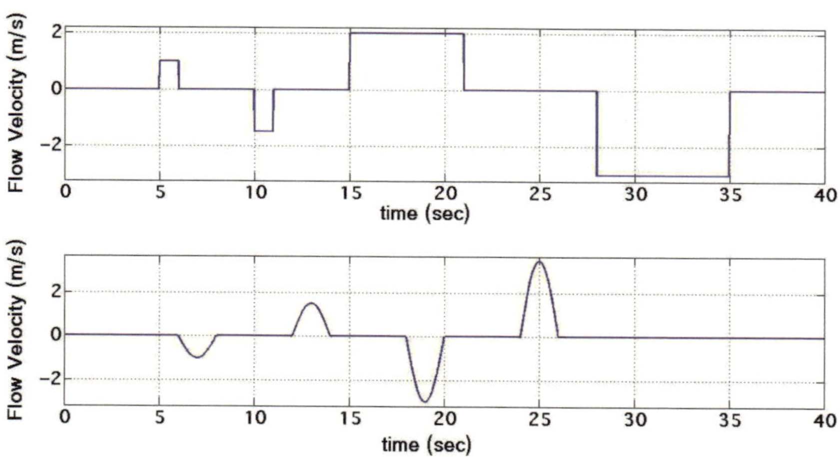

Fig. 11. Flow disturbance models used in simulation.

intelligent controller (FLC based on ANFIS), both of which were tested in a closed-loop system in the presence of the two aforementioned disturbance models and $\dot{Z}_{set}\frac{m}{s} = \{0.10, 0.15, 0.30\}$.

6.1.2 PID Control

The tests were carried out with a classical PID controller of the form :

$$u(t) = K_p e(t) + K_d \frac{de}{dt} + K_I \int_0^t e(\tau) d\tau + u_0 \tag{32}$$

The standard PID controller was designed in accordance with the Ziegler-Nichols tuning criteria (Ziegler & Nichols, 1993). The best value of gains were found to be $K_p = 20.4$, $K_i = 250$ and $K_d = 0.1285$ for proportional, integral and derivative gains, respectively. It is noteworthy that the term u_0 in (32) is to compensate for the gravitational force applied to the robot (vertical pipe).
The PID controller was designed such that the closed-loop control system would be stable and also meet given specifications associated with the following (Ang et al., 2005):

1) Stability Robustness ;
2) Tracking performance at transient, including rise time, overshoot and settling time;
3) Regulation performance at steady state;
4) Robustness against environmental uncertainty.

The response of the closed-loop system using a classical PID controller is shown in Figs. 12 to 17.

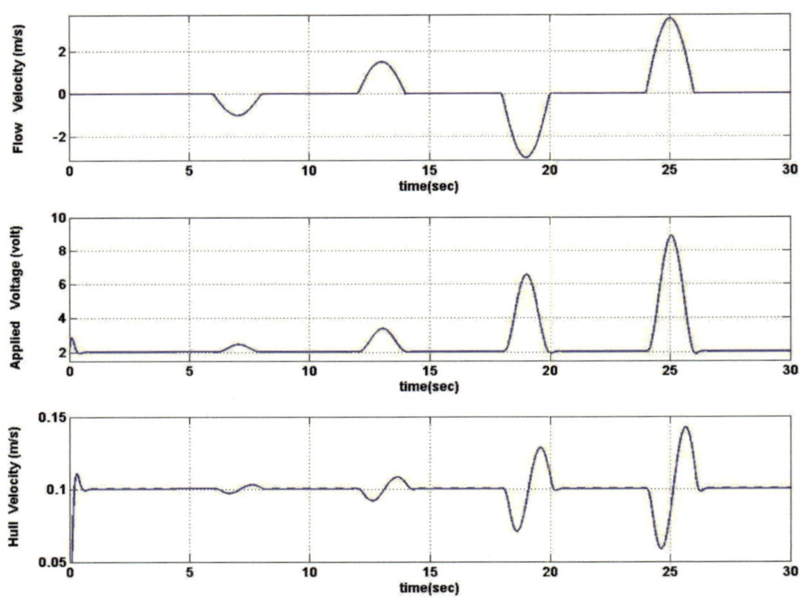

Fig. 12. Response of the closed-loop system with sinusoidal flow disturbance for $\dot{Z}_{set} = 0.10$ m/s using PID.

6.1.3 Fuzzy Logic Controller

The FLC was further optimized using ANFIS based on the following procedure:

- *Training:* A human expert was trained to accomplish the control task within a HITL real-time simulator in the presence of the flow disturbances explained above.
  One should note that for training purpose, we only used the following operating condition:
  - Step Changes in flow disturbance;
  - $\dot{Z}_{set} = 0.15\frac{m}{s}$.

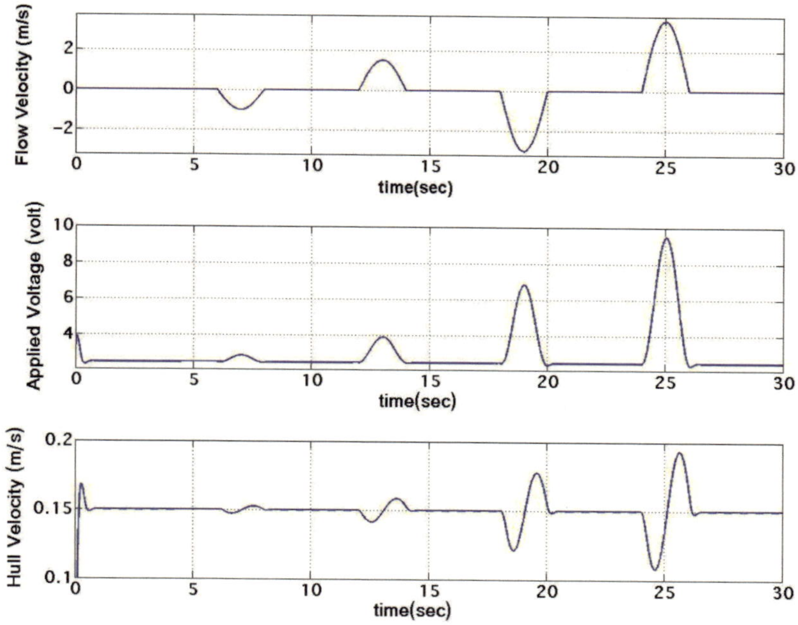

Fig. 13. Response of the closed-loop system with sinusoidal flow disturbance for $\dot{Z}_{set} = 0.15\frac{m}{s}$ using PID.

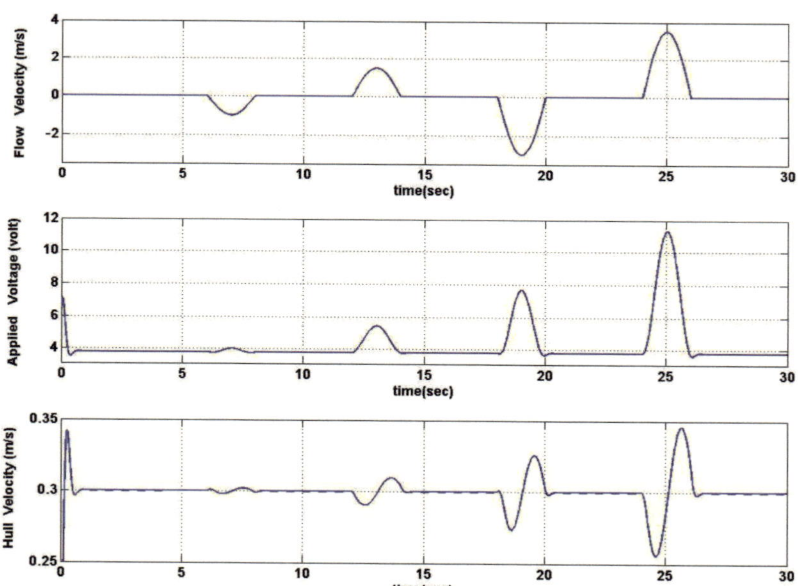

Fig. 14. Response of the closed-loop system with sinusoidal flow disturbance for $\dot{Z}_{set} = 0.30\frac{m}{s}$ using PID.

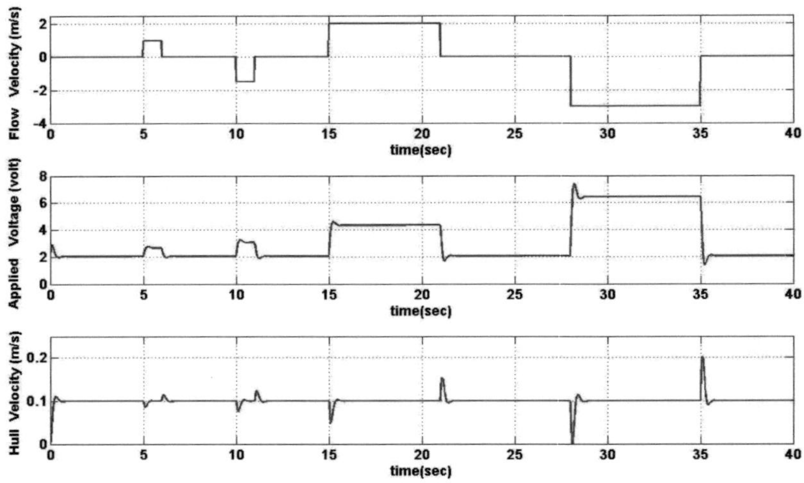

Fig. 15. Response of the closed-loop system with pulse flow disturbance for $\dot{Z}_{set} = 0.10\frac{m}{s}$ using PID.

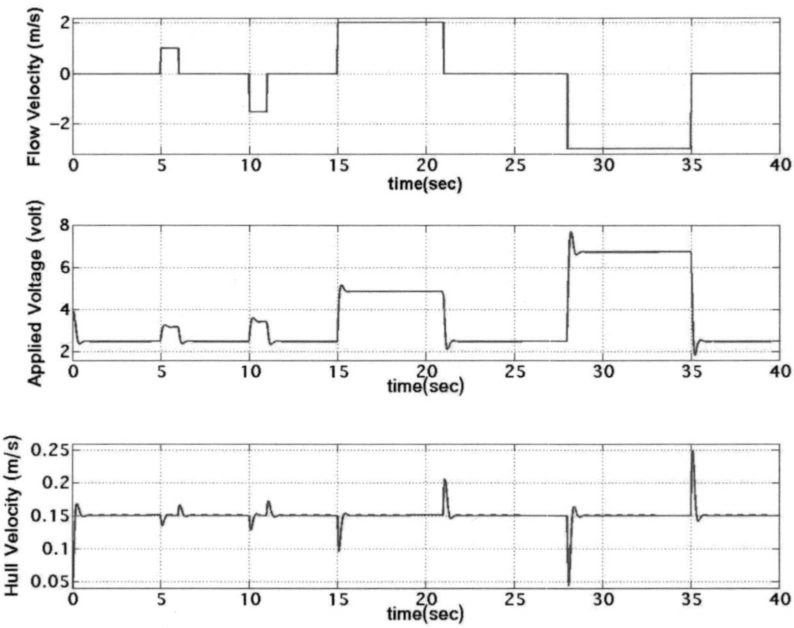

Fig. 16. Response of the closed-loop system with pulse flow disturbance for $\dot{Z}_{set} = 0.15\frac{m}{s}$ using PID.

The final training data (after a few learning trials) is shown in Fig. 18.

We will show through simulation that the FLC tuned based on ANFIS for one particular operating condition is capable of completing the servoing task under various operating

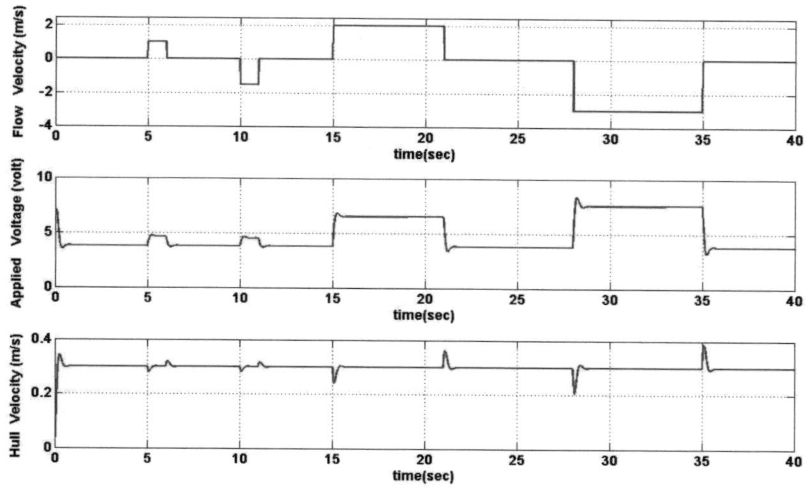

Fig. 17. Response of the closed-loop system with pulse flow disturbance for $\dot{Z}_{set} = 0.30\frac{m}{s}$ using PID.

conditions (i.e. velocity set-points and flow disturbance models, e.g. sinusoidal).

- *Tuning FLC using ANFIS :* Next we used the above acquired data to tune the parameters of the FLC in ANFIS. The inputs to the FLC were the *error,* $e(t)$ and the *rate of change of error,* $\dot{e}(t)$ (see section 5.2.1) and are presented in Fig. 19. The error tolerance in ANFIS was set at 10^{-6} and was reached after 97 epochs on average. The trend in epochs is depicted in Fig. 20. Also the modified MFs and the pertaining control surface after tuning are shown in Figs. 21 and 22, respectively.

  It is noteworthy that as the initial guess, the premise parameters are set to some (arbitrary) non-zero values as listed in Table 3 (INITIAL VALUES). Furthermore, for the consequent parameters, zero is taken as the initial guess. During the optimization process in ANFIS, both sets of values (premise and consequent parameters) are updated. The values of $\{a_l, c_l\}$ for $l = 1,2$ and $\{p_i, q_i, r_i\}$ for $i = 1,2,...,9$ after utilizing the ANFIS are listed in Tables 3 (FINAL VALUES) and 4.

The response of the closed-loop systems using optimized FLC via ANFIS is depicted in Fig. 23 to Fig. 28.

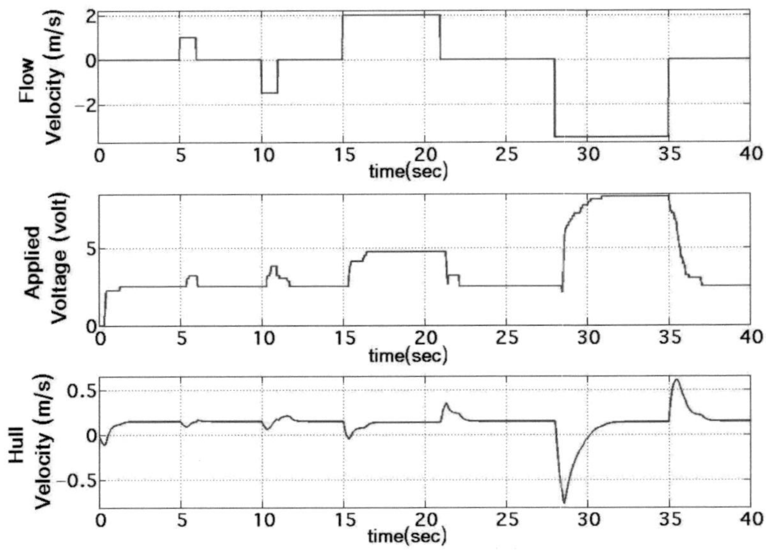

Fig. 18. The human operator's response to the flow disturbances within the HITL simulator.

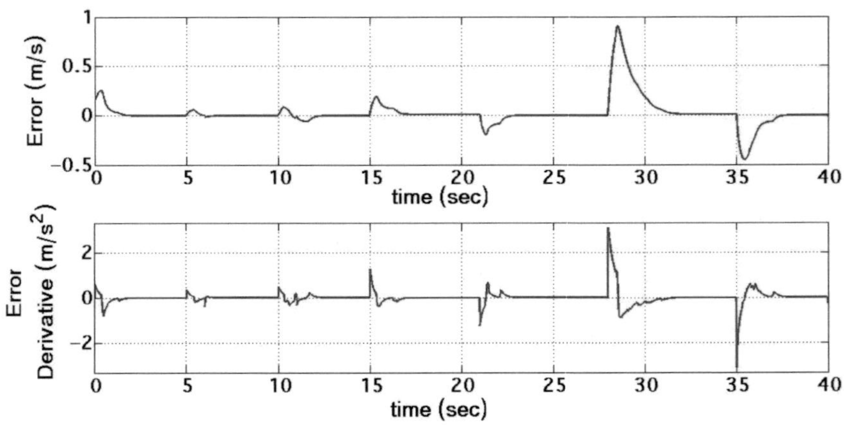

Fig. 19. Input data used in training ANFIS with step flow disturbance .

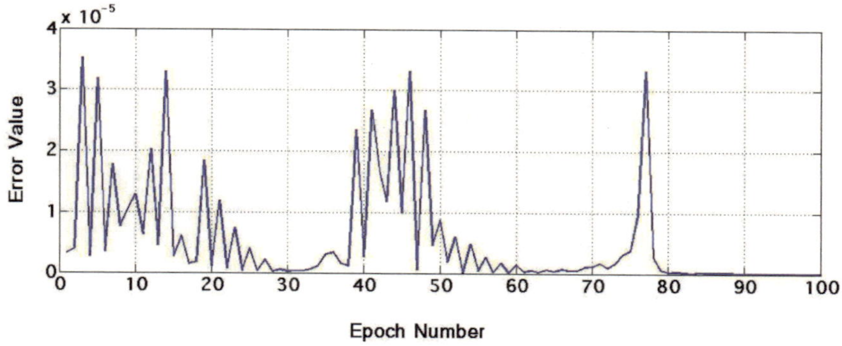

Fig. 20. Epoch evolution using ANFIS for sigmoidal membership function.

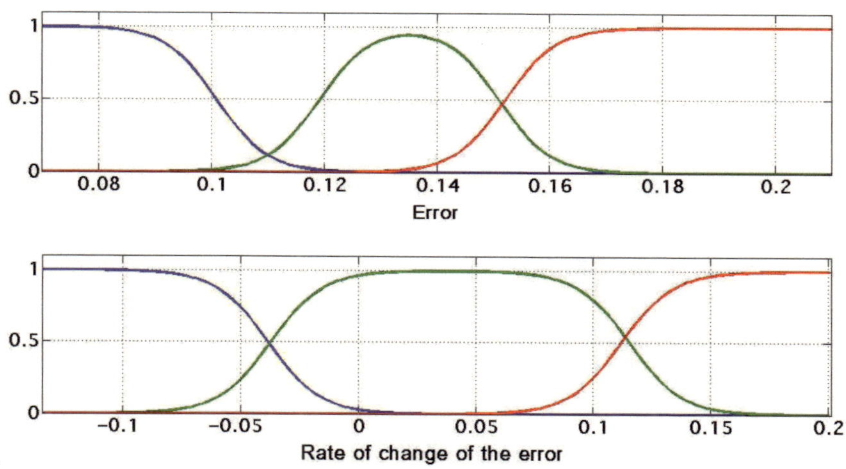

Fig. 21. MF's on the two inputs of the system : error and the rate of change in error, after tuning.

	MF	a_1	c_1	a_2	c_2
INITIAL VALUES	$\mu_{A_{11}}(e)$	222.4	0.041	-222.4	0.11
	$\mu_{A_{21}}(e)$	222.4	0.108	-222.4	0.18
	$\mu_{A_{31}}(e)$	222.4	0.1756	-222.4	0.24
	$\mu_{A_{12}}(\dot{e})$	89.33	-0.2185	-89.33	-0.05
	$\mu_{A_{22}}(\dot{e})$	89.33	-0.05	-89.33	0.12
	$\mu_{A_{32}}(\dot{e})$	89.33	0.12	-89.33	0.29
FINAL VALUES	$\mu_{A_{11}}(e)$	222.4	0.04	-222.4	0.10
	$\mu_{A_{21}}(e)$	222.4	0.12	-222.4	0.15
	$\mu_{A_{31}}(e)$	222.4	0.15	-222.4	0.24
	$\mu_{A_{12}}(\dot{e})$	89.33	-0.2185	-89.33	-0.03
	$\mu_{A_{22}}(\dot{e})$	89.33	-0.04	-89.33	0.12
	$\mu_{A_{32}}(\dot{e})$	89.33	0.1118	-89.33	0.29

Table 3. The initial value of the premise parameters.

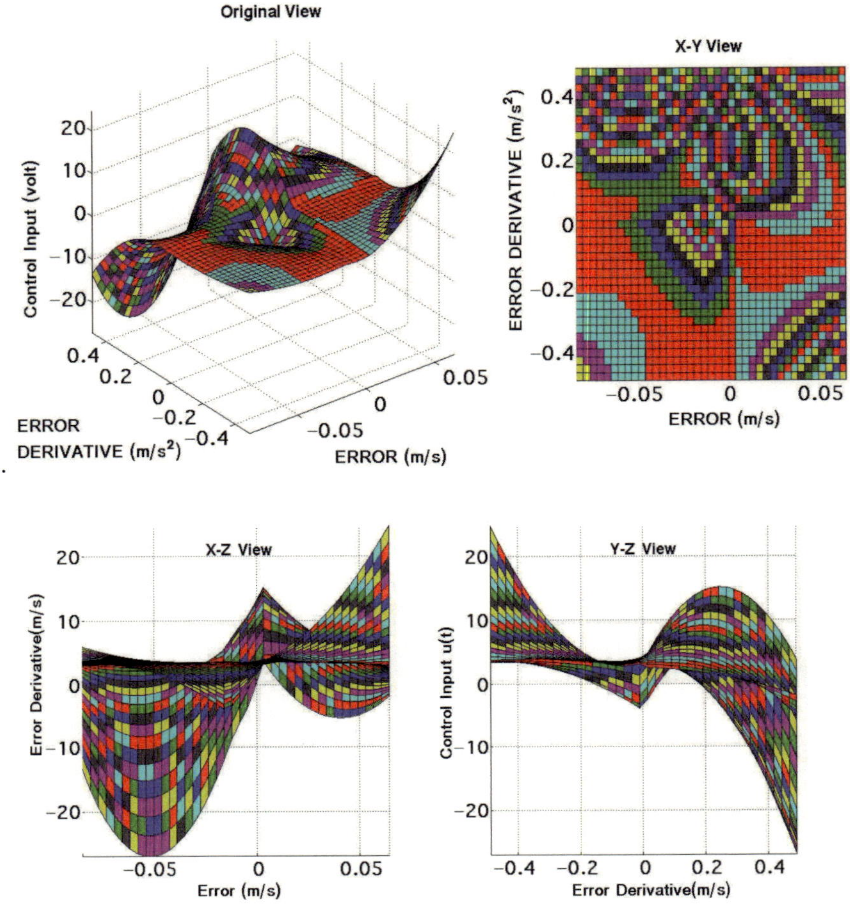

Fig. 22. Control surface of the FLC using three product-of-sigmoidal MF's, four views.

i	p_i	q_i	r_i
1	2.46	2.195	2.582
2	-0.129	1.607	2.728
3	-2.274	2.49	2.849
4	-1.602	2.343	3.03
5	10.36	2.962	0.9316
6	3.308	-1.373	2.846
7	8.068	1.932	1.277
8	8.917	1.853	1.116
9	2.43	0.072	2.497

Table 4. The final value of the consequent parameters.

6.2 Experimental Results

6.2.1 Experimental Setup

The real robotic pipe crawler was placed in an empty transparent PVC pipe of 6 inches in diameter for experimentation. The controller's objective was to track the reference input in form of a time-varying linear velocity set point. The trainee attempts to maneuver the real robot so that it follows the velocity set point visually represented in real time. The same approach was followed as in simulation to tune the FLC parameters.

6.2.2 Data Acquisition for the Real System

The position and essentially the velocity of the robot inside the pipe were captured through the optical encoder mounted on the motor. This data, in turn, was fed into the processing unit (PC) through a Q4 DAQ board from Quanser (Quanser, 2009).
Using a joystick as the haptic interface, the trainee, continuously receiving visual feedback on robot's motion on the monitor screen, applies a suitable voltage to the real system so that the linear velocity of the robot $\dot{z}(t)$ follows the desired trajectory $\dot{Z}_{set}$.

6.2.3 Tuning FLC

The data of the final trial (after a few times of training) was fed into ANFIS to tune the FLC. One should note that the FLC utilized in the experimental set up holds the same structure (i.e. type and number of MF's) used in simulation. After tuning the FLC through ANFIS, two experiments were conducted with a different $\dot{Z}_{set}$ for each, where the objective was to steer the robot along the pipe while $\dot{z}(t)$ follows the set point $\dot{Z}_{set}$ closely. The results are shown in Figs. 29 and 30 along with the control signal deployed. The human-analogous controller succeeded in carrying out the servoing task.

6.3 Discussion

6.3.1 System Performance

We have shown through simulations that not only is the FLC tuned by human data capable of accomplishing the proposed servoing task, but it also posed the following three major advantages over conventional PID controllers:

i. *Response Time:* the $(5-95)\%$ rise time t_r of the closed-loop system was decreased significantly. Table 5 summarizes the results for three different velocity set-points.

$\dot{Z}_{set} \frac{m}{s}$	FLC	PID	Improvement
0.1	0.04	0.2	400%
0.15	0.047	0.180	283%
0.3	0.086	0.158	83.72%

Table 5. Comparison of rise time t_r (in *second*) between FLC and PID.

ii. *Energy Consumption:* The control signal $u(t)$ used over the course of simulation has been decreased by using the proposed FLC. Tables 6 and 7 summarize the diffenerence in energy consumption when the robot is subjected to different flow disturbance models.

iii. *Actuator Saturation :* The control signal stays well below the saturation limit, i.e. $[-12, +12] volts$ when implementing the FLC in the closed-loop system. While using the PID controller, we reach the saturation limit at high amplitude external disturbances.

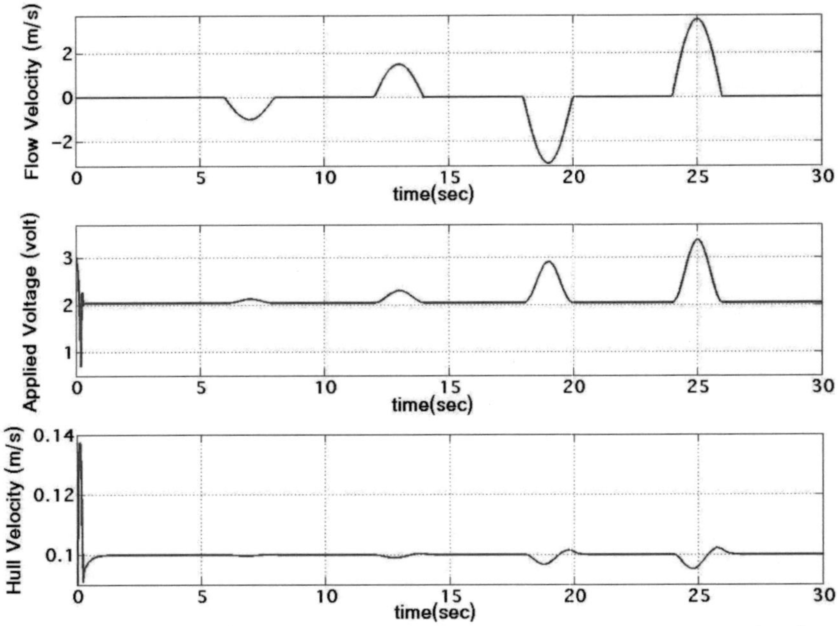

Fig. 23. Response of the closed-loop system with sinusoidal flow disturbance for $\dot{Z}_{set} = 0.10\frac{m}{s}$ using FLC.

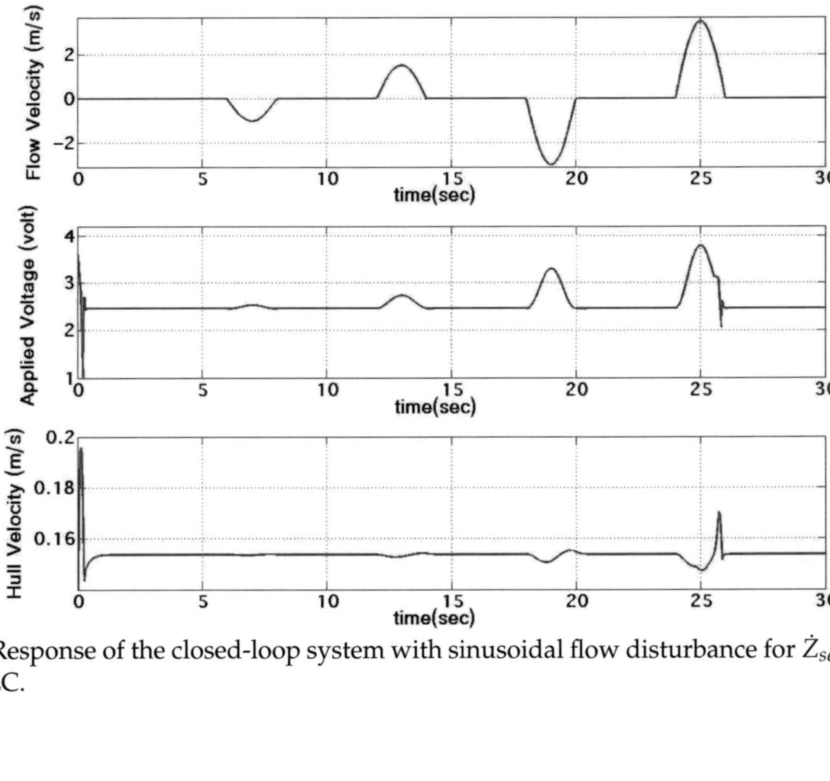

Fig. 24. Response of the closed-loop system with sinusoidal flow disturbance for $\dot{Z}_{set} = 0.15\frac{m}{s}$ using FLC.

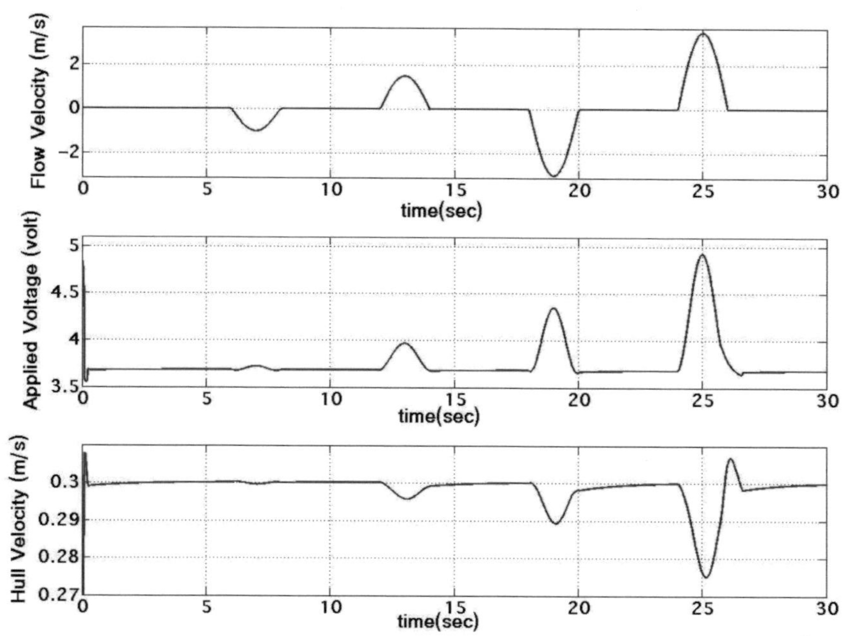

Fig. 25. Response of the closed-loop system with sinusoidal flow disturbance for $\dot{Z}_{set} = 0.30\frac{m}{s}$ using FLC.

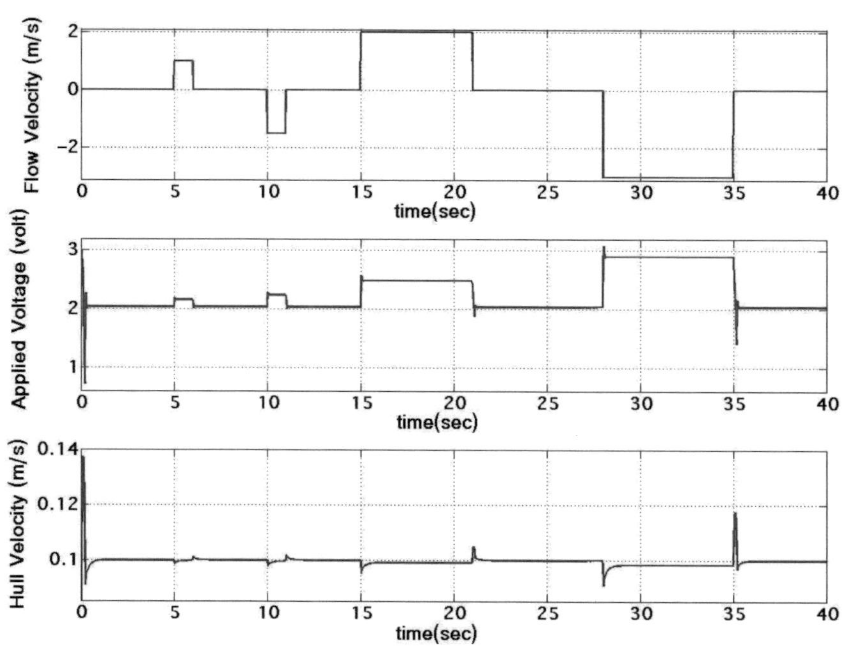

Fig. 26. Response of the closed-loop system with pulse flow disturbance for $\dot{Z}_{set} = 0.10\frac{m}{s}$ using FLC.

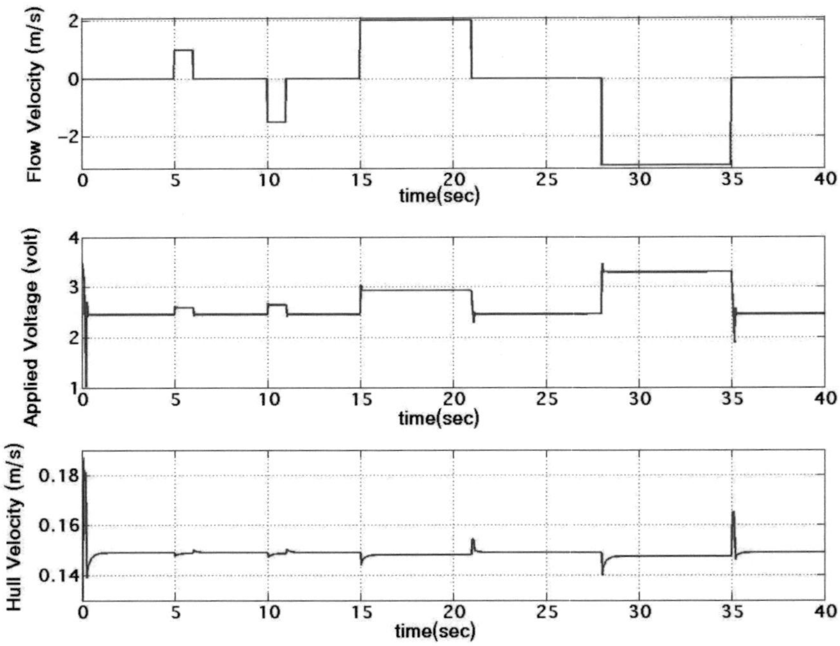

Fig. 27. Response of the closed-loop system with pulse flow disturbance for $\dot{Z}_{set} = 0.15\frac{m}{s}$ using FLC.

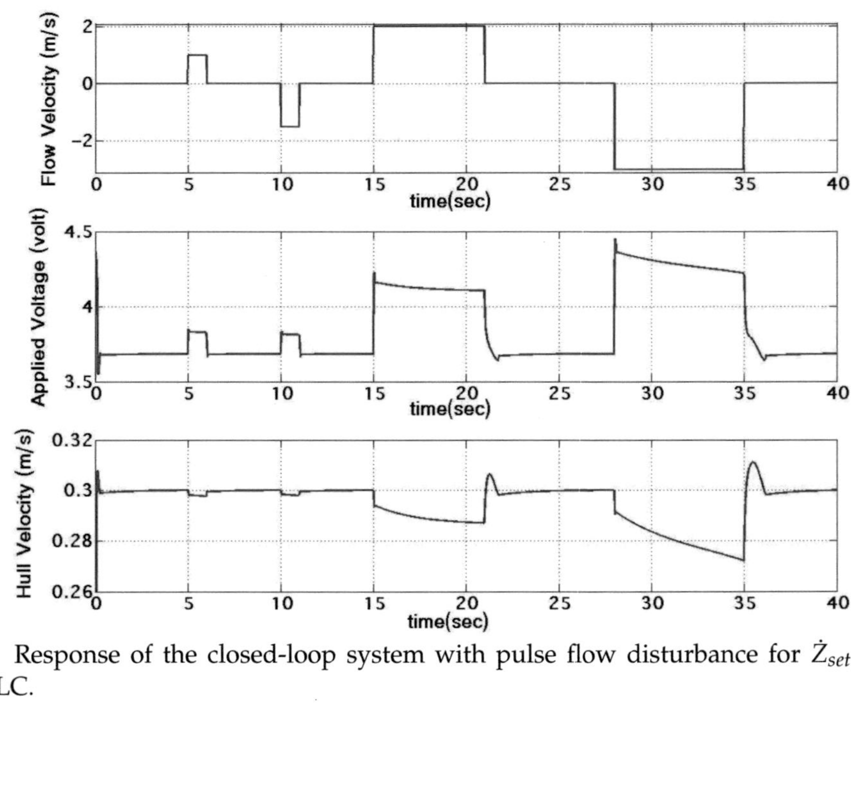

Fig. 28. Response of the closed-loop system with pulse flow disturbance for $\dot{Z}_{set} = 0.30\frac{m}{s}$ using FLC.

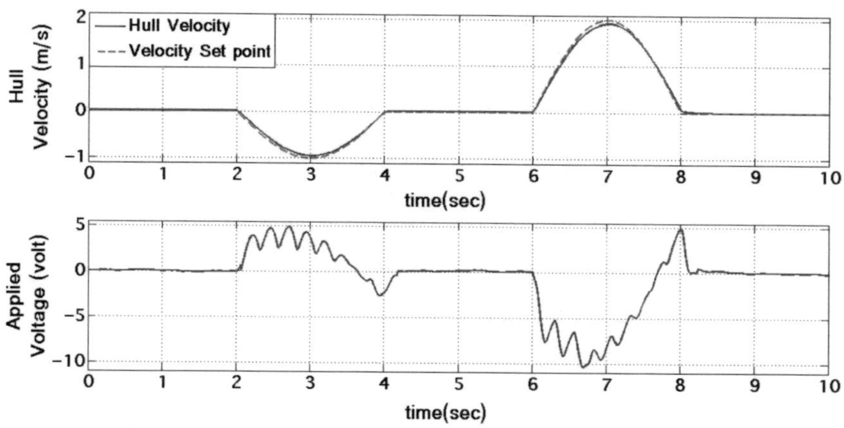

Fig. 29. Experimental results with sinusoidal reference trajectory of the velocity set point, $\dot{Z}_{set}$ using FLC.

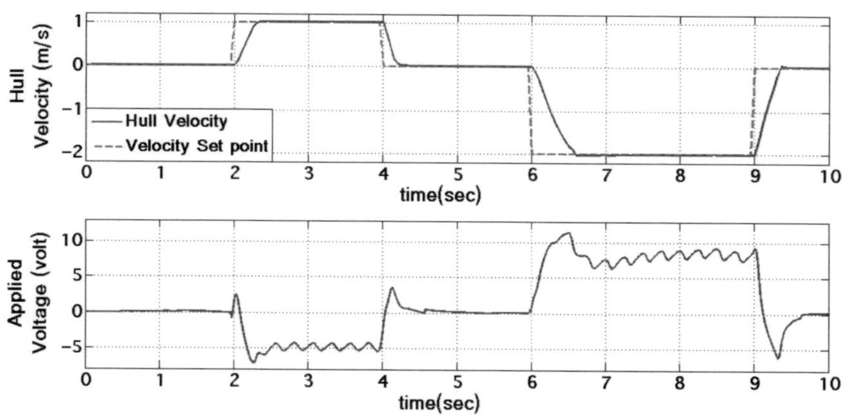

Fig. 30. Experimental results with step reference trajectory of the velocity set point, $\dot{Z}_{set}$ using FLC.

6.3.2 The Effect of MF's on the FLC Performance

As discussed previously (see section 5.2.1), we have chosen three *product-of-sigmoidal* MF's on each of the controller's inputs. This particular choice for MF's proved itself as a reasonable approximation function to relate the inputs (error and its derivative in robot's velocity) and output (voltage applied to the PMDC) , namely $u(t) = f(e(t),\dot{e}(t))$. As an alternative, trapezoidal MF's can be also employed. Fig. 31 compares the control surface of the FLC when using *product-of-sigmoidal* and *triangular* MF's. It is noteworthy that, the same set of human data were used for tuning both systems. From Fig. 31 one can readily deduce that a system with sigmoidal MF's results in a better approximation of the relationship $u(t) = f(e(t),\dot{e}(t))$. In technical terms, one can observe from Fig. 31 that higher control input would be required for large error in the system when using product-of-sigmoidal MF's.

The control surface on the right hand-side of Fig. 31 is not following the same expectation. The two peaks of the surface arise close to the origin (where both inputs to the system are approaching zero) and as we deviate to the edge of the surface, the control signal decreases. This

$\dot{Z}_{set} \frac{m}{s}$	FLC	PID	Improvement
0.1	83.77	94.46	12.76%
0.15	100.64	111.88	11.17%
0.3	149.13	164.75	10.47%

Table 6. Comparison of energy expenditure (*volt.s*) between FLC and PID for sinusoidal flow disturbance.

$\dot{Z}_{set} \frac{m}{s}$	FLC	PID	Improvement
0.1	90.30	127.64	41.35%
0.15	107.07	144.51	34.97%
0.3	154.06	195.72	27.04%

Table 7. Comparison of energy expenditure (*volt.s*) between FLC and PID for pulse flow disturbance.

phenomenon results in a poor performance of the closed-loop system compared to that when using product-of-sigmoidal MF's. Fig. 32 shows the control input required to servo the robot when subjected to flow disturbances using triangular and product-of-sigmoidal MF's. As can be seen from the figure, higher overshoots are observed in the FLC with triangular MF's.

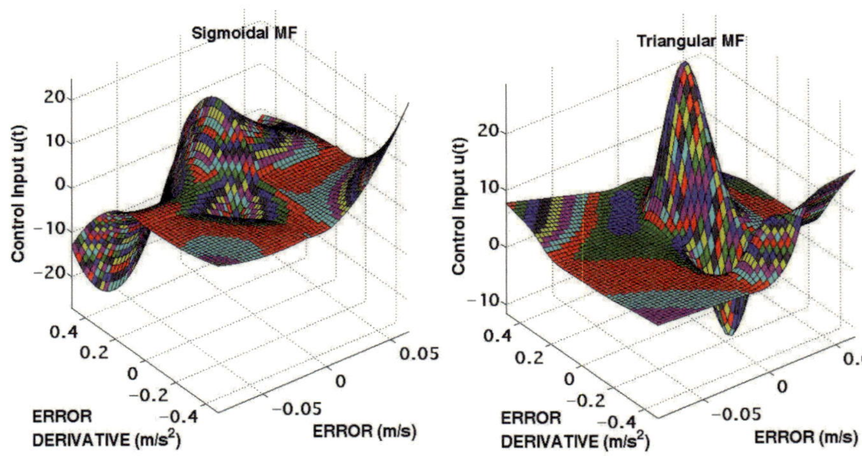

Fig. 31. Comparison between two control surfaces using triangular (left) and sigmoidal MF's (right).

7. Conclusion and Future Works

7.1 Conclusion

We addressed the design and development of a pipe crawler for water pipe inspection along with the detailed derivation and analysis of its governing dynamics equations. Also we implemented a new method for tuning the parameters of a Fuzzy-Logic based controller to regulate

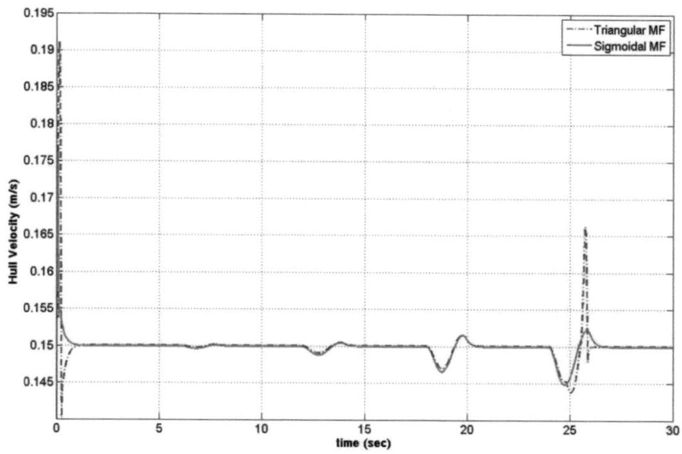

Fig. 32. Compariosn between the response of the closed-loop system using FLC with triangular and sigmoidal MF's.

the pipe crawler's velocity. The controller was designed based on a TSK model whose parameters were tuned using ANFIS. This was carried out by incorporating human analogous control strategy, obtained via a real-time HITL virtual reality simulator, into the ANFIS composition. It was shown that the proposed controller, tuned based on a specific operating condition, was capable of performing the velocity control task under various unrehearsed operating conditions as well. The controller was implemented in both simulation and experiments. It was shown that the proposed FLC outperformed a conventional PID controller.

7.2 Future Work

The future work is twofold: (1) developing a Hardware-In-the-Loop (HIL) simulation system, as depicted in Fig. 33, to control the motion of the robot when located in an empty pipe (or duct) in a dry lab where a motorized flow simulator, mechanically connected to the robot crawler via a force coupling device, will be employed to simulate the effect of hydrodynamic forces exerted on the robot as it were moving inside a live pipe, and (2) using a force/vision haptic interface for HITL control. The haptic device used in data acquisition was a regular joystick commonly used in video games. In this way, the output limitation affects the precision of human action over the course of control task. With a specialized force/vision feedback control interface, the operator will be able to control the robot motion more accurately, and consequently the optimized human analogous FLC would yield better results.

8. References

Ang, K., Chong, G., Li, Y., Ltd, Y. & Singapore, S. (2005). PID control system analysis, design, and technology, *IEEE transactions on control systems technology* **13**(4): 559–576.

Bradbeer, R., Harrold, S., Luk, B., Li, Y., Yeung, L. & Ho, H. (2000). A mobile robot for inspection of liquid filled pipes, *Workshop on Service Automation and Robotics, City University of Hong Kong*.

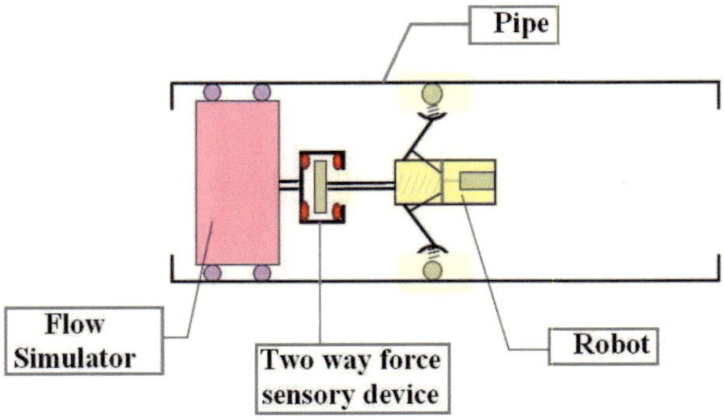

Fig. 33. The proposed HIL simulation system.

Chaudhuri, T., Hamey, L. & Bell, R. (1996). Historical Perspective:" From Conventional Control to Autonomous Intelligent Methods", *IEEE CONTROL SYSTEMS MAGAZINE* **16**: 78–87.

Dimeo, R. & Lee, K. (1995). Boiler-turbine control system design using a genetic algorithm, *IEEE Transaction on Energy Conversion* **10**(4): 752–759.

Eiswirth, M., Frey, C., Herbst, J., Jacubasch, A., Held, I., Heske, C., Hötzl, H., Kuntze, H., Kramp, J., Munser, R. et al. (2001). Sewer assessment by multi-sensor systems, *IWA 2nd World Water Congress*, pp. 15–19.

Ghafari, A., Meghdari, A. & Vossoughi, G. (2006). Intelligent Control of Powered Exoskeleton to Assist Paraplegic Patients Mobility using Hybrid Neuro-Fuzzy ANFIS Approach, *IEEE International Conference on Robotics and Biomimetics, 2006. ROBIO'06*, pp. 733–738.

Glass, S., Levesque, M., Engels, G., Klahn, F. & Fairbrother, D. (1999). UNDER-WATER ROBOTIC TOOLS FOR NUCLEAR VESSEL AND PIPE EXAMINATION, *Framatome ANP, Lynchburg Virginia* .

Griffiths, G. (2003). *Technology and applications of autonomous underwater vehicles*, CRC Press.

Grigg, N. (2006). Assessment of Water Distribution Pipes, *Infrastructure Systems* **12**(3): 147–153.

Gummow, R. & Eng, P. (2000). Corrosion Control of Municipal Infrastructure Using Cathodic Protection, *Materials Performance* .

Horodinca, M., Doroftei, I., Mignon, E. & Preumont, A. (2002). A simple architecture for in-pipe inspection robots, *Int. Colloquium on Mobile and Autonomous Systems*, Vol. 10, pp. 25–26.

Hunt, K., Sbarbaro, D., Żbikowski, R. & Gawthrop, P. (1992). Neural networks for control systems: a survey, *Automatica (Journal of IFAC)* **28**(6): 1083–1112.

Jang, J. (1993). ANFIS: Adaptive-network-based fuzzy inference system, *IEEE transactions on systems, man, and cybernetics* .

Koji, K. (1999). Underwater inspection robot–AIRIS 21, *Nuclear Engineering and Design(Switzerland)* **188**(3): 367–371.

Kulpate, C. (2006). Design and development of a stereo scanner for condition assessment of water pipes, *Technical Report UR/CSIR-01-2006*, University of Regina.

Lee, C. (1990). Fuzzy logic in control systems: fuzzy logic controller. II, *IEEE Transactions on systems, man and cybernetics* **20**(2): 419–435.

Mamdani, E. et al. (1974). Application of fuzzy algorithms for control of simple dynamic plant, *Proc. Iee* **121**(12): 1585–1588.

Miwa, Y., Satoh, S. & Hirose, N. (2002). Remote-Controlled Inspection Robot for Nuclear Facilities in Underwater Environment, *Conference: 10. international conference on nuclear engineering-ICONE 10, Arlington-Virginia (United States), 14-18 Apr 2002.*

Nauck, D., Klawonn, F. & Kruse, R. (1997). *Foundations of neuro-fuzzy systems*, John Wiley & Sons, Inc. New York, NY, USA.

Nguyen, T., Yoo, H., Rho, Y. & Kim, S. (2001). Speed control of PIG using bypass flow in natural gas pipeline, *IEEE International Symposium on Industrial Electronics, 2001. Proceedings. ISIE 2001*, Vol. 2.

Nickols, F., Bradbeer, R. & Harrold, S. (1997). An ultrasonically controlled autonomous model submarine operating in a pipe environment, *Proceedings of 4th International Conference on Mechatronics and Machine Vision in Practice,(M 2 VIP97)*, pp. 142–147.

Quanser (2009). Consulting co., `http://www.quanser.com`.

Ratanasawanya, C., Binsirawanich, P., Yazdanjo, M., Mehrandezh, M., Poozesh, S., Paranjape, R. & Najjaran, H. (2006). Design and Development of a Hardware-in-the Loop Simulation System for a Submersible Pipe Inspecting Robot, *Electrical and Computer Engineering, 2006. CCECE'06. Canadian Conference on*, pp. 1526–1529.

Roh, S. & Choi, H. (2005). Differential-drive in-pipe robot for moving inside urban gas pipelines, *IEEE Transactions on Robotics* **21**(1): 1–17.

Shiho, M., Horioka, K., Inoue, G., Onda, M., Leighty, W., Yokoo, K., Ono, S., Ohashi, K. & Hirata, M. (2004). Proposal for Environmental Observation System for Large Scale Gas Pipeline Networks Using Unmanned Airship, *AIP Conference Proceedings*, Vol. 702, p. 522.

Takagi, T. & Sugeno, M. (1985). Fuzzy identification of systems and its applications to modeling and control, *IEEE transactions on systems, man, and cybernetics* **15**(1): 116–132.

Turing, A. (1950). Computing machinery and intelligence, *Mind* **59**(236): 433–460.

Wang, L. (1993). Stable adaptive fuzzy control of nonlinear systems, *IEEE Transactions on Fuzzy Systems* **1**(2): 146–155.

Wang, L. (1994). *Adaptive fuzzy systems and control: design and stability analysis*, Prentice-Hall, Inc. Upper Saddle River, NJ, USA.

Wang, L. & Mendel, J. (1992). Fuzzy basis functions, universal approximation, and orthogonalleast-squares learning, *IEEE Transactions on Neural Networks* **3**(5): 807–814.

Zadeh, L. (1973). Outline of a new approach to the analysis of complex systems and decision processes, *IEEE Transactions on Systems, Man, and Cybernetics* **3**: 28–44.

Ziegler, J. & Nichols, N. (1993). Optimum settings for automatic controllers, *Journal of dynamic systems, measurement, and control* **115**: 220.

APPENDIX

A. Dynamics Model of the Proposed Pipe Inspecting Robot (Kinetic Energy of the Robot's Wheel)

In order to derive the dynamics model of our proposed system, three coordinate frames, as shown in Figure 3, are taken into consideration which are as follows:

- Inertial reference system - This is the earth-fixed coordinate system defined by a right-handed, orthogonal X_i, Y_i, Z_i coordinate frame. The Z_i axis is parallel to the pipe's centerline and is pointed upward. The X_i and Y_i directions and the location of the origin can be selected arbitrarily (within the requirement of orthogonality). The inertial frame is denoted by letter i.

- Body-fixed coordinate frame - This is the body-fixed coordinate system that is defined by a right-handed, orthogonal coordinate system attached to the robot's hull. Its origin lies on the hull's centre of gravity. The Z_B axis is set parallel to the Z_i and is pointed upward. The X_B and Y_B axes remain perpendicular to the Z_B axis within the requirement of orthogonality. The body-fixed frame is denoted by letter B.
- Wheel-fixed coordinate frame - This is the right-handed orthogonal reference frame fixed to the wheel with its origin located on the intersection point between the Y_B axis and the wheel center. The Z_W axis is perpendicular to the wheel's plane pointing towards the vehicle's direction of motion. It makes an angle of δ with respect to the the Z_B axis. The X_W and Y_W axes remain perpendicular to the Z_W axis within the requirement of orthogonality. The wheel-fixed frame is denoted by letter W.

One should note that the extension of robot's arms is not considered in the dynamics model for simplicity. Therefor distance b remains at constant . Transformations from the wheel- and body-fixed coordinate frames to the inertial reference frame are described here. The general orientation of the wheel-fixed frame (hereinafter called wheel frame) represented in the inertial frame can be utilized through a number of successive rotations called Euler Angles. The relative rotation between the wheel frame and inertial frame can be represented as follows:

$$\mathbf{R}^i_W = \mathbf{R}^i_B \times \mathbf{R}^B_W \tag{A-1}$$

where " $\times$ " denotes matrix product and $\mathbf{R}^i_B$ and $\mathbf{R}^B_W$ are as listed below:

$$\mathbf{R}^i_B = \begin{bmatrix} C_\phi & -S_\phi & 0 \\ S_\phi & C_\phi & 0 \\ 0 & 0 & 1 \end{bmatrix} \tag{A-2}$$

$$\mathbf{R}^B_W = \begin{bmatrix} C_\delta C_\theta & -C_\delta S_\theta & S_\delta \\ S_\theta & C_\theta & 0 \\ -S_\delta C_\theta & S_\delta S_\theta & C_\delta \end{bmatrix} \tag{A-3}$$

Hence:

$$\mathbf{R}^i_W = \begin{bmatrix} C_\phi C_\delta C_\theta - S_\phi S_\theta & -C_\phi C_\delta S_\theta - S_\phi C_\theta & C_\phi S_\delta \\ S_\phi C_\delta C_\theta + C_\phi S_\theta & -S_\phi C_\delta S_\theta + C_\phi C_\theta & S_\phi S_\delta \\ -S_\delta C_\theta & S_\delta S_\theta & C_\delta \end{bmatrix} \tag{A-4}$$

In (A-1) through (A-3), ϕ, θ, and δ denote the rotational angle of the robot's body with respect to the inertial frame, the rotational angle of the wheel with respect to the body frame, and the inclination angle of the wheels, respectively. One should note that the following notation is used in long equations; $C_a = cos(a)$, and $S_a = \sin(a)$.
Total kinetic energy of each robot's wheel can be calculated as follows:

$$T_{wheel} = \frac{1}{2} m (\mathbf{v}^i_W)^T (\mathbf{v}^i_W) + \frac{1}{2} (\mathbf{!}^i_W)^T \mathbf{I}^i_W (\mathbf{!}^i_W) \tag{A-5}$$

where $\mathbf{v}^i_W, \mathbf{!}^i_W$, and $\mathbf{I}^i_W$, m denote the linear velocity of the origin of the wheel frame, the angular velocity of the wheel frame ,the wheel's inertial tensor all represented in the inertial frame and the wheel's mass, respectively.These terms are described below in more detail.
One can write:

$$(\mathbf{v}^i_W) = (\mathbf{v}^i_B) + (\mathbf{v}^i_{B,W}) + (\mathbf{!}^i_B) \times (\mathbf{r}^i_{B,W}) \tag{A-6}$$

where $\mathbf{v}^i_B$ denotes the velocity of the origin of the body frame represented in the inertial frame, $\mathbf{v}^i_{B,W}$ denotes the relative velocity of the wheel frame and body frame represented in the inertial frame, $!^i_B$ denotes the angular velocity of the body frame represented in the inertial frame, and $\mathbf{r}^i_{B,W}$ denotes the vector connecting the origin of the body frame to the origin of the wheel frame represented in the inertial frame. One can readily conclude:

$$(\mathbf{v}^i_B) = \begin{bmatrix} 0 & 0 & rS_\delta\dot{\theta} \end{bmatrix}^T \tag{A-7}$$

and :

$$(!^i_B) = \begin{bmatrix} 0 & 0 & \dot{\phi} \end{bmatrix}^T \tag{A-8}$$

With the assumption that the robot's arms are fixed, namely $b = Constant$ (see Figure 3), one can conclude: $\mathbf{v}^i_{B,W} = \mathbf{0}^T$. One can also write:

$$\begin{aligned} (\mathbf{r}^i_{B,W}) &= (\mathbf{R}^i_B)(\mathbf{r}^B_W) \\ &= (\mathbf{R}^i_B)\begin{bmatrix} 0 & b & 0 \end{bmatrix}^T \\ &= \begin{bmatrix} -bS_\phi & bC_\phi & 0 \end{bmatrix}^T \end{aligned} \tag{A-9}$$

After substituting (A-7-A-9) into (A-6), one gets:

$$\mathbf{v}^i_W = \begin{bmatrix} -bC_\phi\dot{\phi} & -bS_\phi\dot{\phi} & rS_\delta\dot{\theta} \end{bmatrix}^T \tag{A-10}$$

Correspondingly, the angular velocity of the wheel frame represented in the inertial frame can be calculated as follows:

$$(!^i_W) = (!^i_B) + (!^i_{B,W}) \tag{A-11}$$

where $!^i_{B,W}$ denotes the relative angular velocity between the wheel frame and that for the body frame represented in the inertial frame. One can write:

$$\begin{aligned} (!^i_{B,W}) &= (\mathbf{R}^i_B)(!^B_W) \\ &= (\mathbf{R}^i_B)\begin{bmatrix} -S_\delta\dot{\theta} & 0 & -C_\delta\dot{\theta} \end{bmatrix}^T \\ &= \begin{bmatrix} -C_\phi S_\delta\dot{\theta} & -S_\phi S_\delta\dot{\theta} & -C_\delta\dot{\theta} \end{bmatrix}^T \end{aligned} \tag{A-12}$$

By substituting (A-8) and (A-12) in (A-11) one can write:

$$!^i_W = \begin{bmatrix} -\dot{\theta}C_\phi S_\delta & -\dot{\theta}S_\phi S_\delta & \dot{\phi} - \dot{\theta}C_\delta \end{bmatrix}^T \tag{A-13}$$

With the assumption that the wheel assembly has a symmetric mass distribution about its axis of rotation and with the assumption of small wheel's inclination angle, δ, its inertia tensor expressed in the inertial frame can be calculated as follows:

$$\mathbf{I}^i_W = \mathbf{R}^i_B \begin{bmatrix} I_{WX} & 0 & 0 \\ 0 & I_{WY} & 0 \\ 0 & 0 & I_{WZ} \end{bmatrix} (\mathbf{R}^i_B)^T \tag{A-14}$$

Where the diagonal of the inertia matrix given in (A-14) denotes the moment of inertia of the wheel around the X, Y, and Z axes of the wheel frame, respectively. One should note that $I_{WX} = I_{WY}$.

By substituting (A-10), (A-13) and (A-14) in (A-5) , one can derive the following for the kinetic energy of the angled-wheel:

$$T_{AW} = \left\{ \left(mr^2 + I_{WZ} \right) \left(\frac{bC_\delta}{b+r} \right)^2 + (mr^2 + I_{WX}) S_\delta^2 \right\} \frac{\dot{\theta}}{2} \qquad \text{(A-15)}$$

One should note that the kinematics constraint of $\dot{\phi} = \left(\frac{rC_\delta}{b+r} \right) \dot{\theta}$, applicable under no slippage condition on robot wheels, was utilized to derive (A-15) as well. (A-15) can be further simplified for small inclination angles of the wheels, δ in which case one can assume; $\sin(\delta) \approx 0$ and $\cos(\delta) \approx 1$ as follows:

$$T_{AW} = \left(\frac{b}{b+r} \right)^2 (mr^2 + I_{WZ}) \frac{\dot{\theta}^2}{2} \qquad \text{(A-16)}$$

Assuming the wheel as a cylinder with the radius of r and height of h , $I_{WZ} \cong \frac{mr^2}{2}$ and upon substituting in (A-16) yields:

$$T_{AW} = \frac{3}{4} \frac{mr^2 b^2}{(b+r)^2} \dot{\theta}^2 \qquad \text{(A-17)}$$

16

Development of mobile robots based on peristaltic crawling of an earthworm

Hayato Omori, Taro Nakamura,
Tomohide Iwanaga and Takeshi Hayakawa
Chuo University, Department of Precision Mechanics
Japan

1. Introduction

An earthworm moves by peristaltic crawling. It produces extension and contraction waves along the anteroposterior direction, from the anterior to posterior, by varying the thickness and length of its segments. This locomotion mechanism is in contact with a large area on the ground when the robot moves. Furthermore, it requires less space than other locomotion mechanisms, e.g. bipedal, wheel-based or meandering snake-like locomotion. Therefore, this mechanism is attractive not only for rescue and exploration robots operating in constrained environments but also for locomotive endoscope robots in the field of medical engineering. In recent years, some robotic and biomedical engineering researchers have developed peristaltic crawling robots based on earthworm locomotion. A prototype microrobot was developed, which moved using an external magnetic field (Saga et al., 2004). This robot consisted of individual cells made of natural latex rubber tubes with a water-based magnetic fluid sealed inside, corresponding to an earthworm's segments. This peristaltic crawling robot was developed for rescue applications (Saga et al., 2004). The robot is actuated by an artificial rubber muscle that is driven by a pneumatic system (Nakamura et al., 2003). Because this artificial muscle is reinforced longitudinally by Kevlar fibres, the actuator not only contracts in the longitudinal direction, but when internal pressure is applied, it also expands in the radial direction. This robot achieves peristaltic crawling by employing this muscle for segment contraction and expansion. Zou et al. (2005) proposed an earthworm-like microrobot for intestinal inspection. This microrobot consists of several extensor units that are actuated by a micro DC motor and a screw-driven device. Actuating the extensor units in an appropriate sequence allows the microrobot to propel itself. The above robot studies apply various actuation methods to achieve locomotion by peristaltic crawling. However, the motion velocity of these robots is very low and inefficient. Furthermore, only movement through a straight pipe has undergone experimental testing. The influences of various conditions on peristaltic crawling locomotion patterns of earthworm robots must be examined systematically to allow stable adjustment to environmental variations.
In this study, we developed two types of prototype robots based on peristaltic crawling of an earthworm and then introduce one of these as an earthworm robot. One is called the 'flexible-type robot', and consists of flexible units; the other is called the 'rigid-type robot',

which is able to bend between units. We proposed the flexible-type robot for moving passively in narrow spaces such as curved pipes, and the rigid-type robot for moving along given routes on a plane surface. In this paper, we first explain peristaltic crawling of an earthworm, which was analysed for comparison with the locomotion of the proposed robots. Using the flexible-type robot, we performed experiments with forward and turning movements on a plane surface, as well as in a curved pipe, to confirm whether the flexible units suit both environments. We propose two locomotion patterns for turns. The rigid-type robot was modelled to evaluate its position and attitude, and a simulation analysed its movement for given routes on a plane surface. We paid particular attention to the actual movement of an earthworm in dirt and developed an underground explorer robot. This unit was made of parallel links, and we performed experiments in dirt, which showed good performance.

2. Peristaltic crawling motion of an earthworm

2.1 The structure of an earthworm and its peristaltic motion mechanism

The inner structure of an earthworm is shown in Fig. 1. The earthworm consists of numerous segments divided by septa and a coelom containing the alimentary canal and nerve circuits. The inner wall of the body is composed of two muscle layers. The outside layer is called the circular muscle and the inside one is called the longitudinal muscle. When the circular muscle is actuated in a radial direction, the segment becomes thinner and extends in the axial direction. When the longitudinal muscle contracts in the axial direction, the segment gets thicker and shorter.

Fig. 2 shows the locomotion pattern of an earthworm during peristaltic crawling. The earthworm propagates a longitudinal wave from the anterior to posterior part using contraction muscles in consecutive segments. The locomotion pattern can be described as follows:

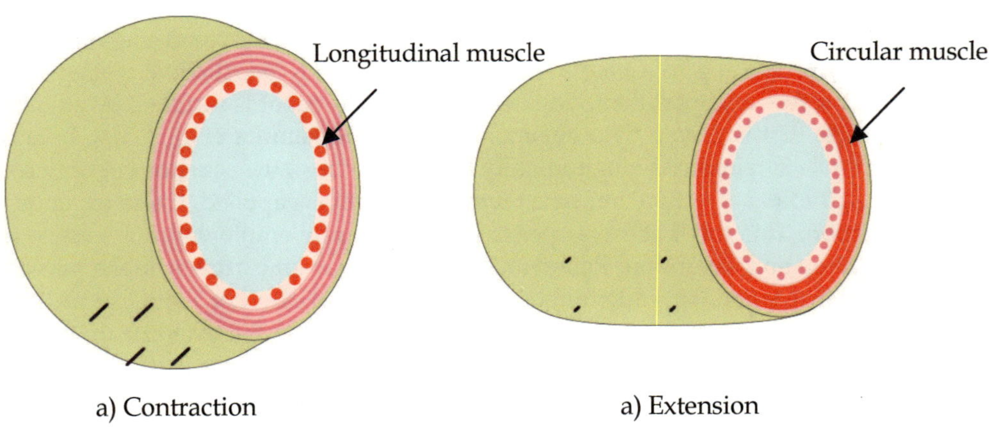

Fig. 1. Inner structure of an earthworm

1) The anterior segments of the earthworm are contracted by the longitudinal muscle. In this case, because the thicker segments are in contact with the surface during locomotion,

the friction between the segments and the ground is increased. Therefore, the segments remain in contact with the surface.
2) The contraction continuously propagates to the rear end. This movement pulls the rear segments in the direction of movement.
3) The anterior segments of the earthworm are extended in the axial direction by the circular muscle. In this case, since the friction between the segments and the moving surface is reduced, thinner segments can move smoothly. Furthermore, because the rear segments remain in contact with the surface, the thinner segments can move forward.
4) The extended segments continuously propagate in the rear direction. The forward segments are pushed in the direction of movement.

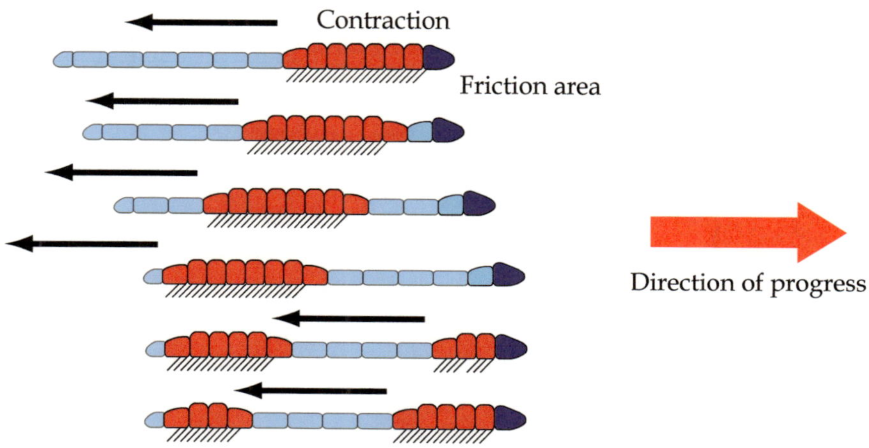

Fig. 2. Locomotion pattern of an earthworm

2.1 Measurement of an actual earthworm's peristaltic crawling

The motion of an actual earthworm was compared with the peristaltic crawling motion of the locomotion robot. A 140-mm-long and 4-mm-thick earthworm was used. Marks were placed on the earthworm's segments every 19 mm (Fig. 3). The locomotion was videotaped with a digital video camera and analysed using a movement analysis software (MOVIAS Pro: Nac Image Technology Inc.). The results obtained are shown in Fig. 4.

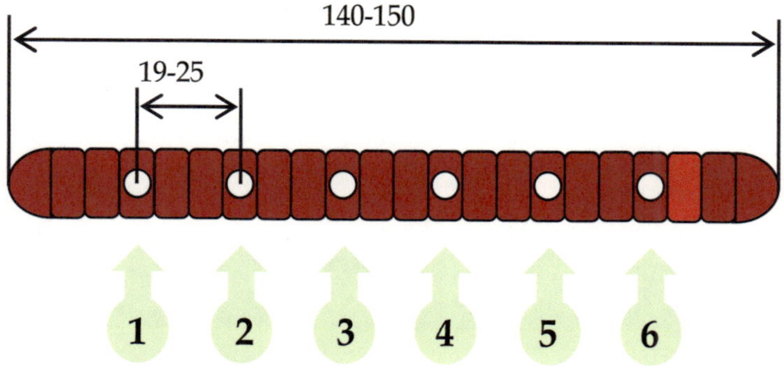

Fig. 3. Earthworm segments and markers

This figure shows that contraction begins from the anterior part of the earth-worm and continuously propagates towards the posterior part. The anterior segments contract again, after propagation to the posterior part is complete. The average velocity of the earthworm was 16.1 mm/s. The peristaltic crawling robot was designed by taking these results into consideration.

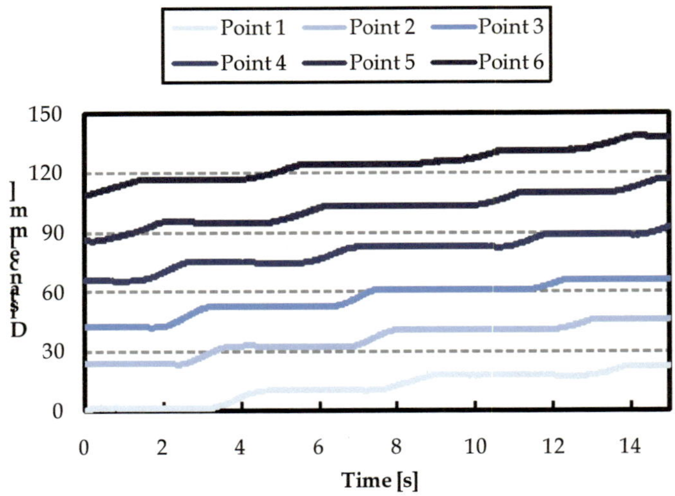

Fig. 4. Peristaltic crawling motion of the actual earthworm

3. Flexible-type robot

3.1 Configuration of a Robot with 7 Units

The peristaltic crawling robot of flexible type was developed by referring to an actual earthworm's motion. Fig. 5 shows the configuration of the robot. The complex mechanism of an actual earthworm was simplified. A unit of the robot is equal to a segment of an actual earthworm. Since peristaltic crawling requires at least 3 units, it has 7 units. Each of them can extend and contract in the axial direction, which is necessary to realise peristaltic crawling. A unit is actuated by servomotors controlled by the pulse signal of PIC. Rubber sheets are attached to each belt for generating friction between the robot and the ground or inner wall of a tube.

Fig. 5. Flexible-type robot

The length of 7 units is 386 mm when fully extended and 295 mm when fully contracted. The robot weights 341 g. Marks are placed on each unit, which are captured in a video, and then the movements of the robot are analysed.

3.2 The Structure of a Flexible Unit

The structure of a unit is shown in Fig. 6. It contains 2 servomotors, whose rotation is transmitted to the cranks and plate A is shifted back and forth. Therefore, the unit expands in the radial direction during contraction due to the curvature of the belts made of flexible plates (Fig. 7).

One of the features of the unit is flexibility. Plate A can be tilted by changing the angle of the servomotor (Fig. 7). A contraction, an extension and a tilt plate are smoothly realised by adopting a leaf spring as a crank. The tilt angle ranges from −20 to 20 degrees. Therefore, the robot has flexibility for environment shape (Fig. 8).

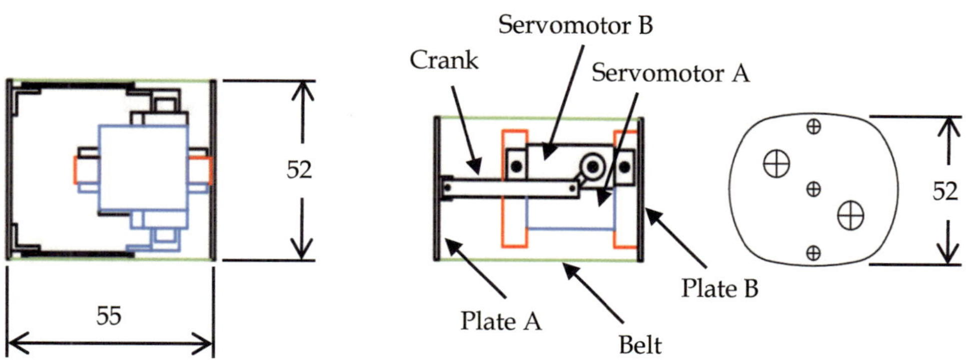

Fig. 6. Structure of a unit

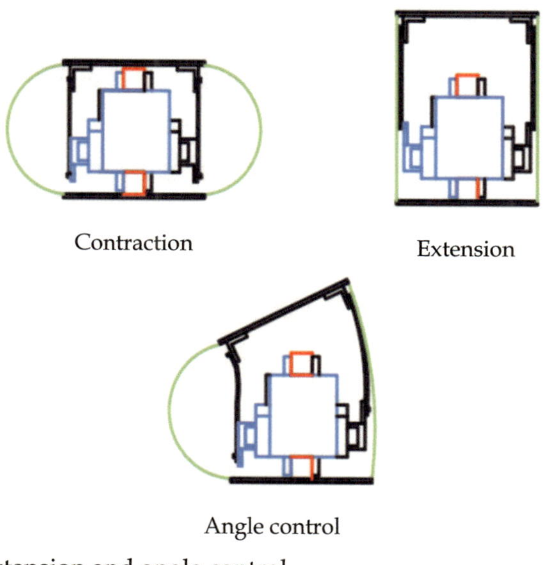

Fig. 7. Contraction, extension and angle control

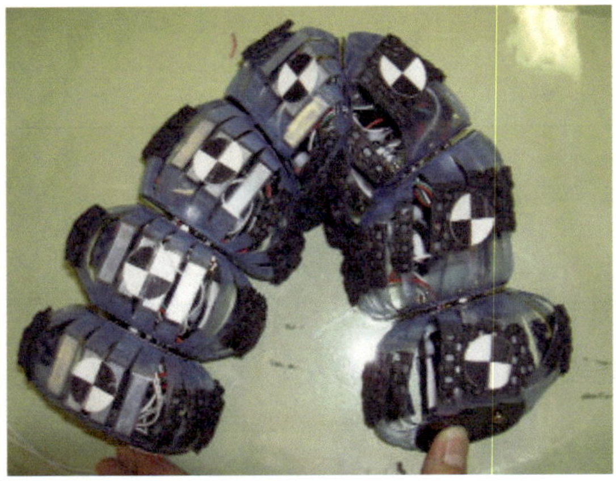

Fig. 8. Units of the robot

4. Experimental results and discussion

4.1 Configuration

In the experiments, the robot was actuated on a plane surface and in a tube. Vinyl chloride, whose friction coefficient is 0.5, was used for the experiments. The transparent tube has an inner diameter of 78 mm and a thickness of 3 mm.

4.2 Forward movement on the ground and comparison with an actual earthworm

The forward movement of the robot on the ground was compared with that of an actual earthworm. Fig. 9 shows the locomotion pattern I that contains 7 motions. Four contracted units were present in all motions, and these propagated along the anteroposterior direction between motions. The period of peristaltic crawling was 0.2 s, which is the time between motions. Fig. 10 shows the movement tracks of the robot.

Fig. 9. Pattern I

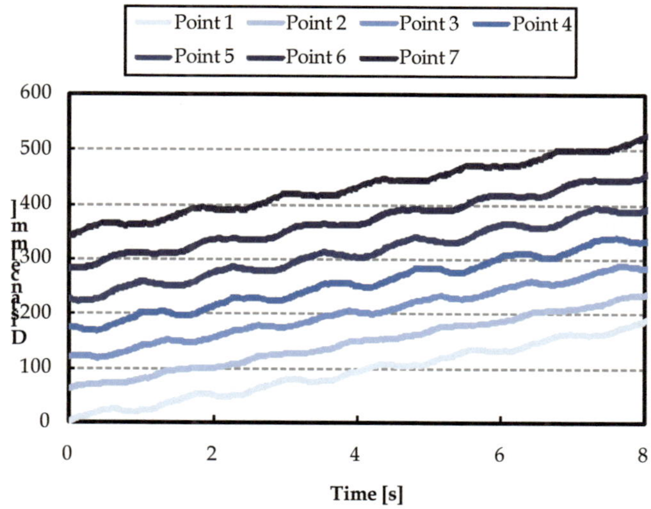

Fig. 10. Movement tracks of flexible robot

Fig. 10 shows that the extension and contraction waves propagate from the anterior to posterior part, resulting in peristaltic crawling of the robot. Compared to Fig. 4, Fig. 10 shows that the robot moved backward since the extension of the anterior part and the contraction of the posterior part occurred simultaneously and that the anterior part was pulled by the posterior part.

4.3 In the tube

The angle of the tube was set at 0 deg, 90 deg and −90 deg. These angles indicate that the robot moves in a horizontal direction, in a vertical upward direction and downward direction, respectively. The locomotion pattern I and a peristaltic crawling period of 0.2 s were used in the tube experiment.

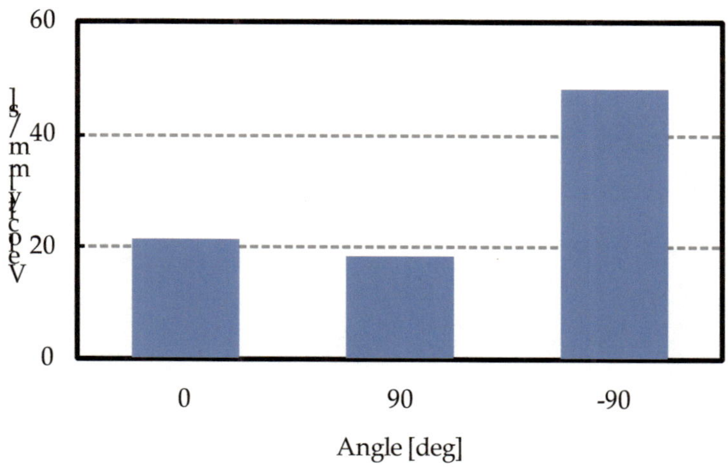

Fig. 11. Velocity in the tube

Fig. 12 shows the movement tracks of the robot at 0 deg, which were analysed to confirm whether the robot moved with peristaltic crawling. The velocity change due to change in the angle of the tube is shown in Fig. 11. Fig. 11 shows that a small difference in the velocity was observed at 90 deg, which was affected by the gravity. It was less than the difference we had expected. The velocity at −90 deg was twice that at 0 deg.

4.3 In the Curved tube

The flexibility of the robot in a curved tube was examined. Fig. 12 shows the shape of the curved tube, which was made of straight tubes and curved tubes. The robot moves straight, turns 90 deg to the left, moves straight again and then moves up vertically. The locomotion pattern I was used in this experiment.

Fig. 13 shows the photos captured when the robot moved in the curved tube. The robot took 150 s to move from A to B. We observed that the robot was able to move in the curved tube when the locomotion pattern was used for forward movement and not for turning. Therefore, we confirmed the robot had flexible units.

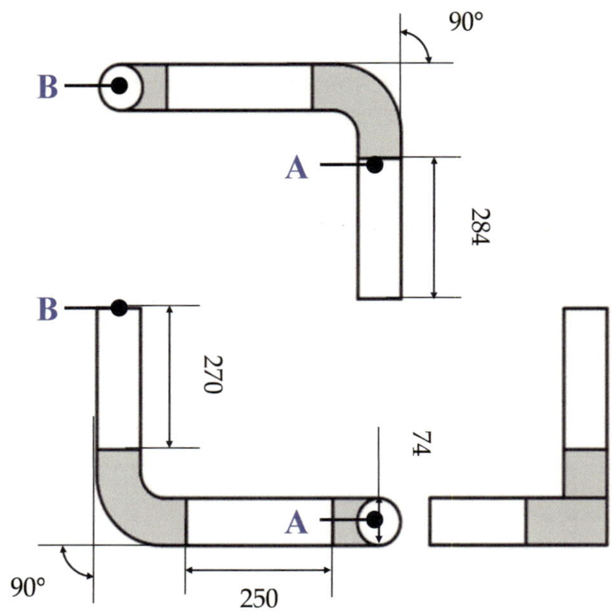

Fig. 12. Shape of the curved tube

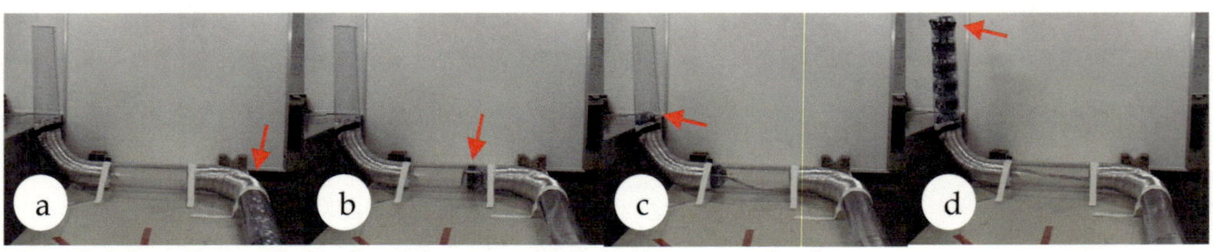

Fig. 13. Movement of the flexible-type robot in the curved tube

5. Turning experimental results and discussion

Turning patterns on the ground were examined. The robot could be turned by having the differences in extension and contraction between right and left segments. The period of peristaltic crawling was 0.2 s. The Y-coordinate, the X- coor-dinate, and θ are defined (Fig. 16). The X- coordinate is at both the anterior and posterior parts, and θ the rotation angle from Y to X.

5.1 Turn I and Turn II

Two patterns were examined to determine which side of the robot needed actuating with peristaltic crawling while turning.

Fig. 14 show turns I and II, respectively. In turn I, the sides of the segments opposite to the turning direction were all extended and those in the same direction were actuated with peristaltic crawling. In turn II, the sides of segments opposite to the turning direction were actuated with peristaltic crawling and those in the same direction were all contracted.

Fig. 15 shows movement tracks of turns I and II. The top track is first unit from the posterior part. Fig. 16 shows transition of the rotational angle.

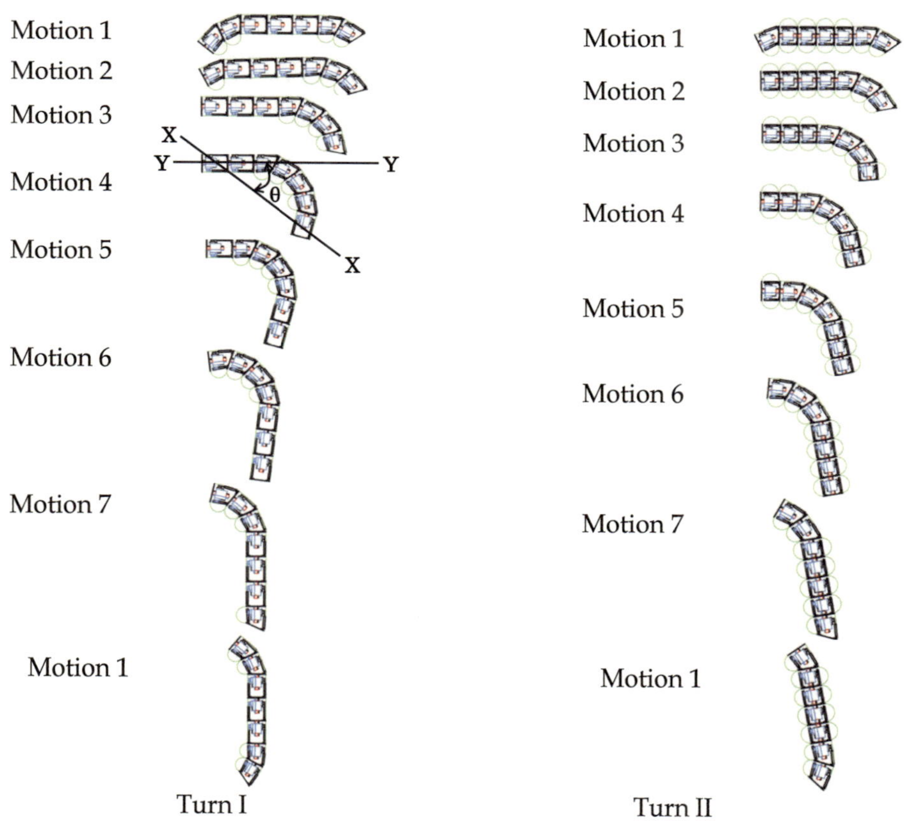

Fig. 14. Turn I and II

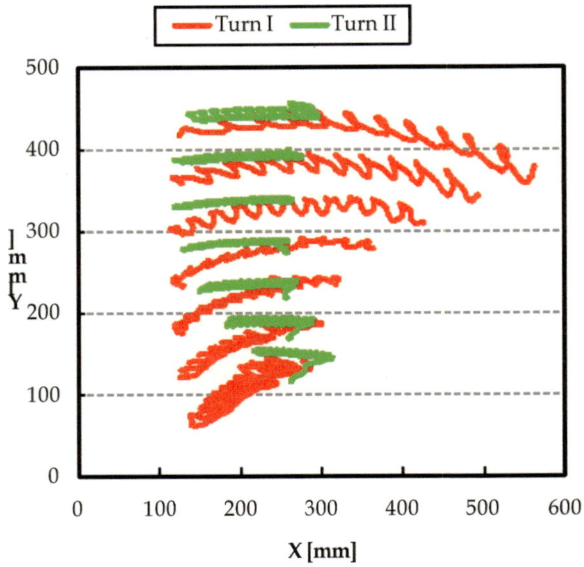

Fig. 15. Movement tracks of Turn I and II

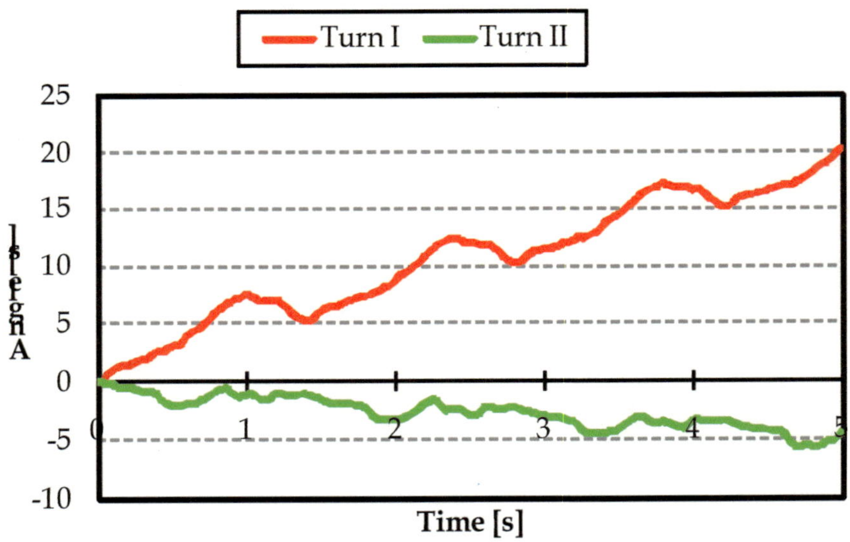

Fig. 16. Transition of the rotational angle

Figs. 15 and 16 show that the robot could turn right using turn I whereas the robot turn left using I. In turn I, the robot moved backward at around 3 deg for every 7 motions, which was because of the short period of peristaltic crawling and increase in the opposite force during expansion of a unit. When the right side of the segments touched the ground in turn II, the robot applied brakes and turned to left by peristaltic crawling.

6. Rigid-type robot

6.1 Configuration of a Robot with 4 Units

Fig. 17 shows the configuration of the rigid-type robot. It consists of several units that can extend and contract in the axial direction. The units are connected with a rotational degree of freedom. Hence, it is possible to bend between units. Since peristaltic crawling requires at least three contraction segments, the robot also must consist of at least three units. However, in practice, the robot requires at least four units to change its attitude. The length of 4 units is 279 mm when fully extended and 231 mm when fully contracted. The robot weights 283 g.

Fig. 17. Rigid-type robot

6.2 The Structure of a Rigid-type Unit

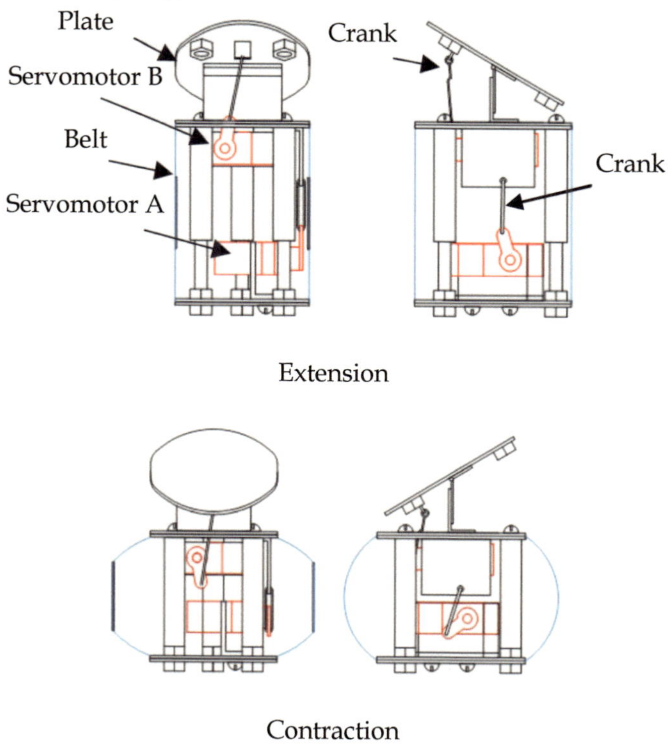

Fig. 18. Rigid-type robot

Fig. 18 shows functions of the rigid-type robot. The units use two servomotors. Servomotor A provides contraction and extension of the unit. The swing motion of the servomotor is converted into a slide motion by a crank mechanism. The unit expands in the radial direction during contraction due to the curvature of belts made from flexible plates. Servomotor B provides bending between the units. A rubber sheet is attached at each belt to increase the friction between the units and the ground when the belts contact the moving surface. Each unit is 65 mm long and of 40 mm in diameter, when fully extended, and 50 mm long and 75 mm in diameter, when fully contracted. It has a mass of 70 g. The bending angle ranges from −30 to 30 degrees.

7. Modelling of the robot and simulation

The developed robot is modeled to evaluate its position and attitude, and a simulation analyses its movement over given routes in 2-dimensional space.

7.1 Setup coordinates

The robot and its setup coordinates are shown in Fig. 19. The robot can be divided into a contraction part and a bending part modeled as a slide joint and a rotational joint, respectively. The rotational angle, anterior segment, and posterior segment are paid particular attention—the coordinates of each unit are set from the posterior to the anterior segment, as Σ1-Σ5. li is a distance between the origins of Σi and Σi+1, and θi is the rotational angle at Σi+1. Further, Σ0 is the world coordinate system.

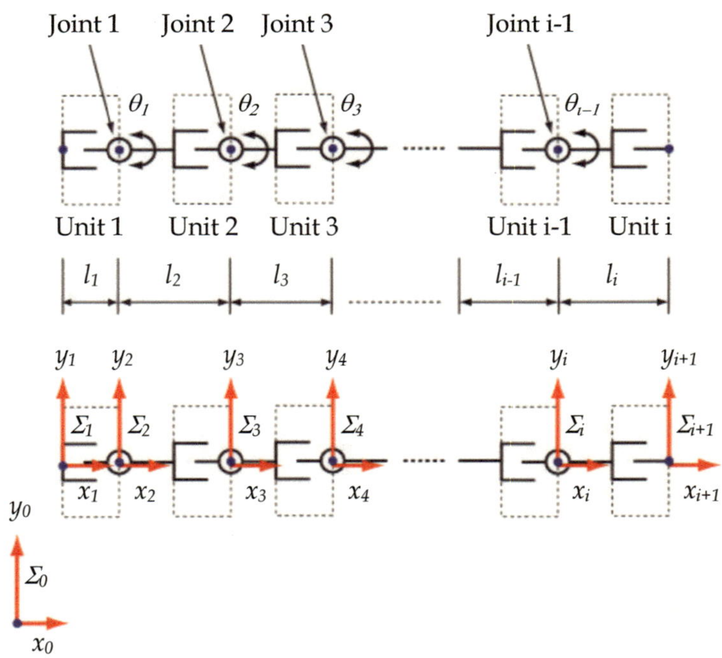

Fig. 19. A set-up of the coordinates of the rigid-type robot

7.2 Locomotion patterns for peristaltic crawling

Fig. 20 shows the pattern of each unit movement required for straight line peristaltic crawling. As shown in this figure, peristaltic clawing is a locomotion pattern whereby contractions of two units propagate along the anteroposterior direction. When the units contract, they come in contact with ground and gain friction. Therefore, the colored contraction unit in each motion (Fig. 20) maintains the position and attitude of the previous motion. This is defined as a basic unit. Hence, when Unit i is the basic unit, Σi should be set as the position and attitude of the previous motion. A contraction unit that has not been colored in Fig. 20 is defined as semi-basic unit. When the attitude of the units is changed to alter the robot's direction, the position and the attitude of the basic and semi-basic units must be fixed. When extension units are between contraction units, such as in Motion 1, the units cannot change their attitude.

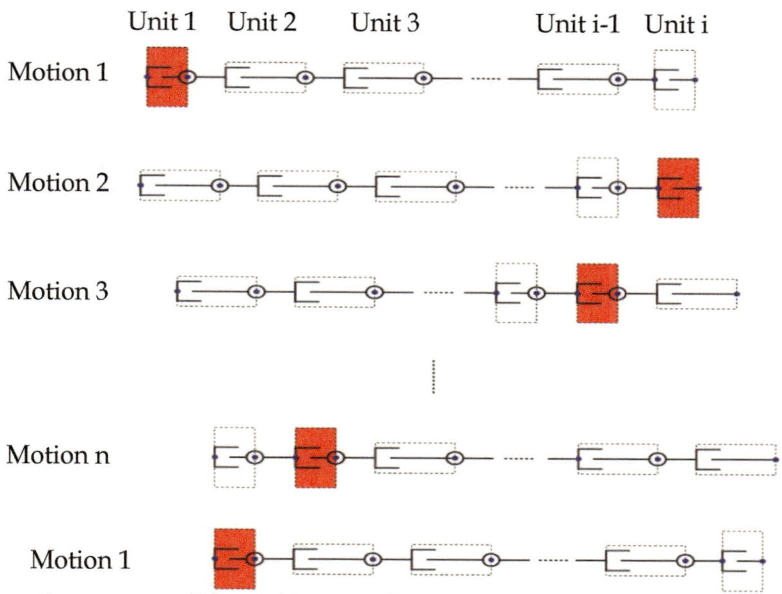

Fig. 20. Locomotion pattern of the rigid-type robot

7.3 Homogeneous transformation matrix of the robot

The attitude of the robot observed from the basic unit in each Motion is derived. It is transformed to world coordinate systems to derive a homogeneous transformation matrix of the robot. For instance, a homogeneous transformation matrix for Motion 1 is shown in (1) to (4). The transformation matrix that gives the relationship between the world coordinates Σ0 and Σi is shown in (5).

$$^{1}T_{2} = \begin{pmatrix} c\theta_1 & -s\theta_1 & l_1 \\ s\theta_1 & c\theta_1 & 0 \\ 0 & 0 & 1 \end{pmatrix} \tag{1}$$

$$ {}^{2}T_{3} = \begin{pmatrix} c\theta_2 & -s\theta_2 & l_2 \\ s\theta_2 & c\theta_2 & 0 \\ 0 & 0 & 1 \end{pmatrix} \tag{2} $$

$$ \vdots $$

$$ {}^{i-1}T_{i} = \begin{pmatrix} c\theta_{i-1} & -s\theta_{i-1} & l_{i-1} \\ s\theta_{i-1} & c\theta_{i-1} & 0 \\ 0 & 0 & 1 \end{pmatrix} \tag{3} $$

$$ {}^{i}T_{i+1} = \begin{pmatrix} 1 & 0 & l_i \\ 0 & 1 & 0 \\ 0 & 0 & 1 \end{pmatrix} \tag{4} $$

$$ {}^{0}T_{i} = \begin{pmatrix} c\phi & -s\phi & x_i \\ s\phi & c\phi & y_i \\ 0 & 0 & 1 \end{pmatrix} \tag{5} $$

7.4 Setting routes and evaluation functions

A simulation is performed using 4 units of the robot's derived position and attitude. The contraction pattern of units for robot locomotion is used for the pattern in Fig. 20. The robot can move along the given routes by changing the angle of the joint between units. First, locomotion routes are decided. The given routes that the robot moves along are expressed by the following functions of (6) and (7), representing Route I and Route II. The robot's tail is set at the origin. Route I shows advancement along a straight line, Route II shows a 45-degree turn.

$$ y=0 \tag{6} $$

$$ \begin{cases} y = 0 & (x < 300) \\ y = x & (x \geq 300) \end{cases} \tag{7} $$

Next, the evaluation function for motions is discussed. The difference between the robot attitudes and the route in each Motion is examined. Then, the evaluation functions shown in (8) and (9) for Route I and II are set. Each evaluation function shows the distance from the origin of Σ1–Σ5 (in world coordinate systems) to the route. The attitude of the robot in each Motion, when these calculation results are the smallest, is defined as the optimal Motion. Here, the range of the angle of each joint is from −30 to 30 degrees (counterclockwise rotation is assumed to be positive), with a resolution of 5 degrees.

$$ E_0 = \sum_{i=1}^{5} |0 - y_i| \tag{8} $$

$$\begin{cases} E_0 = \sum_{i=1}^{5} |0 - y_i| & (x < 300) \\ E_1 = \sum_{i=1}^{5} |x_i - y_i| & (x \geq 300) \end{cases} \tag{9}$$

7.5 Simulation results and discussion

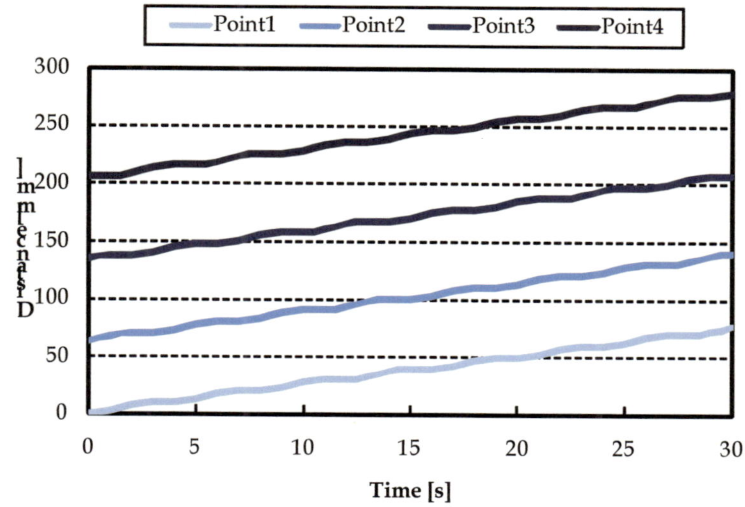

Fig. 21. Simulation results of movement tracks of the robot (Route 1)

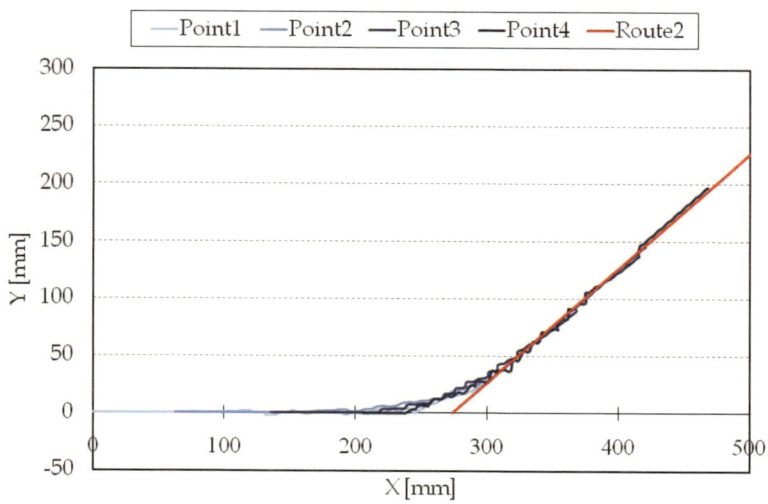

Fig. 22. Simulation results of movement tracks of the robot (Route 2)

The simulation results for each route are shown in Fig. 21 and Fig. 22. First, consider the simulation results for Route 1. If we look at Fig. 21 (the relationship between the advancement distance and time), it confirms that the robot advances by repeated propagation of contractions from the anterior to posterior part. Thus, the simulation results accurately expressed peristaltic crawling motion. We next discuss the simulation results for Route 2. Fig. 21 shows movement tracks of the robot in 2-dimensional planes. Although this figure shows that the robot's tracks deviate a little from the route, the robot does follow the given route without large deviations. This robot is good at moving in narrow spaces, for straight advancement movement, such as Route 1. Thus, it is desirable that the robot passes more close to bending and curve paths, such as Route 2. We think that the robot can be made to follow these tracks more faithfully by increasing the joint size.

8. Experimental results and discussion

Experiments of the mobile robot using peristaltic crawling motion were performed based on the simulation results. A video of the robot moving on the plane was taken from directly overhead and the tracks of the points marked on each unit were analyzed. The experimental results are shown in Fig. 23-25. From Fig. 23, we can see that the motions of the robot are qualitatively similar to those of the actual earthworm shown in Fig. 4, because the longitudinal wave from the front to the back of is propagated by contracting the units, There is, however, a little slipping caused by the relationship between inertia of the units and its friction force. The tracks are shown for Route 2. Here, the angle of the servomotors is controlled using only the values obtained by simulation results, and no feedback control is applied. Fig. 24 shows that, while the robot's tracks are not quite as close to the path as in the simulation, the robot moves around the route. In particular, the robot starts changing its attitude before reaching the point that the curve function actually changes. The peristalsis crawling robot is moved by using the difference of the friction of the contraction units and the extension units.

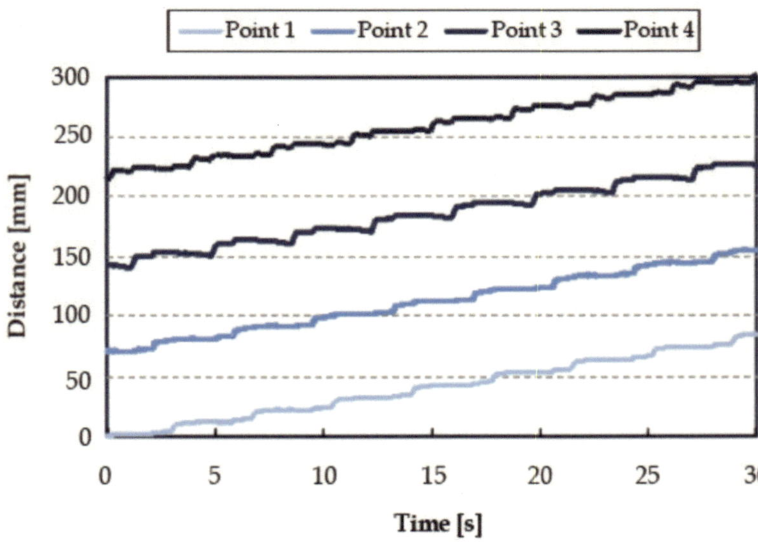

Fig. 23. Experimental results of movement tracks of the robot (Route 1)

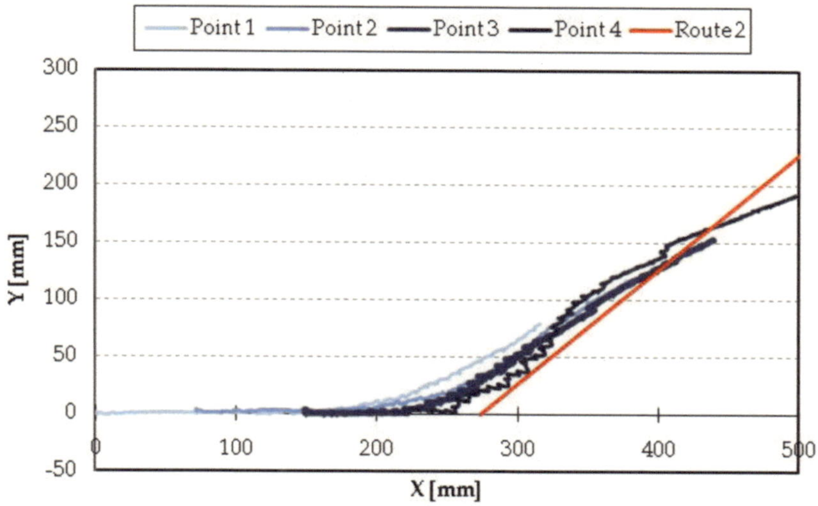

Fig. 24. Experimental results of movement tracks of the robot (Route 2)

Therefore, if the basic and semi-basic units that constitute the contraction units slip, there is a possibility of influencing the distance moved and the bending angle. Also, the robot's weight prevented the belts from achieving full expansion in the radial direction (Fig. 18). Hence, there is a possibility that enough frictional force difference cannot be achieved.

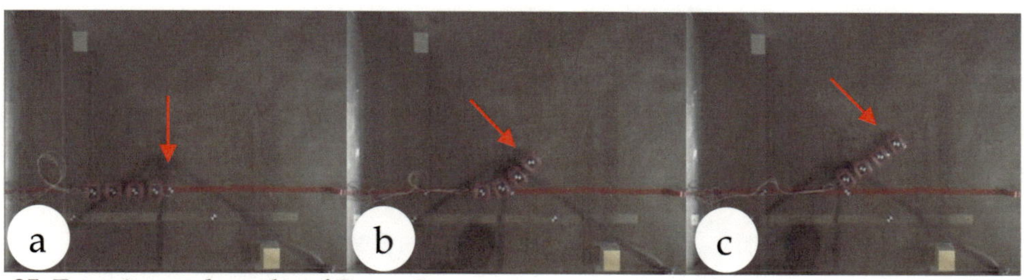

Fig. 25. Experimental results of Route 2

9. Parallel link robot

9.1 Robot

An underground robot is required to be controlled in 3 dimensions and generative force is large. To meet these requirements, we adopted a parallel link mechanism. Figure 26 shows the structure of a single unit (segment). A parallel link mechanism controls the anterior part in three dimensions and generates a large extraction and contraction force. Figure 27 shows the expansion belts; a sheet is placed between the belts to provide dust protection. Figure 28 shows the robot of 4 units. The length of four units is 300 mm when fully extended and 220 mm when fully contracted. The robot weighs 380 g.

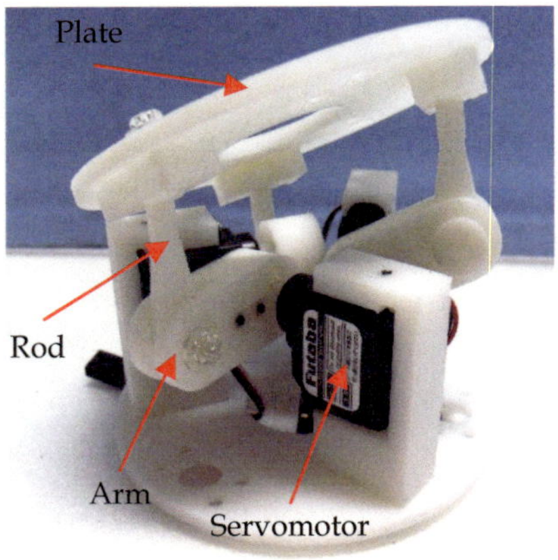

Fig. 26. Internal construction of a single unit

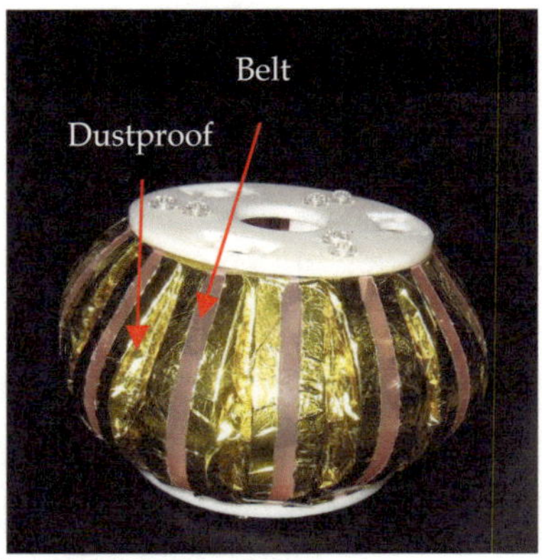

Fig. 27. Expansion belts and dustproof covering

Fig. 28. Prototype of peristaltic crawling robot

9.2 Experiments in Dirt

Fig. 29. Experimental setup

Fig. 30. Moving in dirt

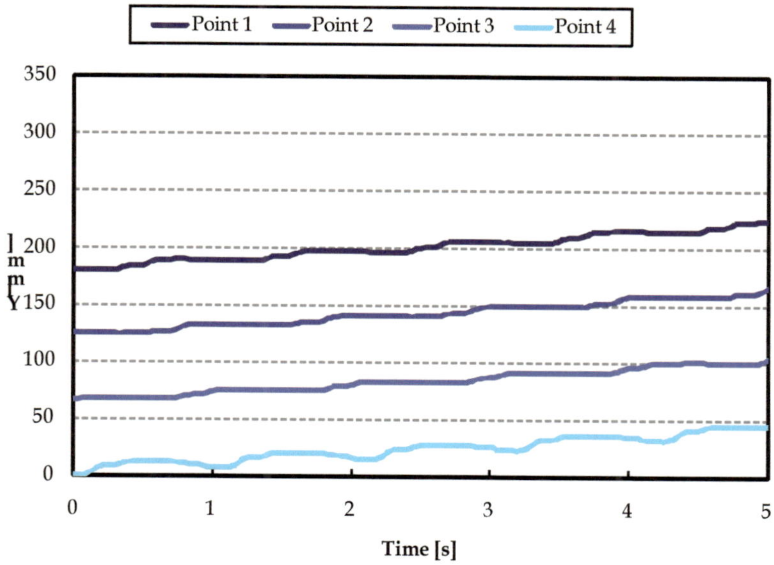

Fig. 31. Locomotion track of the robot in dirt (90°)

The robot was equipped with a dustproof covering so that we could conduct the experiments in dirt. Figure 29 shows the experimental setup. As, the prototype robot did not have an excavation unit, we first set the tube used for the climbing experiments. We then placed black dirt around the tube, and finally withdrew the tube. Through experiments in dirt, we confirmed that the robot can move vertically both upwards and downwards in perforated dirt (Fig.30). Figure 31 depicts the movement tracks of the robot at 90° in dirt. We confirmed that the robot was able to move in the dirt.

9. Conclusion

In this study, we developed two types of peristaltic crawling robots based on an actual earthworm, and introduced one of the applications.
1, Flexible-type of robot
2, Rigid-type of robot
Talking of the flexible-type robot; the movement of the flexible type-robot on the ground and in a narrow pipe was compared to that of an actual earthworm, and it was confirmed that the movement of the robot resembled that of an actual earthworm. In addition, the robot could move in a curved tube. Thus, we successfully developed a peristaltic crawling earthworm robot with flexibility. In turn I, the robot could turn, whereas in turn II, the robot could not turn to the right direction.
Talking of the rigid-type of robot; We proposed a locomotion strategy in 2-dimensional space, based on peristaltic crawling. The robot was modeled to determine its positions and attitude. We performed a simulation of its movements on given routes. The simulation results showed good performance. Experiments with the mobile robot, using peristaltic crawling, were carried out, based on the simulation results. Although the robot did not track the path as closely as the simulation, the robot moved around the routes.
An application introduced was constituted of several parallel links for meeting the requirements of an underground robot. We had experiments in perforated dirt and the experiments were successful.
In the future, these robots will be improved and several sensors will be attached for obtaining surrounding data. In addition, it is necessary that the robots have excavation mechanisms for an underground explorer.

10. References

Mangan, E.V.; Kingsley, D.A.; Quinn, R.D. & Chiel, H.J. (2002). Development of a peristaltic endoscope, *Proceedings of IEEE International Conference on Robotics and Automation*, pp. 3282-3287, Washington DC, May 2002

Menciassi, A.; Gorini, S.; Pernorio, G. & Dario, P. (2004). A SMA actuated artificial earthworm, *Proceedings of IEEE International Conference on Robotics and Automation*, pp. 347-352, LA, April 2004, New Orleans

Zuo, J.; Yan, G. & Gao, Z. (2005). A micro creeping robot for colonoscopy based on the earthworm, *Journal of Medical Engineering &Technology*, Vol. 29, 1, pp. 1-7

Kim B,; Lee MG,; Lee YP, et al. (2006). An earthworm-like micro robot using shape memory alloy actuator, *Sensors and actuators. A, Physical*, Vol. 125 No. 2: pp. 429-437, 0924-4247

Norihiko, S. & Taro N. (2004). A Prototype of Peristaltic Robot Using Pneumatic Artificial Muscle, *Intelligent autonomous system*, No. 8, pp. 85-95

Norihiko, S. & Taro, N. (2004). Development of peristaltic-crawling robot using magnetic fluid on the basis of locomotion mechanism of earthworm, *Smart material and structure*, Vol. 13, No. 3, pp. 566-569

Taro, N.; Takashi, K.; Tomohide, I. & Yoichi, M. (2006). Development of a Peristaltic Crawling Robot Using Servo Motors Based on the Locomotion Mechanism of Earthworms, *Proceedings of IEEE International Conference on Robotics and Automation*, pp. 4342-4344

Taro, N.; Takashi, K.; Tomohide, I. & Yoichi, M. (2006). Development of a Peristaltic Crawling Robot Based on Earthworm Locomotion, *Journal of Robotics and Mechatronics*, Vol.18, No. 3, pp .299-304

Alexander, R. M. (1992). Exploring Biomechanics, Animals in Motion, *W.H. Freeman and Company*, Scientific American Library, 071675035X, New York

Sugi. H. (1977). Evolution of Muscle Motion, the University of Tokyo Press, pp. 72, 1977 (in Japanese

Taro, N. & Tomihide, I. (2008). Locomotion Strategy for a Peristaltic Crawling Robot in a 2-Dimensional Plane, *Proceedings of IEEE International Conference on Robotics and Automation*, pp. 238-243, USA, May 2008, Pasadena

Hayato, O,; Takeshi, H. & Taro, N. (2008). Locomotion and Turning Patterns of a Peristaltic Crawling Earthworm Robot Composed of Flexible Units, Proceedings of IEEE International Conference on Robotics and Automation, pp. 1630-1635, France, September 2008, Nice

17

Dynamically Feasible Probabilistic Motion Planning in Complex Environments for UAVs

Emre Koyuncu
Istanbul Technical University, Controls and Avionics Laboratory
Turkey

Gokhan Inalhan
Istanbul Technical University, Aeronautical and Astronautical Faculty
Turkey

Abstract

In this work, we consider the problem of generating practically implementable path plan for flying unmanned aerial vehicles in 3D Complex environments. This problem is complicated by the fact that, generation of the dynamically and geometrically feasible flight trajectories for agile maneuver profiles requires search of nonlinear state space of the aircraft dynamics. This work suggests a two step feasible trajectory generating approach. In the first step, the planner explores the environment through a randomized reachability tree search using an approximate line segment model. The resulting connecting path is converted into flight way points through a line-of-sight segmentation.

After this first step we explain two different methods to create Dynamically Feasible Path, first one that we called Modal-Maneuver Based PRM Planner is suitable for agile unmanned aerial vehicles that their maneuvers can be define with distinct modes. This allows significant decreases in control input space and thus search dimensions, resulting in a natural way to design controllers and implement trajectory planning using the closed-form flight modes. In this approach the resulting connectivity path and the corresponding milestones are refined with a single query Probabilistic Road Map (PRM) implementation that creates dynamically feasible flight paths with distinct flight mode selections and their modal control inputs. In our second approach that we called Probabilistic B-Spline Planner, every consecutive way points are connected with B-Spline curves and these curves are repaired probabilistically to obtain a geometrically and dynamically feasible path. This generated feasible path is turned in to time depended trajectory with using time scale factor considering the velocity and acceleration limits of the aircrafts.

1. Introduction

Practical usage of Unmanned Air Vehicles has underlined two distinct concepts at which these vehicles are instrumental. First are the routine operations such as border or pipeline monitoring for which manned systems are expensive and inefficient. Second are scenarios such as an armed conflict reconnaissance or nuclear spill monitoring, in which there is a high risk for

human life loss as the proximity to the scenario increases. In this work, we consider a specific case of the second type of scenarios which involves flying through a complex and dense city-like environment rather this be for reconnaissance or monitoring.
When the path planning problem is complicated with dynamic constraints on vehicle along with geometrical constraints, problem becomes challenging due to complexity of dimension. Therefore one cannot simply generate a trajectory that is dynamically feasible (feasible in the sense that it would be trackable by a control system in the flight envelope and actuator limits), relying on the path planning. If the flight trajectory is generated without checking the velocity and acceleration bounds, lower level layers will have hard time searching for angular velocities and angle of attack history that are in the feasible set. Therefore by sharing the dynamic feasibility checks between path planner and lower level layers, these layers covers their disadvantages and provides both dynamically and geometrically feasible flight trajectories for complex environments. Although many kinodynamic motion planning methods that declares generating *dynamically feasible path* have been developed, they rarely can be used in practice especially for the aerial vehicles because of computational complexities. General kinodynamic motion planners require at least exponential time in that dimension of the state space of dynamical systems which is usually at least twice the dimension of the underlying configuration space (Frazzoli et al., 2002). In practice, kinodynamic planners are implementable only for systems that have small state-space dimensions. For example, the work presented in (Frazzoli et al., 2002) suggests a path-planning relaxation which defines a class of maneuvers from a finite state machine, and uses a trajectory based controller to regulate the unmanned vehicle dynamics into these feasible trajectories. However, the trajectories to be controlled are limited to the trajectories generated by the finite state machine and the computational challenges of generating real-time implementable flight trajectories in 3D complex environments still remains as a challenge. Demonstration of path planner solution for flight in the 3D crowded *MelCity* model and landing to base is seen in Fig. 1 .
In our approach, we suggest a real-time implementable two-step path planner strategy. As the first step, 3D environment and the passages are rapidly explored using an RRT based planner. From this geometrically feasible but not dynamically feasible path, line-of-sight critical milestones are extracted. Although these milestones allow point-to-point flyable flight path segmentation, it does not necessarily correspond to a fast agile and continuous motion plan. To address this, as a second step, we will explain two different methods to create Dynamically Feasible Path. First one that we called Modal-Maneuver Based PRM Planner is developed for agile unmanned aerial vehicles that their maneuvers can be define with distinct modes. This allows significant decreases in control input space and thus search dimensions, resulting in a natural way to design controllers and implement trajectory planning using the closed-form flight modes. In this approach the resulting connectivity path and the corresponding milestones are refined with a single query Probabilistic Road Map (PRM) implementation that creates dynamically feasible flight paths with distinct flight mode selections and their modal control inputs. In our second approach, Probabilistic B-Spline Path Planner, every consecutive way points are connected with C^2 continuous B-Spline curve. In face of geometrically and dynamically unfeasibility, generated path is probabilistically reshaped to eliminate the collisions and dynamically unfeasibility thanks to local support property of the B-Spline curves and at the end the *time scale* is adjusted to allow dynamic achievability considering the velocity and acceleration limits of the aircrafts.
Rest of this chapter is organized as follows. In Section 2, framework of the dynamically feasible path planning and literature survey is given. In section 3, first step, finding geometrically

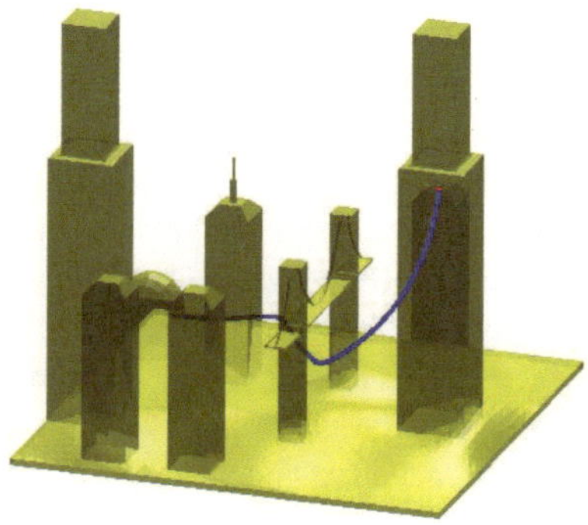

Fig. 1. UCAV flight demonstration in the 3D complex city-like environment and landing to its base

connectivity path method is explained and in Section 4, details of the following step that two generating dynamically feasible trajectory methods is explained. In Section 5, system architecture of the ITU CAL mobile robotic testbed is presented and its experiments are demonstrated in Section 6. Simulation results and its computational time tables are also given in Section 6. The conclusions are discussed in Section 7.

2. Framework of the Dynamically Feasible Path Planning Algorithms

For developing a real-time implementable planner, motion planning researches have been focused on sampling based approaches that rapidly search either the configuration or the state space of the vehicle. In the last few decades, sampling-based motion planning algorithms have shown success in solving challenging motion planning problems in complex geometries while using a much simpler underlying dynamic model in comparison to an air vehicle. Roadmap-based planners, like well-known Probabilistic Road Mapping (PRM) method as mentioned in (Kavraki et al., 1996), are typically used as multi-query planners (i.e. simultaneous search of the environment from different points) that connect these multiple queries using a local planning algorithm. PRM planners converge quickly toward a solution path, if one exists, as the number of milestones increases. This convergence is much slower when the paths must go through narrow passages. For complex environments, some extended algorithms are suggested for PRM like planners in (Hsu et al., 2003) and (Boor et al., 1999). Tree-based planners build a new roadmap for each query and the newly produced samples are connected to the samples that are already exists in the tree as in (Hsu et al., 1999), (Hsu, 2000), and (LaValle & Kuffner, 1999). Rapidly-Exploring Random Tree (RRT) is the most popular representative of tree-based planners that is an exploration algorithm for quickly searching high-dimensional spaces that have both global and differential constraints. Sampling-based planners, especially tree-based planners (RRT and single-query PRM variants), have been adapted to solve dynamically feasible paths that accommodate kinodynamic constraints. Kinodynamic planning refers to problems in which the motion must satisfy nonholonomic and/or dynamic constraints . The main philosophy behind kinodynamic planning is searching a higher dimensional state space that captures the dynamics of the system (LaValle & Kuffner, 1999), (Hsu et al., 2002).

Gradual motion planning methods -our approach can be represented in this class- are recently proposed to solve complex path planning problem in cluttered environments. These methods first solve a relaxed form of the problem and then the approximate solution is refined to solve the original problem with a repairing method. In (Hsu et al., 1998), a roadmap is initially generated by allowing some penetration into the collision workspace. Later, milestones are carried to collision-free space. In Iterative Relaxation of Constraints (IRC) method (Bayazit et al., 2005), first a relaxed version of problem is solved and then this coarse solution is used as a guide to solve original problem iteratively. The strategy of using an approximate solution to obtain a collision-free path is also used in Lazy PRM (Bohlin & Kavraki, 2001) and C-PRM (Song & Amato, 2001).

In comparison, our method utilizes both the probabilistic and the deterministic aspects to obtain a real-time implementable planner strategy. In the first step, the algorithm rapidly explores the complex environment and the passages using an RRT planner because of its well quick spreading ability. In this part, our strategy focuses only finding an obstacle-free path that can be tracked from the initial point to the goal point with line segments in the configuration space. Dynamic constraints of the vehicle are completely disregarded to decrease the computational time. This coarse obstacle-free path will be called as *connectivity path*. After finding the connectivity path, this path is filtered with the line-of-sight implementation to eliminate the points that cause long detours. Remained points that we call as *way points* naturally appear in entering and exiting regions of the narrow passages that are formed between the obstacles. An advantage of this refinement is that we can use these way points as guider-milestones that point out hard regions and directions of the next coming hard regions in the environment.

In explained first method, Modal-Maneuvering PRM planner, milestones are created for each flight segment using randomized flight mode selection. This exploits all the full flight-envelope and capability of the vehicle, and once the planner samples enough number of milestones near one of the way-points, it continues with these milestones to reach other way points. One distinct feature of this planner in comparison with other probabilistic planning methods is the reduced input space selection. Specifically, in each query, instead of choosing all input variables randomly, our planner chooses maneuver mode and its parameters that are constrained by vehicle dynamics. In addition, the size of the search is limited to only partial flight segments. Both of these factors contribute significantly in reduced computation time.

In explained second strategy that we will call Probabilistic B-Spline Path Planner, every way point is connected with a B-Spline curve, then collisions and dynamic feasibility cases are checked on curve. These forth-order B-spline curve presents C^2 continuous flight path. If the generated curve is not feasible, probabilistic repairing methods are achieved by randomized waypoint expansion on the connecting line path and the unit flight time is expanded within controllable regime considering to feasible velocity and acceleration interval. Since B-Spline curves have local support property, these repairing processes can be made on local path segments of interest.

B-Spline curves have been used in many dynamic path planning and control problem implementations. In (Komoriya & Tanie, 1989) , dynamic trajectory is generated with the minimum travel time for two-wheeled-driven type mobile robot. In (Munoz et al., 1994) visibility-based path is modified to continuous feasible path via B-Spline curves. Using the well known local support property of B-Spline curves, real-time path modification methods are proposed for multiple mobile robots in (Paulos, 1998) and robot manipulators in (Dyllong & Visioli, 2003). Constant acceleration time-scalable path generation method for the unmanned helicopters flying in the urban environments is used in our earlier work in (Koyuncu & Inalhan, 2008) that we will use similar method but this time for the unmanned combat aerial vehicles.

3. Finding Geometrically Connectivity Path: First Step

In real-time applications, planners should be able give a reliable answer in minimal permitted time slot. In motion planning problem, especially in complex environments, it is hard to say when planners should stop searching or change the searching strategy (i.e. switching to a more complex planner etc.). Moreover, finding an obstacle free geometrical path does not necessarily mean that a dynamically feasible path can be implemented by the vehicle exists. Although geometrical paths can be implemented via point to point navigation by the helicopter like vehicles with a inefficient manner but this flight strategy is not applicable for agile combat vehicle operations in under-threat environments. For the vehicles that have complex dynamics like combat aerial vehicles, directly searching in high dimensional state-space -as kinodynamic planners do- consumes long computational time to find a feasible path. Specifically, we observed that before the major feasible path planning phase, defining the geometrical obstacle free path and trackable way points significantly accelerates the searching ability and decreases the total computational time of planner.

For finding connectivity path, RRT algorithm is used because of its rapid spreading ability. RRT is a considered as being an efficient algorithm to search even high dimensional spaces. However, one of the important drawbacks of using RRT as a stand-alone planner is biasing of the distribution of milestones towards the obstacle regions if the configuration space has large obstacles. Bi-directional RRT method shows performance more than single tree approach but it has also discontinuity problem on the connection points of the paths. Therefore, we choose to use single Goal-Biased RRT (LaValle & Kuffner, 1999) approach that converges to goal configuration rapidly. We tested performance of the algorithm in different complex environments and to conserve both rapid converging to solution and spreading abilities of the RRT, we chose the 50% percent goal biasing value. In this phase, we are only motivated by good property of the RRT algorithm to obtain connectivity path. Our strategy does not focus on dynamically feasibility in this part of the path planner. Therefore, RRT algorithm is only used for searching configuration-space of the vehicle with primitive maneuvers that includes level and climbing flight and changing instantaneous heading direction. Construction of connectivity path algorithm is given as Goal Biased RRT Algorithm.

In Algorithm I, to find the connectivity path, Goal Biased RRT method is used that one single tree is extended from the initial point. Each loop attempts to extend the τ tree first toward the random selected point m_{rand}, and second toward the goal point by adding new points. To expand the tree, nearest point already within the τ tree to the sampled random point (in Line 3) and the nearest point to the goal point is selected (in Line 9) respectively in every one loop. *Generate* function generates new points m_{new} on the direction of the selected nearest points m_{near} at random selected distances as shown in Line 4 and 10. If direction angles exceed predefined limits, max direction angles are selected. These boundaries should be chosen

Algorithm 1: Goal Biased RRT Algorithm

input : initial configuration q_{init} and goal configuration q_{goal}
output: *connectivity path*

1 $\tau \leftarrow q_{init}$ and $i \leftarrow 1$
2 **repeat**
3   **Select** random point m_{rand} in C and its neighbor point m_{near} in τ
4   **Generate** m_{new} is gone with trajectory e_{new} from m_{near} toward m_{rand}
5   **if** e_{new} *is in* C_{free} **then**
6         $\tau \leftarrow m_{new}$ and $i+1$
7   **if** m_{new} *is in end region* **then**
8   **break** with *success*
9   **Select** neighbor point m_{near} of q_{goal} in τ
10   **Generate** m_{new} is gone with trajectory e_{new} from m_{near} toward q_{goal}
11   **if** e_{new} *is in* C_{free} **then**
12         $\tau \leftarrow m_{new}$ and $i+1$
13   **if** m_{new} *is in end region* **then**
14   **break** with *success*
15   **if** $i = N$ *max iteration number* **then**
16   **break** with fail
17 **until** *end region is reached with success*
18 **Select** *connectivity path* can be gone back from *end region* to initial point in τ

according to vehicle's kinematic boundaries. If new generated point and trajectory is within obstacle-free configuration (checked in Line 5 and 11) then m_{new} is added τ tree as shown in Line 6 and 12. If τ tree reaches *end region* anytime, algorithm returns *connectivity path. End region* can be obtained within a tolerable capture region as explained in (Kindel et al., 2000). A solution of the algorithm in a complex city-like environment is illustrated in Fig. 2.
Because of the RRT's extending strategy and our simplifications, undesirable detours are frequently seen in obtained *connectivity path*. Since we only consider finding the obstacle-free region; we can simply remove the points that cause these detours. In this phase of our strategy, *connectivity path* is refined by Line-of-Sight Filter algorithm that erases points that result in useless fluctuations with using a line-of-sight arguments. As can be seen in Fig.3, remaining points generally appear in nearby entering and exiting field of the narrow passages and inherently hard regions. Hence, these guard points also indicate where hard regions are beginning, what the direction of the next-coming hard region. These points also give a sense of agile maneuvering that are needed to fly over these points.
In this part of algorithm, a simple iteration checks if the selected point $\mathrm{m}_{\mathrm{visib}}$ can connect with the previous points in *connectivity path* with a line segment without colliding with any obstacle. If the line segment collides with an obstacle, in other words, if the current point cannot be connected to the selected point, last connectible point is added to the *way point* sequence and the subsequent search continues from this point. This algorithm runs until the last point of *connectivity path* is reached with a line segment. A solution is illustrated in Fig.3.

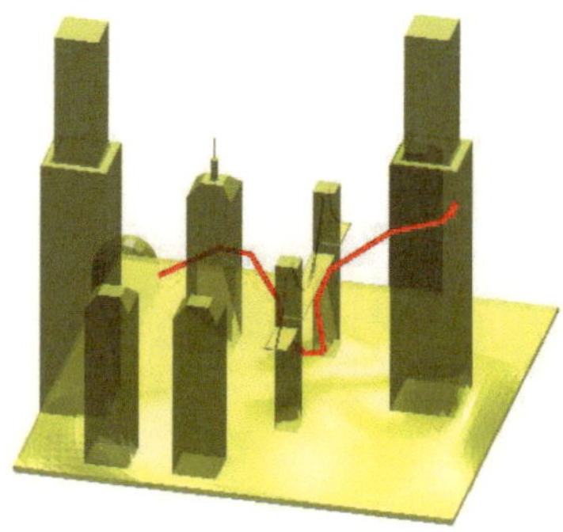

Fig. 2. Demonstration of the RRT based Finding Connectivity Path Algorithm

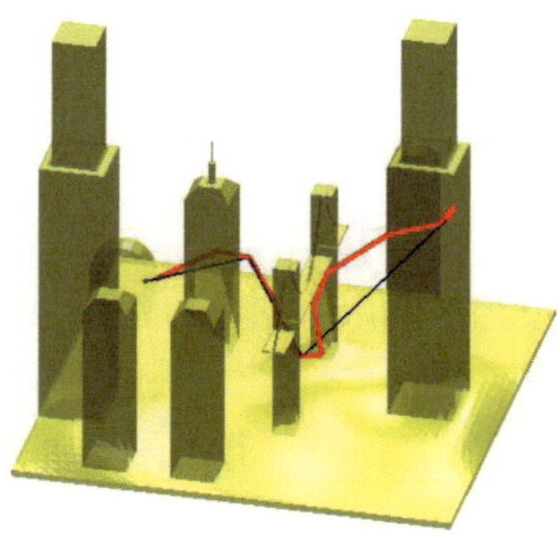

Fig. 3. Demonstration of the refining Connectivity Path with the Line-of-Sight Filter Algorithm

4. Generating Dynamically Feasible Trajectory: Second Step

In the first step, generated connectivity path with straight line segments result in a simple and implementable piecewise flight plan. However, this flight plan is not a fast agile and continu-

Algorithm 2: Line-of-Sight Filtering

input : *connectivity path*
output: *way point set* WP
$m_{visib} \leftarrow m_1$ and $m_i \leftarrow m_2$
repeat
  Generate line ℓ_{visib} from m_{visib} to m_i
  if ℓ_{visib} *is collide with* C_{obst} **then**
    $WP \leftarrow m_{i-1}$
  else
    $i+1$
until *last point of connectivity path is reached*

ous motion plan - a desirable feature in many complex unmanned aerial vehicles applications. After obtaining the way points - we will call remaining points as *way point* set - on the environment, many deterministic and sampling based path planner methods can be used to find the dynamically feasible path between the way points.

On this point, we suggest two methods to create *Dynamically Feasible Path*, first one that we called *Mode Based PRM Planner* is suitable for agile unmanned vehicles that their maneuvers can be define with distinct modes.This mode-based structure is also well suited designing a systematic hybrid flight control system. Our second path planner method is called *Probabilistic B-Spline Planner*. This algorithm generates continuous flight paths using B-Spline fitting strategy. In our earlier work, we used this strategy to generate constant acceleration time-scalable path for the unmanned helicopters (have hovering and instantaneously changing heading angle etc. abilities) flying in the urban environments in (Koyuncu & Inalhan, 2008). We will explain the similar method that generates C^2 continuous flight path but this time it will also be suitable for the agile aircrafts (unmanned combat aerial vehicles).

4.1 Modal-Maneuver Based Probabilistic Road Mapping Method

Our Mode-Based PRM approach (Koyuncu et al., 2008) is based on the simple idea of exploiting the full flight envelope of the air vehicle through distinct flight modes from which almost any maneuver can be created. This mode-based structure is especially well suited both creating flight paths and also designing a systematic flight control system. However, this structure does not necessarily solve the critical problem of finding a) the possible fly-through passages and b) the necessary mode selections to utilize these passages. These two points and the corresponding solution method are the main focus of this work.

One distinct feature of this planner in comparison with other probabilistic planning methods is the reduced input space selection. Specifically, in each query, instead of choosing all input variables randomly, our planner chooses maneuver mode and its parameters that are constrained by vehicle dynamics. In addition, the size of the search is limited to only partial flight segments. Both of these factors contribute significantly in reduced computation time. It is noted by (Frazzoli et al., 2002) that, in general kinodynamic motion planners require at least exponential time in that dimension of the state space of dynamical systems which is usually at least twice the dimension of the underlying configuration space. Because of this consideration, in practice kinodynamic planners are implementable only for systems that have small state-space dimensions. Thus, for the air vehicles that we focus on, it is hard and time-consuming to obtain a feasible path using a standard kinodynamic planner. We address this problem by directing the search not to the expensive state-space, but to only a subset of the input space as required by the flight modes (and their resulting controlled state-space selections). Thus, for almost every flight mode, the input space is either two or three dimensional, with the most complex mode 3D spin, being four dimensional.

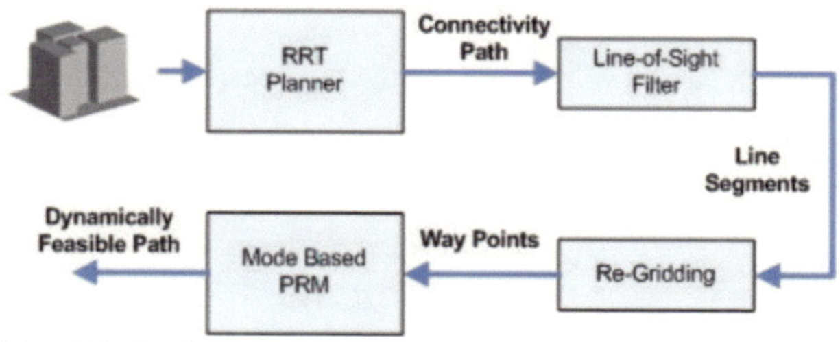

Fig. 4. Dynamically Feasible Path Generating Process with Mode Based PRM Planner

Motion planning of agile vehicles from a control perspective has been mainly studied under the topic of hierarchical hybrid systems. Basically, the flight path of an aircraft can be divided into modes, and these modes serve as reference blocks to a control system, and the control system regulates these modes with given specifications by possibly using nonlinear control laws (Ghosh & Tomlin, 2000). Note that this modal approach can also be used for trajectory optimization for multiple vehicles (Inalhan et al., 2002). (Frazzoli et al., 2002) suggested a path-planning system which defines a class of maneuvers from a finite state machine, and uses a trajectory based controller to regulate the agile vehicle dynamics into these feasible trajectories. This approach has got the advantage of generating both feasible and optimal trajectories in an environment with an obstacle while ensuring robust tracking of these trajectories. However, the trajectories to be controlled are limited to the trajectories generated by the finite state machine. Similar approaches have also been developed in (Schouwenaars et al., 2004). Although these works provide asymptotic tracking, the modes as governed by the state-machines do not exploit the full flight envelope and the generated maneuvers are not really tailored toward an environment, which demand tactical advantage through exploitation of the vehicle's full capability - a feature that exists in sample based motion planning algorithms and we exploit this feature while creating dynamically feasible paths using the probabilistic roadmap approach over flight modes. We illustrate the control of a complex agile maneuver over such modes in Fig. 5.

Multi Modal control framework consists of decomposition of the arbitrary maneuvers into set of maneuver modes and associated maneuver parameters. The main aim of the work was the help to reduce complexity of the both planning and control part. Complexity of maneuver planning part has been reduced by reducing the dimension of the problem (modal sequence has strictly lower dimension than state space description) and control part was relaxed by designing specific controllers for each mode and switch between them in order track maneuver mode sequence instead of designing a single controller for maneuver tracking over full flight envelope. In this work we shall only focus on side of the path planning, discussion of the switched control layer can be found on (Ure & Inalhan, 2009) and (Ure & Inalhan, 2008)

Basically, main idea is to divide an arbitrary flight maneuver into smaller maneuver segments (called maneuver modes) and associated maneuver parameters (called modal inputs). If the maneuver modes are found properly, one can describe any maneuver by giving the maneuver mode sequence. This idea makes use of the fact that, 12 states of the conventional aircraft are

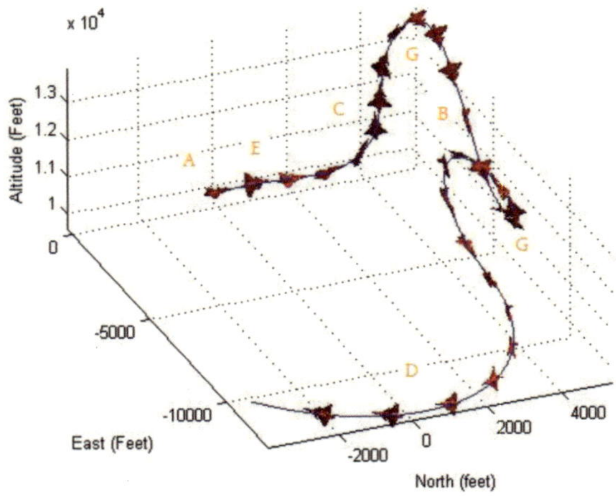

Fig. 5. Flight Modes of an Aggressive Maneuver of F-16 over Flight Using the Full-Envelope Dynamic Model

not independent during all maneuvers and one does not need to give all the state trajectory of the aircraft to define a maneuver. Demonstration of those modes table (Table 1) is seen in below and these modes and their generations are explained in (Ure & Inalhan, 2008) and one of applications is demonstrated in (Koyuncu et al., 2009). These same modes table approach will be used during design of the algorithms.

	Mode	State Constraints	Modal Inputs
q_0	Level Flight	$\dot{h}=0, (\dot{\varphi},\dot{\theta},\dot{\psi})=0$	V_T, α
q_1	Climb/Descent	$(\dot{\varphi},\dot{\theta},\dot{\psi})=0$	$V_T, (\dot{h}, \theta_w)$
q_2	Roll	$(\dot{\theta},\dot{\psi})=0$	$V_T, \int P_w dt$
q_3	Longitudinal placeLoop	$(\dot{\varphi},\dot{\psi})=0$	$\left(V_T, r_{loop}\right), \dot{\theta}$
q_4	Lateral placeLoop	$\dot{h}=0, (\dot{\varphi},\dot{\theta})=0$	$\left(V_T, r_{loop}\right), \dot{\psi}$
q_5	3D Mode	$\{\}$	$V_T, P, Q, R/$ $V_T, \varphi_w, \theta_w, \psi_w$
q_6	Safety	$\{\}$	$\{0,1\}$

Table 1. Flight Modes and Modal Inputs

Previous step provided us to obtain a flight path with *way-points* that generally appear as long straight flight segments that occasionally enter and exit passages. As a result of the underlying randomized exploration toward the goal region, the tendency of the path is toward larger passages and toward direct paths that lead to the goal region. Although this provides a reasonably good approximation, it does not necessarily correspond to a flyable trajectory as it is based on a simple point mass model with velocity and heading.

This part of our planning strategy is an extension of single-query PRM algorithm. It iteratively builds a tree-shaped roadmap to connect the way-points one-by-one. In every inner loop, it

Algorithm 3: Mode-Based PRM Planner

```
   input : way point vector
   output: dynamically feasible path
1  repeat
2      Tree ← reached points
3      goal points ← other way points
4      repeat
5          Select a milestone m_{rand} from Tree probabilistically
6          Select a maneuver mode uniformly at random
7          Select modal inputs according to selected milestone
8          Create a trajectory segment e_{new} with selected modal inputs and maneuver mode
9          Extend m_{rand} with e_{new} to m_{new}
10         if e_{new} is in C_{free} then
11             if m_{new} is in approaching field of any goal points then
12                 Compute weight value w
13                 reached points ← (m_{new}, w)
14                 if size(reached points) is enough then
15                     exit with success
16             else
17                 Tree ← m_{new} and i + 1
18         if i = N max iteration number then
19             exit with failure and f + 1
20     until success or failure
21     if f = M max failure number then
22         Erase prior path segment and go back prior interest region to recalculate
23     else
24         f ← 0
25 until last way point is reached
```

first selects at random a milestone as in Line 5, maneuver-mode and its modal inputs from Table 1 as seen in Line 6 and 7 and then generates a trajectory with selected maneuver mode and modal inputs from selected milestone as shown in Line 8. If this trajectory does not collide with any obstacle (checked in Line 10), its end point is added to tree as a new milestone as shown in Line 17 and its modal inputs are stored. If newly generated milestones fall in nearby region of any goal points, the planner assigns a weight value to these milestones according to their approaching angles as depicted in Line 12.

Milestone Selection; The planner selects an existing milestone in the Tree at random according to direct proportion to their values. A milestone which has higher weight value has a greater chance of being selected by planner, in the other words; milestones which can be propagated with more smooth trajectories have greater chance to continue. These weight values are assigned by *approaching field*. This milestone selection technique pushes our planners to side of kinodynamic planners that use single-query PRM method. Differently, RRT based kinodynamic planners select existing milestones which are nearest (metrically) to randomly-selected states (within all space). PRM like kinodynamic planners can select a milestone in tree according to their values that are charged by planners before. Hence, planner can make decision about which milestones should be chosen more densely. Therefore, our planner uses alike method to select milestones.

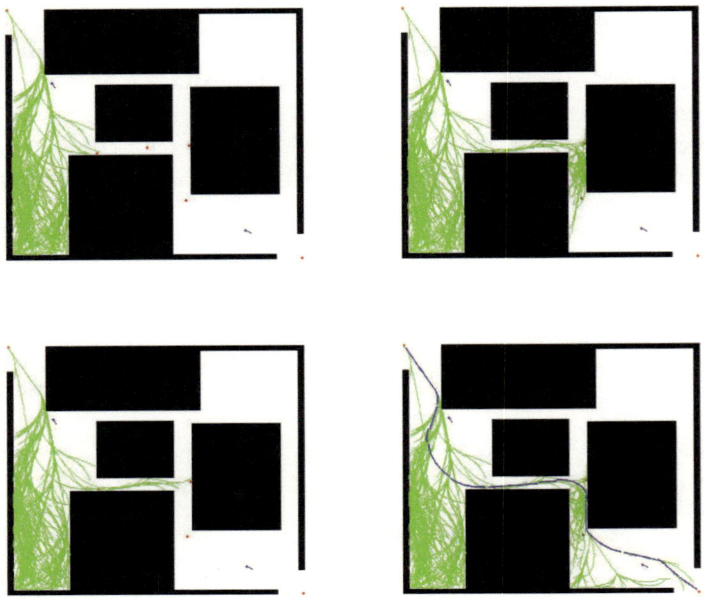

Fig. 6. 2D Demonstration of Mode-Based PRM Searching

Modal-Input selection; Our planner only chooses modal inputs of the distinct maneuver modes instead of choosing control inputs from high dimensional control-input space. These distinct modes can exploit the full flight envelope and almost every flight paths can be created with their combinations. In every loop of algorithm, after the milestone selection, planner first chooses flight maneuver-mode and then chooses its modal inputs according to weight value of the selected milestone. For example, considering to selecting level-flight mode, the milestones which have close angles with line-of-sight to next-coming goal in a small interval (therefore, it is assigned with higher weight values by *approaching field*) can be propagated with longer straight flight paths. Hence, if this milestone is selected by planner, higher velocity rates (in constant time, longer distance rates) are mostly preferred as modal input and then more agility is obtained.
Computing Weight Values; If new generated collision-free milestone falls in nearby of any way points in distinct distance metric, according to its angles and distance, it is assigned with the specific weight value. For deciding these values, every way points are enclosed with *approaching fields* includes distinct regions. If the angles of the milestone (felt in the region) are within specific rate interval of the region, it is enumerated with the respective weight value. Thus, it is aimed that; to charge the milestones which have angles closer to angle rates that can carry it easily (i.e. with smoother curve) to next-coming way point with higher values. The planner more densely selects the milestones are charged with higher value. Hence, it is

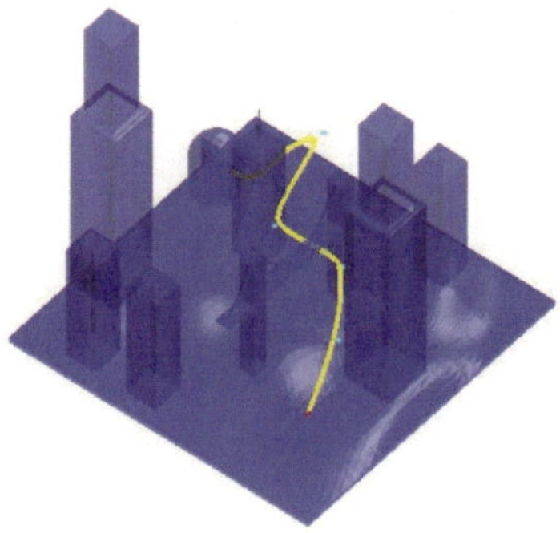

Fig. 7. Complete Solution of Mode Based PRM Planner

intended to create more smooth flight segments. This tunneling effect provides a straightforward solution to the classical narrow passage effect seen in standard PRM methods.
In the main loop, the inner PRM path segment loops try to connect the way-points one-by-one until the ultimate goal region is attained. During this iteration, if the inner loop returns an increased number of failures, prior path segment is erased as depicted in Line 22 and PRM searching is run from prior interest region. Snap-shots from the evolving PRM iterations and the completed solution are given in Fig. 6.

4.2 Probabilistic B-Spline Planning Method

In our second way, we still desire that generated path must be continuous on the way points. During the path generation phase, trajectory generation method should allow reshaping to supply collision avoidance and dynamic feasibility. Therefore, local support is also a desirable property on the path generation method. Local support means that the paths only influence a region of the local interest. Thus, obstacle avoidance and dynamic-feasibility repairing can be achieved without changing the whole shape of the generated path. B-Spline approach can supply these main requirements. An overview of B-Spline can be found in (Piegl & Tiller, 1997) .
Basically, output $C(u)$ can be defined in terms an k order B-Spline curve;

$$C(u) = \sum_{i=0}^{n} P_i N_{i,k}(u) \quad 0 \leq u \leq u_{max} \tag{1}$$

The coefficients P_i in Eq. 1 are called control points that will represent way points and pseudo way points in our approach.
The B-Spline basis functions $N_{i,k}$ are given by the Cox De Boor recursion;

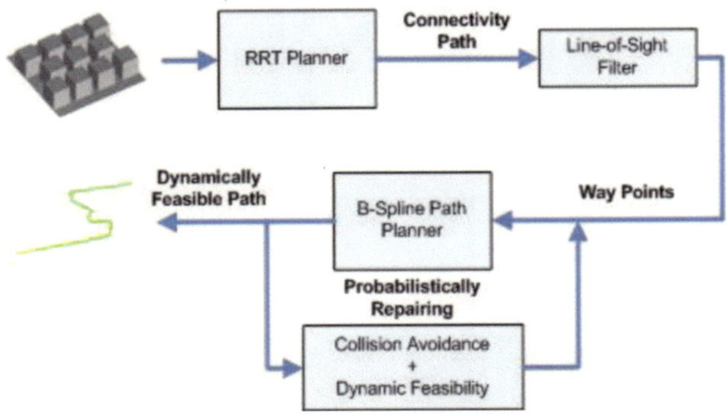

Fig. 8. Dynamically Feasible Trajectory generating using Probabilistic B-Spline Planner

$$N_{i,0}(u) = \begin{cases} 1, & u_i \le u \le u_{i+1} \\ 0, & otherwise \end{cases} \tag{2}$$

$$N_{i,k}(u) = \frac{u - u_i}{u_{i+k-1} - u_i} N_{i,k-1}(u) + \frac{u_{i+k} - u}{u_{i+k} - u_{i+1}} N_{i+1,k-1}(u) \tag{3}$$

A B-Spline curve can be constructed from Bezier curves joined together with a prescribed level of continuity between them. A nondecreasing sequence of real numbers $U = \left[u_0 \ \ldots \ u_{max} \right]$ is called the knot vector. Frequently, the knot points are referred to as the break points on curve (Piegl & Tiller, 1997). B-Spline basis function $N_{i,k}$ is zero outside the interval $[u_i, u_{i+k}]$ and non-negative for all values of k, i and u.
Derivatives of B-Spline curve exist on the knot vector span. Since, the kth-order B-spline is actually a degree $(k-1)$ polynomial, produced curve can be differentiated $k-1$ times.

$$C(u)^{(j)} = \sum_{i=0}^{n} P_i N_{i,p}^{(j)}(u)\, 0 \le u \le u_{max} \tag{4}$$

A valuable characteristic of the B-Spline curves is that the curve is tangential to the control polygon (formed by the control points) at the starting and ending point if some modifications are supplied. This characteristic can be used in order to define the starting, ending and transition directions of the curve by inserting an extra *pseudo control points* in directions which are defined according to way points' orientations assigned in the first step as explained in (Nikolos et al., 2003).
In this strategy, we choose generate forth-order path B-Spline (cubic polynomials) to obtain continuous inertial velocity and acceleration. All dynamically feasible trajectory planning process illustrated in Fig. 8. In our earlier work (Koyuncu & Inalhan, 2008), we used constant acceleration time-scalable path generation method for the unmanned helicopters(have hovering and instantaneously change-heading abilities) flying in the urban environments that we used to generate third-order segmented (between the way point) B-Spline curves. It was good approximation to obtain rapid path planner algorithm that generates continuous inertial velocity and constant acceleration. In this approach, we will generate single cubic B-Spline curve. Complete process of that method seen in Fig. 8.
For generating B-Spline trajectory pass through way points, two *pseudo control points* are inserted after and before the every waypoint(except initial and last way point) in direction of the their tangent vector that these tangent directions are assigned during the path planning

step. Note that; for the first way point, only further pseudo way point and for the last way point, only back pseudo way point should be added to way point set. The distance value between the way-point and the added pseudo-way points will define the transition velocity and acceleration on the way points of the path. Hence, C^2 continuity, in other words, continuous velocity and acceleration transition is naturally achieved on the way points.

For generate cubic B-Spline curves, we use specific nonuniform knot vector form $\mathbf{U} = [\ 0\ \ 0\ \ 0\ \ 0\quad U_{mid}\quad 1\ \ 1\ \ 1\ \ 1\ ]$ to obtain the coincidence between the first and last control points and the first and the last ends of the generated B-Spline curve respectively. Detailed information about this effect can be found in (Piegl & Tiller, 1997) as *open uniform knot vector* effect. $\mathbf{U_{mid}}$ is represents middle *knot vector* that is initially uniformly distributed in $(0,1)$ interval -number of points depends on the number of control points- and algorithm can add new knot points to the vector without preventing its uniform form. We choose using arbitrary $[0,1]$ interval for parameter u such that it represents unit-time scale (Vazquez et al., 1994). This property is later used to allow dynamic feasibility via time scaling (i.e. expanding the time horizon of the maneuver). Overall B-Spline path planning algorithm can be demonstrated in Algorithm 3.

This algorithm tries to find dynamically feasible B-Spline curve passes through on the way points with their heading angles and runs until the last way point is connected with a feasible path. Initially, m number way points - generated in the first step- are added in *control point* set $\mathbf{P}$ as seen in Line 2. Then, for every way point, except first and last way point, *back* pseudo way point $gp_{i,b}$ and *further* pseudo way point $gp_{i,f}$ is located on random selected distance d from way points on their heading tangent directions and these *pseudo way point* set $\mathbf{gp}$ is also added in *control point* set $\mathbf{P}$ that is demonstrated in Line 3 to 6 . Different from other way points, for the first way point, only *further pseudo way point* $gp_{1,f}$ is located and for the last way point, only *back pseudo way point* $gp_{m,b}$ is located. Thus, algorithm initially begins with $3m-2$ control points where m indicates that number of way points but note that the algorithm can add new control points during to implementation to repair the B-Spline. As initial form, open uniform knot vector form that is chosen in unit interval [0,1] is used in our implementation as shown in Line 8. As depicted in Line 9, in a loop, B-Spline basis function is generated via u parameter and then collision and dynamic feasibility is checked on every discrete point of the curve as shown in Line 10. Since velocity and acceleration on the path is a function of time, for each point of the trajectory we have to check if the instantaneous velocity and acceleration is within the limits of the flight envelope. This Dynamic feasibility check is done by checking the first and the second derivatives of the B-Spline curve which gives the velocities and accelerations of the aircraft respectively. If these velocity and acceleration values are within the limitations of the aircraft (*flight envelope*) using chosen *unit time scale*, generated path segment is accepted as dynamically feasible. One of the most critical step in the path planning layer is to determine the velocity and accelerations on the trajectory; if the aircrafts velocity constraints are not taken into account, lower layers(maneuver planners, low-level control layers etc.) would not be able to find feasible references from this generated trajectory. This concept is used for giving a sense on *Dynamic Feasibility* as Path Planners do.

If feasibility cannot be obtained during the path planning, repairing methods are implemented hierarchically. Firstly, location pseudo control points are slid on the same tangent directions as shown in Line 12 and the algorithm decreases the u value from current interest curve segment $-k/2$ curve segment where k is represents order that is four in this implementation. Note that, local support property of the B-Splines allows local control over the generated spline. Specifically, this control is over the curve segments with $\pm k/2$ polygon spans around the

Algorithm 4: Dynamically Feasible Trajectory Generating with B-Spline

input : way point set $\mathbf{g} = [g_1 \dots g_m]$
output: *dynamically feasible path*

TimeScale $\leftarrow$ unit-time scale
$\mathbf{P} \leftarrow \mathbf{g} = [g_1 \dots g_m]$ as control point set
foreach *element* g_i *of the* **g do**
  Insert back pseudo way point $gp_{i,b}$ to the random selected distance d from g_i on its negative heading direction
  Insert further pseudo way point $gp_{i,f}$ to the random selected distance d from g_i on its positive heading direction
$\mathbf{P} \leftarrow \mathbf{gp} = [gp_{1,f}\ gp_{2,b}\ gp_{2,f} \cdots gp_{m-1,b}\ gp_{m-1,f}\ gp_{m,b}]$ as control point set
$\mathbf{U} \leftarrow [0\,0\,0\,0\,\mathbf{U_{mid}}\,1\,1\,1\,1]$
for $u \leftarrow 0$ **to** 1 **do**
  Evaluate B-Spline basis function
  Check collision and dynamic feasibility
  if *collision occurs or point of the spline is not dynamically feasible* **then**
    Change locations of the pseudo way-points $gp_{i,b}, gp_{i,f}$ of the interest curve segment according to u
    Set u value as indicates that local interest curve segment $-k/2$ segment
    $m_1 + +$
    if $m_1 > M_1$ **then**
      Change locations of the way points g_i and its pseudo way-points $gp_{i,b}, gp_{i,f}$ of the interest curve segment according to u in its small region
      Set u value as indicates that local interest curve segment $-k/2$ segment
      $m_2 + +$ and $m_1 = 0$
      if $m_2 > M_2$ **then**
        $\mathbf{P} \leftarrow P_{new}$ as new control point around unfeasibility and update knot vector
        Set u value as indicates that local interest curve segment $-k/2$ segment
        $m_3 + +$ and $m_1, m_2 = 0$
        if $m_3 > M_3$ **then**
          Change *TimeScale* to insert min/max values of velocity and acceleration in dynamically feasible region
          Set u value as 0
          $m_4 + +$ and $m_1, m_2, m_3 = 0$
          if $m_4 > M_4$ **then**
            break with *fail*
Set *TimeScale* as generated path can be implemented as possible as in optimal time
return B-Spline*Trajectory*

displaced or newly added point. Therefore, when any changes is made on spline, instead of evaluate all spline over and over again, u value is decreased by value interval that spans local interest. Note that, all the repairing steps are tried with predefined threshold iteration times

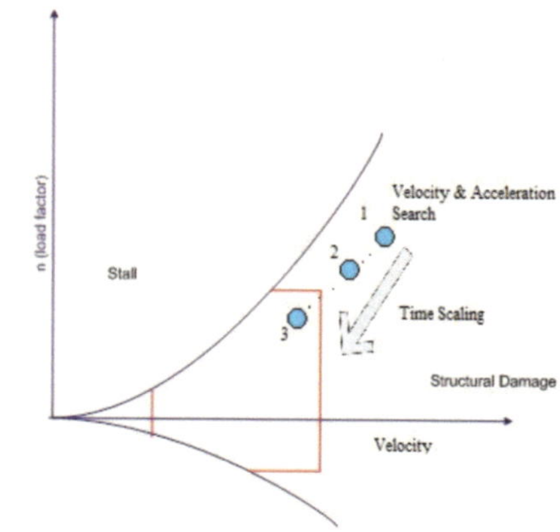

Fig. 9. Feasible Velocity - Acceleration Search on Flight Envelope

illustrated as M_is in the algorithm. After predefined number of trials, if the spline cannot be repaired, the way point and its pseudo way points within local interest are carried to a new collision-free locations and these locations are chosen as small random-selected distance away from the prior locations as seen in Line 16.

If the B-Spline still cannot be repaired, new control point P_{new} is added in control point vector **P** around the region in which collision or infeasibility has occurred(Line 20). Since we know the infeasible knot value and its interval in the knot vector, new knot point is added to the midpoint of the infeasible knot interval. Hence, only a limited interval of the knot vector **U** is updated. Reader should remember, only $\pm k/2$ polygon spans around the displaced or newly added point will be effected with this change.

If all these processes can not repair the path, the *time scale* value is scaled in Line 24 to reallocate the min-max velocity and acceleration interval of the trajectory(time depended path) within the dynamically feasible interval that can be achieved by the aircraft (falls into limits of flight envelope). For example in Fig. 9, search begins in a point outside the flight envelope and in two steps it is moved into limits of flight envelope by time scaling. Finding the feasible velocity-acceleration by this method is similar to what authors had done for finding the feasible modal inputs in the (Koyuncu et al., 2008). Note that other than flight envelope check there is not any dynamic model involved in path planning layer. All the other feasibility problems (actuator saturation, attitude discontinuity etc.) are left to low-level layers.

The end result is a time expanded or shortened flight path. Dynamic feasibility of the all generated spline should be checked from the beginning considering to newly changed time scale. After the generating B-Spline, *TimeScale* is also set again to fly over the all path in optimal time interval. Dynamically Feasible Path solution of the B-Spline Planner Algorithm in the *MelCity* model for a UCAV is demonstrated in Fig. 10

In (Koyuncu et al., 2009), one of application of this approach is demonstrated. In this work (Koyuncu et al., 2009), with adding one more interval maneuver planning layer, generated dynamically feasible trajectory is refined with modal-maneuvers that are briefly explained in previous section. This additional step gave a change to use advantages of the multi modal maneuvering approach.

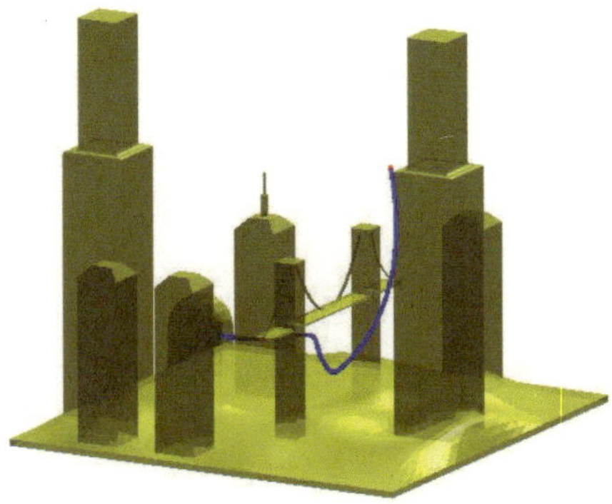

Fig. 10. Dynamically Feasible Path solution of the B-Spline Planner Algorithm in the complex 3D environment for UCAV

5. Mobil Robot Testbed for Algorithm Verification Tests

For 2D workspace verification tests of the algorithms, we set up a mobile robotic testbed in Controls and Avionics Lab (CAL) at the Istanbul Technical University. On this testbed, we implemented simplified 2D versions of the our algorithms to observe the practical implementability. 2D verification is good implementation method for such that 3D algorithms since its have rapid prototyping property. Similar approach is seen in (Clark et al., 2003), that such 3D Multi-robot space system tests have been implemented in 2D MARS test-platform.

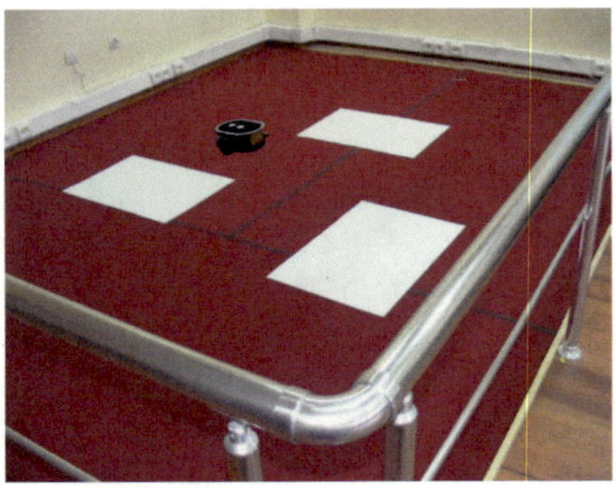

Fig. 11. ITUCAL Robotic Testbed

5.1 Mobile Robot Platform

The robotic testbed platform consists of a 4m x 3m movement platform with autonomous mobile robots. The testbed robots build in CAL are circular shaped and have two independently driven wheels which allows the robots to rotate on its center points. The robots are equipped with a 400 MHz Linux computer (Gumstix) and 8-bit microcontroller (Robostix) to perform own tracking controllers and communication.

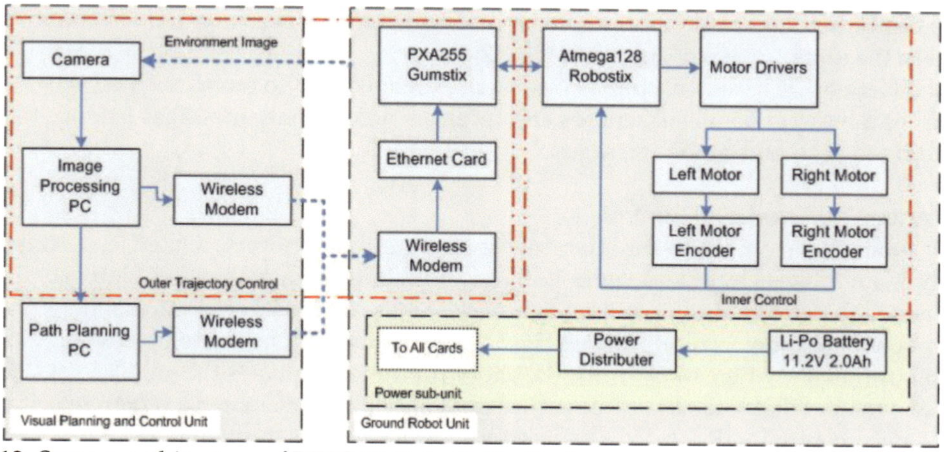

Fig. 12. System architecture of ITUCAL Robotic Testbed

5.2 Visual Positioning

The configurations and color-ids(for multi-robot applications) of the robots are detected by an overhead vision positioning system that is mounted on top of the platform. Every robot have two spots on its top side and one of them takes place center of the robot while other one takes place on heading of the robot with its own id color. Configurations of the robot consists of position and heading direction is tracked thanks to these center and heading spots. General appearance of the testbed can be seen in Fig. 11.

Visual positioning application runs on a centralized PC and uses two CCTV camera to cover the platform. Taken images are processed to detect both configurations of the robots and the obstacles. Processed images are turned in to point-cloud matrix to perform in the path planner computer. All visual positioning application is coded on MATLAB and visual positioning unit can give us configurations of the robot and obstacles in 15 Hz while path planning unit gives solution in > 3 Hz on this application specifications. Example processed images can be seen in Fig. 13

Fig. 13. Example processed environment image, detection obstacles and robots

5.3 Network Communication

Two centralized computers (visual positioning and path planner) and robots' computers runs on the system at the same time. These all computers are connected through a LAN network. Communication among the centralized PCs is performed with the physical ethernet cable while Centralized PCs and robots are connected with wireless network. Data communication between the units are demonstrated in Fig. 12.

Testbed Network is based on a publish/subscribe architecture. To broadcast messages, sender publishes a message to all subscribers and receivers accepts only messages belongs to them according to head-tags of the messages.

5.4 Performing Low-Level Control

Every robots have ability to run own low-level control algorithms. Outer loop control algorithm, a nonlinear trajectory controller, runs on robots' own embedded Linux computers (Gumstix). To perform this algorithm, reference path is received from Path Planner PC while current configurations is received from Visual Positioning PC via wireless network. According to position and orientation errors, trajectory controller evaluates the angular velocities of both two motors that leads the robot to track reference path. Evaluated angular velocities are sent to microcontroller (Robostix) as reference control variables through UART port. Robostix also counts the pulses of the optic encoders of the motors to evaluate current angular velocities. Received reference angular velocities and current angular velocities are compared and PWM signals are generated via PID controllers (as inner control loop) and then these PWM signals are sent to motor drivers. These both application is coded in C performs on Gumstix and Robostix. All these control architecture demonstrated in Fig. 14.

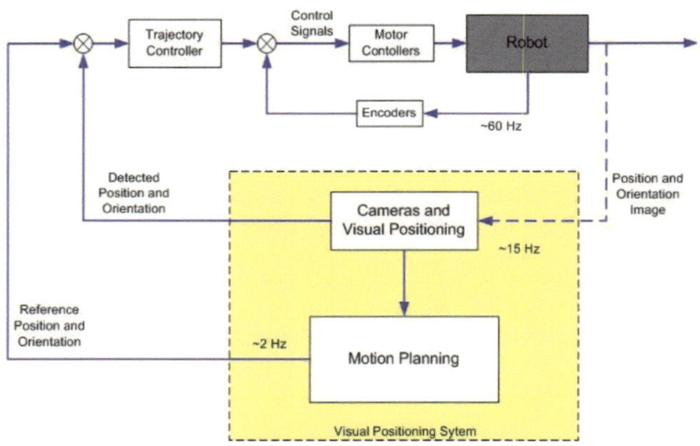

Fig. 14. Control Architecture of the ITUCAL Robotic Testbed

6. Experiments and Simulation Results

6.1 Physical Hardware Demonstrations

To demonstrate the applicability of the algorithms on physical systems, robot experiments have been implemented. In first group experiments on the ITU CAL Robotic Testbed, simplified version of the Modal-Maneuver Based PRM is used. On this application, two independently controlled primitive maneuver modes -straight forward mode and turning mode- are used on robots. Example tunnel problem solving tests can be seen in Fig. 15. On this experiment, the path planner evaluated its solution in 357 ms. Visual positioning system published current configurations in 7 Hz and motor controllers run at 60 Hz. Robot completed its all motion in 57 s.

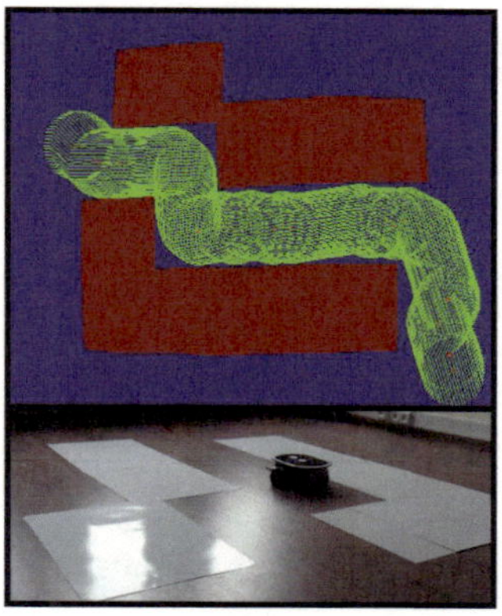

Fig. 15. Modal-Maneuver Based PRM experiments on Robotic Testbed

In the second group experiments in the ITU CAL Robotic-Testbed, Probabilistic B-Spline Based Trajectory Planner is implemented. A capture seen in Figure 16 is taken from one of the experiments of the Probabilistic B-Spline Planner in robotic testbed. Evaluated path is tracked by a nonlinear control algorithm runs on robots computer. This experiment took 29 s and the path planner evaluated its solution in 278 ms while visual positioning system published current configurations in 15 Hz and motor controllers run at 60 Hz.

Fig. 16. Probabilistic B-Spline Trajectory Planner experiments on Robotic Testbed

6.2 Simulations of 3D Environments

To illustrate the applicability the algorithms on 3D complex environments in varying ratio of obstacle-space, performance of the algorithms is tested for 3D single-narrow-passage problem, city-like environment, mostly-blocked environment and MelCity model environment that has volume 2^3 times greater than the others. All the experiments were conducted on a 3.00 GHz Intel Pentium(R) 4 processor and the average results are obtained over 50 runs.

In the first group simulations, Modal-Maneuver Based PRM algorithm for aerial vehicles is tested and computational times of the all phases of the algorithm are illustrated in Table 2

As can be seen in results, total times mostly based on Mode-Based Planner phase. As anticipated, increasing blocked space also increases the solution time as seen in Mostly-Blocked

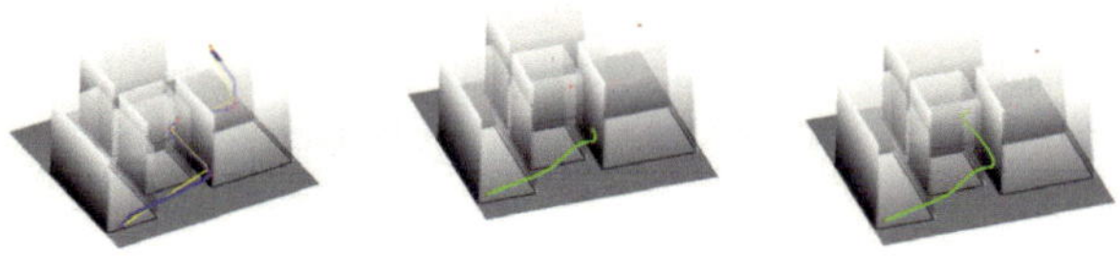

Fig. 17. Simulation example; Modal-Maneuver Based PRM Path Planer Construction Steps for City-Like Environment

		Connectivity Path Planner	Filtering Phase	Mode-Based Planner	Total Time
Single-Narrow Passage	time	0.350 s	0.032 s	32.706 s	33.089 s
	std	0.231 s	0.008 s	17.699 s	17.732 s
City-Like Environment	time	0.712 s	0.036 s	42.397 s	43.145 s
	std	0.942 s	0.008 s	24.478 s	24.639 s
Mostly-Blocked Environment	time	1.376 s	0.042 s	222.229 s	223.648 s
	std	1.132 s	0.011 s	273.451 s	273.134 s

Table 2. Modal-Based PRM Path Planer Construction Times (Seconds)

environment test. However, this increasing rate does not grow exponentially according to percentage of obstacle space. Note that in this approach, modal inputs of the independently controlled modes directly obtained that can be used by low level control layers. Therefore, this approach may be seen slower than other path planner methods, but this method significantly decrease task-load of the low-level layers and should be compared with kinodynamic approaches.

In the second group simulations, we tested the performance of Probabilistic B-Spline Trajectory Planning method on 3D environments. The computational times of steps of the algorithm are illustrated in Table 3 for 3D single-narrow-passage problem, city-like environment, mostly-blocked environment and MelCity model environment that has volume 2^3 times greater than the others.

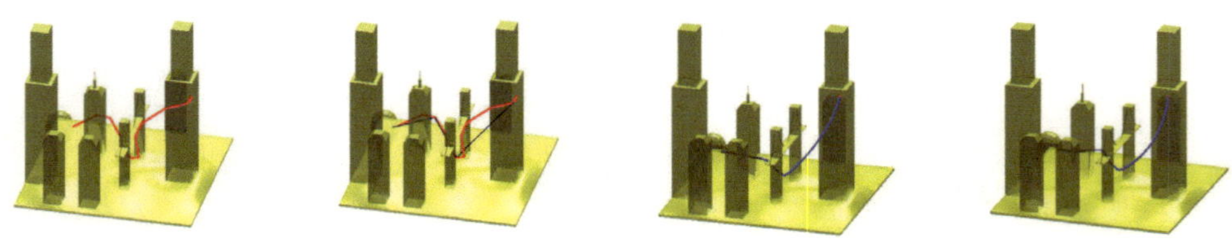

Fig. 18. Simulation example; Probabilistic B-Spline Path Planer Construction Steps for MelCity Model Environment

On this approach, increasing complexity of the environment, as shown in Table 2, mainly increases computational time of the *connectivity path* that is implemented with a simplified version of RRT. Since repairing part of the algorithm is visited much more in planning complex

		Connectivity Path Planner & Filtering	B-Spline-Based Planner	Total Time
Single-Passage Problem	avr	0.440 s	0.206 s	0.646 s
	std	0.287 s	0.011 s	0.293 s
City-Like Environment	avr	0.977 s	0.326 s	1.303 s
	std	0.935 s	0.254 s	0.930 s
Mostly-Blocked Environment	avr	3.930 s	2.182 s	5.837 s
	std	2.504 s	3.347 s	3.912 s
MelCity Model Volume; 2^3x	avr	3.306 s	0.538 s	3.844 s
	std	1.528 s	0.650 s	1.212 s

Table 3. Mode-Based Path Planer Construction Times (Seconds)

environments, computational time of the B-Spline based planner phase is also rises. However, this rising rate does not grow exponentially and computational times mostly based on Finding Connectivity Path phase. The complete solution times suggest that our method will be applicable for real-time implementations as the solution time is favorably comparable to implementation times.

7. Conclusion

Trajectory design of an air vehicle in dense and complex environments, while pushing the limits of the vehicle to full performance is a challenging problem in two facets. The first facet is the control system design over the full flight envelope and the second is the trajectory planning utilizing the full performance of the aircraft. In this work, we try to address the mostly second facet via the generating dynamically feasible trajectory planning. Hence, a real-time implementable two step planner strategy is implemented for obtaining 3D flight-path generation for an Unmanned Aerial Vehicles in 3D Complex environments. Thus simplifications on the problem improved the real time implement ability.

In our approach, initially, simplified version of the RRT planner is used for rapidly exploring the environment with an approximate line segments. The resulting connecting path is converted into flight way points through a line-of-sight segmentation.

In second step, we explained two different methods to generate dynamically feasible trajectory. First one that we called Modal-Maneuver Based PRM Planner is developed for agile unmanned aerial vehicles that their maneuvers can be define with distinct modes. This allows significant decreases in control input space and thus search dimensions. In this approach the resulting connectivity path and the corresponding milestones are refined with a single query Probabilistic Road Map (PRM) implementation that creates dynamically feasible flight paths with distinct flight mode selections and their modal control inputs. In our second approach, remaining way points are connected with cubic (C^2 continuous) B-Spline curve and this curve is repaired probabilistically to obtain a geometrically (prevents collisions) and dynamically feasible (considers velocity and acceleration constraints) path. At the end, the *time scaling* approach allow dynamic achievability considering the velocity and acceleration limits of the aircrafts. Resulting strategy is tested on real-time physical hardware system using ITU CAL mobile robot testbed for 2D environments and simulations for 3D complex environments. Computational times showed satisfactory results to used for real time implementation for UAVs operations in challenging urban environments.

One of the limitations of the algortihm is on very narrow passages, which require aircraft to tilt considerably to avoid collision. In the problems we have examined distance between obstacles are far wider compared to wing span of the aircraft so we didn't include this case. One of the possible future works is to handle these extreme cases. Moreover, extension of the algorithms presented to UAV fleets is another natural application of this work.

8. References

Bayazit, O. B., Xie, D. & Amato, N. M. (2005). Iterative relaxation of constraints: a framework for improving automated motion planning, *Intelligent Robots and Systems, 2005. (IROS 2005). 2005 IEEE/RSJ International Conference on* pp. 3433–3440.

Bohlin, R. & Kavraki, L. E. (2001). A randomized algorithm for robot path planning based on lazy evaluation, *Handbook on Randomized Computing, Kluwer Academic Publishers, p.221-249 (2001)* pp. 221–249.

Boor, V., Overmars, M. H. & van der Stappen, A. F. (1999). The gaussian sampling strategy for probabilistic roadmap planners, *IEEE International Conference on Robotics & Automation* p. 6.

Clark, C. M., Rock, S. & Latombe, J.-C. (2003). Dynamic networks for motion planning in multi-robot space systems, p. 8.

Dyllong, E. & Visioli, A. (2003). Planning and real-time modifications of a trajectory using spline techniques, *Robotica* **21**(5): 475–482.

Frazzoli, E., Dahleh, M. A. & Feron, E. (2002). Real-time motion planning for agile autonomous vehicles, *AIAA Journal of Guidance and Control* **25**(1): 116–129.

Ghosh, R. & Tomlin, C. (2000). Nonlinear inverse dynamic control for mode-based flight, *AIAA Guidance, Navigation and Control Conference* .

Hsu, D. (2000). Randomized single-query motion planning in expansive spaces, *PhD Thesis* p. 134.

Hsu, D., Jiang, T., Reif, J. & Sun, Z. (2003). The bridge test for sampling narrow passages with probabilistic roadmap planners, *IEEE International Conference on Robotics & Automation* .

Hsu, D., Kavraki, L. E., Latombe, J.-C., Motwani, R. & Sorkin, S. (1998). On finding narrow passages with probabilistic roadmap planners, *International Workshop on Algorithmic Foundations of Robotics* pp. 141 – 153.

Hsu, D., Kindel, R., Latombe, J.-C. & Rock, S. (2002). Randomized kinodynamic motion planning with moving obstacles, *International Journal of Robotics Research* **21**(2): 233 – 255.

Hsu, D., Latombe, J.-C. & Motwani, R. (1999). Path planning in expansive configuration spaces, *International Journal Computational Geometry and Applications* **4**: 495–512.

Inalhan, G., Stipanovic, D. & Tomlin, C. (2002). Decentralized optimization, with application to multiple aircraft coordination, *Decision and Control, 2002, Proceedings of the 41st IEEE Conference on* **1**: 1147–1155 vol.1.

Kavraki, L., Svestka, P., Latombe, J. & Overmars, M. (1996). Probabilistic roadmaps for path planning in high-dimensional configuration spaces, *Robotics and Automation, IEEE Transactions on* **12**(4): 566 – 580.

Kindel, R., Hsu, D., claude Robert, J. & Latombe, S. (2000). Randomized kinodynamic motion planning with moving obstacles, *The International Journal of Robotics Research* **21**(3): 233–255.

Komoriya, K. & Tanie, K. (1989). Trajectory design and control of a wheel-type mobile robot using b-spline curve, *Intelligent Robots and Systems '89. The Autonomous Mobile Robots and Its Applications. IROS '89. Proceedings., IEEE/RSJ International Workshop on* pp. 398 – 405.

Koyuncu, E. & Inalhan, G. (2008). A probabilistic b-spline motion planning algorithm for unmanned helicopters flying in dense 3d environments, *Intelligent Robots and Systems, 2008. IROS 2008. IEEE/RSJ International Conference on* pp. 815 – 821.

Koyuncu, E., Ure, N. K. & Inalhan, G. (2008). A probabilistic algorithm for mode based motion planning of agile unmanned air vehicles in complex environments, *Int. Federation of Automatic Control(IFAC'08) World Congress* .

Koyuncu, E., Ure, N. K. & Inalhan, G. (2009). Integration of path/maneuver planning in complex environments for agile maneuvering ucavs, *Proc. 2th Int. Symposium on Unmanned Aerial Vehicles (UAV'09)* .

LaValle, S. & Kuffner, J. (1999). Randomized kinodynamic planning, *Robotics and Automation, 1999. Proceedings. 1999 IEEE International Conference on* **1**: 473 – 479 vol.1.

Munoz, V., Ollero, A., Prado, M. & Simon, A. (1994). Mobile robot trajectory planning with dynamic and kinematic constraints, *Robotics and Automation, 1994. Proceedings., 1994 IEEE International Conference on* pp. 2802 – 2807 vol.4.

Nikolos, I., Valavanis, K., Tsourveloudis, N. & Kostaras, A. (2003). Evolutionary algorithm based offline/online path planner for uav navigation, *Systems, Man, and Cybernetics, Part B, IEEE Transactions on* **33**(6): 898–912.

Paulos, E. (1998). On-line collision avoidance for multiple robots using b-splines, *University of California Berkeley Computer Science Division (EECS) Technical Report* **98-977**.

Piegl, L. A. & Tiller, W. (1997). *The NURBS BookâĂŐ*, Springer-Verlag New York, Inc.

Schouwenaars, T., Feron, E. & How, J. (2004). Hybrid model for receding horizon guidance of agile autonomous rotorcraft, *IFAC Symposium on Automatic Control* .

Song, G. & Amato, N. (2001). Randomized motion planning for car-like robots with c-prm, *Intelligent Robots and Systems, 2001. Proceedings. 2001 IEEE/RSJ International Conference on* **1**: 37 – 42 vol.1.

Ure, N. & Inalhan, G. (2008). Design of higher order sliding mode control laws for multi modal agile maneuvering ucavs, *2nd Int. Symposium on Systems and Controls in Aerospace* .

Ure, N. & Inalhan, G. (2009). Design of a multi modal control framework for agile maneuvering ucavs, *IEEE Aerospace Conference* .

Vazquez, G. B., Sossa, A. H. & de Leon, S. J. L. D. (1994). Auto guided vehicle control using expanded time b-splines, *Systems, Man, and Cybernetics,Humans, Information and Technology, IEEE International Conference on* **3**: 2786–2791 vol. 3.

18

A Model-Based Synthetic Approach to the Dynamics, Guidance, and Control of AUVs

Kangsoo Kim and Tamaki Ura
NTT Communication Science Laboratories, Nippon Telegraph and Telephone Corporation
Institute of Industrial Science, The University of Tokyo

1. Introduction

In this article, we present a model-based synthetic approach applied to the dynamics, guidance, and control of Autonomous Underwater Vehicles (AUVs). The feature of vehicle dynamics is one of the most important concerns in designing and developing an AUV, while the guidance and control are the key issues in fulfilling the vehicle performance. Our approach deals with these individual but closely interrelated issues in a consistent way based on the model-based simulations.

In our research, as the dynamic model of an AUV, we employ a set of equations of motion describing the coupled 6-D.O.F. behaviour in 3-D space. In the linearized form of the equations of motion derived on the basis of the small perturbation theory, to complete the dynamic model of an AUV we have to determine so called stability derivatives or hydrodynamic coefficients of the AUV. In general, determination of the stability derivatives requires quite amount of effort and time, since they are functions of the fluid dynamic loads depending on the vehicle motion (Etkin, 1982; Lewis et al., 1989). There are many, well-established approaches for determining stability derivatives of the air vehicles (McRuer et al., 1990; Etkin, 1982) or the marine vehicles (Lewis et al., 1989), which are based on either experiment or theoretical prediction. While the experimental approach allows direct measurement of the fluid dynamic forces and moments acting on the vehicle, it requires large amount of time, labour, expense, as well as the experimental facility. On the other hand, nowadays a few state-of-the-art techniques are available in predicting the stability derivatives theoretically. Most of them are however specialized in deriving the stability derivatives for the dynamics of conventional airplane or ship, hard to be directly applied to the modelling problems related to a specific AUV dynamics. In this respect, we present a technique of deriving the dynamic model of an AUV mainly on the basis of the CFD (Computational Fluid Dynamics) analyses, which is applicable to any kind of vehicle moving in a fluid environment. In our approach based on this technique, we determine some stability derivatives dominating the dynamics of an AUV by differentiating the hydrodynamic loads obtained from the CFD analyses.

The derived dynamic model is directly applied to the model-based design of the motion control systems of a vehicle. Two PID type low-level controllers are employed to let a

vehicle follow the desired trajectories in the longitudinal and the lateral sections, represented by time sequences of the depth (altitude) and the heading.
As an intelligent high-level control of AUVs, a strategy of optimal guidance in current disturbance is presented. Suppose that a vehicle is to transit to a given destination in a region of environmental disturbance. Then it is quite natural that navigation time of the vehicle should change according to the selection of a specific trajectory. The optimal guidance proposed in this research is the minimum-time guidance in sea current environments, letting a vehicle reach a destination with the minimum travel time. When the power consumption of an AUV is controlled to be constant throughout the navigation, the navigation time is directly proportional to the total energy consumption. Released from the umbilical cable, an AUV has to rely on restricted energy stores during the undersea mission. Therefore for an AUV, minimizing navigation time offers an enhanced potential for vehicle safety and mission success rate.
We present a numerical procedure deriving the optimal heading reference by tracking which a vehicle achieves the minimum-time navigation in a given sea current distribution. The proposed procedure for implementing the optimal guidance is systematic and works in any deterministic current field whether stationary or time-varying. Moreover, unlike other path-finding algorithms such as Dynamic Programming (DP) or Generic Algorithm (GA) (Alvarez et al., 2004), our procedure does not require computation time increase for the time-varying problem.
In real environments of AUV navigation, there are some factors which can cause the failure in realizing the proposed optimal guidance strategy (Kim & Ura, 2008). Some examples of such factors are environmental uncertainties in sea currents, severe sensor noises, or temporally-faulty actuators. As a fail-safe strategy in realizing the optimal or minimum-time navigation, we present the concept of quasi-optimality. Basic idea of the quasi-optimal navigation is quite simple. It consists of repetitive revisions of the optimal heading reference in respond to the on-site request of the optimal guidance revision for preventing from the failure in on-going optimal navigation. The quasi-optimal navigation has practical importance since in real sea environments, there actually are several possible actions which deteriorate the realization of the optimal navigation.

2. An AUV "R-One"

In this article, we practice our strategy in dynamics, guidance, and control on an AUV "R-One". The R-One is a long-range cruising type AUV, developed by the Institute of Industrial Science (IIS), the University of Tokyo. Figure 1 shows overall layout of the R-One.

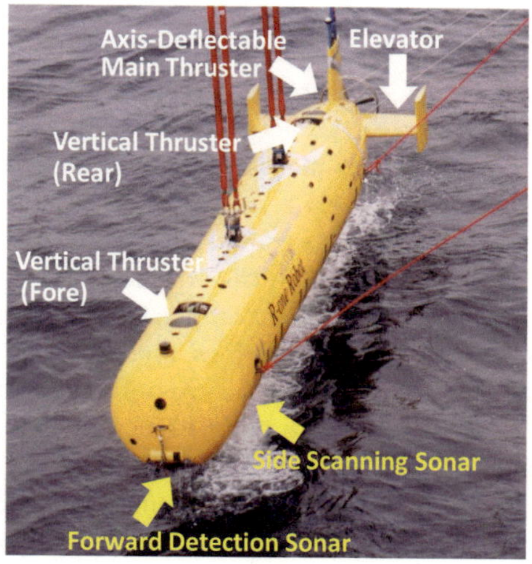

Fig. 1. Overall layout of the long-range cruising type AUV R-One.

Figure 2 shows the coordinate system and the actions of actuators installed in the R-One. While the axis-deflectable main thruster keeps and changes the vehicle's kinematic states in horizontal plane, two elevators and two vertical thrusters play the same role in vertical plane.

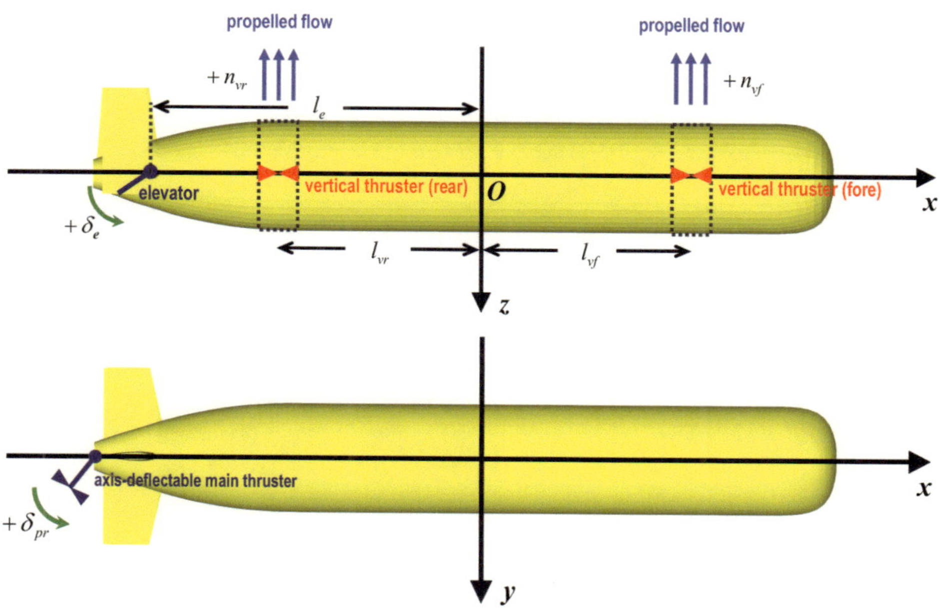

Fig. 2. Coordinate system and actuator actions in describing the dynamics of R-One. The coordinate system takes its origin at the center of gravity of the vehicle. n_{vf} and n_{vr} are the rpms of fore and rear vertical thrusters. δ_e is the elevator deflection.

3. Modelling Vehicle Dynamics

3.1 Equations of Motion for Vehicle Dynamics

The equations of motion describing the vehicle motion mathematically can be derived from the conservation law of the linear and the angular momenta with respect to the inertial frame of reference. If the axes of reference frame are nonrotating however, it should be noted that as the vehicle rotates, mass moments and products of inertia will vary, thus the time derivatives of them appear explicitly in the equations of motion (McRuer et al., 1990; Etkin, 1982). This increases the mathematical complexity which causes serious interference in treating the equations numerically as well as analytically. This is why the most of equations of motion of a rigid body in 3-D space are defined with respect to the body-fixed frame of reference. In (1), equations of motion describing the 6-D.O.F. motion of an AUV are shown. The equations are defined with respect to the body-fixed frame of reference shown in Fig. 3, in which the origin is taken at the vehicle's center of gravity.

$$m(\dot{U} + QW - RV) = -(m - \rho\nabla)g\sin\Theta + X \tag{1a}$$

$$m(\dot{V} + RU - PW) = (m - \rho\nabla)g\cos\Theta\sin\Phi + Y \tag{1b}$$

$$m(\dot{W} + PV - QU) = (m - \rho\nabla)g\cos\Theta\cos\Phi + Z \tag{1c}$$

$$I_{xx}\dot{P} - I_{xz}\dot{R} - I_{xz}PQ + (I_{zz} - I_{yy})QR = \rho\nabla g z_B \cos\Theta\sin\Phi + L \tag{1d}$$

$$I_{yy}\dot{Q} + (I_{zz} - I_{zz})RP + I_{xz}(P^2 - R^2) = \rho\nabla g z_B \sin\Theta + M \tag{1e}$$

$$-I_{xz}\dot{P} + I_{zz}\dot{R} + (I_{yy} - I_{xx})PQ + I_{xz}QR = N \tag{1f}$$

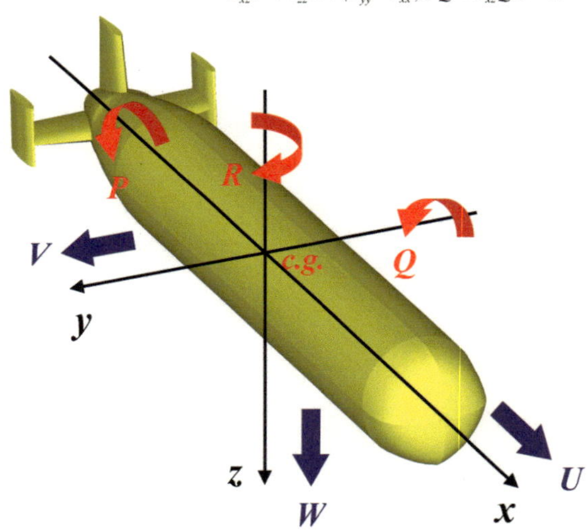

Fig. 3. Body-fixed coordinate system with linear and angular velocity components.

In (1), U, V, W, and P, Q, R are x, y, z components of linear and angular velocities. ∇, M, and I represent displacement, mass, and mass moments or products of inertia of a vehicle. ρ and g are constants expressing water density and gravitational acceleration. Hydrodynamic forces and moments are represented by X, Y, Z, and L, M, N, each of which is the component in the direction of x, y, z. Φ, Θ, and Ψ are so called Euler angles to be defined in the coordinate transformation between the body-fixed and the inertial frames of reference. z_B is the z-directional displacement of the buoyancy center of the vehicle. Eqs. (1) are quite similar to those describing the dynamics of the aircraft. It should be noted that however, terms expressing the hydrostatic forces and moments do not appear in the equations for aircraft dynamics.

The equations of motion are frequently linearized for use in stability and control analysis as remarked in Etkin (1982) or McRuer et al. (1990). The equations are linearized on the basis of the small perturbation theory in which it is assumed that the motion of the vehicle consists of small deviations from a reference condition of steady motion. Equations (2) are the linearized form of the Eqs. (1), in which *u, v, w, p, q, r, ϕ, θ,* and *ψ* denote small amounts of velocities and displacements perturbed from their reference values which are expressed by their uppercase letters.

$$m(\dot{u}+qW_0)=-\theta(m-\rho\nabla)g\cos\theta_0+X \tag{2a}$$

$$m(\dot{v}+rU_0-pW_0)=\phi(m-\rho\nabla)g\cos\theta_0+Y \tag{2b}$$

$$m(\dot{w}-qU_0)=-\theta(m-\rho\nabla)g\sin\theta_0+Z \tag{2c}$$

$$I_{xx}\dot{p}-I_{xz}\dot{r}=\phi\rho\nabla g z_B\cos\theta_0+L \tag{2d}$$

$$I_{yy}\dot{q}=\theta\rho\nabla g z_B\cos\theta_0+M \tag{2e}$$

$$-I_{xz}\dot{p}+I_{zz}\dot{r}=N \tag{2f}$$

$$\dot{\phi}=p+r\tan\theta_0 \tag{2g}$$

$$\dot{\theta}=q \tag{2h}$$

$$\dot{\psi}=r\sec\theta_0 \tag{2i}$$

To complete the linearized equations of motion for the use in stability and control analysis, hydrodynamic loads are expanded and linearized on the assumption that they are functions of the instantaneous values of the perturbed velocities, control inputs, and their derivatives. Thus the expressions of the hydrodynamic loads are obtained in the form of a Taylor series in these variables, which is linearized by discarding all the higher-order terms. For example, X is expanded as

$$X=X_u u+X_w w+X_{\dot{u}}\dot{u}+X_{n_m}n_m \tag{3}$$

where

$$X_u=\left.\frac{\partial X}{\partial u}\right)_0,\quad X_w=\left.\frac{\partial X}{\partial w}\right)_0,\ \text{etc.}$$

The subscript zero indicates a reference condition where the derivatives are evaluated. In (3), the derivatives such as X_u or X_w are called stability derivatives. By expanding all the external hydrodynamic loads introducing stability derivatives of dynamical correlations, the equations of motion (2) are expressed by means of the stability derivatives as

$$(m-X_{\dot{u}})\dot{u}+mW_0\dot{\theta}-X_u u-X_w w+\theta(\rho\nabla-m)g\cos\theta_0=X_{n_m}n_m \tag{4a}$$

$$(m-Z_{\dot{w}})\dot{w}-Z_{\dot{q}}\dot{q}-Z_u u-Z_w w-(mU_0+Z_q)q-\theta(\rho\nabla-m)g\sin\theta_0=Z_{n_{Vf}}n_{Vf}+Z_{n_{Vr}}n_{Vr}+Z_{\delta e}\delta_e \tag{4b}$$

$$-M_{\dot{w}}\dot{w}+(I_{yy}-M_{\dot{q}})\dot{q}-M_u u-M_w w-M_q q-\theta\rho\nabla g z_B\cos\theta_0=-Z_{n_{Vf}}l_{vf}n_{Vf}+Z_{n_{Vr}}l_{vr}n_{Vr}+Z_{\delta e}l_{\delta_\varepsilon}\delta_e \tag{4c}$$

$$(m-Y_{\dot{v}})\dot{v}-Y_{\dot{r}}\dot{r}-Y_v v-(mW_0+Y_p)p+(mU_0-Y_r)r=Y_{\delta_{pr}}\delta_{pr} \tag{4d}$$

$$-L_{\dot{v}}\dot{v}+(I_{xx}-L_{\dot{p}})\dot{p}-(I_{xz}+L_{\dot{r}})\dot{r}-L_{v}v-L_{p}p-L_{r}r-\phi\rho\nabla g z_{B}cos\theta_{0}=L_{\delta_{pr}}\delta_{pr} \tag{4e}$$

$$-N_{\dot{v}}\dot{v}+(I_{xz}+N_{\dot{p}})\dot{p}+(I_{zz}-N_{\dot{r}})\dot{r}-N_{v}v-N_{p}p-N_{r}r=N_{\delta_{pr}}\delta_{pr} \tag{4f}$$

$$\dot{\phi}=p+rtan\theta_{0} \tag{4g}$$

$$\dot{\theta}=q \tag{4h}$$

$$\dot{\psi}=rsec\theta_{0} \tag{4i}$$

3.2 Evaluation of Stability Derivatives by CFD Analyses

As noticeable in (4), within the framework of small perturbation theory, constructing dynamic model is reduced to the determination of the stability derivatives defined in the linearized equations of motion. Some stability derivatives in (4) are to be evaluated by using the techniques proposed in the flight dynamics or ship manoeuvrability (Etkin, 1982; Lewis et al., 1989). But since the configuration and layout of an AUV are generally quite different from those of aerial vehicle or surface ship, not all of stability derivatives appearing in (4) are to be determined by such techniques. Moreover, it is generally not easy to evaluate the stability derivatives deemed to dominate the calculated vehicle motion, for they are closely related to the damping and the energy transfer accompanied by the fluid flow (Lewis et al., 1989). The most commonly and widely employed approaches to evaluate the dominant stability derivatives are the wind tunnel test for aerial vehicles and the towing tank test for marine vehicles. Experimental approaches are however, implemented with huge experimental facility and many workforces, which require quite amount of expense even when the test is for a single model. In this article, we present a model-based approach for evaluating the dominant stability derivatives. In the approach, dominant stability derivatives are evaluated by means of the hydrodynamic loads, obtained as the results of CFD analyses. The basic idea of evaluating stability derivatives by the proposed approach is quite simple. When we are to evaluate the value of X_u defined at a reference speed of U_0 for example, we conduct CFD analyses repeatedly with the cruising speed of $U_0(1\pm\eta)$, where U_0 is the reference cruising speed and η is the perturbation ratio of U_0. By taking central difference approximation of X with respect to u by using the X values obtained at $U_0(1\pm\eta)$, we can derive X_u defined at U_0. However, while the majority of dominant stability derivatives are to be evaluated by this technique, there are other stability derivatives which are not. For such stability derivatives, estimation formulae proposed in the field of flight dynamics are modified and applied (McRuer et al., 1990; Etkin, 1982).

Figure 4 shows the grid system for evaluating the hydrodynamic loads by CFD analyses. In our CFD analyses, we used a solver called "Star-CD" (http://www.cdadapco.com/), developed by CD-adapco. The Star-CD is based on the finite difference numerical scheme and thus works with a structured grid system. In the aftbody of the R-One, geometric feature of the body surface is quite complicated due to the existence of tail fins. To generate a computationally robust structured grid system adapting to the geometric feature of the vehicle, we employed a grid generation technique called multi block method (Thomson, 1988). In the multi block method, entire grid system is subdivided into several local subgrid blocks, each of which shares the interfacing grids with the adjacent subgrid blocks.

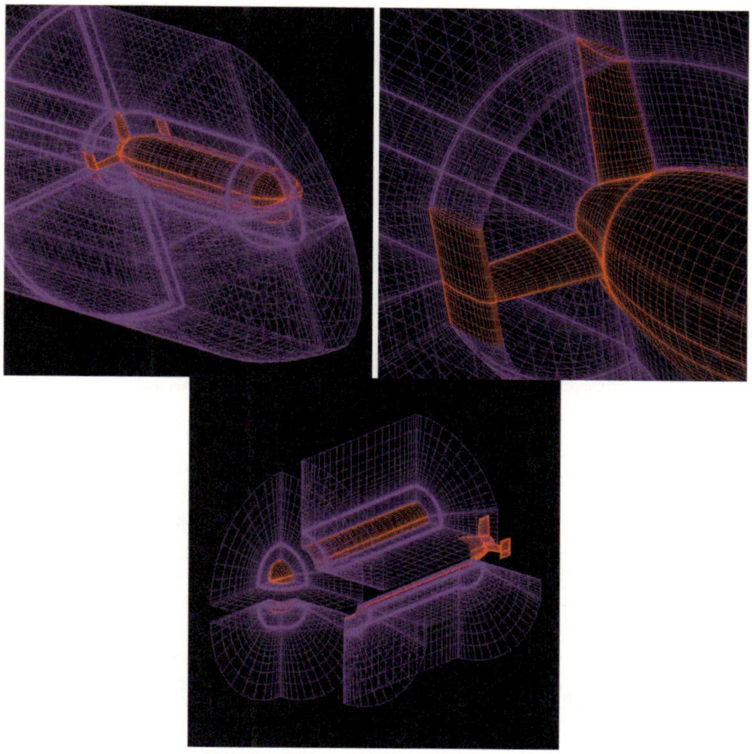

Fig. 4. Grid system for the CFD analyses of flow field around the body surface of the R-One. Due to the complicated surface geometry in the aftbody, entire grid system is completed by assembling individually generated subgrid blocks.

Figure 5 shows the pressure distribution with few selected streamlines along the body surface of R-One. By integrating the pressure over the entire body surface, hydrodynamic loads are obtained.

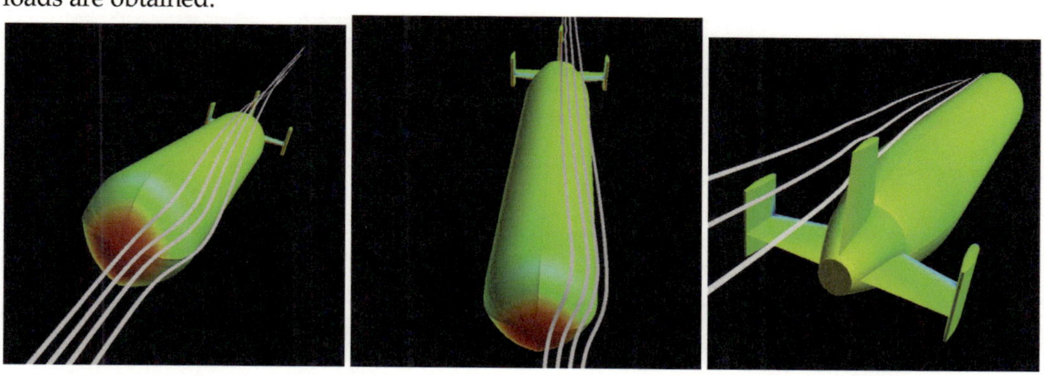

Fig. 5. Visualized results of a CFD analysis: Pressure distribution with few selected streamlines over the body surface of R-One.

By substituting all stability derivatives in (4) with their corresponding numerical values, dynamic model of R-One is completed. It is generally known and also noticeable from (4) that according to the coupling relation, linearized equations of motion are to be split into two independent groups: the longitudinal equations including surge, heave, and pitch, and the lateral equations including sway, roll, and yaw (McRuer et al., 1990; Etkin, 1982). In Table 1, longitudinal and lateral stability derivatives appearing in (4) are summarized.

$X_{\dot{u}}$	(kg)	-237.65	Z_u	(kg/s)	-64.41
$Z_{\dot{w}}$	(kg)	-2819.64	Z_w	(kg/s)	-2819.64
$Z_{\dot{q}}$	(kg·m)	-25547.00	Z_q	(kg·m/s)	-11360.06
$M_{\dot{u}}$	(kg·m)	0.00	M_u	(kg·m/s)	0.00
$M_{\dot{w}}$	(kg·m)	-1928.80	M_w	(kg·m/s)	870.36
$M_{\dot{q}}$	(kg·m^2)	-153400.00	M_q	(kg·m^2/s)	-39351.25
X_u	(kg/s)	-728.73	Z_{δ_e}	(kg·m/s^2)	-3168.10
X_w	(kg/s)	33.36	M_{δ_e}	(kg·m^2/s^2)	-10974.31
X_q	(kg·m/s)	0.00			

Table 1a. Stability derivatives in the longitudinal equations for R-One.

$Y_{\dot{v}}$	(kg)	-237.65	Y_r	(kg·m/s)	3931.06
$Y_{\dot{r}}$	(kg)	-2819.64	L_v	(kg·m/s)	-515.99
$L_{\dot{v}}$	(kg·m)	-25547.00	L_p	(kg·m^2/s)	-1165.25
$L_{\dot{p}}$	(kg·m^2)	0.00	L_r	(kg·m^2/s)	1500.45
$L_{\dot{r}}$	(kg·m^2)	-1928.80	N_v	(kg·m/s)	-4054.37
$N_{\dot{v}}$	(kg·m)	-153400.00	N_p	(kg·m^2/s)	0.00
$N_{\dot{p}}$	(kg·m^2)	-728.73	N_r	(kg·m^2/s)	-13704.63
$N_{\dot{r}}$	(kg·m^2)	33.36	$Y_{\delta_{pr}}$	(kg·m/s^2)	-399.04
Y_v	(kg/s)	-1809.97	$L_{\delta_{pr}}$	(kg·m^2 /s^2)	0.00
Y_p	(kg·m/s)	0.00	$N_{\delta_{pr}}$	(kg·m^2 /s^2)	1677.38

Table 1b. Stability derivatives in the lateral equations for R-One.

3.3 Vehicle Motion Simulation

Equations (5a) and (5b) represent state-space forms of the longitudinal and the lateral equations of motion for R-One, completed by assigning the numerical values in Table 1 to corresponding stability derivatives in (4).

- *Longitudinal Equations of Motion for R-One:*

$$\begin{bmatrix} \dot{u} \\ \dot{w} \\ \dot{q} \\ \dot{\theta} \end{bmatrix} = \begin{bmatrix} -0.1571 & 0.0072 & 0 & 0.0145 \\ -0.0103 & -0.4725 & 0.2456 & 0.0610 \\ 0.0001 & 0.0108 & -0.2420 & -0.0156 \\ 0 & 0 & 1 & 0 \end{bmatrix} \begin{bmatrix} u \\ w \\ q \\ \theta \end{bmatrix} + \begin{bmatrix} 0.0554 & 0 & 0 \\ 0 & 0.0027 & -0.2169 \\ 0 & -0.0001 & -0.0684 \\ 0 & 0 & 0 \end{bmatrix} \begin{bmatrix} n_m \\ n_v \\ \delta_e \end{bmatrix} \tag{5a}$$

- *Lateral Equations of Motion for R-One:*

$$\begin{bmatrix} \dot{v} \\ \dot{p} \\ \dot{r} \\ \dot{\phi} \end{bmatrix} = \begin{bmatrix} -0.2097 & 0.0053 & -0.5388 & 0.0112 \\ -4.7444 & -11.2192 & 16.1215 & -23.6516 \\ -0.1185 & 0.0643 & -1.1931 & 0.1357 \\ 0 & 1 & 0 & 0 \end{bmatrix} \begin{bmatrix} v \\ p \\ r \\ \phi \end{bmatrix} + \begin{bmatrix} -0.0388 \\ -0.0948 \\ 0.0634 \\ 0 \end{bmatrix} \delta_{pr} \tag{5b}$$

By solving (5) in time domain with appropriate initial conditions and actuator inputs, motion responses of the R-One are computed. In the Inertial Navigation System (INS) installed in R-One, not only vehicle kinematics but also time sequences of the actuator inputs during an undersea mission are recorded. In Figs. 6 and 7, simulated vehicle trajectories are shown and compared with the actual vehicle trajectories which were recorded during the Teisi knoll survey mission (Ura et al., 2001). By applying the actual actuator inputs taken from the INS record to (5), simulated motion responses are computed producing the simulated vehicle trajectories. As noticeable from Figs. 6 and 7, the dynamic model of R-One implemented by our model-based approach provides the motion responses exhibiting sufficiently good similarity between the simulated and the actual vehicle trajectories.

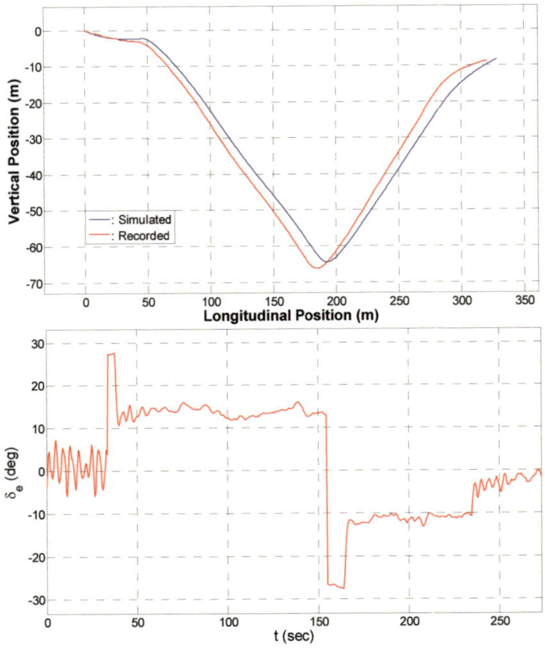

Fig. 6. Simulated and actual vehicle trajectories in vertical plane (top) generated from the corresponding actuator input (bottom).

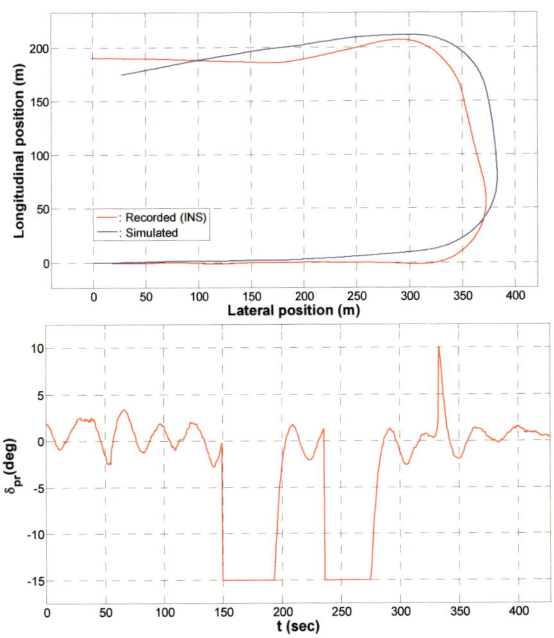

Fig. 7. Simulated and actual vehicle trajectories in horizontal plane (top) generated from the corresponding actuator input (bottom).

4. Tracking Control Design

The controller model selected to be implemented in the motion control system of R-One is based on the PID compensation. Needless to say, PID type controller is the most commonly and widely used controller for most artificial control systems. However in designing a PID controller, precise plant dynamics is an important prerequisite to ensure the acceptably good control performance. Deriving a precise plant dynamics is not easy in some cases. For this reason during the past two decades, a few significant attempts have been made to provide the controller models which do not require precise description of the plant model in its design. Neural network (NN) controllers based on the self-organizing map, or fuzzy logic controllers are the most famous ones in such attempts (Haykin, 1999). In order to derive a practically useful controller by NN or fuzzy however, we have to ensure huge diversity in training data. This is also a very difficult task in a real world problem, for in general we do not have any useful guideline to decide whether the prepared training data is biased or not. Moreover, since we have derived a fine dynamic model as shown in the previous section, we do not have any particular reason for choosing NN- or fuzzy-based controller model instead of the PID-based controller.

To change or keep the kinematic states of the vehicle, two independent low-level controls were implemented in the R-One: the depth (altitude) control for the longitudinal motion and the heading control for the lateral motion. Configurations of the depth and the heading controls are depicted in Fig. 8 and Fig. 9.

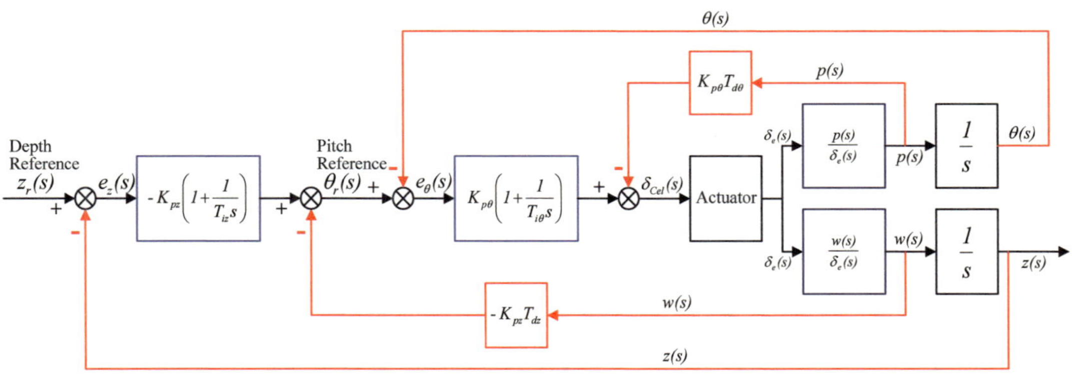

Fig. 8. Configuration of the depth (altitude) control system for R-One.

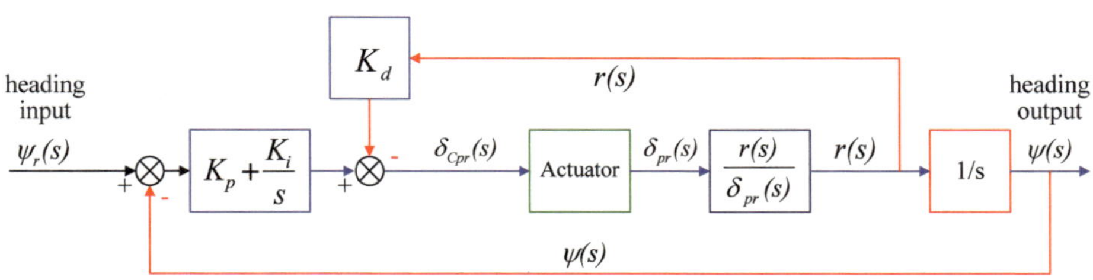

Fig. 9. Configuration of the heading control system for R-One.

In the model-based or computer-aided control system design, mathematical model describing the system behaviour should be prepared. To build up the mathematical models for the control systems shown in the Figs. 8 and 9, transfer functions of $p(s)/\delta_e(s)$, $w(s)/\delta_e(s)$, and $r(s)/\delta_{pr}(s)$ are extracted from (5). Then, the PID-tuning is carried out to determine the optimal values of controller gains from the standpoint of the system robustness and the swiftness in response. In determining the optimal gain values, we used the model-based control system design tool called "SISO Design Tool", offered by "Control System Toolbox" included in "Matlab" (The Mathworks Inc., 2001).

An example exhibiting the performance of the designed control systems is shown in Fig. 10. In the figure, it is clearly seen that the designed heading controller let the vehicle follow the heading reference with sufficient swiftness and small overshoot.

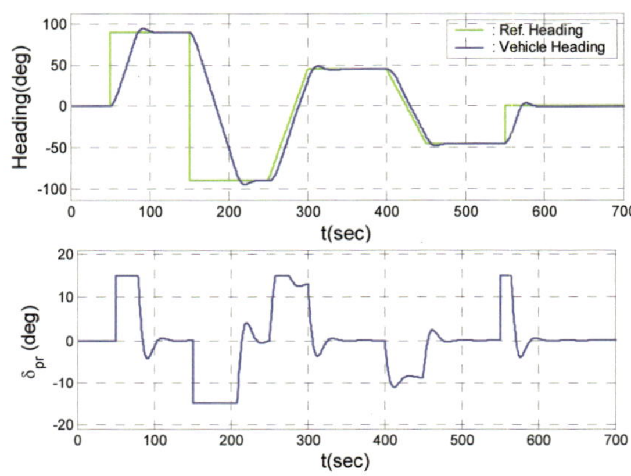

Fig. 10. Simulated heading tracking control (top) with corresponding actuator input (bottom).

5. Optimal Guidance of AUV

5.1 Background

The sea environment contains several disturbances such as surface wave, wind, and sea current. Among them, the sea current is known to be the most important disturbance for the dynamics of an underwater vehicle directly interacting with the vehicle motion (Lewis et al., 1989). Considering the guidance problem to make a vehicle transfer to a given destination in a region of sea current, it is quite natural that there happens navigation time difference according to the selection of an individual navigation trajectory. The problem of the minimum-time vessel guidance in a region of current flow interested people so long years ago as ancient Greece (Bryson & Ho, 1979). But since the problem requires minimization technique of the functionals, it hardly had been treated mathematically until the advent of the calculus of variations. On the basis of this mathematical tool, Bryson and Ho derived the minimum-time guidance law of a surface vessel in a region of a surface current flow. Though it is an optimal controller of explicit form described as an ordinary differential equation, to obtain its solution is not easy since it actually is so called a two point boundary value problem. As an ad hoc approach for the minimum-time navigation problem in a linearly varying, shear flow like current distribution presented by Bryson and Ho, a graphical solution finding technique has been presented (Lewis & Syrmos, 1995). As is naturally expected however, such an approach is a problem specific one, lacking in universality in its applicability. Papadakis & Perakis (1990) treated the minimal time routing problem of a vessel moving in a region of wave environment. In their approach, by subdividing the navigation region into several subregions of different sea states, the path for the optimal routing is obtained by the dynamic programming approach. Aside from the difficulties in constructing a numerical solution procedure for their approach, it has a problem of significant solution dependency on the features of regional subdivision. As a completely discrete and nonlinear approach, the cell mapping technique was applied to derive the minimal time tracking trajectory to catch a moving target in a deterministic vortex field (Crespo & Sun, 2001). Though the cell mapping is quite suitable for the optimal control of strongly nonlinear systems, it also has the problem of significant solution dependency on the subdivision.

In this research, we propose a solution procedure to obtain the numerical solution of the optimal guidance law, achieving the minimum time navigation of an AUV in current fields. It is a simple but consistently applicable algorithm to the optimal navigation problem defined in any current filed, if only the distribution of which is deterministic. In applying the proposed procedure, a simulated navigation along any feasible trajectory terminated at the destination is required. We call this navigation the "reference navigation", because it is utilized as the reference in realizing the optimal navigation. As a fault-tolerable strategy in putting the proposed optimal navigation into practice, concept of the quasi-optimal navigation is introduced. The basic idea of the quasi-optimal navigation is quite simple, consisting of the on-site feedbacks of the optimal guidance revisions.

5.2 The Optimal Guidance Law

In this optimal guidance problem, we employed the guidance law presented by Bryson and Ho (1979), shown in (6).

$$\dot{\psi} = sin^2\psi \frac{\partial v_c}{\partial x} + \frac{1}{2}\left(\frac{\partial u_c}{\partial x} - \frac{\partial v_c}{\partial y}\right) sin2\psi - cos^2\psi \frac{\partial u_c}{\partial y} \qquad (6)$$

In (6), ψ represents the vehicle heading, and u_c, v_c are x, y components of the sea current velocity. Though Bryson and Ho derived (6) on the assumption of the stationary flow condition, we have shown that it is also valid for time-varying currents, like tidal flows (Kim & Ura, 2009).

Equation (6) is a nonlinear ordinary differential equation of an unspecified vehicle heading $\psi(t)$. Though the solution of (6) seems to be readily obtainable by a suitable numerical scheme such as Runge-Kutta, there still remains a significant shortfall. While (6) defines an initial value problem, its solution cannot be obtained with an arbitrarily assigned initial heading. If we solve (6) with an arbitrary initial value of ψ, the vehicle following the solution of (6) as the heading reference does not arrive at the destination. It is because (6) is a two point boundary value problem in fact, correct initial value of which constitutes a part of the solution.

5.3 Numerical Solution Procedure

To obtain the solution of the two point boundary value problems, an iterative solution procedure is generally used such as "shooting" or "relaxation" (Press et al., 1992). Starting from an initial guess, solutions generated by these schemes are repeatedly adjusted to eliminate the discrepancies between the estimated and the desired boundary conditions at both endpoints. But these schemes strongly rely on the initial guess, inappropriate assignment of which may lead to a local solution or divergence. In this study, we present a numerical procedure to obtain the solution of (6), called "AREN", named as an acronym of Arbitrary REference Navigation". In applying the AREN, first we need to make a simulated navigation in which the vehicle travels following an arbitrary guidance law. In this navigation, traveling time for the vehicle to reach the destination should be recorded. This navigation we call the "reference navigation", and the corresponding traveling time the "reference navigation time", denoted by t_{f_ref}. The only requirement for the reference navigation is that the vehicle has arrived at the destination at the final state.

In AREN, to search for the correct initial heading numerically, the interval of $0 \sim 2\pi$ is divided by equally spaced *N-1* subintervals as

$$\psi_0^{(i)} = i\Delta\psi \quad \text{for} \quad i=0,1,...N-1 \tag{7}$$

where $\Delta\psi = 2\pi/N$.

In (7), $\psi_0^{(i)}$ is *(i)*-th initial heading trial and $\Delta\psi$ is its increment, i.e. the interval of initial heading trials. Next, for an initial heading trial $\psi_0^{(i)}$, we solve (6) in time domain by a time marching scheme, which produces a simulated navigation starting from $\psi_0^{(i)}$. The navigation having produced here is called *(i)*-th "trial navigation" adjoining to $\psi_0^{(i)}$. Once the trajectory produced by a trial navigation passes through the destination, it is the optimal navigation since only if the optimal guidance law (6) with the correct initial heading let a vehicle reach the destination. Therefore, *N* trial navigations are possible candidates for the optimal navigation. In practice however, discretization error in initial heading trials causes the convergence error at the destination, so that the optimal navigation should be identified in an approximate manner.

For a navigation trajectory, we define the "minimum distance" which is the shortest distance between the destination and a trajectory. In Fig. 11, $l_{min}^{(k-1)}$, $l_{min}^{(k)}$, and $l_{min}^{(k+1)}$ represent the minimum distances between the destination and the trajectories generated by *(k-1)*-th, *(k)*-th, and *(k+1)*-th trial navigations, respectively.

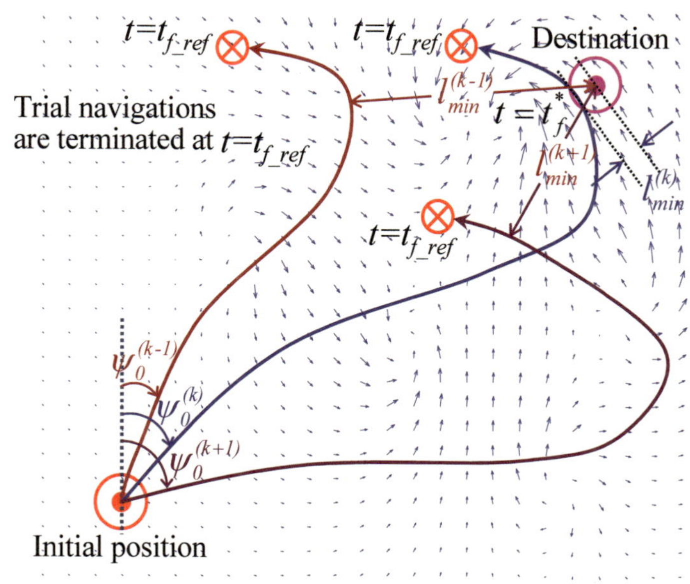

Fig. 11. Trial navigations starting from discrete initial headings. The approximate optimal navigation converging to destination is determined based on the minimum distances. Note that trial navigations are continued until $t = t_{f\_ref}$.

If the minimum distance of *(k)*-th trial navigation is smaller than any other minimum distance, and thus satisfies (8), we choose it as the optimal navigation because on its trajectory vehicle approaches the destination with the smaller deviation than any other trial navigation.

$$l_{min}^{(k)} \leq l_{min}^{(i)} \quad \text{for } i=0,1,...N\text{-}1 \tag{8}$$

But in choosing the optimal navigation among the trials, there still remains a serious drawback. We have no idea how long we have to continue a trial navigation so as to determine its true minimum distance. Here comes the use of the reference navigation time prepared beforehand. It is apparent that the reference navigation is non-optimal, since it is the one following an arbitrary guidance law, only achieving the vehicle's arrival at the destination. Therefore, reference navigation time must be larger or equal to that of the optimal navigation as

$$0 < t_f^* \leq t_{f\_ref} \tag{9}$$

where t_f^* represents navigation time by the optimal guidance. It should be noted here that by the minimum principle, once we have started a trial navigation with an initial heading close to optimal one, vehicle never fails to pass by the vicinity of the destination at the time smaller than t_{f_ref}. In other words, the reference navigation time qualifies to be the upper limits of the simulation times of the trial navigations in order to identify an optimal navigation among the trials, on the basis of the minimum distances. In Fig. 11, *(k)*-th trial navigation is picked up as the optimal navigation, since on its trajectory the vehicle marks the smallest minimum distance.

6. Optimal Navigation Examples

6.1 The Reference Navigation

As mentioned in the previous section, to implement the optimal guidance for vehicle navigation by AREN, it is necessary to make a reference navigation beforehand. The simplest guidance satisfying the vehicle's arrival at the destination may be the Proportional Navigation (PN). In PN, vehicle heading is continuously adjusted to let the line of sight direct toward the target. In our work, we employ PN as the reference navigation.

6.2 Optimal Navigation in a Shearing Flow

The first numerical example in this research is an optimal navigation in a current disturbance of the linear shear flow, taken from Bryson and Ho (1979). The current velocity in this problem is described by

$$u_c(x, y) = 0 \quad (10a)$$

$$v_c(x, y) = -U_c x/h \quad (10b)$$

where U_c and h are set to be 1.544 m/s and 100 m, respectively. Starting from the initial position at $(x_0, y_0) = (-186\ m, 366\ m)$, R-One is subjected to move the destination at the origin in this example. Due to its simplicity, the current distribution of (10) allows derivation of the analytic optimal guidance law expressed as

$$\frac{x}{h} = csc\,\psi - csc\,\psi_f \quad (11a)$$

$$\frac{y}{h} = \frac{1}{2}\left[csc\,\psi_f(cot\,\psi - cot\,\psi_f) + cot\,\psi(csc\,\psi_f - csc\,\psi) + log\frac{csc\,\psi_f - cot\,\psi_f}{csc\,\psi - cot\,\psi} \right] \quad (11b)$$

where ψ_f is the vehicle heading at the final state.

Navigation trajectories are shown in Fig. 12. In the reference navigation by PN, significant adverse drift happens at initial stage, since within the region of $|x| > 100\ m$ current flow speed exceed the advance speed of the vehicle. Optimal guidance detours the vehicle across the upper half plane, taking advantage of the favorable current flow. Navigation times by PN and optimal guidance are 353.7 s and 739.2 s respectively, implying 52 % decrease in navigation time by the optimal guidance.

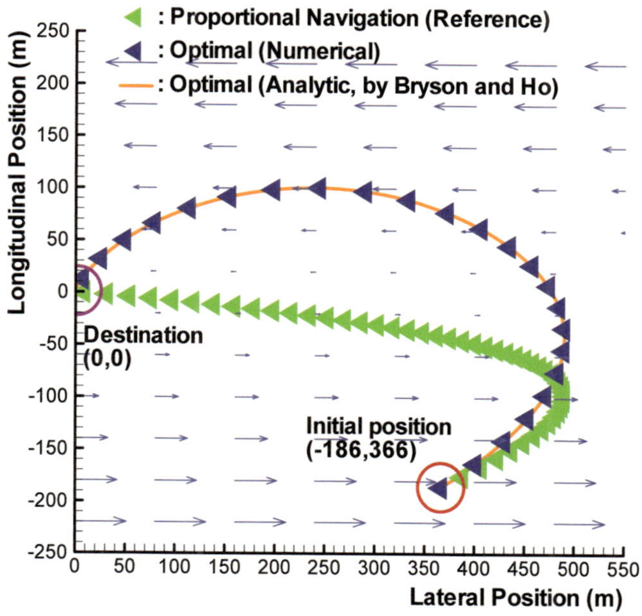

Fig. 12. Navigation trajectories in a linear shearing current flow.

6.3 Optimal Navigation in a Time-Varying Flow

The next numerical example is an optimal navigation in a time-varying current flow. In actual sea environments, for the most of currents the direction and the magnitude of their velocities change continuously like tidal flows. As mentioned previously, the optimal guidance law (6) is also valid for time-varying current flows as well as the stationary ones. Therefore, once the flow velocity distribution in a navigation region is described as a function of the position and time, our numerical scheme equally works and realizes the minimal time navigation.

Navigation trajectories of the R-One in an artificially made time-varying current are depicted in Fig. 13. As shown in the figure, near the middle of the navigation region, vehicle following the guidance of PN is temporarily failing to proceed toward the destination due to severe drift caused by local current flow of adverse direction. Occurrence of such a disadvantage is prevented from in the optimal navigation. By following the optimal guidance, the vehicle proceeds taking advantage of the favorable flows and avoids passing through the gradually changing region of strong adverse current flow. In Fig. 13, it should be noted that at 623.0 s, having released from the severe drift, R-One being controlled under the guidance of PN is about to restart toward the destination. At the same instant however, the optimal guidance has already made the vehicle arrive at the destination.

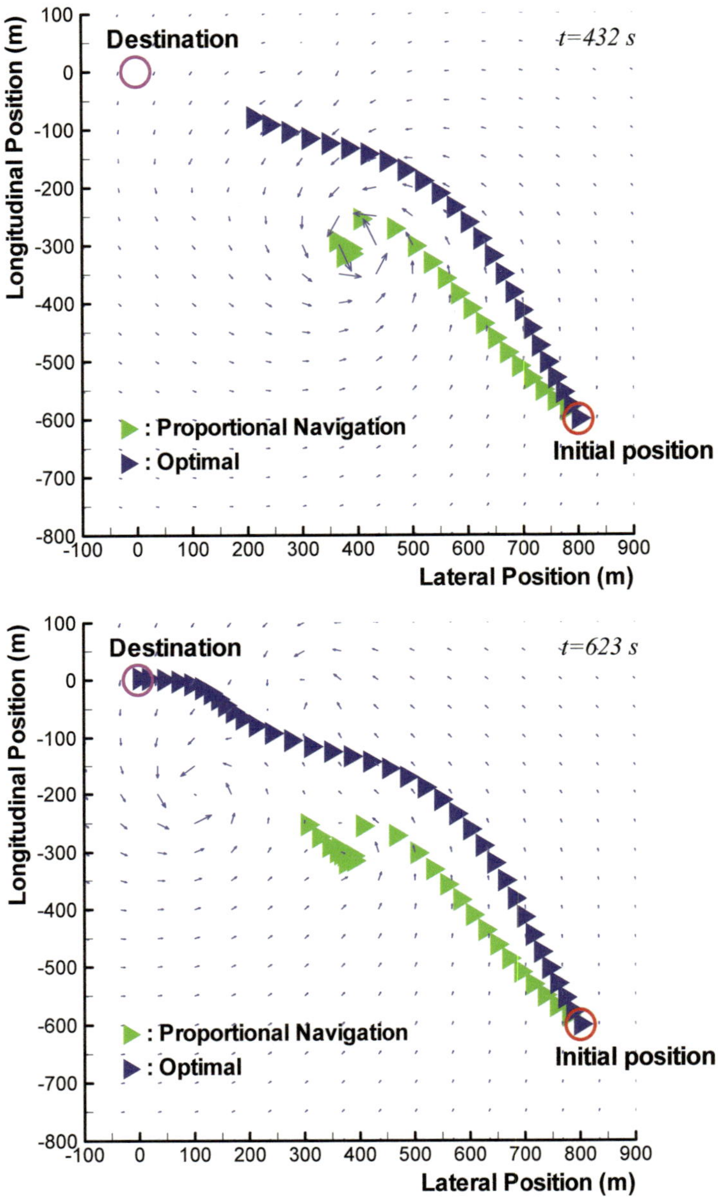

Fig. 13. Navigation trajectories in a time-varying current flow at t = 432 s (top), and t = 623 s (bottom).

6.4 Quasi-Optimal Navigation

The quasi-optimal navigation is a fail-safe strategy introduced to react to the failure in realizing the optimal navigation due to environmental uncertainties or temporal malfunctions in sensors and actuators. Suppose that a vehicle has failed in following the optimal trajectory due to one of the abovementioned problem. Not only has the navigation lost the optimality, but by following the current optimal guidance, vehicle is even unable to arrive at the destination. According to the Bellman's principle of optimality (Lewis & Syrmos, 1995), once we have failed in tracking the optimal trajectory, the best policy we can take from then on is reconstructing and following the new optimal trajectory at the present state. The resulting trajectory is not optimal since it already has included the past non-optimal interval. But it is evidently the best trajectory we can take under present situation, so that we call the corresponding navigation the quasi-optimal navigation. In Fig. 14, optimal and quasi-optimal navigation trajectories in a time-varying current are shown.

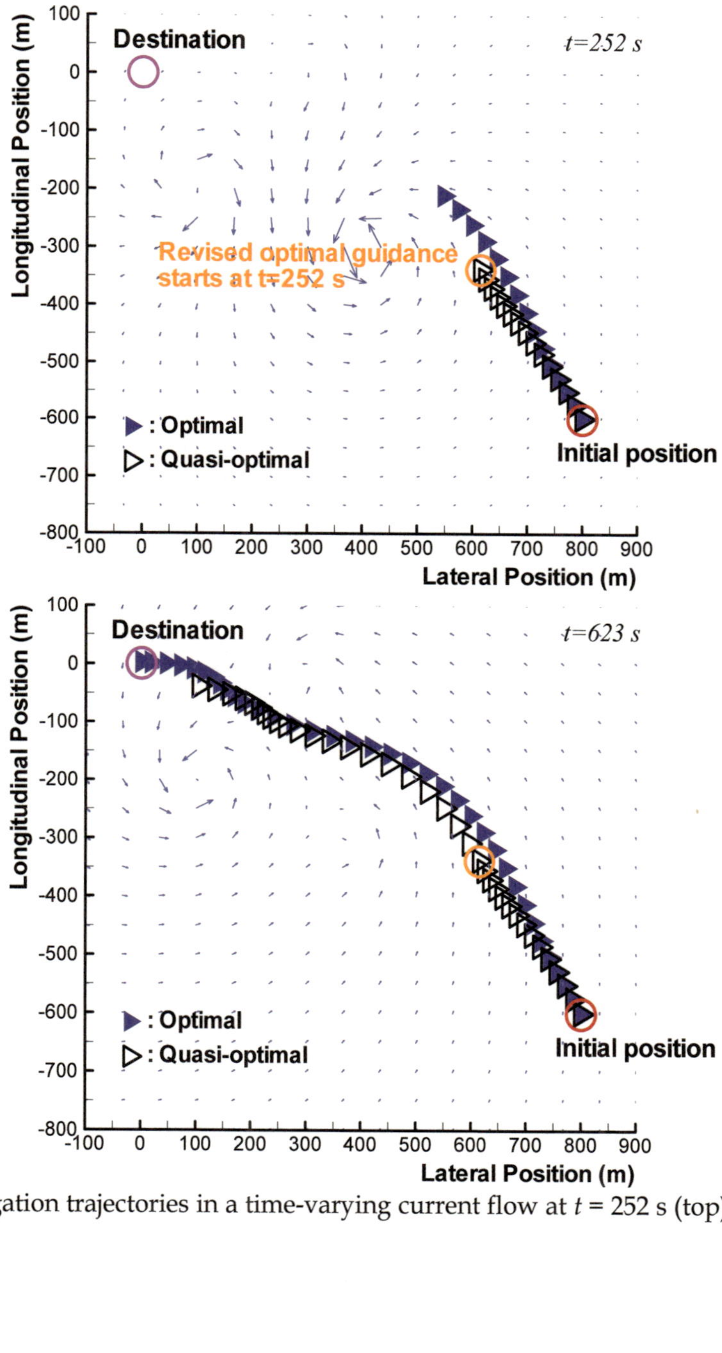

Fig. 14. Navigation trajectories in a time-varying current flow at t = 252 s (top), and t = 623 s (bottom).

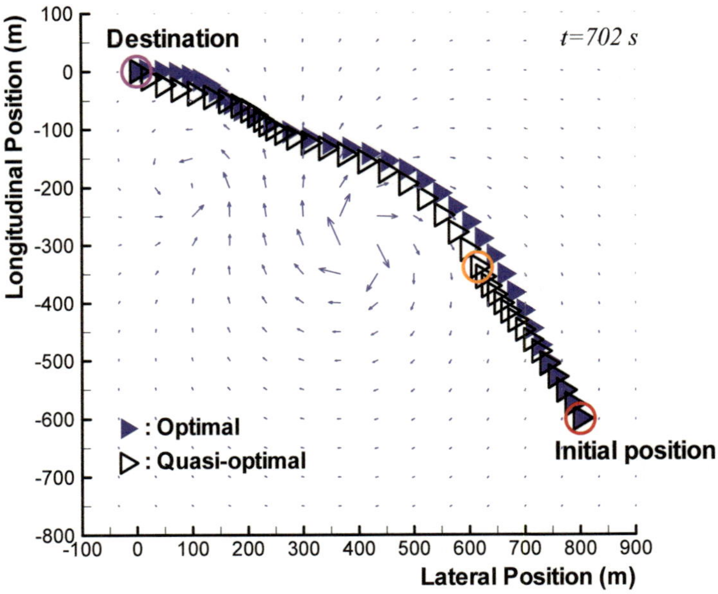

Fig. 14. Navigation trajectories in a time-varying current flow at t = 702 s (continued).

The current distribution in this example is the same one that we took in the previous one. In this example, while the optimal navigation is performed with the exact information on the current flow distribution, assuming a situation of incorrect localization due to sensor failure, mismatched current flow information is fed to the vehicle guidance controller in the quasi-optimal case. The time interval during which mismatched current flow information is taken in the quasi-optimal navigation is 0.0 s ~ 252.0 s. Starting at 252.0 s, revised optimal guidance on the basis of the correct current flow information achieves the quasi-optimal navigation. Fig. 15 shows the time sequence of the vehicle headings during the optimal and the quasi-optimal navigations.

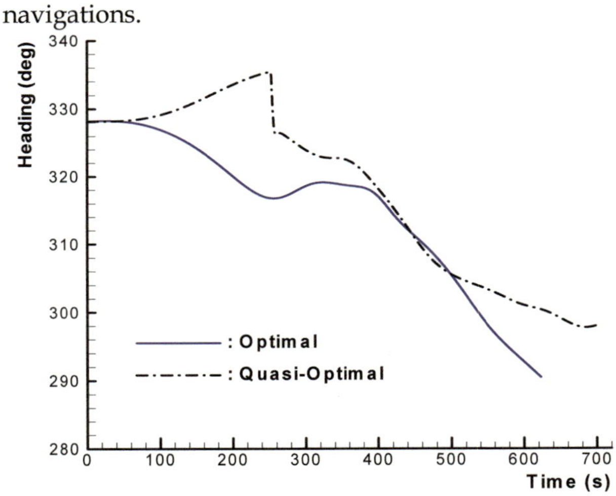

Fig. 15. Time sequence of vehicle headings.

As expected, the performance of the quasi-optimal navigation is not as high as that of the optimal one. While the optimal guidance completes the navigation at 623 s, the quasi-optimal one continues it until 702 s. Note that in Fig. 15, abrupt heading change occurs at 252.0 s during the quasi-optimal navigation.

7. Conclusions

In this article, a model-based synthetic approach to the following three main research fields in AUV design and development has been presented.

- Dynamic system modeling of AUV
- Motion control design and tracking control application
- Optimal guidance of AUV in sea currents

In the dynamic system modeling of the AUV R-One, we evaluated the hydrodynamic loads by using CFD analyses. Then by differentiating a hydrodynamic load with respect to a perturbed amount in vehicle kinematics, corresponding stability derivative were obtained. Using the stability derivatives evaluated, we built up the dynamic model of the R-One, which is characterized to be 6-D.O.F. (3 longitudinal (surge, heave, pitch) + 3 lateral (sway, roll, yaw)), linear, and Multiple-Input Multiple-Output (MIMO).

Depth and heading control systems are designed by employing controller models which are based on the PID compensations. In the PID-tuning, simulation models for the depth and the heading controls are exploited in determining the optimal gains.

Finally, concerning the guidance problem of AUVs moving in sea environmental disturbances, a procedure for obtaining the numerical solution of the optimal guidance law to achieve the minimum time navigation has been presented. The optimal heading is obtained as the solution of the optimal guidance law, which is fed to the heading control system as the heading reference. Reduced computational cost is one of the outstanding features of the proposed procedure. Numerical calculations of the optimal navigation examples presented in this article are completed within 10 minutes, on a single core 2.4 GHz windows XP platform. Moreover, unlike other path-finding algorithms such as dynamic programming or generic algorithm, our procedure does not require computation time increase for the time-varying problem.

As a fail-safe strategy in realizing the optimal navigation, the concept of the quasi-optimal guidance has been proposed. The fact that there actually are several possible actions lessening the chance of achieving optimality emphasizes the practical importance of the quasi-optimal navigation strategy.

We have not considered the problem of unknown or nondeterministic currents. Our approach cannot be applied to an entirely unknown environment. For a sea region including partially or coarsely defined current however, an estimated distribution can be built via interpolation and extrapolation. The estimation possibly contains more or less uncertainty. Notably it is the quasi-optimal strategy that can cope with the environmental uncertainty. When the uncertainty in the estimation is considerably significant, convergence may not be guaranteed, however.

8. References

Etkin, B. (1982). *Dynamics of Flight - Stability and Control*, John Wiley & Sons, ISBN 0-471-08936-2, New York

McRuer, D.; Ashkenas, I. & Graham, D. (1990). *Aircraft Dynamics and Automatic Control*, Princeton University Press, ISBN 978-0691024400, New Jersey

Lewis, E. V. (Ed., 1989). *Principles of Naval Architecture Vol III, Motions in Waves and Controllability*, The Society of Naval Architects and Marine Engineers, ISBN 0-939773-02-03, New Jersey.

Haykin, S. (1999). *Neural Networks*, Prentice Hall, ISBN 0-13-908385-5, New Jersey

The Mathworks Inc. (2001). *Control System Toolbox User's Guide*, The Mathworks Inc., Natick

Bryson, A. E. & Ho, Y. C. (1975). *Applied Optimal Control*, Taylor & Francis, ISBN 0-89116-228-3, Levittown

Lewis, F. L. & Syrmos, V. L. (1995). *Optimal Control*, John Wiley & Sons, ISBN 0-471-03378-2, New York

Press, W. H.; Teukolsky, S. A.; Vetterling, W. T. & Flannery, B. P. (1992). *Numerical Recipes in Fortran*, Cambridge University Press, ISBN 0-521-43064-X, Cambridge

Kim, K. & Ura, T. (2008). Optimal and Quasi-Optimal Navigations of an AUV in Current Disturbances, *Proceedings of IEEE/RSJ 2008 International Conference on Intelligent Robots and Systems (IROS2008)*, pp. 3661-3667, Nice, September 2008

Ura, T. ; Obara, T. ; Takagawa, S. & Gamo, T. (2001). Exploration of Teisi knoll by Autonomous Underwater Vehicle R-One Robot, *Proceedings of MTS/IEEE OCEANS2001*, pp. 456-461, Honolulu, November 2001

Alvarez, A.; Caiti, A. & Onken, R. (2004). Evolutionary Path Planning for Autonomous Underwater Vehicles in a Variable Ocean. *IEEE Journal of Oceanic Engineering*, Vol. 29, No. 2, (April 2004) 418-429

Thomson, J. F. (1988). A composite grid generation code for general 3D regions – the Eagle code. *AIAA Journal*, Vol. 26, No. 3, (March 1988) 1-10

Papadakis, N. A. & Perakis, A. N. (1990). Deterministic Minimal Time Vessel Routing. *Operations Research*, Vol. 38, No. 3, (May 1990) 426-438

Crespo, L. G. & Sun, J. Q. (2001). Optimal Control of Target Tracking with State Constraints via Cell Mapping. *Journal of Guidance, Control, and Dynamics*, Vol. 24, No. 5, (September 2001) 1029–1031

Kim, K. & Ura, T. (2009). Optimal Guidance for Autonomous Underwater Vehicle Navigation within Undersea Areas of Current Disturbances. *Advanced Robotics*, Vol. 23, No. 5, (April 2009) 601–628

19

Towards multimodal interface for interactive robots: challenges and robotic systems description

Burger Brice[1,2,3], Ferrané Isabelle[23] and Lerasle Frederic[13]
[1] CNRS ;LAAS [2] IRIT; UMR CNRS/Université de Toulouse [3] Université de Toulouse; UPS
France

1. Introduction and framework

Making robots to assist people in human-centred environments is a goal to which the robotics community has aspired to for many years (Fong *et al.*, 2003). This field of research is a deep challenge because robots moving out of laboratories have to gain more social skills in order to improve peer-to-peer interaction with a more or less novice user in public, domestic or industrial areas. In contrast to today's specialized service robots these robotic assistants could be well used in such areas and for a variety of tasks like elderly people care, or helping handicapped people as well as assistance in factories or offices. Such prospects require both spatial and transactional intelligence. The former is based on environment perception capabilities. For a robot, this means "being able to understand and navigate in its environment; locating objects and knowing how to manipulate them". The later is based on human perception capabilities. This means "being able to meaningfully communicate with a human user". While the first issue has received much attention in the past, relatively few recent robotic systems are equipped with multimodal user interfaces that permit to control the robot using natural means like human body motion, speech even tactile senses.
RHINO and MINERVA (Thrun *et al.*, 2000) were the first robots to be deployed in a public area, but they do not emphasize the interaction part so much even if they understand speech. Though recent demonstrators embed more advanced human-robot interfaces (frequently based on speech), their capabilities to perceive close human motion remain surprisingly fairly limited (Bennewitz *et al.*, 2005; Maas *et al.*, 2006; Siegwart *et al.*, 2003) or omitted (Breazal *et al.*, 2001; Breazal *et al.*, 2004). To detect persons, ROBOX (Siegwart *et al.*, 2003) relies on two sensors and dedicated algorithms. First, the laser scan data is fed to a motion detector. Secondly, colour images enable the identification of skin colour as well as the subsequent 2D detection and tracking of human faces using simple heuristics.
BIRON (Maas *et al.*, 2006) and ALPHA (Bennewitz *et al.*, 2005) consider also information from heterogeneous sources namely microphone for speaker localization, and vision for frontal face detection... which both lead to intermittent cues. Part of the aforementioned systems (Maas et al., 2006; Siegwart *et al.*, 2003; Thrun *et al.*, 2000) and beyond *i.e.* HERMES

(Bischoff & Graefe, 2004), MAGGIE (Gorostiza et al., 2006), WAKAMARU (Harte & Jarvis, 2007), PEARL (Pineau *et al.*, 2003), etc. focus also on 2D laser data, even radio frequency data *e.g.* ROBOVIE (Kanda *et al.*, 2004) to detect and track humans. To our greater view, using such sensors to perceive human motion seems questionable and vision technologies should be privileged due to their good price-performance ratio and the rich information they encompass.

Perceiving human motion from on-boarded vision is a key-point of human-robot interaction (HRI). On one hand, any interactive robot needs to maintain estimates, or beliefs as probabilistic distributions, about its human user's kinematic (and beyond his/her state) to make effective decisions during interaction. On the other hand, body movements are important in any communication as 65% of the information in a HRI act is non-verbal (Davis, 1971). Visual gestures show human thoughts, replay complements, accent and adjust verbal information. Therefore, vision-based gesture interpretation is valuable in environments where the speech-based communication may be garbled or drowned out. Moreover, the mutual assistance between the robot's speech and vision capabilities enables a user to robustly specify location references in verbal statements. Combined with pointing gestures, such prominent commands open up the possibility of intuitively indicating objects and locations e.g. to make the robot change its direction/position or to mark object. Yet, it can be argued that vision techniques for human perception and natural language processing have mostly been studied rather independently because they constitute research areas in themselves (Pro-danov & Drygajlo, 2003; Skubic *et al.*, 2004; Triesch et al., 2001; Waldherr *et al.*, 2000). Spatial intelligence based on environment perception capabilities has led to numerous and complementary multimodal interfaces in order to label places (Theobalt *et al.*, 2002) or objects (Bischoff & Graefe, 2004) but goes beyond the chapter scope.

Possibly combined with speech for multimodal communication, gesture recognition (GR) has recently received attention in the robotics community. When designing such interface, several requirements must be taken into considerations. First, the visual system must cope with uncontrolled real world environments (background clutter, changes in lighting conditions, presence of several individuals). It must be person independent i.e. many users should be able to operate it, without the necessity for retraining the system. Finally, the system must work at an acceptable speed as on-boarded processing is limited and human users would not accept several seconds for simple gesture recognition tasks. However, hardly any existing interface integrating speech and gesture inputs fulfils all the requirements stated above while symbolic and pointing 3D gestures usually co-occur with speech in a natural interaction scenario. The most advanced multimodal interfaces are probably those presented in (Rogalla *et al.*, 2004; Yoshizaki *et al.*, 2002), even if 2D gestures are considered, and especially (Stiefelhagen *et al.*, 2004) in which the framework trend is similar to ours. In the latter work, a constraint based multimodal system for speech and 3D pointing gestures was developed. Though, monomanual hand gestures are pre-supposed while upper human body extremities are tracked separately, inducing inevitably tracking failures when they overlap. Finally, evaluations are performed on a small data set and independently of any robotic key-scenario. In brief, though remarkable progress has been made, the human motions perceptual capabilities of assistant robots remain fairly limited.

From these considerations, and to best fit these new challenges, a novel framework is proposed here in order to:

- perceive the user's motion namely (i) his/her body placing thanks to the 3D head position which is a commonly stated simplification, (ii) gestures tracking and recognition,
- perform a late-stage fusion with verbal cues to define a multimodal interface,
- integrate and evaluate real-time processes on a mobile robot called JIDO. So, given a targeted scenario, the robustness and the usability of this framework, in the context of a realistic robotic service task is finally investigated.

The tasks we chose are motivated by an "object exchanging" scenario, involving novice human users in the loop: JIDO interprets multimodal orders given by its user in order to move locally, pick a pointed object up, and safely manage to exchange this object with him/her. The primary motivation here is to show that thanks to our interface JIDO is able to interpret multimodal commands like "come to me", "stop", "take this bottle", "bring it to me", "go there", etc. and responds to them by performing the corresponding actions.

In the following, each component integrated on JIDO and involved in the humanrobot interaction process, namely gesture recognition, speech understanding and late stage multimodal fusion are described in section 2. Next, for a global evaluation, live experiments carried out on JIDO in the context of interactive manipulation tasks are described. Finally, this leads us to give some prospects and future work to be done on this topic.

2. Description of JIDO and its multimodal interface

This section gives first some considerations about the JIDO platform and the targeted scenario. Our multimodal interface is embedded on a robot called JIDO which is equipped with a 6-DOF arm, a pan-tilt stereo system at the top of a mast and two laser scanners (Fig. 2). The embedded functionalities are managed thanks to the "LAAS" layered software architecture (Fig. 1) and detailed in (Alami *et al.*, 1998). Such functionalities enable JIDO to:

build maps and navigate in indoor environments. The embedded functionalities are represented in orange boxes.

recognize and manipulate objects thanks to the modules depicted in blue and red boxes. A standard procedure is: the "Hueblob" module extracts the 3D position of the object thanks to blob detection, then "MHP" computes a trajectory which is executed by the "Xarm" module.

perceive humans, namely (i) detection/recognition and view-based tracking from the "ICU" modules, (ii) control of the pan-tilt unit mounted stereo head from the "PTU" module, (iii) 3D gestures tracking and interpretation from the "GEST" modules, (iv) speech utterances interpretation from the "RECO" module, (v) merging of speech and gestures results from the "FUSION" module. The last three modules are briefly described here below.

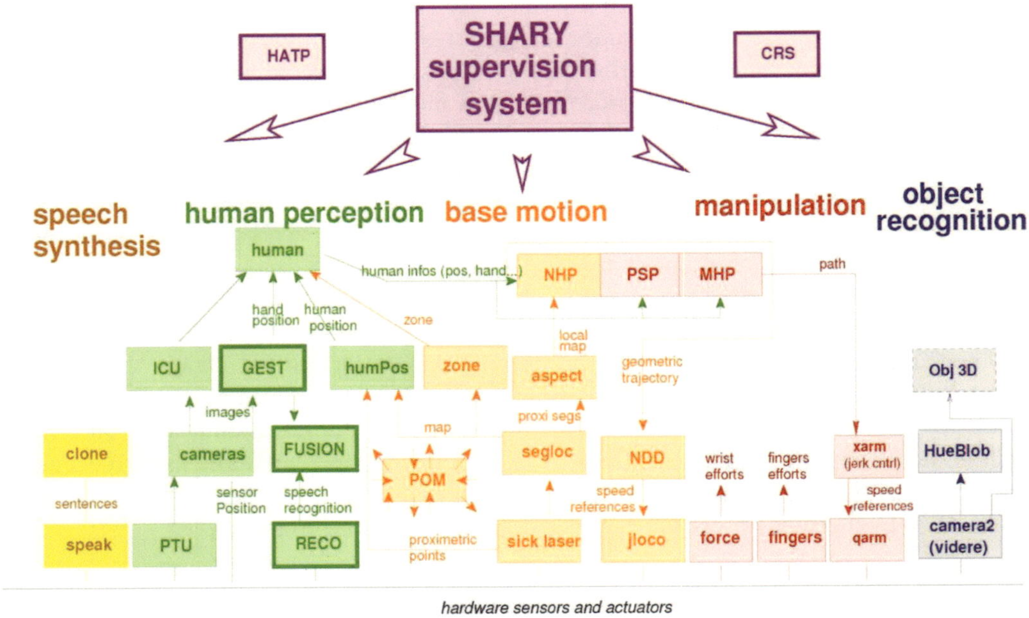

Fig. 1. The JIDO software architecture.

Dynamic gesture recognition is carried out in two phases by the **"GEST" module.** First, the upper extremities of the human body are tracked in 3D in stereoscopic video stream thanks to interactively distributed particle filters devoted to the human's hands and head (Qu *et* al., 2007). Recent investigations concern the second phase, *i.e.* the classification of legitimate gestures. These gestures are here assumed to start and end in the same natural/rest position (the hands lying along the thighs). Given an isolated gesture segment, classification outputs the class the gesture belongs to among a vocabulary composed of: - 7 symbolic gestures defined by their motion templates, namely: "calling out" (with one or two hands), "introducing oneself", "come to me" (with one or two hands), "stop", "go away".

- 5 deictic gestures depending on the coarse pointed direction relatively to the user who performs the gesture *i.e.* "in front of", "bottom left", "bottom right", "top left", "top right". The pointing direction is calculated by the connecting line between the centre of the head and the hand in 3D.

Each gesture is here straightforward modelled by a dedicated HMM. The features used as model's inputs are derived from tracking the 3D positions of both hands relatively to the head to achieve invariance with respect to the person location. For evaluations, we acquired a gesture database consisting of 772 video shots of gestures carried out by 11 different people in front of JIDO. From these experiments, about 70% of the examples are correctly classified. The most prominent error was a failure to recognize "stop" and "introducing oneself" (i.e. "hello") gestures which we can attribute to a poor set of motion template for this gesture. Another observation is that bi-manual gestures are better classified than their mono-manual counterparts as their motion templates are more discriminating.

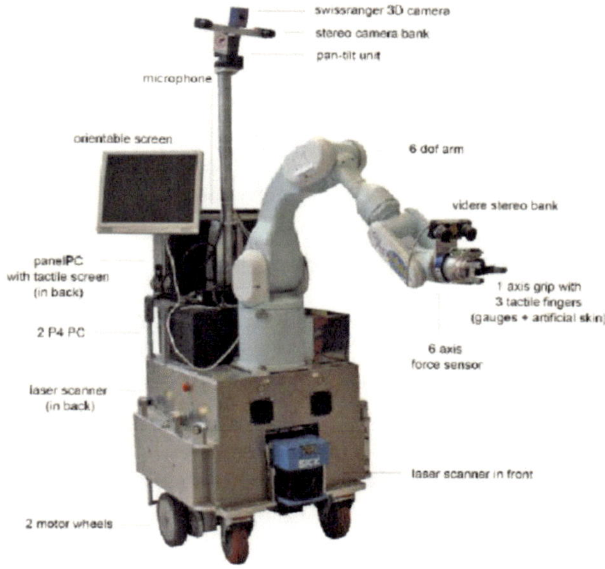

Fig. 2. JIDO and its actuators/sensors.

The "RECO" module performs a classical and speaker-independent speech recognition process, briefly described here. Thirty-nine-parameter vectors are extracted from the audio stream: 12 Mel Frequency Cepstral Coefficients (MFCC), the log energy and their delta and acceleration. The Julius engine is used to decode speech using a set of acoustic and phonetic models for French. This set comprises 37 phonemes and 2 pauses (short/long). It is HMM-based (3-state models with 32 Gaussians per state) and was first trained on about 31 hours of Broadcast News recorded on French radios, in the scope of a completely different transcription task. These models were re-estimated on a thirty-minute audio corpus, directly recorded on the robotic platform in the same conditions as our other experiments. The lexicon (246 words/428 pronunciations drawn up from the French lexical database BDLEX (Perennou & De Calmes, 2000) and the language model were specifically designed for our multimodal experiments. Deictic and anaphoric sentences as well as some language flexibility were taken into account. Context free grammars were designed for different types of user requests like starting interaction with the robot, asking for guidance, global robot or robot's arm movements, or object exchanges. This represents an overall set of 2334 different well-formed sentences, although the experiments described in the robotic scenario only involved a little subset of them, focusing on local movement and deictic requests for object exchange. Nevertheless, the evaluations carried out in our experimental conditions were made on 1200 various sentences, covering all request types and uttered by 16 different speakers (including 7 non native). The Word Error Rate (WER) is around 7%, while the Sentence Error Rate (SER) is equal to 19%. As word recognition errors have an immediate impact on sentence recognition and a less immediate one on sentence interpretation, when errors occur on words that are not significant in the interaction context, the sentence Interpretation Error Rate (IER) is 6% lower (13%) than the SER.
The "FUSION" module merges gesture recognition results and speech interpretation thanks to late-stage and hierarchical fusion ones. The speech is used as the main channel and

actions needing a gesture disambiguation are identified by the "RECO" module. Following a rule-based approach, the command generated by "RECO" is completed. Thus, for human-dependent commands *e.g.* "viens ici" ("go there"), the human position and the pointed direction are characterized thanks to the 3D visual tracker. Late-stage fusion consists of fusing the confidence scores for each N-Best hypothesis produced by the speech and vision modalities according to (Philipp *et al.*, 2008). The associated performances are reported thanks to the targeted robotic scenario detailed here below.

3. Targeted scenario and robotics experiments

These "human perception" modules encapsulated in the multimodal interface have been undertaken within the following key-scenario (Table 1). Since we have to deal with robot's misunderstandings, we refer to the human-human communication and the way to cope with understanding failure. In front of such situations, a person generally resumes his/her latest request in order to be understood. In our scenario, although no real dialogue management has been implemented yet, we wanted to give the robot the possibility to ask the user to repeat his/her request each time one of the planed step fails without irreversible consequences. By saying "I did not understand, please try again." (*via* the speech synthesis module named "speak"), the robot resume its latest step at its beginning. The multimodal interface runs completely on-board the robot. From this key-scenario, several experiments were conducted by several users in our institute environment. They asked JIDO to follow their instructions given by means of multimodal requests: by first asking JIDO to come close to a given table, take over the pointed object and give it to him/her. Figure 3 illustrates the scenario execution. For each step, the main picture depicts the current H/R situation, while the sub-figure shows the tracking results of the GEST module. In this trial, the multimodal interface succeeds to interpret multimodal commands and to safely manage objects exchanges with the user.

#	The human user command	The JIDO action	Demonstrated modules	Comments
1.	"Hello, I am here" accompanied with a symbolic gesture	Local motion towards the user	RECO, GEST, FUSION, MHP, PSP	JIDO moves and stops in front of the user
2.	"Hi JIDO it's Brice"		RECO, ICU	The user must be identified to be granted to interact with JIDO
3.	"Come to me" with a pointing gesture	Location motion towards the pointed location on the floor	GEST, RECO, FUSION, MHP	The command execution requires the 3D location of the user
4.	"Stop" with a symbolic gesture"	Stop of the robot	GEST, RECO, FUSION	This command is performed while the robot is moving
5.	"Take this object" with a pointing gesture	Grasping of the object	GEST, RECO, FUSION, Hue-Blob, MHP, Xarm	Jido searches for an object the user points to, then picks it up
6.	"Go to my left side" with a pointing gesture	Local motion according to the user location	GEST, RECO, FUSION, MHP, PSP	The command execution requires the 3D location of the user
7.	"Give the object to me"	Object handling	GEST, RECO, FUSION, MHP, Xarm	The command execution requires no gesture recognition but only the hand tracking
8.	"Go away" with a symbolic gesture	Local motion to go away from the user	GEST, RECO, FUSION, MHP	

Table 1. Excerpt of an interaction scenario between a human user and JIDO.

Fig. 3. From top-left to bottom-right, snapshots of a scenario involving speech and gesture recognition and data fusion: current H/R situation -main frame-, "GEST" module results - bottom right then bottom left-, other modules ("Hue Blob", "ICU") results -top-.

Given this scenario, quantitative performance evaluations were also conducted. They refer to both (i) the robot capability to execute the scenario, (ii) and potential user acceptance of the ongoing interaction scenario. The less failures of the multimodal interface will occur, the more comfortable the interaction act will be for the user. The associated statistics are summarized in Table 2 which synthesizes the data collected during 14 scenario executions.
Let us comment these results. In 14 trials of the full scenario execution, we observed only 1 fateful failure (noted fatal) which was due to a localisation failure and none attributable to our multimodal interface. Besides, we considered that a run of this scenario involving more than 3 failures is potentially unacceptable by the user, who can be easily bored by being constantly asked to re-perform his/her request. These situations were encountered when touching the limits of our system like for example when the precision of pointing gestures decreases with the angle between the head-hand line and the table. In the same manner, short utterances are still difficult to recognize especially when the environment is polluted with short sudden noises.

#	"RECO"	"GEST"	"FUSION"	Others	Comments
1.	0	1	0	0	
2.	0	0	0	1 ICU	Face recognition
3.	1	3	1	0	The distance to the robot makes this gesture hard to track
4.	3	2	2	0	Computing time sometimes too long when the robot is moving
5.	0	0	0	2 HueBlob	The bottle is not always seen
6.	0	0	0	0	The left is not really on the left...
7.	0	0	0	2 MHP (1 fatal)	Hand too far, localisation failure
8.	2	4	1	0	

Table 2. Modules' failure rates during scenario trials.

Apart from these limitations, the multimodal interface is shown to be robust enough to allow continuous operation for the long-term experimentations that are intended to be performed.

4. Conclusion

This article described a multimodal interface for a more natural interaction between humans and a mobile robot. A first contribution concerns gesture and speech probabilistic fusion at the semantic level. We use an open source speech recognition engine (Julius) for speaker independent recognition of continuous speech. Speech interpretation is done on the basis of the N-best speech recognition results and a confidence score is associated with each hypothesis. By this way, we strengthen the reliability of our speech recognition and interpretation processes. Results on pre-recorded data illustrated the high level of robustness and usability of our interface. Clearly, it is worthwhile to augment the gesture recognizer by a speech-based interface as the robustness reaches by cue proper fusion is much higher than for single cues. The second contribution concerns robotic experiments which illustrated a high level of robustness and usability of our interface by multiple users. While this is only a key-scenario designed to test our interface, we think that the latter opens in increasing number of interaction possibilities. To our knowledge, quite few mature robotic systems enjoy such advanced embedded multimodal interaction capabilities.

Several directions are currently studied regarding this multimodal interface. First, our tracking modality will be made much more active. Zooming will be used to actively adapt the focal length with respect to the H/R distance and the current robot status. A second envisaged extension is, in the vein of (Richarz *et al.*, 20006; Stiefelhagen *et al.*, 2004), to incorporate the head orientation as additional features in the gesture characterization as our robotic experiments strongly confirmed by evidence that a person tends to look at the pointing target when performing such gestures. The gesture recognition performances and the precision of the pointing direction should be increased significantly. Further investigations will aim to augment the gesture vocabulary and refine the fusion process, between speech and gesture. The major computational bottle-neck will become the gesture recognition process. An alternative, pushed forward by (Pavlovic *et al.*, 1999), will be to privilege dynamic Bayesian networks instead of HMMs which implementation requires linear increasing complexity in terms of the gesture number.

5. Acknowledgements

The work described in this chapter was partially conducted within the EU Project CommRob "advanced Robot behaviour and high-level multimodal communication" (URL www.commrob.eu) under contract FP6-IST-045441.

6. References

Alami, R.; Chatila, R.; Fleury, S.; Ingrand, F. (1998). An architecture for autonomy. *Int. Journal of Robotic Research (IJRR'98)*, 17(4):315—337.

Benewitz, M.; Faber, F.; Joho, D., Schreiber, M.; Behnke, S. (2005). Towards a humanoid museum guide robot that interacts with multiple persons. *Int. Conf. on Humanoid Robots (HUMANOID'05)*, pages 418—423, Tsukuba, Japan.

Bischoff, R.; Graefe, V. (2004). HERMES - a versatile personal robotic assistant.*IEEE*, 92:1759—1779.

Breazal, C.; Brooks, A.; Chilongo, D.; Gray, J.; Hoffman, A.; Lee, C.; Lieberman, J. (2004). Working collaboratively with humanoid robots. *ACM Computers in Entertainment*, July.

Breazal, C.; Edsinger, A.; Fitzpatrick, P.; Scassellati, B. (2001). Active vision for sociable robots. Trans. *On Systems, Man, and Cybernetics*, 31:443—453.

Davis, F (1971). Inside Intuition What we know about non-verbal communication. *Mc Graw-Hill book Co.*

Fong, T; Nourbakhsh, I.; Dautenhahn, K. (2003). A survey of socially interactive robots. *Robotics and Autonomous Systems (RAS'03)*, 42: 143—166.

Gorostiza, J.; Barber, R.; Khamis, A.; Malfaz, M. (2006). Multimodal human-robot Framework for a personal robot. *Symp. on Robot and Human Interactive Communication (RO-MAN'06)*, pages 39—44, Hatfield, UK.

Harte, E.; Jarvis, R. (2007). Multimodal human-robot interaction in an assistive technology context. *Australian Conf. on Robotics and Automation*, Brisbane, Australia.

Isard, M.; Blake, A. (1998). I-CONDENSATION: unifying low-level and high-level tracking in a stochastic framework. *European Conf. on Computer Vision (ECCV'98)*, pages 893—908, Freibourg, Germany.

Kanda, T.; Ishiguro, H.; Imai, M.; Ono, T. (2004). Development and evaluation ofInteractive humanoide robots. *IEEE*, 92(11): 1839—1850.

Maas, J.F.; Spexard, T.; Fritsch, J.; Wrede, B.; Sagerer, G. (2006). BIRON, what's the topic ? A multimodal topic tracker for improved human-robot interaction. *Symp. on Robot and Human Interactive Communication (RO- MAN'06)*, Hatfield, UK.

Pavlovic, V.; Rehg, J.M.; Cham, T.J. (1999). A dynamic Bayesian network approach to tracking using learned switching dynamic models. *Int. Conf. on Computer Vision and Pattern Recognition (CVPR'99)*, Fort Collins, USA.

Pérennou, G.; De Calmes, M. (2000). MHATLex: Lexical resources for modeling the french pronunciation. *Int. Conf. on Language Resources andEvaluations,* pages 257–264, Athens, Greece.

Philipp, W.L.; Holzapfel, H.; Waibel, A. (2008). Confidence based multimodal fusion for person identification. *In ACM Int. Conf. On Multimedia,* pages 885-888, New York, USA.

Pineau, J.; Montemerlo, M.; Pollack, M.; Roy, N.; Thrun, S. (2003). Towards robotic assistants in nursing homes: challenges and results. *Robotics and Autonomous Systems (RAS'03),* 42:271–281.

Prodanov, P.; Drygajlo, A. (2003). Multimodal interaction management for tour-guide robots using Bayesian networks. *Int. Conf. on Intelligent Robots and Systems (IROS'03),* pages 3447–3452, Las Vegas, USA.

Qu, W.; Schonfeld, D.; Mohamed, M. (2007). Distribution Bayesian multiple-targettracking in crowded environments using collaborative cameras. *EURASIP Journal on Advances in Signal Processing.*

Rabiner, L. (1989). A tutorial on hidden markov models and selected applicationsin speech recognition. *IEEE,* 77(2): 257–286.

Richarz, J.; Martin, C.; Scheidig, A., Gross, H.M. (2006). There you go ! Estimatingpointing gestures in monocular images from mobile robot instruction.*Symp. on Robot and Human Interactive Communication (RO-MAN'06),*pages 546–551, Hartfield, UK.

Rogalla, O.; Ehrenmann, M.; Zollner, R.; Becher, R.; Dillman, R. (2004). Usinggesture and speech control for commanding a robot. *Book titledAdvances in human-robot interaction,* volume 14, Springer Verlag.

Siegwart, R.; Arras, O.; Bouabdallah, S.; Burnier, D.; Froidevaux, G. ; Greppin, X. ;Jensen, B. ; Lorotte, A. ; Mayor, L. ; Meisser, M. ; Philippsen, R. ; Piguet,R. ; Ramel, G. ; Terrien, G., Tomatis, N. (2003). RoboX at expo 0.2: alarge scale installation of personal robots. *Robotics and Autonomous Systems (RAS'03),* 42:203–222.

Skubic, M.; Perzanowski, D.; Blisard, S.; Schutz, A.; Adams, W. (2004). Spatial language for human-robot dialogs. Journal of Systems, *Man andCybernetics,* 2(34):154–167.

Stiefelhagen, R.; Fugen, C.; Gieselman, P.; Holzapfel, H.; Nickel, K., Waibel, A. (2004). Natural human-robot interaction using speech, head pose and gestures. *Int. Conf. on Intelligent Robots and Systems (IROS'04),* Sendal, Japan.

Theobalt, C.; Bos, J.; Chapman, T.; Espinosa, A. (2002). Talking to godot: dialogue with a mobile robot. *Int. Conf. on Intelligenr Robots and Systems (IROS'02),* Lausanne, Switzerland.

Thrun, S.; Beetz, M.; Bennewitz, M.; Burgard, W.; Cremers, A.B.; Dellaert, F.; Fox, D.; Halnel, D.; Rosenberg, C.; Roy, N.; Schulte, J. Schulz, D. (2000). Probabilistic algorithms and the interactive museum tour-guide robot MINERVA. *Int. Journal of Robotics Research (IJRR'00).*

Triesch, J.; Von der Malsburg, C. (2001). A system for person-independent hand posture recognition against complex background. *Trans. On Pattern Analysis Machine Intelligence (PAMI'01),* 23(12):1449-1453.

Waldherr, S.; Thrun, S.; Romero, R. (2000). A gesture-based interface for humanrobot interaction. *Autonomous Robots (AR'00),* 9(2): 151–173.

Yoshizaki, M.; Kuno, Y.; Nakamura, A. (2002). Mutual assistance between speech and vision for human-robot interface. *Int. Conf. on Intelligent Robots and Systems (IROS'02)*, pages 1308 – 1313, Lausanne, Switzerland.

20

Robotic Localization Service Standard for Ubiquitous Network Robots

Shuichi Nishio[1], JaeYeong Lee and Wonpil Yu[2], Yeonho Kim[3], Takeshi Sakamoto[4], Itsuki Noda[5], Takashi Tsubouchi[6], Miwako Doi[7]
[1]ATR Intelligent Robotics and Communication Laboratories, Japan
[2]Electronics and Telecommunications Research Institute, Korea
[3]Samsung Electronics Co., Ltd., Korea
[4]Technologic Arts Inc., Japan
[5]National Institute of Advanced Industrial Science and Technology, Japan
[6]University of Tsukuba, Japan
[7]Toshiba Research and Development Center, Japan

1. Introduction

Location and relevant information is one of the essentials for every robotic actions. Navigation requires position and pose of robot itself and other positions such as goals or obstacles to be avoided. Manipulation requires to know where the target objects is located, and at the same time, to know positions and poses of robot arms. Robot-human interaction definitely requires where the person to interact with is located. For making interaction even effective, such by having eye contacts, further detailed information may be required. As such, robots need to acquire various location related information for its activity. This means that components of robots need to frequently exchange various types of location information. Thus, a generic framework for representing and exchanging location information that is independent to specific algorithms or sensing device are significant for decreasing manufacturing costs and accelerating the market growth of robotic industry. However, currently there exists no standard mean to represent and exchange location or related information for robots, nor any common interface for building localization related software modules. Although localization methods are still one of the main research topics in the field of robotics, the fundamental methodology and elements necessary are becoming established (17).

Although numbers of methods for representing and exchanging location data have been proposed, there exist no means suitable for robotic services that aim to serve people. One of the industrial robot standards defined in International Organization for Standardization (ISO) defines a method to define position and pose information of robots (6). Another example is the standards defined in Joint Architecture for Unmanned Systems (JAUS) (16) where data formats for exchanging position information are defined. However, these standards only define simple position and pose information on fixed Cartesian coordinate systems and are neither sufficient nor flexible enough for treating complex information required for service robots and modern estimation techniques.

Probably the most widespread standard on positioning is for the Global Positioning System (GPS) (12). GPS provides absolute position on the earth by giving 2D or 3D coordinate values in latitude, longitude and elevation. Although the GPS itself is a single system, the terminals that people use to receive the satellite signals and perform positioning varies. Thus, there are variations in how GPS receivers output the calculated position data. One of the most commonly used format is the NMEA-0183 format defined by National Marine Electronics Association (NMEA) (13). However, as NMEA format only supports simple absolute positions based on latitude and longitude, it is not sufficient for general robotics usage. Another related field is Geographic Information System (GIS). GIS is one of the most popular and established systems that treats location information. In the International Organization for Standardization, many location related specifications have been standardized (for example, (7)). There already exist versatile production services based on these standards such as road navigation systems or land surveying database. However, current GIS specifications are also not powerful enough to represent or treat information required in the field of robotics.

In this paper, we represent a new framework for representing and exchanging Robotic Localization (RoLo) results. Although the word "localization" is often used for the act of obtaining the position of robots, here we use for locating physical entities in general. This means that the target of localization is not just the robot itself, but also includes objects to be manipulated or people to interact with. For robotic activities, mere position is not sufficient. In combination with position, heading orientation, pose information or additional information such as target identity, measurement error or measurement time need to be treated. Thus, here the verb "locate" may imply not only measuring position in the spatio-temporal space.

Our framework not only targets the robotic technology available today but also concerns of some near future systems currently under research. These include such systems as environmental sensor network systems (14), network robot systems (4) or next-generation location based systems. Figure 1 illustrates a typical network robotic service situation where localization of various entities is required. Here, a robot in service needs to find out where a missing cellular phone is by utilizing information from various robotic entities (robots or sensors) in the environment. These robotic entities have the ability to estimate the location of entities within their sensing range. Thus, the problem here is to aggregate the location estimations from the robotic entities, and to localize the cellular phone in target.

Since 2007, the authors have been working on the standardization of Robotic Localization Service. This is done at an international standardization organization Object Management Group (OMG). OMG is an consortium widely known for software component standards such as CORBA and UML. As of May 2009, the standardization activity on Robotic Localization Service (RLS) (15) is still ongoing and is now on its final stage. The latest specification and accompanying documents can be found at: `http://www.omg.org/spec/RLS/`.

In this following sections, we will describe elements of the new framework for representing and exchanging localization results for robotic usage. We first present a new method for representing position and related information. Items specific to robotics use such as mobile coordinate systems and error information are described. We describe several functionalities required for exchanging and controlling localization data flow. Finally, some example usages are shown.

2. Data Architecture

In this section, we present a new method for representing location data and related information that is suitable for various usages in robotics, which forms the core part of the proposed

Fig. 1. Example of a typical robotic service situation requiring location information exchange (from (15))

framework. Modern robotic algorithms for position estimation or localization require more than simple spatial positioning. Various types of information related to the measurement performed are also required for precise and consistent results. One obvious example can be seen when combining outputs of laser range finder (LRF) and odometer installed in a mobile robot. When the robot turns around, LRF measurement values quickly change. If the two sensors are not temporally synchronized, the combined output will result in a complete mess (18). In order to obtain precise results, measurement time and error estimation is crucial for integrating measurements from multiple sensors. Pose information is also important. When grasping an object or talking to a person nearby, robots always need to obtain in which direction they or their body parts are heading. When sensors in use can perform measurements of multiple entities at once, target identity (ID) information is also required. For example, when vision systems are used to locate people in an environment, the system needs to track and distinguish people from each other. As such, there are numbers of information to be treated in combination with simple spatial location. In order to make various robotic services treat and process these versatile information easily and effectively, our idea is to represent these heterogeneous information within a common unified framework. As stated before, the proposed framework is defined by extending the existing GIS specifications such as ISO 19111(7). In the following sections, we describe three extensions required for robotics usage. And then we describe a generic framework for composing structured robotic localization results.

2.1 Relative and Mobile Coordinate Reference Systems

In general, spatio-temporal locations are represented as points in space. Cartesian coordinate is one typical example where location is defined by a set of numeric values that each represent the distance from the origin, when the location is projected to each axis that defines the coordinate system. As described in this example, locations are defined by a combination of information: a coordinate value and the coordinate system where the coordinate value is based on.

Before going further, let us clarify the terms used in the following sentences. A *coordinate system* (CS) is a system for assigning an n-tuple of scalar values to each point in an n-dimensional space. Mathematically, a scalar is in general an element of commutative ring, but we do not apply this restriction here. Instead, each of the tuple element is allowed to be taken from arbitrary set of symbols, as explained later. Normally this set consists of rational numbers. A *coordinate value* denotes the n-tuple of scalars assigned with respect to a CS. In this document, we will assume that every coordinate value is related to a CS, through which it was assigned. That is, every coordinate values are typed to be a member of an element set that belongs to a certain CS. Note that, there exists no uncertainty with coordinate values themselves. The uncertainty (or error) with the observation or estimation, if any, is represented by another accompanying value. We will call this an *error value* (or error in short).

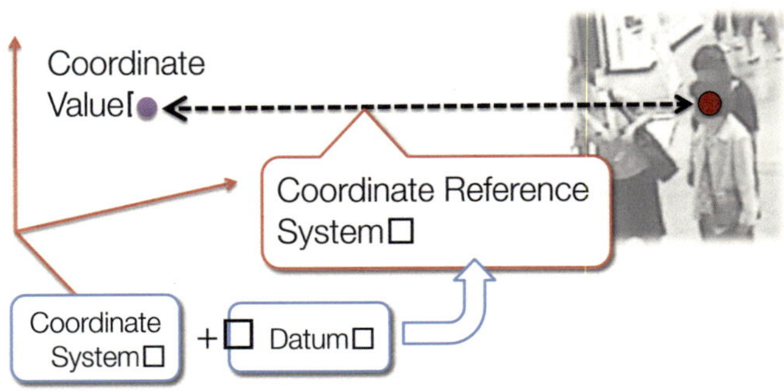

Fig. 2. Coordinate Reference System (CRS)

Figure 2 shows how a position in real world is mapped to a coordinate value defined over a certain CS. This mapping is typically done through some observation or sensing. However, in order to obtain the measurement values in a consistent manner, some rule must be defined on how to associate the coordinate values to the real world position. This mapping or grounding rule is called *datum*. With both the CS and datum defined, the mapping function from real world position to coordinate values can be defined. In other words, in order to define a perform an observation of real world phenomena, a combination of CS and datum is required. This combination is called *coordinate reference system* (CRS).

The basic idea in GIS specifications is that every CSs used for representing position data are fixed relative to the earth (i.e. referenced). There exists descriptions for relative coordinate systems in GIS standards (e.g. *Engineering CRS*), but they are hardly used and the usage is not clear. In robotics usage, however, CSs are not always fixed, and in many cases they are also mobile. That is, its relation to the earth changes by time. Although it may not be impossible to express every data in some global, fixed CSs, in most cases, it is much convenient to treat data in a relative form. There exists a GIS specification recently published (8) which specifies a method for describing moving entities. However this method is mainly aimed for car navigation that assumes the localized objects to move alongside some predefined roads and is not easy to use in robotics usage.

Especially in mobile robots, CRSs defined on a moving robot change its relation with other CRSs in time. For example, imagine that there are two rooms, room A and room B, and a mobile robot equipped with a 3-degree-of-freedom hand. When this robot grasp and move some objects from room A to room B, at least one CRS that represents the area including two rooms and one CRS that moves along as robot navigates are required. In some cases, each room may also have an individually defined coordinate space, related to the 'global' coordinate space representing both rooms in common. Moreover, in order to represent the gripper location at the end of the robotic hand, several CSs must be defined over the robotic hand each related to other coordinate systems by some means such as Denavit-Hartenberg convention (1). The object to be gripped and moved by the robot may also hold some CRSs that indicate the position or the pose of the object. When the object is carried by the robot, these CSs also shift in space as the robot moves.

As can be seen from this example, not all the CRSs used need to be grounded to the earth or the 'global' CRS in use. Requiring users to strictly define datums for every CRS in use is not realistic. Also, some mechanism for easily process CRSs on a moving platform and for transforming coordinate values on it to some static CRSs on demand is required. In the proposed framework, a relative coordinate reference system is defined as a CRS where the relation with the fixed world may be not known at some instant or users have no interest in referencing it to other CRSs. A mobile CRS is defined as a relative CRS with an dynamic datum referring to output of different localization output (figure 3).

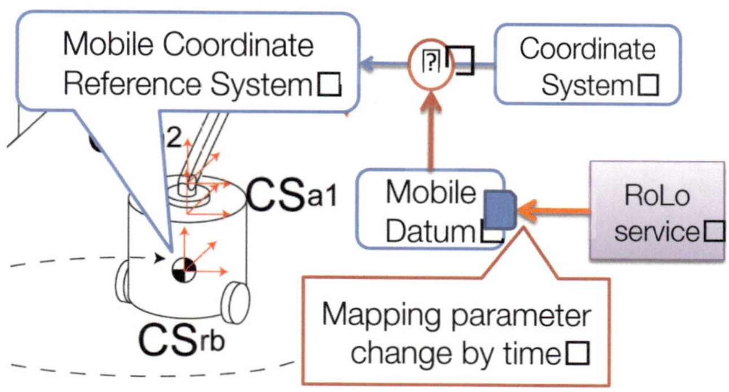

Fig. 3. Mobile CRS

2.2 Identity Representation

Identity (ID) information which is assigned for localized targets, can also be treated as a value on some CS. For example, MAC addresses used in Ethernet communication protocols can be represented as a coordinate value on a two-dimensional CS, vendor code and vendor-dependent code (5). Electric Product Code (EPC) (3) used for identifying RF tags is another example of identification systems defined by multi-dimensional coordinate system. There also exists some ID systems, such as family names, that is usually not explicitly defined over some mathematical structure.

In general, sensors hold their own ID system, and each observed entity are assigned an ID from this local ID system. This is because, at least on the initial stage, there are no means to identify an observed entity and assign it a global ID. Thus, when multiple sensors are in use, there exist multiple local ID systems independent to each other, and it becomes necessary to properly manage and integrate these ID systems (ID association problem). Also as previously described, ID assignments are probabilistic, just like other location information.

From these considerations, we can say that ID information requires representation or access methods similar to other types of location information. Thus, we propose to treat ID information in the same manner as spatial positions or other location information, as a value on a CS.

Since in GIS specifications CSs cannot handle axis defined over a set of symbols or discrete set of numbers, we extend this point. Note however, that some operations such as comparison is not always defined over this axis, as symbols in ID systems do not form an ordered set in general. Also, transformation between ID CSs will likely to be defined as a conversion table, not an mathematical operation.

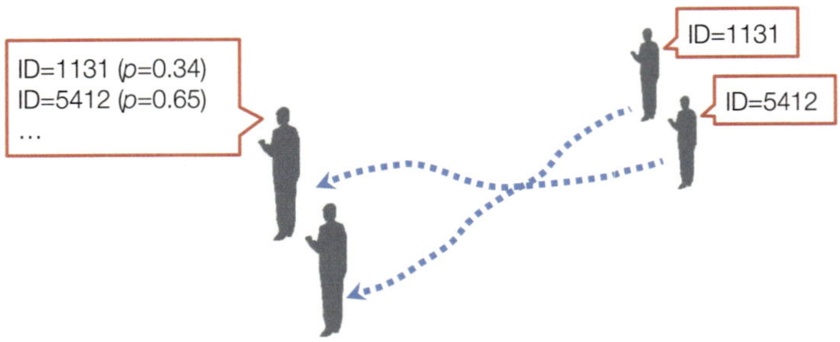

Fig. 4. Identity Association Problem

2.3 Error Representation

Error (or uncertainty, reliability) information plays a important role in robotic algorithms. This is especially important when localization results are used for further processing such as sensor fusion or hi-level estimation. Thus, measurement or estimation errors are one of the most essential features required for robotics usage. In GIS specifications, only static information of expected error concerning inter-coordinate transformation can be stated. Thus, here we extend the GIS specification in the following points:

- Localization results can be attributed an error information.
- Allow errors to be represented in various forms.

Just like every location information is associated with some coordinate (reference) system that defines what the information represents, every error information is associated with some error type. Modern measurement or estimation techniques require versatile forms of error representation depending on what computation methods are used or what devices are used. These include reliability, covariance or probability distribution. Distributions are typically approximated by finite combination of simple distributions, such as a single Gaussian distribution or mixture of Gaussians. Distributions may also be represented by random sampling such as by Monte Carlo method. Thus, rather than fixing how error information are represented to a single way, it is better to define a framework that allows multiple forms of error representation and allows users to extend necessary forms if necessary. Figure 5 shows some predefined error types that are commonly used in current localization techniques. We have designed the framework to be extendable so that users can extend their own error type if necessary.
In some cases, a single error information may be related to multiple position data. In such cases, a special structure for describing such relation is necessary. This structure is described in the next section.

2.4 Describing Complex Data Structure

Up to now, we have defined necessary elements for describing individual information required in robotics usage. The next step is to provide a mean to describe complex data struc-

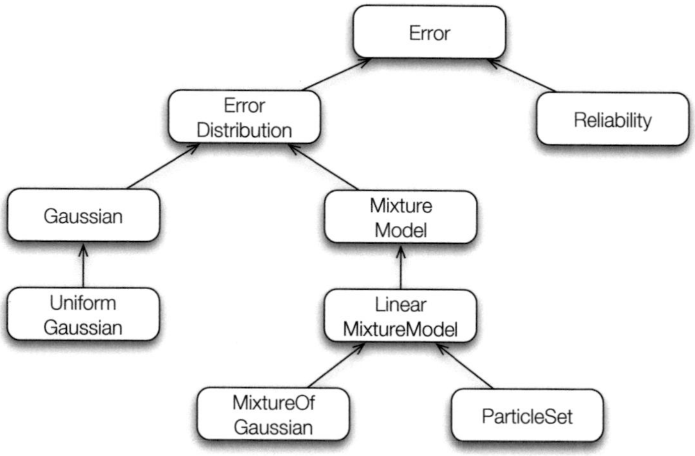

Fig. 5. Hierarchy of predefined error types

ture that consists of multiple measurements or estimation results. This combination is often required as it is quite rare that only a single type of measurement is obtained as a result of localization; in most practical cases, measurements such as position, velocity, ID and pose are given out. Also multiple sensing equipments are often used in order to increase robustness of the measurement or resulting output. This is also true if we assume a situation as described in Figure 1, where numbers of environmental sensors or robots are utilized and combined to perform a single task.

Figure 6 shows an example of an combined data definition. Here three types of information are combined: measurement time, target ID and spatial position. As such, users can define their own data structure that contains arbitrary numbers of necessary data elements. In this case, each element contains an error information. However note that errors are not mandatory, and if unnecessary, users can safely omit the error part both in definition and in the actual data set.

In defining a data structure, CSs that each values are based on, are kept independent to each other and individual values remain in its original form. This means that, definition of each values are kept in their original form, and the containing structure defines their relation with each other elements. In other words, multiple information are represented in combination to be suitable for certain usage, still remaining the 'meaning' of each individual values. Note that, this specification does neither oblige the users to specify information of some 'meaning' nor restrict the 'meaning' of information expressed by RoLo Data Specification. For example, the spatial coordinate in the above example may represent the centroid of the robotic body, or it may represent the position of a robotic arm. The meaning of each coordinate information contained in RoLo Data Specification definitions are out of the scope of this specification. Only the users and provider of the output module needs to agree in how each coordinate information will be interpreted.

Normally, error information is associated with one main location data. However in certain cases, there is a need to hold an integrated error among multiple data. For example, in a typical Kalman filter usage, multiple measurements such as spatial position and velocity are used to form a single state vector. When the elements of the state vector are not independent, which is the usual case, errors include cross effect among multiple measurements and

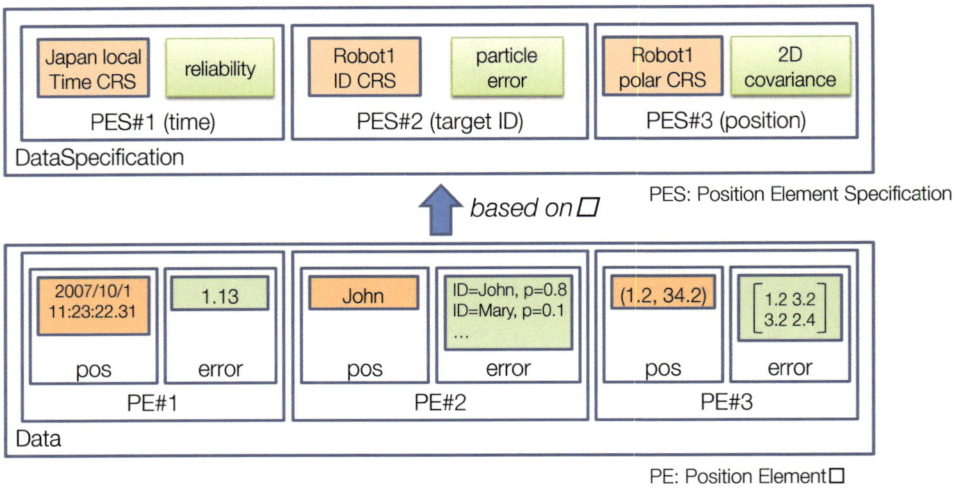

Fig. 6. Sample of RoLo Data Specification

are often represented as a covariance matrix. In such case, the Error Element Specification instance specifies which main information slot the error is related to, and the actual error data is contained by Error Element instances (Figure 7).

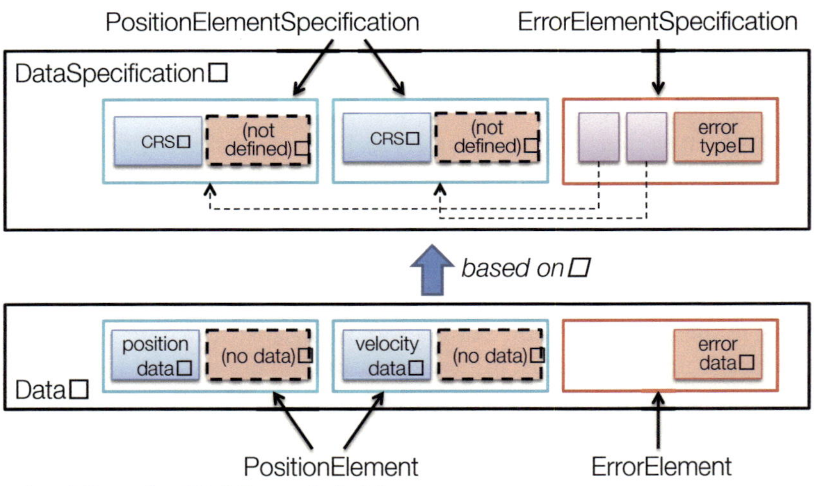

Fig. 7. Sample of Complex RLS Error definition

2.5 Don't-Cares

Consider that you want to develop a database system with RLS interface which accumulates results from numbers of people tracking modules. These modules are measurement modules corresponding to sensors of identical type installed in different locations (Figure 8). Being installed in different location means that each camera output is bound to a different CRS, *i.e.*, same coordinate system but different datum. As the sensor hardware are the same, and the DB system will not see the datum for each data, you may want to develop a generic interface that accepts data from all of the sensor modules. However, as stated later, RLS interfaces can only be bound to a finite number of data specifications. Thus you cannot make a generic interface which may accept infinite variations of coordinate reference systems. Would you give up and go on with the tedious development of interfaces for every newly added sensor modules?

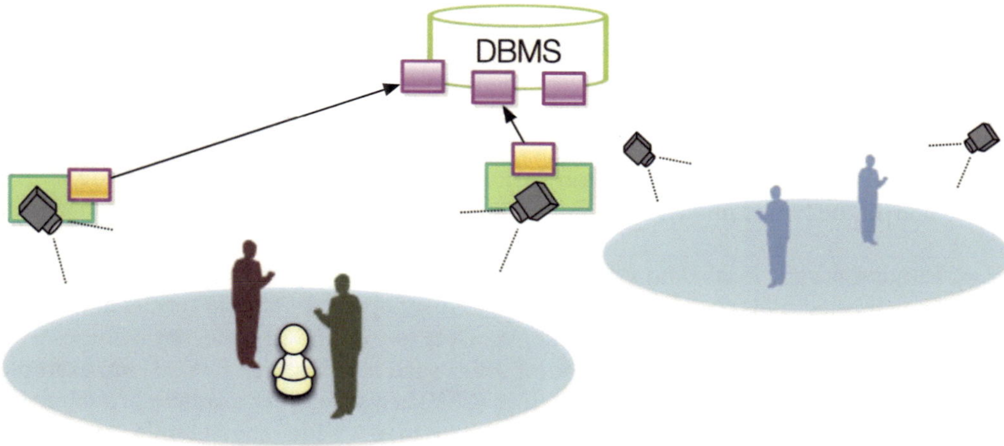

Fig. 8. Example situation where don't-care elements are required

As such, there are often cases that you are not interested in some part of the data specification. You want to just pass-through those parts. That is, you don't care whatever elements are specified in the uninterested parts of the data. *Don't-care elements*, similar to the usage in automata theory, are prepared for such usage. In the above example, you specify a data specification for the database's input stream ability with a coordinate reference system that contains a don't-care datum. This way you can specify only the specification parts you (the module) are interested, and leave the other parts unspecified.

Issues similar to the above example can quite often be seen, so the use of don't-cares will increase the flexibility and usability of the service. However, this use of don't-cares may require notice as they are quite likely to result in high computation requirements not suitable for systems with limited resources. Also, don't-cares may lead to ambiguous, useless specifications that break the idea of having specifications for data. Therefore in the specification, some rules are defined to prohibit misleading usages and to avoid unnecessary costs.

3. Data Format

In the previous section, we have showed a framework for defining and holding complex localization data. However, when exchanging information amongst modules, knowledge on data structures is not enough. For example, data may be exchanged in XML, in comma-separated values or in some binary format. Roughly speaking, the relation between data specification and data formats are similar to Saussurian relation between signifié and signifiant. The mapping from data specification to data format is basically arbitrary. That is, the same data may be represented in numbers of ways. Thus, in order to exchange data between different systems or modules, we need to specify how data is represented in addition to how it is structured and what it means. Here, data formats are means to indicate how data is represented in a machine readable form.

Generally, when defining a specification, there are two approaches about data formats. One is to fix the data format to be used, and another is to define a meta-specification for describing the variety of formats. Specifications such as NMEA-0183 (13) or JAUS (16) are example of the first form. Fixing the way data are represented can lead to a simple and rigid specification. However, the extendability is in general low. On the other hand, some specifications such as ASN.1 (10) allows several data formats to be used for data exchange. This provides flexibility and extendability, but often leads to difficulty in implementation.
On designing RLS specification, there were three requirements: 1) Make the standard to be extendable so that the standard can be used with forthcoming new technologies and devices. 2) Allow existing device/system developers and users to easily use the new standard. 3) Maintain minimum connectivity between various modules. As for 1) and 2), we decided to include both of the two approaches for data formats. As for 3), we prepared the 'common data format'.
For exchanging robotic localization data, we can think of roughly two categories of data formats. The first category is about data formats that are independent to data specifications. This category is for the first requirement given above. Data formats in this category includes those specified by some systematic rules. Encoding rules such as XER(9) or PER(11) are examples of such data formats. Comma Separated Values (CSV) is also another example of such that is widely used. These rules generally have the ability to map a wide range of data structures to data representations such as XML or bit sequences. Based on the defined rules, data formats specific to the data structure in use can be generated. In this sense, we can think that this category of data formats are independent to data specification. Another category, for the second requirement, is about formats bound to some specific data specification. That is, formats are defined over some target data specification. Most of the sensor device outputs in market these days uses formats of this type. Usually, reference manuals for these devices describe data structure and how they are given out in combination.
In the RLS specification, both categories of data format are supported. In order to clarify what data format is used, data format information is provided implicitly or explicitly in the access interface. Details are described in the next section. In some cases, users or developers do not need to care about data formats. For example, when sending messages with arguments between C++ objects, there is no need to think about formats. The compiler will take care of the actual data packing. The same can be said for distributed systems with IDL definitions. However in some cases such as receiving raw outputs from sensors or reading data files, data format information is essential.

3.1 Common Data Formats

Think of a situation where two foreigners meet together. When both of them cannot speak others' mother language, how shall they communicate with each other? Often in such case, they will choose to talk in a third language that both can speak and understand, such as English, even if not fluent. It is always better to have something than nothing.
Common data formats are aimed to be something like the third language in this example. They are for maintaining minimum connectivity between heterogeneous RLS modules. The RLS specification defines three common data formats each accompanied with two data specifications. These combinations were chosen from the most frequently used CSs in robotics, Cartesian CS, polar CS and geodetic (GPS) CS. Figure 9 shows an example of common data format definition.

Table 71 - Common data format type II-2 (Spherical Coordinate System, *xyz*-Euler Angle Representation)

Parameter	Format of value	Value type	Unit
Position	$[r, \theta, \varphi]$	Real, Real, Real	meter, radian, radian
Orientation	[yaw α, pitch β, roll γ]	Real, Real, Real	radian, radian, radian
Timestamp	POSIX time	Integer, Integer	second, nanosecond
ID	--	Integer	--

Fig. 9. Example of common data format definition (from (15))

The specification requires every RLS module to support at least one of these common data formats and accompanying data specification. Thus, even in cases where two modules has no common way to exchange data, they can always 'fall back' to these data formats. They will not be able to transmit detailed results, but can exchange the least amount of information. Such fall-backs are assumed to be especially important on near-future network robot usages as in figure 1, as robots need to get as much as possible from variations of modules situated in different environments. We can also see a similar example in today's home appliances. Recent appliances are equipped with advanced plugs such as HDMI and so on for transmitting hi-definition videos or digital audios, but older equipments are not. When connecting newer appliances to older ones, you use traditional connectors such as the yellow RCA plug. You may not get the best out of it, but at least there's something on the screen.

4. Interface

In this section, we will describe how RLS modules can be accessed. As stated in the introduction, one of our goal was to make a scalable specification. That is, a specification that can be used not only for complex, large-scaled systems but also for small embedded systems where available resources are limited. However, as our original purpose was to compile a specification which can be used in near-future network robot systems, module interfaces must be able to handle intelligent requests.

4.1 Module Structure

In general, several types of modules are commonly used for treating location data in robotic systems. The simplest form of module is which receives data from sensors, calculates location and outputs the results. However this type of interface strongly depends on sensor interfaces or sensor output formats. Strong dependency on specific products or vendors is not suitable for standardization. Moreover, when a location is calculated, many kinds of resources such as map data, specific to each sensing system, are required. It is impractical to include each of these resources into the standard specification. Thus, we decided to embed and hide the individual device or localization algorithm details inside the module structure.

On the other hand, if we focus on functionalities required to localization modules, we can classify them into roughly three classes (figure 10):

- Calculate localization results based on sensor outputs (measurement)
- Aggregate or integrate multiple localization results (aggregation)
- Transform localization results into different coordinate reference systems (transformation)

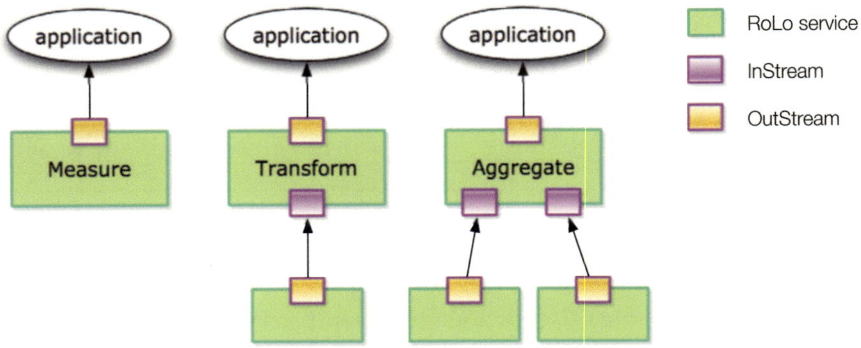

Fig. 10. Three forms of typical RLS module usage

These functionalities differ in their internal algorithms or the number of input / output streams. However, in all of these, the main data to be exchanged is localization results. As we are focusing on the interface of RLS modules, and not on their functionalities, we decided to abstract these different types of modules into a single form of module. This abstract module holds n input streams and a single output stream. By abstracting various types of modules and assuming a uniform interface, complex module compositions such as hierarchical or recursive module connections can be easily realized.

4.2 Module Ability

If each module can represent what it can perform, or provide information on available configurable parameters, a large amount of development efforts can be reduced. By defining the "meaning" of parameters, the ambiguity in functional definition or parameters can be eliminated, resulting in increase of developing efficiency. Moreover, advanced features can be implemented such as verification of inter-module connection, automatic search of specific modules or semi-automatic parameter negotiation between modules. In future network robot environments where sensors or robots are distributed in the environment and cooperate with each other, it becomes essential to register each module's capabilities in repositories and make them searchable.

As a basis for realizing such functionality, we have also defined a facility to describe *ability* of each modules or streams. In definition, each stream owns an ability description, which shows how this stream can perform and be configured. This includes the list of Data Specifications or Data Formats that this stream can handle. Also it describes configurable parameters specific to the module. Each service module also owns an ability description for the service it provides, besides the ability description for its streams. The configurable parameters defined in the ability description can be specified values via the module interface, restricted by the ability description held by the service module or the belonging stream modules.

5. Using Robotic Localization Service

In this section, we will introduce how RLS modules can be used. Typical steps of using RLS Services are as following:

1. (optional) Obtain ability description
   This can be performed by calling the 'getAbility' method toward RLS service or stream.

This step can be omitted if users already have sufficient information such by reading reference manuals.

2. (optional) Set up parameters
   This is done by calling the 'setParameterValues' method. If the default settings are sufficient or if there exists no parameter to be configured, this step can be omitted.
3. Establish connection
   Connection establishment can be initiated from both side; either from the service that outputs data (OUT service) or from the service that accepts data inputs (IN service).
4. (optional) Set initial data
   Robots often require initial data setting. For example, when you bring a mobile robot to a room and power it on, the first thing you need to do is to let the robot know where it is located and to which direction it is heading. By obtaining these, the robot can establish reference from its own mobile CS to the coordinates in the room. Initialization is performed by calling the 'adjust' method.
5. Perform data passing
   Data passing is the main aim for RLS modules. Two types of data passing are defined, *PUSH mode* and *PULL mode*. In PUSH mode, data passing is triggered by OUT service and in PULL mode, IN service triggeres data passing.
6. (optional) Perform adjustment
   Occasionally while passing data, perform adjustment if necessary. Adjustment is an act to provide auxiliary information to the target module for improving localization result.
7. Disconnect from service
   Shut down the connection and disconnect from the target module.

5.1 Scenario 1: Simple Usage

The first scenario is about a very simple navigation robot whose purpose is to reach a predefined goal. The robot repeatedly estimates its position in the space, performs path planning and move. This repetition of seek-and-act is necessary as both estimation and motion imply errors. Figure 11 shows a sample block diagram for this robot and the RoLo data specifications used.

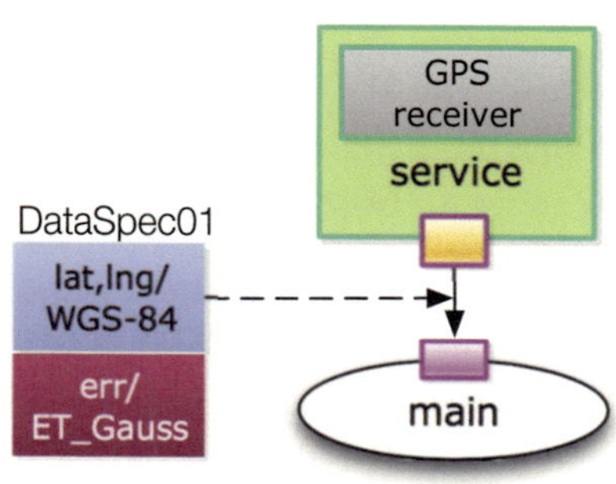

Fig. 11. [Scenario 1] Block diagram and data specification

The robot is equipped with a single a GPS receiver which is embedded in the 'service' module. This module outputs geodetic position (latitude and longitude) and an error modeled as an uniform normal distribution, based on the measurements from the GPS receiver. Note that

here the GPS receiver is embedded in the module, as its interface is not based on RLS specification. However, we can also think of a RLS module which outputs GPS measurement results in NMEA format or some vendor-specific binary data format. These can be realized by coding a wrapper for RLS interface API and by defining some data specifications and data formats describing necessary data structures and formats.
Figure 12 shows a sample code fragment for the 'main' process. In this example, the 'service' module

```
#include <RoLo.hpp>
extern Service *service;

class MyInStream: public InStream
{
public:
  void setData(const Data& d){ calc_n_move(d); }
  ...
};

int main(int argc, char **argv)
{
  MyInStream inStream;
  ...
  OutStream *outStream;
  try {
    service->connect(inStream, outStream);
  } catch (Returncode_t r){  }

  outStream->activate();
  ...
  outStream->deactivate();
  outStream->disconnect();
}
```

Fig. 12. [Scenario 1] Sample code

5.2 Scenario 2: Environmental Sensors

The second scenario describes a vacuum cleaning robot in a room that can return to a charging station when it finishes cleaning or go to the position that is located by a person as shown in Figure 13.
The robot is assumed to be able to calculate its position by combining a position information from an embedded odometer and from several landmarks in the environment. There is no restriction on the landmarks in this scenario and they can be obtained from any type of sensor such as cameras or laser range sensors. It is also assumed in this scenario that the position of a charging station and a pointer can be calculated by using a location sensing system in the environment such as RF beacons.
Figure 14 shows a block diagram that describes a structure of RoLo services and RoLo data specifications for each service modules. In this block diagram, each RoLo data specification shows time, identification, location and error parameters. In Figure 14, 'DataSpec01' is the RoLo data specification for the odometer where 't' represents the time-stamp in UTC format

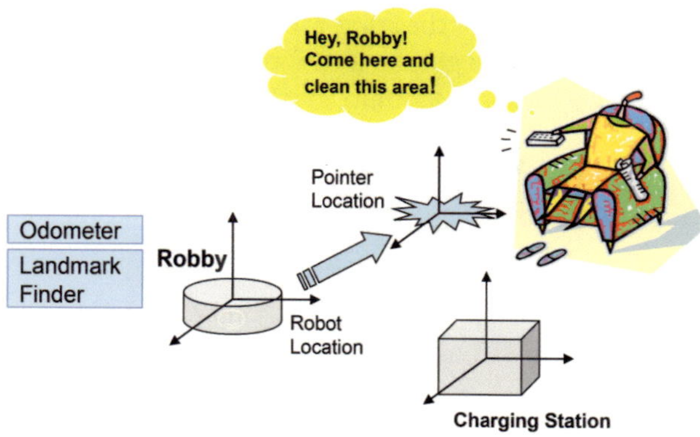

Fig. 13. Scenario 2

and 'x,y,α' the 2D position with a reference CS denoted by 'pCRS1' and 'err' the estimation errors with an error type denoted by 'ET01'. 'DataSpec02' is the RoLo data specification for the landmark finder where 'id' is the identification parameter with a reference coordinate system denoted by 'iCRS2' and 'x,y,z' the 3D position with a position reference CS 'pCRS2' and other parameters are the same as in 'DataSpec01'. 'DataSpec02' is for the robot and the type of parameters are the same as in 'DataSpec01'. 'DataSpec04' is for the beacons and 'id' is the identification parameter with a reference CS denoted by 'iCRS4' and 'd' is the distance between the beacon and the object that is to be localized using the beacons such as the charging station. 'DataSpec05' and 'DataSpec06' have parameters representing time-stamp, 3D position and position errors in the same manner with other RoLo data specifications.

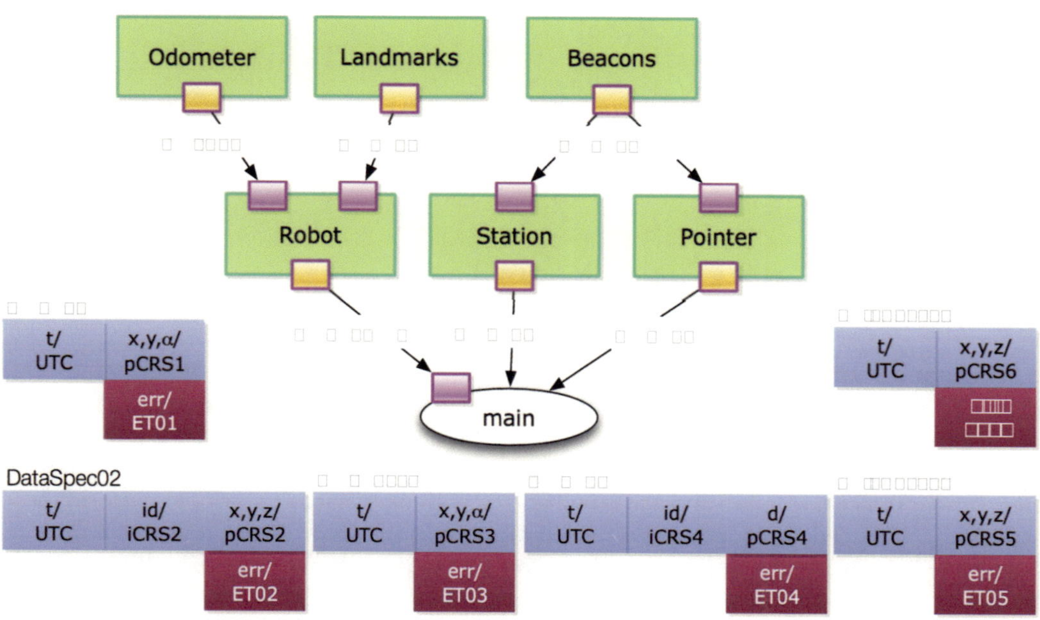

Fig. 14. [Scenario 2] Block diagram and data specifications

Figure 15 show an exemplary sequence diagram for the main part of this scenario. Here, the 'main' application uses PUSH mode to get the location of the robot and PULL mode to get the location of the charging station and the pointer.

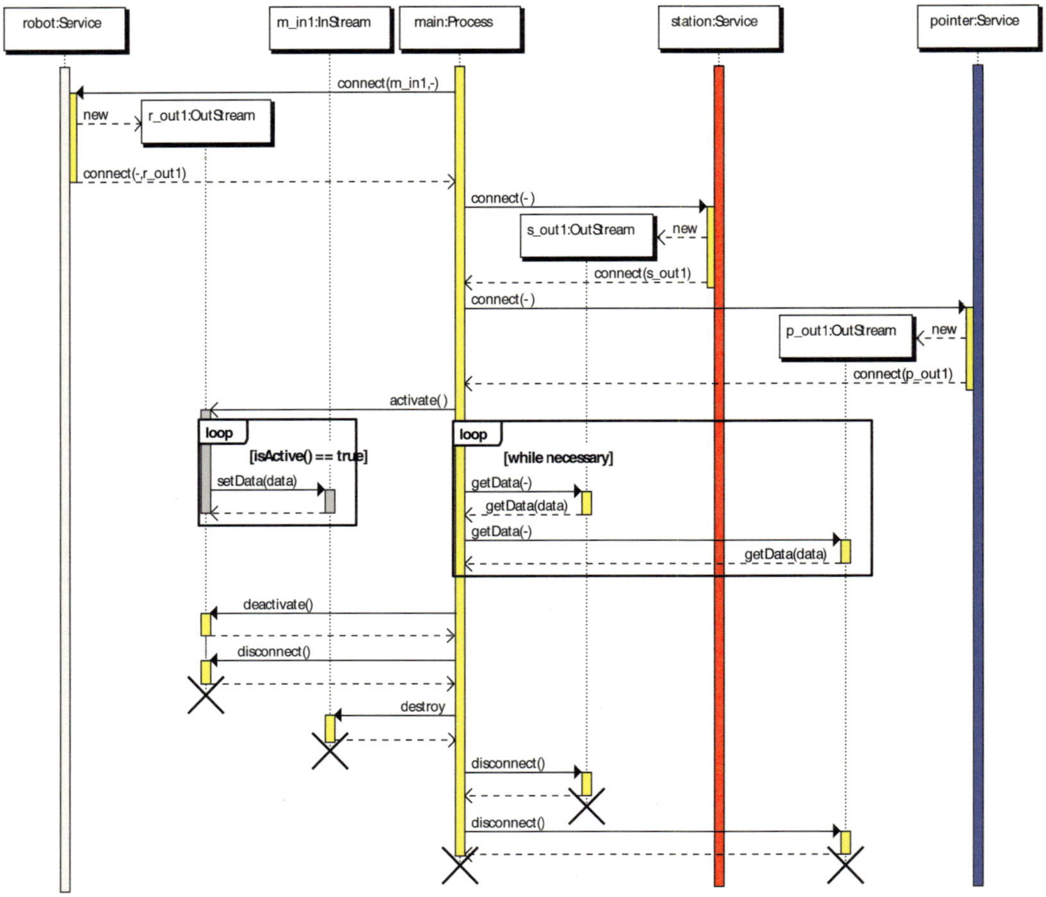

Fig. 15. [Scenario 2] Sequence Diagram (Main Part)

5.3 Scenario 3: Ubiquitous Network Robot

Figure 16 shows a situation where numbers of robots cooperate with each other to assist people over multiple locations. In this example, a robot navigates a person from one environment to another environment. The robots finds a person who needs help, approaches the person, searches for the goal location and starts navigating with the person. In order to perform these tasks, the robot needs to cooperate with sensors and information systems in the environment. Especially in this case, not just those in the surroundings, but also with those in different environments. Also, the robot needs to service the person seamlessly while navigating along two (or more) environments. Such wide-area robotic system is called *Ubiquitous Network Robot* system.

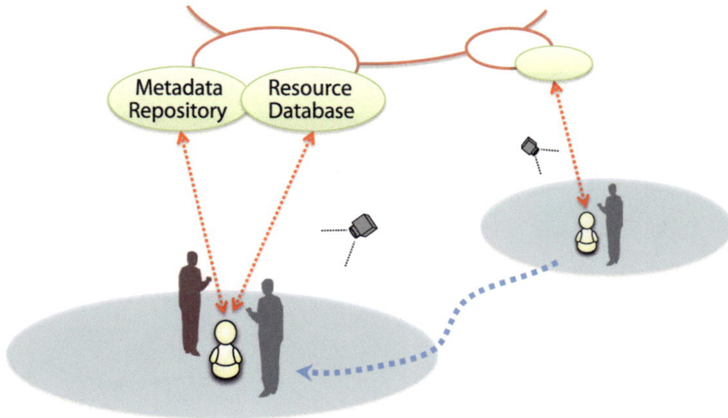

Fig. 16. Scenario 3

Upon entering a new environment, the robot needs to ask the management system there for a list of available RLS modules, and to find out which module is suitable for its use. This search and examination will be performed by using the ability descriptors of each module. Here, we need to assume for additional specifications for the management system: that is, registering and searching ability descriptors for the modules the system manages. Such specification is out of scope of the RLS specification and will be considered for future standardization. One such example is the one implemented in the *Kansai Environmental Platform* (14). Although this environment consisted of only a single environment, a resource management system was implemented experimentally. The steps for robots were as follows:

1. Robot (R) asks resource manager (RM) for a certain type of RLS modules
2. RM searches its database and returns a list of available modules
3. R obtains ability descriptor for each of the modules and examines which to use
4. After decision, R connects to target modules and starts setting up parameters
5. If parameter setup was successful, R activates the modules and start receiving (or sending) data

Figure 17 shows a sample block diagram for scenario 3. As robots move to different environment, the block diagram changes by connecting to modules specific to each environment. As can be seen from these flow and diagram, the robot and the environment need to have a common 'knowledge' for performing device search or parameter setting. For example, if the search request contains devices or CS definitions unknown to the robot, will the robot be able to 'understand' what or how the device can give out? Generally speaking, this requires exchange and bridging of ontologies from heterogeneous systems and currently there are no firm method for performing these. However, if we can limit the target domain, such as to robotic localization, knowledge exchange between heterogeneous systems may become possible. As every data following our specification is related with its formal definition, implicitly or explicitly, our idea is that through accessing a repository of definitions, it will become possible to exchange 'meanings' alongside with data. Such functionality is essential for ubiquitous network robots that provide services seamlessly and continuously in multiple locations. This means that wherever you bring robots, the robots will obtain information necessary from the

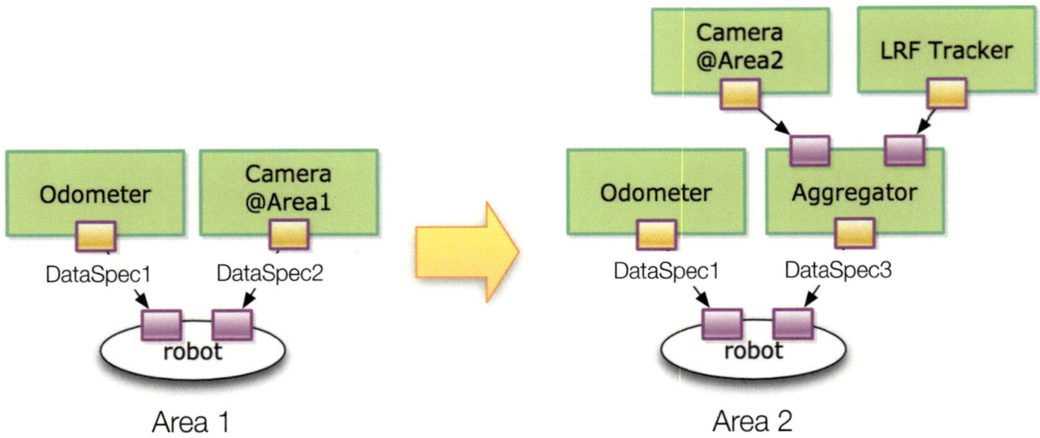

Fig. 17. [Scenario 3] Block diagrams in different areas

environment, automatically adjust to the place and start working. That is, the *Robot Plug-and-Play* (2).

Also note that, although the example here describes the usage for robots, this is not limited to robots only. Applications such as people navigation have similar requirements. Ability search and exchange facilities can thus be said to be one of the key requirements for next-generation location based applications where sensors and information sources local to each environments are actively utilized.

6. Conclusion and Future Works

In this paper, we have presented the concept behind a new framework for treating and exchanging location information in network robotic systems. One of the issues regarding frameworks of this kind is how vendor-specific functionalities or raw data can be represented. Although in this paper we have focused on describing the representation of localization data, our framework includes elements such as data format or vendor-specific parameter specification that can be utilized for treating vendor specific data or interfaces. However note that most of the existing outputs (including raw sensor data) follows some rule. This means these data is defined over some kind of system, and this allows the data to be represented by means of CSs in most of the realistic cases. Although the proposed framework has no ability to treat grammar-based complex data flows, in realistic cases this limitation can be avoided by defining symbolic axes based on codebooks such as the symbolic information described above.

Although the formal standard document has not been published yet, several companies and research projects have started to implement and use this specification. One such system is the *Kansai Environmental Platform* (14). The Kansai Environment is a field-trial environment for robotic services where sensor network is installed in a real world location such as shopping mall. Various types of sensors are installed in the environment to produce information on the environment such as where people is, how people are behaving or how a certain area (space) is used. This is done by performing people tracking and real-time pattern recognition based on statistical processing of the measured trajectories. These results are sent to servicing robots in the environment on request. By utilizing these information, robots can easily provide useful services to people in the environment. Here, RLS specification is extended for passing these position information and, at the same time, also for passing symbolic recognition results to robots.

As can be seen in this implementation, the framework described here can be easily extended to treat and exchange versatile information, such as what a person is doing or what kind of place a person is staying at. Thus, the proposed specification has the possibility to be extended to further generic data exchange for robotics use. This generalization of localization service specification is one of the future issues.

The localization standard is only one of the fundamental frameworks required to build robots with common interfaced modules. Profiling on robotic devices, or descriptions on robotic service flows may be other candidates for standardization. Also as we stated briefly with the module ability description, by defining and exchanging module description metadata, much flexible and dynamic configuration of network robots composed of distributed modules over networks will be available. This will also make the interconnection with heterogeneous network systems such as ubiquitous sensing systems or GIS systems much easier. In the future, we will consider common frameworks for such architecture and metadata repository.

Acknowledgements

Standardization of the Robotic Localization Service has been initially worked by the Robotic Functional Services working group in Robotics Domain Task Force and later by Robotic Localization Service Finalization Task Force, OMG. The authors would like to thank the members of the working groups, especially the ex-co-chairs Olivier Lemaire and Kyuseo Han. The authors would also like to thank the members of joint localization working in Japan Robot Association and Network Robot Forum for their discussion and continuous support. Part of this work was supported by the Ministry of Internal Affairs and Communications of Japan, New Energy and Industrial Technology Development Organization (NEDO) and the Ministry of Economy, Trade and Industry of Japan.

7. References

[1] Denavit, J. & Hartenberg, R. S. (1955) A Kinematic Notation for Lower-Pair Mechanisms Based on Matrices, *J. Applied Mechanics*, ASME 22, pp. 215-221

[2] Doi, M. (2007) Network Robots' Plug and Play, In: *Proc. ICRA 2007 Workshop on Network Robot System: Ubiquitous, Cooperative, Interactive Robots for Human Robot Symbiosis*

[3] EPC global (2008) Tag Data Standard, ver. 1.4

[4] Hagita, N. (2006) Communication Robots in the Network Robot Framework, In: *Proc. Int. Conf. Control, Automation, Robotics and Vision*, pp. 1–6

[5] Institute of Electrical and Electronics Engineers, Inc., Standard for local and metropolitan area networks: overview and architecture. Amendment 2: registration of object identifiers, IEEE Std. 802-1990, 1990.

[6] International Organization for Standardization (1999) Manipulating industrial robots - coordinate systems and motion nomenclatures. 9787:1999

[7] International Organization for Standardization (2007) Geographic information - Spatial Referencing by Coordinates, 19111:2007

[8] International Organization for Standardization (2008) Geographic information – Schema for moving features, 19141:2008

[9] International Telecommunication Union Telecommunication Standardization Sector (2001) XML Encoding Rules (XER), ITU-T Rec. X.693

[10] International Telecommunication Union Telecommunication Standardization Sector (2002) Abstract Syntax Notation One (ASN.1): Specification of basic notation, ITU-T Rec. X.680
[11] International Telecommunication Union Telecommunication Standardization Sector (2002) Specification of Packed Encoding Rules (PER), ITU-T Rec. X.691
[12] Misra, P. & Enge, P. (2001) Global Positioning System: Signals, Measurements, and Performance, Ganga-Jamuna Press
[13] National Marine Electronics Association (2002) NMEA 0183 Standard for Interfacing Marine Electronic Navigational Devices, Version 3.01
[14] Nishio, S.; Hagita, N.; Miyashita, T.; Kanda, T.; Mitsunaga, N.; Shiomi, M. & Yamazaki, T. (2009) Sensor network for structuring people and environmental information, In: Cutting Edge Robotics 2009, Kordic, V., Lazinica, A. and Merdan, M. (Eds.), 383-402, IN-TECH, ISBN 978-3-902613-46-2, Vienna
[15] Object Management Group (2009) Robotic Localization Service specification (Beta3), document no. dtc/2009-06-04
[16] SAE International (2009) JAUS/SDP Transport Specification, AS5669, rev. A
[17] Thrun, S., Burgard, W. and Fox, D. (2005) Probabilistic Robotics, MIT Press, Cambridge, MA
[18] Ueda, T., Kawata, H., Tomizawa, T., Ohya A. and Yuta, S. (2006) Mobile SOKUIKI Sensor System – Accurate Range Data Mapping System with Sensor Motion –, In: *Proceedings of Third International Conference on Autonomous Robots and Agents*, pp.309–314

Robot Localization Using Distributed and Onboard Sensors

Dražen Brščić and Hideki Hashimoto
The University of Tokyo
Japan

1. Introduction

This chapter is dealing with the introduction of mobile robots inside environments with ubiquitously distributed sensors, specifically with the problem of robot tracking in such environments. There have been many researches recently on environments where both sensors distributed in the space and actuators such as mobile robots are used to support human users in various ways (Saffiotti et al, 2008; Hasegawa & Murakami, 2006). While there are several names that are often used for such spaces, here we refer to them as Intelligent Spaces (iSpaces) (Lee & Hashimoto, 2002). Intelligent Spaces have the purpose to provide human with both informative and physical services. Fig. 1 shows the concept of iSpaces. It relies on a network of devices to obtain information about the space and its state and to act inside it.

The primary function required by the Intelligent Space is the ability to track the position of objects in the space, most importantly that of humans and mobile robots. The issue of determining the robot's location has been dealt with a lot in the robotics community, mainly using onboard sensors. In Intelligent Spaces distributed sensors provide additional sources of information that can be used in the localization task. The issue of combining onboard and distributed sensors for robot localization has not achieved much attention yet.

A variety of sensors have been utilized for tracking in ubiquitous systems like iSpace (Hightower & Borriello, 2001). However, the tracking characteristics vary greatly from sensor to sensor, and this influences their choice. Some commonly used types of tracking systems are camera systems, tracking systems based on a combination of transmitters or receivers of ultrasound or electromagnetic waves, floor pressure sensors, distance sensors like laser range finders, etc. While all of these types of tracking systems have both advantages and limitations, here we use laser range finders as both distributed and onboard sensors for tracking in Intelligent Space.

Laser range finders are devices that determine the distance to an object or surface using laser rays. They give accurate measurements and recently some comparatively low priced sensors appeared. Due to the nature of applications of Intelligent Space, the tracking system is required to have a relatively high accuracy. It is also desirable that human users do not have to wear any special tags. Plus, the sensors should be easy to install and work with, and

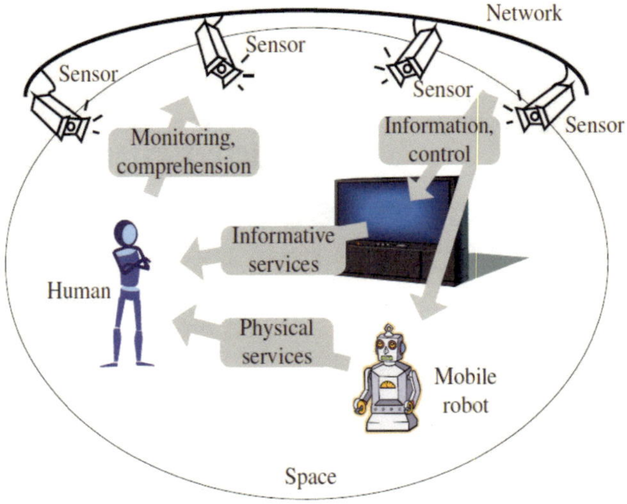

Fig. 1. Intelligent Space concept.

should be easy to apply for both human and robot tracking. Laser range finders satisfy all these requirements.

The research on position estimation using laser range finders is large. However, most of this work is on the self-localization of mobile robots equipped with onboard sensors. This also includes the extensive research on simultaneous localization and mapping (SLAM) (Thrun et al., 2005). Some recent works include also the detection and tracking of humans in the vicinity of robots, also using a laser scanner mounted on the robot (Schulz et al., 2001; Wang, 2004).

On the other hand, there are also several works dealing with objects tracking when the laser range finders are distributed in space, as is the case in Intelligent Spaces. In (Fod et al., 2002) static laser range finders were used for tracking people in everyday environments. In (Zhao & Shibasaki, 2005) the positions of pedestrians in large open spaces were estimated. Some researches used laser range finders to learn human motion patterns by extracting the trajectories of humans in the space (Bennewitz et al., 2005; Kanda et al., 2008).

This chapter tries to give the basic characteristics and methods for tracking using both distributed and onboard sensors. In the next section the basic characteristics of using distributed or onboard sensors and the advantages and disadvantages of using each of these sensing approaches are reviewed. Then, a centralized distributed architecture for Intelligent Spaces is proposed which provides flexibility and easy extension of its function.

Based on this architecture, methods for combined sensing using both robot's onboard sensors and sensors distributed in the space are described. The methods are tested in experiments where laser range finders were used as both onboard and distributed sensors.

First, tracking and fusion based on the Gaussian representations of the states (using an extended Information filter) is presented. Here the problem of correlations between the tracked objects which appears when onboard sensors are used is described and a solution based on the use of Covariance Intersection (CI) for onboard sensors is described, which explicitly avoids the correlation problem but gives conservative estimates.

Next, robot tracking based on the use of particle filters is presented. Here an extension of the popular Monte Carlo localization (MCL) algorithm that includes also the sensors distributed in the space is introduced. In addition, modeling of the robot geometry and mapping of the environment is addressed by extending the Rao-Blackwelized particle filter. In general these methods require centralized processing, so the issue of distributed sensing with so called Gaussian particle filters is also discussed.

2. Distributed and mobile sensors and fusion architecture

The application of sensors distributed in an environment to moving object tracking is straightforward and used in many applications. However, while they give very good results for tracking, distributed sensors are not very appropriate for building the map of the space. Since the sensors are not moving, the reading from static objects is always the same and further measurements give virtually no additional information. Attempts to build a map of the space with completely static sensors will in general be affected by calibration and other systematic errors. This is not the case with mobile robots equipped with onboard sensors since they make observations from different positions in space, so this types of errors average out.

Another problem with distributed sensors, especially for tracking with laser range finders, is that the orientation of the tracked object usually cannot be observed directly, since the 2D scan of the tracked object is usually small and lacking distinct features. In localization with onboard sensors this is in general not an issue since the used part of the scan is much larger. However, tracking of moving objects, such as humans, using only onboard sensors is usually more difficult than it is when distributed sensors are used. Even with a known map of the space, localization and moving object tracking can be challenging (Schulz et al., 2001). Also, distributed sensors offer continuous coverage of a tracked area, whereas continuous tracking of an object using onboard sensors has to be specifically implemented. A more psychological reason why distributed sensors are preferred is that they are embedded and not easily noticeable by users, whereas a following robot can make the users feel uncomfortable. These reasons justify why in most works on extraction of human walking patterns distributed sensors are used, e.g. (Bennewitz et al., 2005; Vasquez et al., 2008). The case of tracking humans was considered in our previous work (Brscic & Hashimoto, 2007).

Table 1 summarizes the most important points in sensing using only distributed or only onboard sensors. A combination of both distributed and onboard sensors, as considered here, can overcome the limitations of both types of sensors. Here we concentrate on the problems of robot tracking and mapping.

As explained in the previous section, the Intelligent Space is a complex system with multiple different sensors and actuators, and as such it is an important question how to design its architecture. In (Sasaki & Hashimoto, 2007) a hierarchical architecture of the iSpace as shown in Fig. 2 was proposed. It has a four layer structure consisting of sensors, information servers, applications and actuators. Information servers fuse the data from sensors and provide it to applications, which in turn send the commands to the actuators to realize various services. As discussed in (Sasaki & Hashimoto, 2007) this kind of hierarchical architecture makes it easy to add or change parts of the architecture without affecting the rest of the system, thereby providing the flexibility and modularity to iSpace. Because of that, we assume the use of this architecture throughout the chapter.

	Robot	Human	Map
Distributed	Good (orientation problem)	Good (and continuous)	Poor (non-informative)
Onboard	Good	Difficult	Good

Table 1. Comparison between sensing with distributed and with onboard sensors.

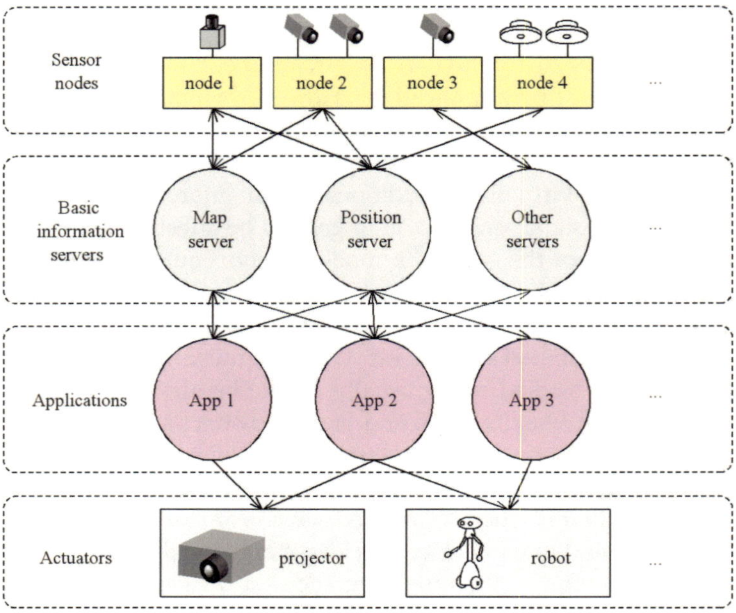

Fig. 2. Hierarchical architecture of Intelligent Space.

An important consequence of using this iSpace architecture is that for the fusion of sensor data a centralized fusion architecture is adopted, with an information server serving as the central processing unit. However, a pure central architecture where all the processing is done solely by the information server is not desirable, since computational and communication burden can be very large. Instead we make use of distributed processing, where the data is first processed on sensor nodes and then sent to the server.

3. Tracking with distributed and onboard laser range finders

We consider a system like the one depicted in Fig. 3. It consists of several distributed laser range finders distributed in the space and a mobile robot equipped with a laser range finder. Distributed sensors can observe the mobile robot, whereas both distributed and onboard sensors can observe other objects that serve as landmarks. It is assumed that the position of the landmarks is initially unknown.

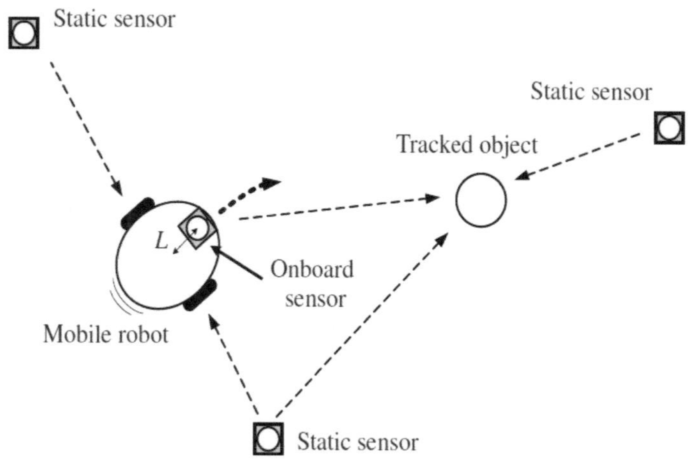

Fig. 3. Tracking system with both distributed and onboard sensors.

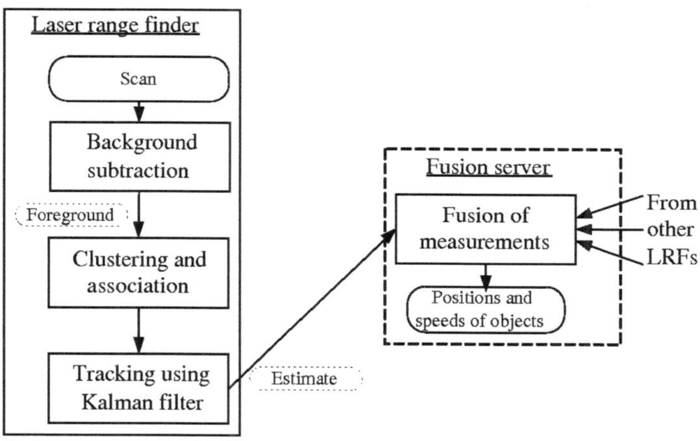

Fig. 4. Flow of object tracking using distributed sensors.

Figure 4 shows the flow of object tracking for the case of a distributed sensor. Each newly taken scan is processed through a number of steps and sent to the fusion server.

By comparing the new scan with the one where no moving object was present (the background scan), the parts that belong to moving objects can be extracted. These are further clustered and the clusters are associated to the tracked object. Finally an estimation algorithm is applied and the estimate is sent to the fusion server.

For onboard sensors the procedure is similar. From the scans landmarks are extracted from scans and used in the estimation of the robot's and landmarks' position, and this estimate is then sent to the fusion server. Any type of landmark can be used here, e.g. lines or corners - in our experiments we used artificial cylindrical landmarks which were easy to extract from the scans.

The laser scans give only the contours of the robot or landmarks, so for the use in the estimation algorithm the object centers are calculated first. This is done by calculating the center of the cluster as the mean value of the points in the cluster. Then the object center

position is approximated as being in the same direction from the sensor as the cluster center, but at a fixed larger distance than the, where the distance is dependent on the type of object (landmark or robot).
This approximation is simple, but can introduce errors if the shape of the robot or landmark are not circular, or in case of noise or misdetections. Also, for objects that are occluded by other objects, as often happens in cluttered or crowded environments, the approximation can give biased results. A different approach to using the object scans will be presented in section 4.

3.2 Estimation using the Information Filter (IF)

The described object tracking using a laser range finder gives as output the range and angle to the detected objects. The measurement model can be described as:

$$\mathbf{z} = \begin{bmatrix} q \\ \tan^{-1}\left(\frac{\Delta y}{\Delta x}\right) + \theta_0 \end{bmatrix} + \mathbf{w} \tag{1}$$

where Δx, Δy and q are given by the following relations:

$$\Delta x = x_0 - x_1, \tag{2}$$

$$\Delta y = y_0 - y_1, \tag{3}$$

$$q^2 = \Delta x^2 - \Delta y^2 \tag{4}$$

The index 0 and 1 stand for the (x, y, θ) coordinates of the sensor (either distributed or onboard) and observed object, respectively, whereas $\mathbf{w}$ represents the measurement noise, which is assumed zero mean Gaussian with covariance matrix $\mathbf{R}$.
Linearizing the measurement equation gives the following measurement matrix for the distributed sensor case:

$$\mathbf{H} = \begin{bmatrix} \frac{-\Delta x}{q} & \frac{-\Delta y}{q^2} & 0 \\ \frac{\Delta y}{q} & \frac{-\Delta x}{q^2} & 0 \end{bmatrix}. \tag{5}$$

For the onboard laser range finder the position of the sensor is also changing with the robot's position, so the measurement matrix becomes:

$$\mathbf{H}_o = \begin{bmatrix} \frac{\Delta x}{q} & \frac{\Delta y}{q^2} & 0 \\ \frac{-\Delta y}{q} & \frac{\Delta x}{q^2} & 1 \end{bmatrix} \mathbf{H} \Bigg], \tag{6}$$

where $\mathbf{H}$ is the measurement matrix for the distributed sensors from eq. (5).

Using the measurement model given above together with the model of the robot and landmarks, which we omit here, it is possible to develop an estimator for the positions of the robot and landmarks. A standard approach is to use an extended Kalman filter (EKF). Here we use the dual form of the EKF – the Extended Information Filter (EIF) (Maybeck, 1979; Thrun et al., 2005). EIF gives the same result as EKF, but it has different computational characteristics which make it easy to be decentralized. Instead of representing the state with the mean $\mathbf{x}$ and a covariance matrix $\mathbf{P}$, it uses the so-called canonical parametrization where the state is given by the information vector $\mathbf{i} = \mathbf{P}^{-1}\mathbf{x}$ and information matrix $\mathbf{I} = \mathbf{P}^{-1}$.
The full description of EIF can be found in the references and we leave it out here. In the distributed version each sensor separately calculates its information contributions:

$$\mathbf{i}_i = \mathbf{H}_i^T \mathbf{R}_i^{-1} \left[\mathbf{z}_i - h(\overline{\mathbf{x}}_i) + \mathbf{H}_i^T \overline{\mathbf{x}}_i \right], \tag{7}$$

$$\mathbf{I}_i = \mathbf{H}_i^T \mathbf{R}_i^{-1} \mathbf{H}_i , \tag{8}$$

where i is the sensor index, the function h is defined by (1) and other values are as defined above. The variables with a bar represent predicted values.
These information contributions are sent to the fusion server, where they are combined into the final fused value using a simple summation:

$$\mathbf{i} = \overline{\mathbf{i}} + \sum_i \mathbf{i}_i , \tag{9}$$

$$\mathbf{I} = \overline{\mathbf{I}} + \sum_i \mathbf{I}_i . \tag{10}$$

3.3 Estimation using Covariance Intersection (CI)

However, as is well known from Kalman filter based solutions to the SLAM problem, the use of onboard sensor results in cross-correlations between the estimates. This means that the positions of different objects cannot be estimated separately, but instead all estimates must be lumped together in a large information vector and matrix. This in turn causes a significant increase in the computation, but also in the communication since the full estimates have to be transferred to and from the sensors.
To alleviate this problem we combine the use of the Information filter with the Covariance Intersection (CI) method. Covariance Intersection (Julier & Uhlmann, 2001) was proposed as a method of fusion of two estimates when the correlation between the estimates is unknown. It can also be shown that it is the optimal method of fusion in that case. It also uses the Gaussian representation as the Kalman and Information filters, so the estimates are given by the mean state $\mathbf{x}$ and the corresponding covariance matrix $\mathbf{P}$, or equally by the information vector and matrix $\mathbf{i}$ and $\mathbf{I}$. Fusion of two estimates using CI is given by: $\mathbf{I} = \omega\mathbf{I}_1 + (1\text{-}\omega)\mathbf{I}_2$ and $\mathbf{i} = \omega\mathbf{i}_1 + (1\text{-}\omega)\mathbf{i}_2$, ω being a parameter between 0 and 1.
It is also possible to write the CI equations in a form appropriate for use in the fusion equations (7) and (8). The vector and matrix representing new information become:

$$\mathbf{i}_i = \omega\left(\mathbf{H}_i^T \mathbf{R}_i^{-1} \left[\mathbf{z}_i - h(\overline{\mathbf{x}}_i) + \mathbf{H}_i^T \overline{\mathbf{x}}_i \right] - \overline{\mathbf{i}} \right) \tag{11}$$

$$\mathbf{I}_i = \omega\left(\mathbf{H}_i^T \mathbf{R}_i^{-1} \mathbf{H}_i - \overline{\mathbf{I}} \right) \tag{12}$$

One good characteristic of CI is that since it treats all information as possibly being correlated, it prevents information reuse. Another good feature is that it is possible to obtain separate estimation for all tracked objects, thus giving uncorrelated estimates even when using onboard sensors. This is explained in (Julier & Uhlmann, 2007), where CI was applied to the SLAM problem to achieve separate estimation of the robot and landmark positions.
Consequently, onboard sensors can be easily included in sensor networks, without introducing correlations between estimates. Using (11) and (12) CI can be implemented for the onboard sensor, whereas distributed sensors use the EIF method (7) and (8) .
A weak point of CI is that, in contrast to the Information filter, it does not utilize all the available information. Hence the obtained estimate is somewhat conservative, as will be shown in the experimental results.
Another problem of CI is that calculation of the ω parameter usually involves optimization of some performance criteria, like the determinant of the fused matrix (Julier & Uhlmann, 2001), which makes the method more computationally involved when compared to EIF.

3.4 Experimental results

We made experiments in our experimental Intelligent Space. Part of the setup is shown in Fig. 5. Four Hokuyo URG-04LX laser range finders were used as distributed sensors and placed at the sides in a way that they cover the whole space (about 5x7 meters). A Pioneer 2-DX robot was equipped with the same laser range sensor in the front as onboard sensor. Ten cylindrical objects were used as landmarks.
Fig. 6 shows the result of estimating the robot and landmark positions after a sample run. The ellipses show the uncertainty in the landmark estimate. Also shown are the positions of distributed sensors.
While the position estimates are very similar for all estimation methods, differences can be seen in the obtained estimate uncertainties. Fig. 7 shows the uncertainty in the robot position estimate during the experiment. As expected, tracking using only distributed sensors gave the largest uncertainty. On the other hand, when the information from the robot onboard sensor was also included using the EIF the best result of all proposed methods was obtained, reflecting the fact that all of the available information was utilized. The result obtained when only the onboard sensor was used lies somewhere in between.
It is obvious that Covariance Intersection gave a larger uncertainty than EIF as expected, and at times it even becomes larger than the estimate using only the onboard sensor. This is however achieved with a much smaller computation and communication load.

4. Geometrical model based method

In the method presented in section 3 all scan points belonging to an object were combined into one point to extract the center of the tracked object. This can be seen as an unnecessary loss of valuable information and a source of errors in the case when the object is occluded. Here we present another method that uses all these scan points and compares them with a geometrical model of the object. This approach results in a considerably different estimation method.

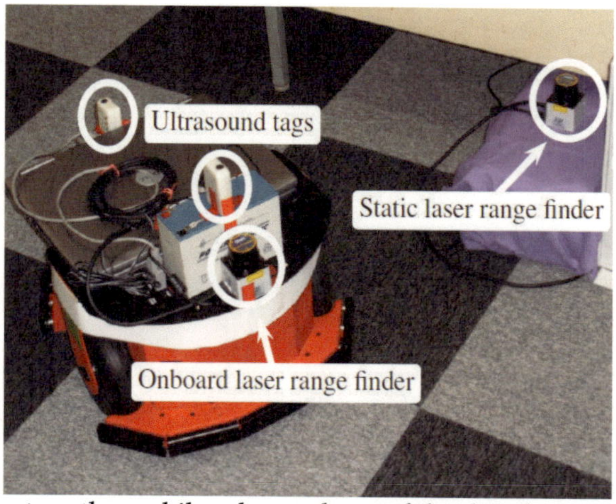

Fig. 5. Experimental setup: the mobile robot and one of the distributed laser range finders.

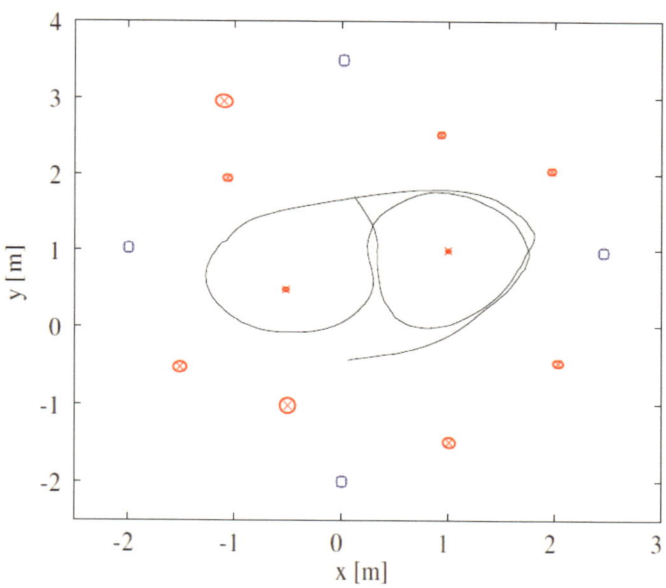

Fig. 6. Estimation result: the obtained robot path is shown, along with the estimated position of landmarks and their uncertainty ellipses; the position of distributed sensors is marked with squares.

4.1 Modeling of the mobile robot and space

Although several choices for the model of the robot exist, we use occupancy grids because of they can represent any kind of shape and are easy to build online. Occupancy grids have been very popular in the robotics community and can be easily combined with laser range finder measurements.

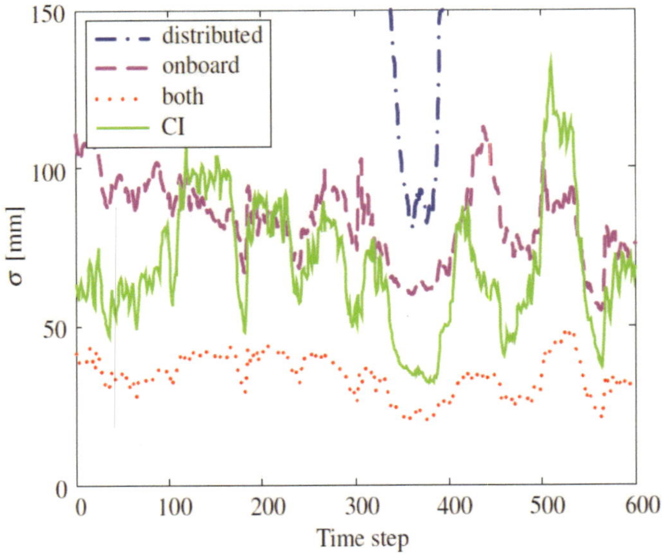

Fig. 7. Obtained robot position uncertainty when using the Information filter for only distributed, only onboard or both sensors, and when CI is used for onboard sensors.

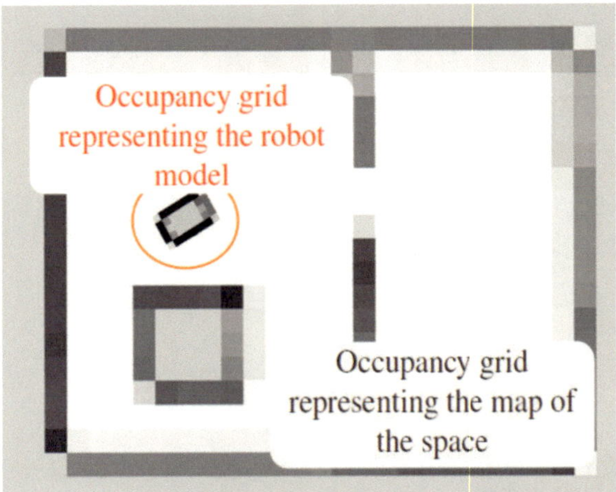

Fig. 8. Representation of the space and mobile robots with occupancy grids.

The idea of occupancy grids is to divide the space in a grid, usually evenly spaced. Each cell in the grid is given a measure representing the probability that the cell is occupied by an object. In order to make the combination easier, we model both the robot's shape and the geometry of the space using occupancy grids (i.e. instead of using landmarks like in the previous section here we use an occupancy grid representation of a map of all static objects in the space).

An example illustrating the overall model is given in Fig. 8. In addition to the occupancy grid representing the map of the space, there is also an occupancy grid that outlines the shape of the robot. The space map is fixed to the world coordinate system, whereas the robot model changes its position with the robot.

4.2 Estimation of the robot position

First we assume that the occupancy grids representing the robot and space models are known. A method how to build them will be explained in the next subsection.

Occupancy grids do not allow a straightforward usage of the Kalman filter or similar methods, like in the previous section. A common approach is to use a particle filter based approach (Arulampalam et al., 2002), and one standard particle filter based method used in robot localization with onboard sensors is Monte Carlo localization (MCL) (Dellaert et al., 1999).

The idea of MCL is to represent the belief on the robot's position by a set of particles, each representing one hypothesis on the current pose - i.e. the robots (x, y) position and orientation θ. At every step the algorithm goes through several functions to update the set of particles, namely motion model sampling, particle weighting based on the measurement and resampling (Arulampalam et al., 2002).

In motion model sampling the position of each particle is updated based on the dynamic model of the tracked robot, given in (7). Next, for each updated particle a weight w is calculated, which describes a belief how well the new laser range finder measurement matches the occupancy grid map, given the updated pose (in our experiments we used the likelihood field method (Thrun et al., 2005)). Finally, the particles are resampled, i.e. a new set of particles is drawn from the current one based on the weights. This is repeated in each step with the new scans to obtain a new set of particles that defines the current belief on the robot pose.

Standard MCL uses only onboard laser range finders, however if the robot is also represented using an occupancy grid it is easy to extend it to distributed laser range finders. The only part that changes is the weighting of the particles. In addition to weighting based on the onboard sensor scan the particles are additionally weighted based on the matching of scans from distributed sensors with the robot model, and these new weights are used instead in the resampling.

4.3 Building the robot model and space map

For the SLAM problem, a well known particle filter that builds up on MCL is FastSLAM (Montemerlo, 2003). This algorithm implements the same three functions in each update as described above for MCL, but in addition it has also a function that updates the occupancy grid. Similar to MCL, FastSLAM can also be easily extended to incorporate measurements from distributed sensors.

When building the model, not only the pose of the robot but also the log odds for each cell in the occupancy grid are unknown. A direct application of the particle filter would result in a huge number of necessary particles, but in FastSLAM the particles instead contain the robot pose and Gaussian representations of cell occupancy probabilities (Montemerlo, 2003). This is possible because of conditional independence of the map cells given the robot pose (this corresponds to the Rao-Blackwellization procedure from statistics, so this method is also commonly referred to as Rao-Blackwellized particle filter).

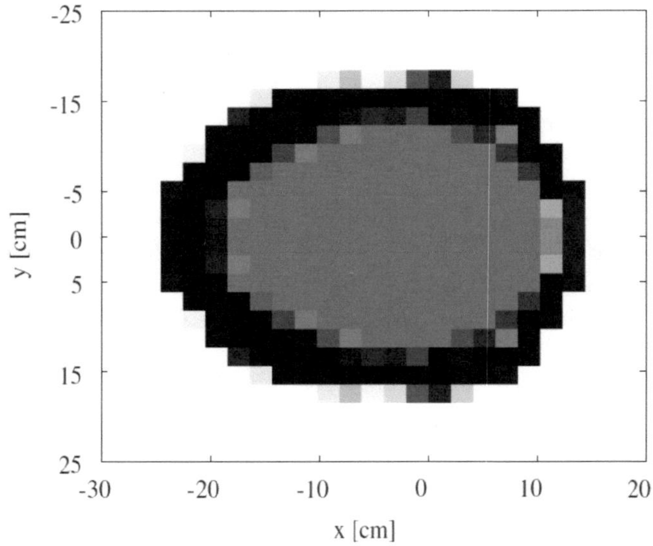

Fig. 9. Obtained occupancy grid representation of the robot model.

Since in our problem in addition to the map of the space we have also the model of the robot, both of these have to be contained in a particle. Due to the conditional independence of the robot models given the robot pose, a Gaussian representation of the cells can be used in the same way as for the map of the space. Therefore, each particle contains the robots pose, an occupancy grid representing the map of the space and an occupancy grid describing the robot model.

In addition to model sampling, particle weighting and resampling, at every step and for each particle both occupancy grids are updated. This is implemented using ray casting, where for each laser ray all the cells that it traverses are updated (Thrun et al., 2005).

We also use an assumption that the left and right side of the robot are symmetric, which is the case with most mobile robots. This simplifies the calculations since we only need half of the occupancy grid to represent the robot model.

4.4 Experimental results

We made experiments in our Intelligent Space room using the same setup as in section 3.4. For the map of the space we used an occupancy grid with cells of size 10x10cm, whereas for the robot model we used 2x2cm cells.

Fig. 9 shows the occupancy grid representing the model of the robot obtained in the experiments using the method described above. This model matches closely the geometry of the real robot (the Pioneer 2-DX shown in Fig. 5), which has a width of about 30 cm, its front is at a distance of about 15 cm from the center, whereas the back is at about 25 cm.

Fig. 10 shows the obtained occupancy grid representing the map of the space.

Finally, we also made a tracking experiment, where the robot model and map of space obtained above were used. A comparison with the EIF method presented in section 3 is shown in Fig. 11. In order to get a measure of the precision of the proposed estimation method, the obtained estimation results are also compared with the measurements from the

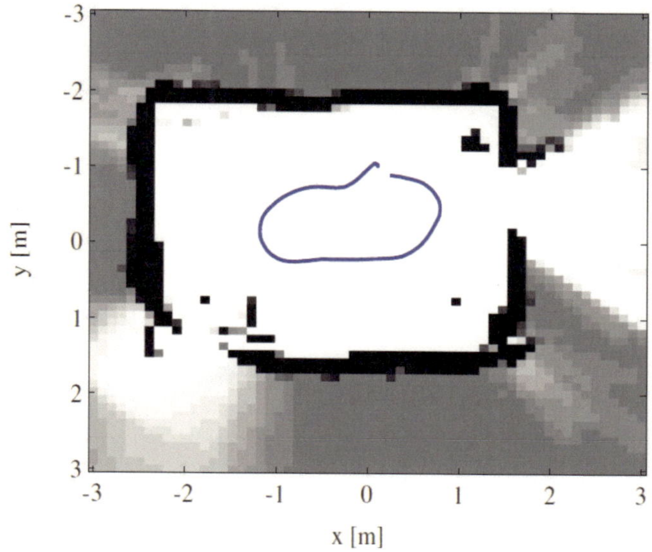

Fig. 10. Robot path and obtained map of the space.

ZPS ultrasound positioning system (by Furukawa Electric, Ltd.) that is installed in our experimental Intelligent Space.
We see that the two results are very close. By observing the differences between the estimates and the ultrasound system results, we noticed that throughout the experiment they mostly stayed bellow 5 cm for both methods. Overall the model based method gave slightly better results compared to EIF, which is expected since instead of an approximation it uses an accurate model of the robot.

4.5 Distributed fusion using Gaussian particle filters

The model based tracking method discussed so far assumes that the data from all sensors are collected in a central place and processed simultaneously. However, as discussed also previously in section 2, centralized fusion is characterized by a large computational and communicational load, especially since all laser scans need to be transferred through the network.
Some approaches to distributed fusion of particle filters have been presented in e.g. (Rosencrantz et al., 2003; Ong et al., 2006), which use approximations of the particle sets, such as Gaussian mixture models. Here we investigate the approximation using simple Gaussian representations. One of the advantages of using particle filters is that they can deal with distinct hypotheses and non-Gaussian states easily, so representing particle sets with Gaussians might not seem justified. However, it can also be argued that due to measurements from distributed sensors the robot position can always be set to just one location, and therefore an unimodal distribution like the Gaussian can give a satisfactory representation.
Using the Gaussian representation of the state with a particle filter corresponds to the "Gaussian particle filters" (GPF) discussed in (Kotecha & Djurić, 2003). A distributed architecture is also assumed, where the sensors calculate new information from the

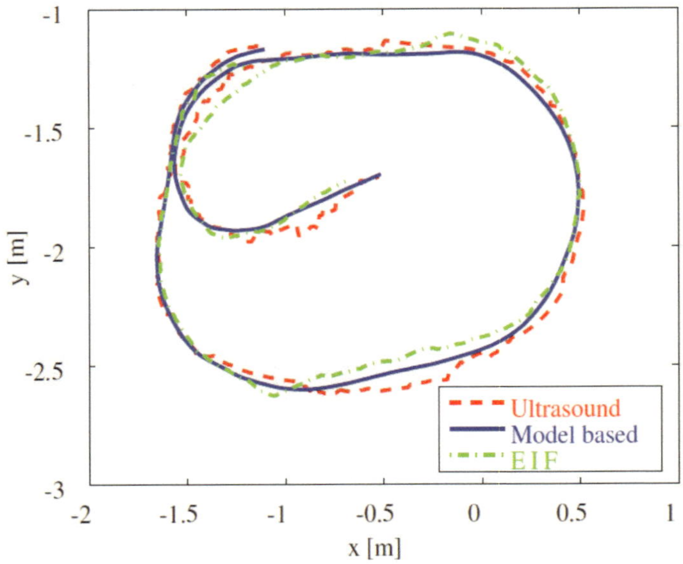

Fig. 11. Robot tracking results using EIF and the model based method shown in comparison to the ultrasound system tracking measurements.

measurements and send it to a fusion server, which in turn calculates the prediction step and feeds it back to sensors.

Since the Gaussian representation is used, the prediction on the fusion server can be performed in the usual way, based on the process model and the Kalman filter equations. The calculation on the sensor nodes is implemented in particle filter form, as follows:

- a set of particles is drawn from the predicted state vector and covariance matrix
- based on the measurements for each particle weights are calculated in the same way as in MCL or FastSLAM
- based on the particle states $\mathbf{x}^{(n)}$ and weights $w^{(n)}$ the updated state $\mathbf{x}$ and covariance $\mathbf{P}$ are calculated using the following equations:

$$\mathbf{x} = \frac{\sum_{n=1}^{N} w^{(n)} \mathbf{x}^{(n)}}{\sum_{n=1}^{N} w^{(n)}} \tag{13}$$

$$\mathbf{P} = \frac{\sum_{n=1}^{N} w^{(n)} \left(\mathbf{x}^{(n)} - \mathbf{x}\right)\left(\mathbf{x}^{(n)} - \mathbf{x}\right)^T}{\sum_{n=1}^{N} w^{(n)}} \tag{14}$$

At each step a new set of particles is drawn from the Gaussian representation of the predicted state. For each of these particles weights are calculated based on the measurement - for distributed sensors this is done based on the robot model, and for onboard sensor on the environment model. In the final step the particles are transformed back to the Gaussian

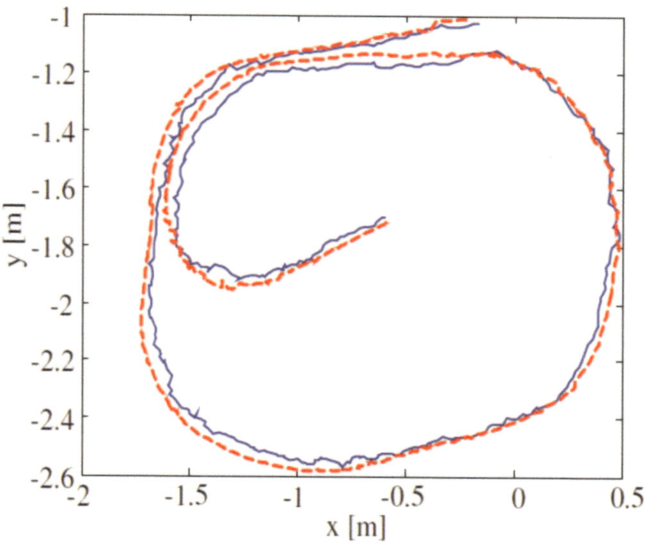

Fig. 12. Comparison of model based robot tracking results using distributed Gaussian particle filters (solid line) and standard particle filters (dashed line)

representation, by calculating their mean and covariance. This is done in the place of the resampling step of standard particle filters and is in general computationally less expensive. Fig. 12 shows the result of tracking a robot using the just described distributed Gaussian particle filter (same data as in Fig. 11 were used) compared with the result obtained using a centralized particle filter. Even though the GPF uses only an approximation of the particles for fusion, the obtained results are very close. The result is not so smooth due to the random nature of the algorithm and the fact that a small number of particles (50) was used.
Since in the distributed architecture all sensors work independently there is no need for all of them to implement the same tracking method. For example, onboard sensors could use EIF or CI as described in section 3, while distributed sensors use model based tracking.

5. Method comparison

The two methods presented here differ in their complexity. The first method determines the object center based on a simple calculation and then applies a Kalman filter based estimation method, whereas the second method uses a particle filter for the estimation. The particle filter methods in general have more computational demands, even though both methods can be made to run in real time on an average computer. A larger difference appears in the model building process, which for the second method is much more involved. For the first method model building corresponds to the inclusion of the estimation of the landmarks' positions, which as we said can be done uncorrelated with the robot position estimation and therefore very efficiently.
Yet, when the accuracies of the methods are compared there is not a large difference, as shown in the experiments. The reason for this is that even though the calculation in the first method is only approximate, the error introduced by the approximation is mitigated by

fusing the results from multiple sensors. Even though the second method uses a more exact algorithm, it does so at a cost of an involved computation.
Nevertheless, the advantage of using the model based method becomes more highlighted when used for tracking the robot in crowded or complex spaces. In these situation the occlusions in the sensor scans increase, thereby introducing an error in the approximation used in the first method. For onboard sensors it is obviously an advantage of using the whole scan instead of landmarks, which can be easily occluded.
Another question that arises in actual applications is the fusion of distributed sensors, since in iSpaces we deal with multiple sensors connected through a network. The first method can easily be implemented in a centralized architecture, with each sensor sending the extracted object center to the fusion node. But due to uncorrelated estimates, it can also be very simply turned into a distributed or decentralized algorithm, while keeping the network load relatively small. On the other hand, an exact solution for the second method admits only centralized processing, which in turn also requires significant communication loads since the whole parts of scan belonging to each object have to be sent. The Gaussian particle filter based estimation presented here gives an approximate way of obtaining a distributed estimation, but the characteristics and conditions under which it can be used still have to be studied.

6. Conclusion

This chapter discussed the issue of robot tracking inside Intelligent Spaces, where both sensors onboard the robot and sensors distributed in the space can be utilized. Two methods for robot tracking were presented, where a robot with an onboard laser range finder and several laser range finders at fixed locations in the space were used. One method uses an approximate relation for calculating the center of the tracked objects, and employs a combination of an extended Information filter and Covariance Intersection for position estimation. The other method extends the idea of occupancy grids to the modeling of the robot, and applies a particle filter based estimator for tracking and building of the robot model and map of the space. The methods are tested in experiments and a comparison is given.

7. References

Arulampalam, S.; Maskell, S.; Gordon N. & Clapp, T. (2002). A Tutorial on Particle Filters for On-line Non-linear/Non-Gaussian Bayesian Tracking. *IEEE Transactions on Signal Processing*, Vol. 50, 174-188

Bennewitz, M.; Burgard, W.; Cielniak, G. & Thrun, S. (2005). Learning Motion Patterns of People for Compliant Robot Motion. *International Journal of Robotics Research*, Vol. 24, No. 1, 31-48

Brscic, D.; Hashimoto, H. (2007). Tracking of Humans Inside Intelligent Space using Static and Mobile Sensors. *Procedings of the 33rd Annual Conference of the IEEE Industrial Electronics Society*, pp. 10-15, Taipei Taiwan, Nov. 2007

Dellaert, F.; Fox, D.; Burgard, W. & Thrun, S. (1999). Monte Carlo Localization for Mobile Robots. *Proceedings of the IEEE International Conference on Robotics and Automation*, pp. 1322-1328, Detroit, MI, USA, May 1999

Fod, A.; Matarić, M. & Sukhatme, G. (2002). A Laser-Based People Tracker. *Proceedings of the IEEE International Conference on Robotics and Automation*, pp. 3024-3029, Washington DC, USA, May 2002

Hasegawa, T. & Murakami, K. (2006). Robot Town Project: Supporting Robots in an Environment with Its Structured Information. *Proceedings of the 3rd International Conference on Ubiquitous Robots and Ambient Intelligence*, pp. 119-123, Seoul, Korea, Oct. 2006.

Hightower, J. & Borriello, G. (2001). Location Systems for Ubiquitous Computing. *IEEE Computer*, Vol. 34, No. 8, 57-66

Julier, S. & Uhlmann, J. (2001). General decentralized data fusion with covariance intersection (CI), In: *Handbook of Multisensor Data Fusion*, Hall, D. & Llinas, J. (Ed.) CRC Press

Julier, S. & Uhlmann, J. (2007). Using covariance intersection for SLAM. *Robotics and Autonomous Systems*, Vol. 55, No. 1, 3-20

Kanda, T.; Glas, D.F.; Shiomi, M.; Ishiguro, H. & Hagita, N. (2008). Who will be the customer?: a social robot that anticipates people's behavior from their trajectories. *Proceedings of the 10th international conference on Ubiquitous computing*, Seoul, Korea, Sep. 2008

Kotecha, J. & Djurić, P. (2003). Gaussian particle filtering. *IEEE Transactions on Signal Processing*, Vol. 51, No. 10, 2592-2601

Lee, J.H & Hashimoto, H. (2002). Intelligent Space - Concept and Contents. *Advanced Robotics*, Vol. 16, No. 3, 265-280

Maybeck, P. (1979). *Stochastic Models, Estimation, and Control*, Academic Press, ISBN: 978-0124807020

Montemerlo, M. (2003). *FastSLAM: A Factored Solution to the Simultaneous Localization and Mapping Problem with Unknown Data Association*, PhD thesis, Robotics Institute, Carnegie Mellon University

Ong, L.; Upcroft, B.; Bailey, T.; Ridley, M.; Sukkarieh, S. & Durrant-Whyte, H. (2006). A decentralised particle filtering algorithm for multi-target tracking across multiple flight vehicles. *Proceedings of the IEEE/RSJ International Conference on Intelligent Robots and Systems*, pp. 4539-4544, Beijing, China, Oct. 2006

Rosencrantz, M.; Gordon, G. & Thrun, S. (2003). Decentralized sensor fusion with distributed particle filters. *Proceedings of the 19th Annual Conference on Uncertainty in Artificial Intelligence*, pp. 493-500, Acapulco, Mexico, Aug 2003

Saffiotti, A.; Broxvall, M.; Gritti, M.; LeBlanc, K.; Lundh, R.; Rashid, J.; Seo, B.S. & Cho, Y.J. (2008). The PEIS-Ecology project: Vision and results. *Proceedings of the IEEE/RSJ International Conference on Intelligent Robots and Systems*, pp. 2329-2335, Nice, France, Sep. 2008

Sasaki, T. & Hashimoto, H. (2007). Hierarchical Framework for Implementation of Intelligent Space, *Proceedings of the 33rd Annual Conference of the IEEE Industrial Electronics Society* , pp. 28-33, Taipei, Taiwan, Nov. 2007

Schulz, D.; Burgard, W.; Fox, D. & Cremers, A. (2001). Tracking multiple moving targets with a mobile robot using particle filters and statistical data association. *Proceedings of the IEEE International Conference on Robotics and Automation*, pp. 1665-1670, Seoul, Korea, May 2001

Thrun, S.; Burgard, W. & Fox, D. (2005). *Probabilistic Robotics,* The MIT Press, ISBN: 978-0-262-20162-9, Cambridge, MA, USA

Vasquez, D.; Fraichard, T.; Aycard, O. & Laugier, C. (2008). Intentional motion on-line learning and prediction. *Machine Vision and Applications,* Vol 19, No. 5-6, 411–425

Wang, C. (2004). *Simultaneous Localization, Mapping and Moving Object Tracking,* PhD thesis, Robotics Institute, Carnegie Mellon University

Zhao, H. & Shibasaki, R. (2005). A novel system for tracking pedestrians using multiple single-row laser-range scanners. *IEEE Transactions On Systems, Man and Cybernetics – Part A: Systems and Humans,,* Vol. 35, No. 2, 283-291

Sensing Terrain Parameters and the Characteristics of Vehicle-terrain Interaction Using the Locomotion System of a Robot

Ilkka Leppänen
Helsinki University of Technology
Finland

1. Introduction

A good perception system is important for a mobile robot that must execute autonomous piloting and navigation in an unknown unstructured environment. Terrain information is needed for path planning, terrain trafficability evaluation, obstacle avoidance and also for automatic locomotion mode control of robots having several modes available. There are two basic principles for the sensing of terrain. The first sensing method is based on touching the ground and the second is based on contactless range measurements. Of course, it is possible to use pre-information that has been catched before the mission started.

With the contactless measuring system, it is possible to model the terrain along a greater distance than sensing by touching the ground, but when operating off-road there are a lot of features that dilute ranging reliability. Vegetation, trees, bush and even snow hinder or prevent scanning real terrain surfaces. Above all, with vision and other range-based systems, it is hard to discover soil characteristics, such as the deformation and compressibility of the soil under pressure and the reactions of soil particles to the horizontal forces, e.g. shear strength or soil strength. Therefore, to get information about soil characteristics, sensing by touching is needed.

The main factors of vehicle-terrain interaction are load-carrying capacity of soil and traction performance of the wheel. Furthermore, geometry of the terrain has a great effect on locomotion. Deformation and compressibility of the soil represent the bearing capacity of the soil. Shear strength of the soil represents the traction performance of the wheel. There are standard methods to measure deformation, compressibility and shear strength of soil, but these methods are not practical for a moving vehicle because they are based on standard devices and procedures requiring in-place measurements (Bekker, 1969 ; Karafiath & Nowatzki, 1978; Wong, 1994).

This chapter describes new real time sensing methods, while driving, for determining terrain geometry parameters such as slope and roughness, and also the characteristics of vehicle-terrain interaction through energy consumption, motion resistance and slippage using the locomotion system as a sensing system so that minimal or no additional sensors are needed. The terrain sensing methods described are generally not dependent on the particulars of the implementation of the locomotion system (such as degrees of freedom or joint and linkage types). The methods have been tested with the WorkPartner service robot that has a multimode locomotion system.

2. Sensing the characteristics of vehicle-terrain interaction

Moving over natural terrain and measuring soil parameters at the same time is difficult (Ojeda et al., 2005). It is really more practical to sense the characteristics of vehicle-terrain interaction, such as the energy consumption and traction or rolling resistance of the wheel, and use them instead of these parameters.

2.1 Sensing load-carrying capacity of soil using energy consumption

Wheel or foot sinkage is a good characteristic to describe load-carrying capacity (Iagnemma et al., 2004), but unfortunately it cannot be measured in a reliable way in a demanding environment. It can be estimated indirectly by using the energy consumption of the wheel or energy consumption of walking. When moving with constant speed, the used energy finally converts to deformation of soil and heating of the wheel. In very soft soil, the wheel also looses energy when bulldozing the ground.

In a fully electrical actuating system, the torque of each wheel can be easily calculated using the current of the electric motor. The power of the wheel is derived from the torque and angular speed of the wheel. The used energy should then be scaled to the traveled distance. In general, the total energy of the propulsion system is used for acceleration of the robot, driving uphill or in some cases pulling a load. In all cases some part of the energy is lost in wheel-soil interaction. To calculate the lost energy correctly, all these essential matters should be considered. Ground slope can be derived from the robot attitude and the wheel positions relative to the robot body. Hence the sum of all the wheels' lost energy in wheel-soil interaction per traveled distance, which can also be called rolling resistance, is

$$F_{roll} = \frac{\sum_{j=1}^{n} E_j}{s} + mg\sin\theta_p - ma_x, \tag{1}$$

where E_j is energy of the wheel j, s is traveled distance of the vehicle, m is mass of the vehicle, g is acceleration due to gravity, θ_p is pitch angle of terrain slope, a_x is longitudinal acceleration of the vehicle and n is the number of the wheel. In rolking (Leppänen et al., 1998) or walking locomotion energy consumption can be calculated by considering all joints.

2.2 Motion resistance in static situations

In off-road operation, it can happen that an obstacle or obstacles force the vehicle to stop or the front wheels sink into the soil, thus preventing the vehicle from driving forward. Then the vehicle speed is zero and energy consumption per traveled distance cannot be determined. In this static situation, motion resistance can be estimated by using wheel torque as follows

$$F_{static} = \sum_{j=1}^{n} \frac{T_j}{r_R} + mg\sin\theta_p, \tag{2}$$

where T_j is torque of the wheel j, r_R is rolling radius of the wheel, m is mass of the vehicle, g is acceleration due to gravity and θ_p is pitch angle of terrain slope.
The static motion resistance can be used to indicate a sudden and difficult obstacle, such as a hole in the ground or if the vehicle body gets stuck.

2.3 Energy consumption of a single wheel in wheel-soil interaction

The wheel generates a drawbar force to the robot body in good conditions. In very soft terrain, a single wheel can produce a resisting force with respect to the body, although the wheel still consumes lot of energy in generating wheel torque. This happens especially when a single wheel sinks into the ground and bulldozes soil. If the pulling or resisting force of the wheel can be measured with the help of forces measurement in the suspension system of the wheel, and if the wheel power is known, it is possible to estimate the lost energy in each wheel-soil interaction. The energy that the wheel consumes in wheel-soil interaction per traveled distance is

$$F_{roll\_w} = \frac{E_w}{s} - F_{DP}, \tag{3}$$

where E_w is energy of a wheel, s is traveled distance of the vehicle and F_{DP} is the drawbar pull of the wheel.
In a static situation, when the vehicle speed is zero, the motion resistance of a wheel can be estimated by using the wheel's torque as follows

$$F_{static\_w} = \frac{T_W}{r_W} - F_{DP}, \tag{4}$$

where T is torque of the wheel, r_W is rolling radius of the wheel and F_{DP} is the drawbar pull of the wheel.
The motion resistance of a single wheel can be used to localize the problematic ground area under the vehicle, which enables more sophisticated locomotion control.

2.4 Motion resistance of the transfer leg's wheel in rolking mode

In rolking mode, the transfer leg's wheel drives to the next support position without propelling the robot and lightened. Then the transfer leg's wheel does not strain soil and the real load carrying capacity of soil remains unknown. However, with the unloaded driving wheel, it is possible to recognize very soft soil that resists the driving wheel. In addition, with the unloaded driving wheel, it is possible to sense terrain obstacles that also resist the wheel. In the transfer phase, the wheel is under speed control and the transfer leg moves the wheel to the next position. Thus, the motion resistance of the transfer leg's wheel is a sum of the wheel force generated by wheel torque and the leg pushing force as determined in Equation (4).

2.5 Sensing drawbar force of the wheel in wheeled mode

The traction performance of the wheel is one of the main criteria in the analysis of the vehicle mobility. If the drawbar pull produced by the wheel with respect to the robot's body can be measured directly, it describes wheel-soil interaction. If the motion resistance of the wheel is lower than the generated tractive force, the wheel pulls the robot's body.

Otherwise, the wheel resists the robot. The pulling or resisting force generated by the wheel can be calculated using force measurement in the suspension system of the wheel. In the case of the wheeled leg, the force affecting in the fixing point of the body can be derived using the leg joint's torques.

2.6 Slip of the wheel

In slippery conditions where the friction coefficient between the wheel and the soil is low or the shear strength of the soil is limited, the wheel slips and therefore the traction force is lower. The longitudinal slip of the wheel, when a driving torque is applied, is usually defined by the equation

$$S = \left(\frac{r_R\omega - v}{r_R\omega}\right) * 100\%, \tag{5}$$

where S is slip, r_R is rolling radius, ω is angular velocity of the wheel and v is forward velocity of the vehicle (Gillespie, 1992; Wong, 2001). It has been observed experimentally that using this kind of slippage value and driving with rubber tires the best traction occurs with a slippage of 15 - 20 % (Wong, 2001).
The angular velocity of the wheel can be measured reliably by an encoder. The speed of the vehicle can be estimated using the average of the wheel speeds so that the wheel that has its speed deviate most from the average of all is not taken into account in speed calculation. Problems, however, occur if all wheels slip at the same time. Then the velocity of the vehicle cannot be determined with pure wheel speed sensor information. In the four-wheel locomotion system, it is possible to use one wheel as a sensor wheel for estimating the vehicle speed. Then a sensor wheel rolls freely and measures vehicle speed and odometry. Unfortunately, the drawbar pull of this sensor wheel is lost, which can be critical in low frictional conditions.
In very slippery conditions, vehicle inertia sensors with GPS (Global Positioning System) navigation can be used for estimating vehicle speed and odometry. Another way is to use a separate sensor wheel in bad conditions. Vision-based sensing methods for estimating the speed of a vehicle are still under research, but might help in the future.
For example, excessive wheel slippage can be used as an indicator that the terrain has become problematic for wheeled driving and a switch of locomotion to rolking could be beneficial.

3. Sensing terrain geometry

A terrain range scanner, based on laser or vision technology, is crucial for autonomous mobile robots, because the terrain information is needed for path planning and obstacle avoidance. Using a vision system or laser-based technology, it is possible to get the geometry of terrain in good environment conditions, but when operating off-road there are a lot of features that dilute ranging reliability. The main emphasis here is put on sensing the geometry of the terrain surface by touching, because the actual terrain surface under snow or vegetation can be detected with the loaded wheel reliably. A negative aspect of the touching method is that it limits the robot's velocity, because it is possible to sense only terrain that is under the robot. Terrain slope and roughness are the main characteristics that express the geometry of terrain.

3.1 Sensing terrain slope

The terrain slope angles, terrain pitch (θ_p) and terrain roll (θ_r), express the inclination of the terrain according to the gravitational vector (g). Terrain slope can be derived from the robot's body inclination and wheel positions utilizing the robot coordinate systems that are shown in Figure 1.

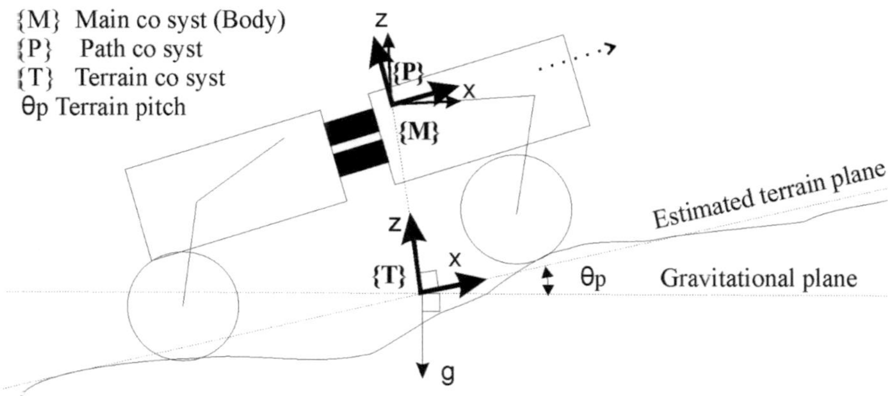

Fig. 1. Relation of the coordinate systems of the wheel-legged robot and terrain pitch.

The estimated terrain plane is derived from the wheel positions in the main coordinate system. It can be obtained by using linear regression in fitting a plane into the set of the wheels' position coordinates. The pose of the main coordinate system with respect to the gravitational vector can be measured by inclination sensors. Using this information, the terrain pitch and roll angles with respect to the gravitational vector can be calculated.

3.2 Sensing terrain roughness

In locomotion studies, the geometry of the terrain surface cannot be divorced from vehicle characteristics, because what is rough for a small vehicle, such as a small conventional car, may be smooth for a large off-road vehicle with large wheels. There is no good and standard way to model terrain roughness, because natural terrain is composed of an unlimited number of features. Typical often-used separate features in modeling terrain are ditch, step, ramp or a separate obstacle with basic parameters, such as height and width. Modeling of the terrain with standard features can be achieved, but collecting enough geometrical information in real operating conditions can be very difficult. When driving at high speed, vibration of the body informs of terrain surface roughness, but it only tells about the ground surface pattern. In addition, soft soil dampens the vibration effectively. At low speed, the suspension system of the vehicle filters the unevenness of terrain.

The wheeled vehicle with a suspension system will align itself according to terrain. If the wheel positions with respect to the vehicle body are known, it is possible to use the sum of the wheel distances to the estimated support plane as an indicator of terrain roughness. Because this terrain roughness indicator does not take the vehicle size into account, another way to estimate terrain roughness is proposed. This method is based on using virtual axles that connect the wheel centers as illustrated in Figure 2.

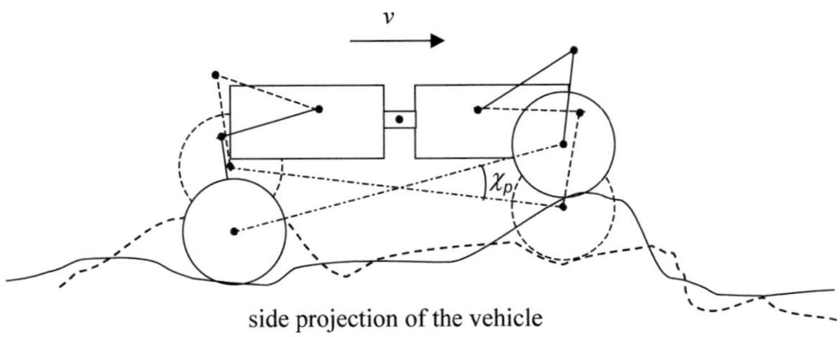

side projection of the vehicle

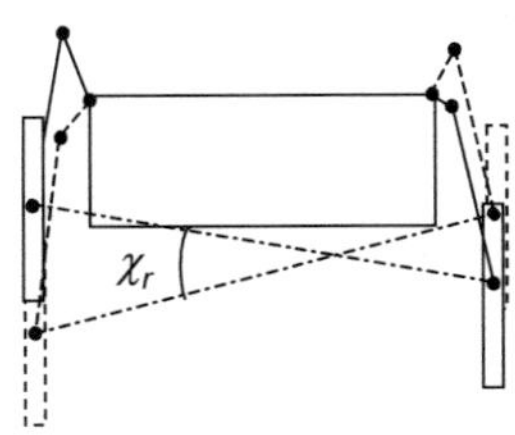

front projection of the vehicle

Fig. 2. Terrain roughness indicator based on the virtual axles and their difference angles in the side and front projection.

This terrain roughness indicator takes the vehicle size, i.e. wheelbase and wheel distance, into account automatically. This indicator does not detect vertical surface parts that are problematic for the wheel. On the other hand, if the value of this indicator is high, it is likely that terrain consists of more vertical surface parts that are problematic for the wheel and need more drawbar force. In wheeled mode, there is another limitation using this method for sensing terrain roughness: if the front or back wheels, or even all the wheels, encounter a hump (or ditch) at the same time, the value of this roughness indicator remains small although the terrain will be difficult for the wheel to negotiate.

4. Experimental verification of the sensing methods

The described sensing methods are generally not dependent on the particulars of the implementation of the locomotion system (such as degrees of freedom or joint and linkage types). However, since the methods have been tested with a particular robot with a multimode locomotion system it is necessary to describe the multimode locomotion system and the robot used for testing.

4.1 The multimode locomotion system

The basic idea is to combine the best properties of both of the two main locomotion principles, wheeled and legged locomotion: high-speed locomotion of the wheel and good terrain negotiating capability of legged locomotion. This is achieved with a multimode locomotion system. The minimum requirement for this is that a wheel is attached with a suspension system that has active horizontal (longitudinal) and vertical degrees of freedom.

From here on the wheel suspension is referred to as a "leg" although it does not need to have a leg-like implementation.
The multimode system can then be used for new hybrid modes to help find an optimal solution for greatly varying ground conditions. In hybrid locomotion the wheel and the leg actuators generate the propulsive force simultaneously. It can also be called rolking (rolling-walking), as named in (Leppänen, 1998) or wheel-walking as named in (Kemurdjian, 1990). Rolking resembles walking and works like the following. When a wheeled leg is in the supporting state, the propulsive force is generated by the leg actuators. When the wheeled leg is in the transferring phase, it is not lifted in the air, but lightened and moved along the ground by touching it all the time, while applying a slight forward torque to the wheel. Rolking offers better propulsion in soft soil and good obstacle negotiation.
The locomotion system offers sensor information that is useful for measuring and calculating the quality of the terrain the robot is moving on, and the characteristics of the interaction between the robot and the terrain. This information includes the strain forces affecting each leg and wheel or foot and also the positions of the ankles, or wheel hubs, relative to the robot body.

4.2 The test robot - WorkPartner

One of the more advanced hybrid vehicles to date is the WorkPartner service robot. WorkPartner is a centaur-like service robot with four wheeled legs and a human-like upper body with two hands and a head (see Figure 3).

Fig. 3. The WorkPartner service robot seen here in winter on a test track, negotiating obstacles.

The size of WorkPartner is such that it is suited to co-operating with humans. It weighs about 270 kg. The actuation system is fully electrical. The "muscles" of the legs are identical electric linear actuators that consist of a 250W motor, a reduction gear and a ball screw. The

energy system is hybrid, which carries the energy in the form of fuel and transforms it into electrical power for robot actuation. For this, the system includes a combination of a small lightweight combustion engine with generator and batteries.
WorkPartner is built on a mobile platform that consists of four legs equipped with wheels and an active body joint (Halme et al., 1999) shown in Figure 4.

Fig. 4. WorkPartner in an early stage of development, where the structure of the legs and front and rear bodies can be seen.

The wheeled leg consists of a 3-dof mammal-type leg and an active wheel. The rounded profile rubber wheel is designed to have two functions: as a foot in the rolking or walking mode and as a wheel in driving mode.
The locomotion system of WorkPartner allows motion with legs only, with legs and wheels powered at the same time or with wheels only. This enables it to move over different types of terrain with minimum strain to the robot and terrain (Virekoski & Leppänen, 2007) and can sense terrain parameters and the characteristics of vehicle-terrain interaction successfully.

4.3 Sensor system of WorkPartner

The WorkPartner sensor system includes sensors for observing the internal state of the robot and perceiving the outside world. Only the sensors for controlling the locomotion of the robot are described here. A more detailed description of other sensors and sensing algorithms can be found in (Selkäinaho, 2002) and (Halme et al., 2003).
Leg joints angles are measured by potentiometers and encoders that are connected to the leg controller. Inclination of the body is measured using gravity-based inclinometers connected to the middle joint controller that also takes care of the energy system. This sensor information is enough for motion control in most situations, but when force control is needed, force measurement should also be available. Implementing force sensors in every actuator is complicated. Therefore an alternative indirect method of measuring the forces through the driving currents of the joint motors is used. The joint gear reductions are relatively small, which allows a moderate accuracy in force measurements to be achieved

this way. Kinematic calculations are then used to calculate the forces in the contact points that the robot has with the environment. This sensor system is then readily available for sensing terrain parameters and the characteristics of vehicle-terrain interaction.

4.4 Energy consumption of WorkPartner

Table I shows the measured energy consumption of WorkPartner in both wheeled and rolking locomotion modes in different terrain conditions. Each terrain type was driven through 3 times and the average is shown. The deviation was ~10% in the worst, softest conditions. The speed value was calculated as an average of the three least slipping wheels, i.e. the wheel that had its speed deviate most from the average of all four was not taken into account in speed (or odometry) calculation. It should also be noted here that the wheels do not slip as freely as in e.g. a car because they are under a speed controlled mode. During the tests for table I it was also visually confirmed that excess slippage did not occur and the robot advanced at all times. If the slippage should be excessive, and by multiple wheels, it would hamper the measurements of real vehicle speed and odometry, and then an external measuring device, e.g. a distance/speed wheel would be needed.

Locomotion mode	Speed [cm/s]	Energy consumption [J/m]	Terrain
wheeled mode	50	130 (wheels only)	Flat hard
wheeled mode	50	600 (wheels only)	Soft sand
wheeled mode	50	600 (wheels only)	Soft snow 15 cm, without ice layer
wheeled mode	50	900 (wheels only)	Soft snow 22 cm, without ice layer
rolking, crawling gait	2	2800	Flat hard
rolking, crawling gait	2	4500	Soft snow, 25 cm
rolking, trot gait	4	3700	Flat hard
rolking, worm gait	4	4000	Flat hard

Table 1. Energy consumption of WorkPartner. Losses in the gears and the linear actuators are included in energy consumption values. See text for more details about speed and energy measurements.

The energy consumption of rolking as well as walking will increase in softer soil. Slip of the support legs doesn't only affect the speed of the robot, but also the energy consumption, as shown in Figure 5. Note that energy consumption of rolking is generally higher than wheel motion, but in some terrain conditions, like deep snow, it is the only way to move.
In this experiment, the robot climbed up a slippery 18 ° slope successfully. Slipping of the support legs causes more internal forces in the robot (Lehtinen, 1994), which results in increasing energy consumption. When the robot succeeded to move from the slippery slope to more frictional and flat terrain, the energy consumption decreased. Energy consumption is a good indication of the softness of the soil.

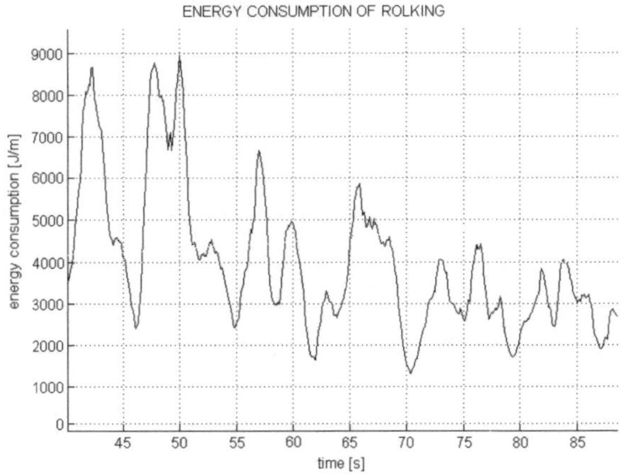

Fig. 5. During this test run the robot climbs up, rolking, a slippery 18 ° slope that causes slip of the support legs. Energy consumption decreases when the robot arrives on flat and level ground.

4.5 Energy consumption and slip of a single wheel in wheel-soil interaction

The energy consumption of a wheel per traveled distance increases if the wheel drives on soft soil or slips a lot under low-frictional condition. To illustrate energy consumption and slip, WorkPartner climbed a slippery 18 ° slope; test results of this climb can be seen in Figure 6. The rear left wheel started to slip and the speed of the robot decreased. In general, all slipping wastes energy.

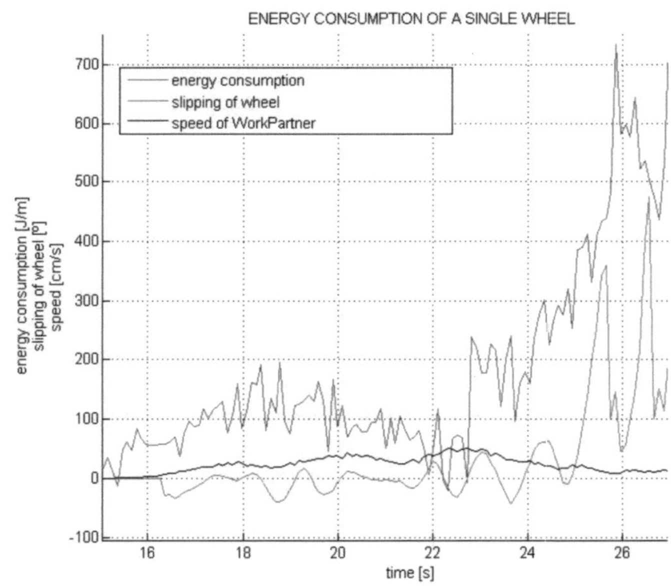

Fig. 6. WorkPartner tried to climb a slippery 18 ° slope with wheels. The rear left wheel started to slip, which also increased the energy consumption of the wheel per traveled distance.

4.6 Motion resistance of the transfer leg's wheel in rolking mode

Terrain trafficability can be estimated with motion resistance of a transfer leg's wheel. For example, when the depth of snow decreases, so does the motion resistance of the transfer leg's wheel also, as can be seen in Figure 7. At the end of this test, WorkPartner moved on firm ground during 214 - 235 seconds, when the average motion resistance of the transfer leg's wheel was about 80 N. The load of the transfer leg's wheel stays almost constant, because the load of WorkPartner is divided so that the support legs carry about 85% of the robot's weight and the transfer legs only about 15 % (Virekoski & Leppänen, 2007).

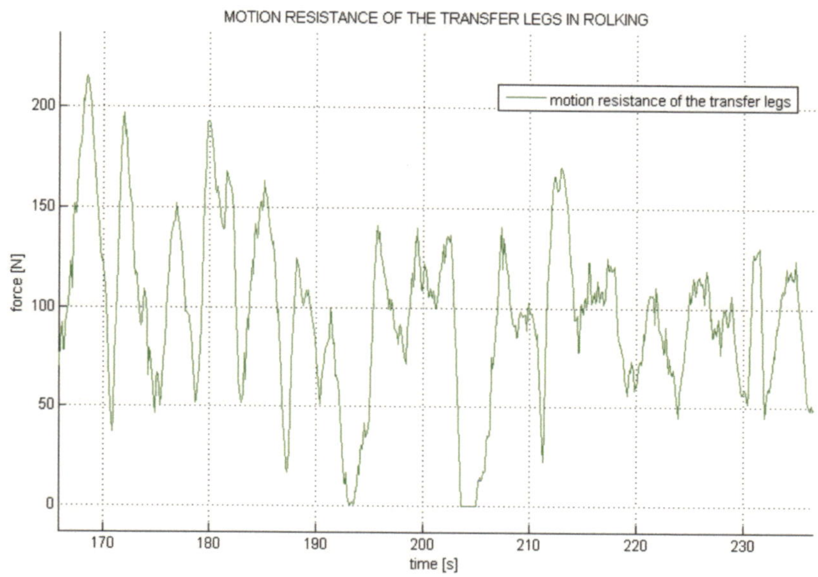

Fig. 7. WorkPartner rolked from soft snow to firm ground. Motion resistance of the transfer legs decreased.

4.7 Sensing terrain slope and roughness

To demonstrate the terrain slope and roughness sensing principles, , WorkPartner climbed over a 30 cm high and 140 cm long ramp with the left side wheels; the results are shown at the bottom of Figure 8, while the ramp is shown at the top. The ramp is so short that the left front wheel climbed first up then down and the robot was level for a short time before the rear wheel did the same. The wheels were in the nominal positions so that the wheel base was 1,45 m and sideways distance between the wheels was 1,0 m. If a wheel of WorkPartner is on an obstacle with a height of the radius if the wheel (0,23 m) and the other wheels are in the nominal positions, the roughness is about 22 °. If the opposite corner wheels are on the same size obstacle (0,23 m), then the roughness will double (44 °). In wheeled mode, there is a limitation using this method for sensing terrain roughness: if the front or back wheels, or even all the wheels, encounter a hump (or ditch) at the same time, the value of this roughness indicator remains small although the terrain will be difficult for the wheel to negotiate. In rolking, this kind of limitation only exists when using a worm-type gait in which each pair of wheels in the frontal plane moves only when the other one is firmly braked to the ground.

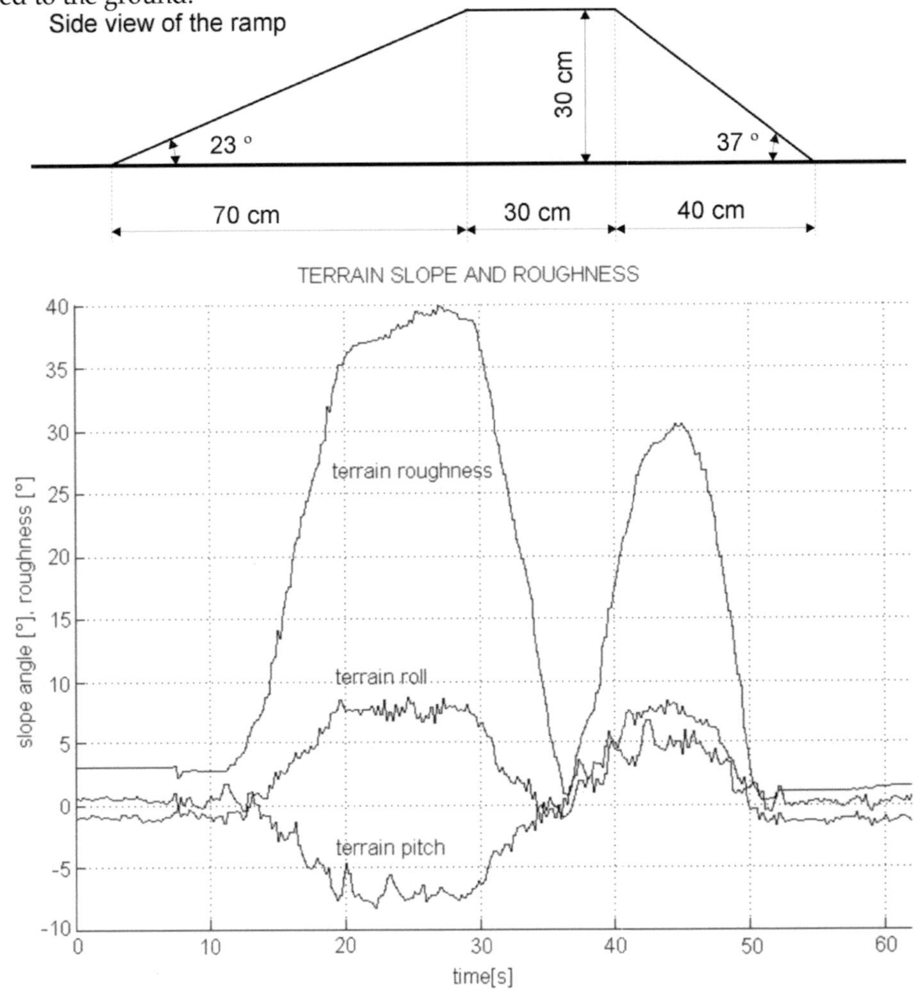

Fig. 8. Terrain slope and roughness. WorkPartner drives over a 30 cm high and 140 cm long ramp with the left wheels. This is a test to demonstrate the terrain geometry parameters only, and the consistency of the driving speed or other parameters is not essential here. The first peak occurs when the front wheel is on top of the ramp (x = 20 - 30 s) and the second with the rear wheel (x = 40 - 50 s). The rear wheel peak is lower because the robot's middle joint is asymmetric and flexes more to the other diagonal.

The roughness sensing method is also valid in rolking mode. A test with WorkPartner rolking over logs, as illustrated in Figure 3, shows that the calculated roughness clearly indicates when the individual front legs go over the log one by one. The result is shown in Figure 9.

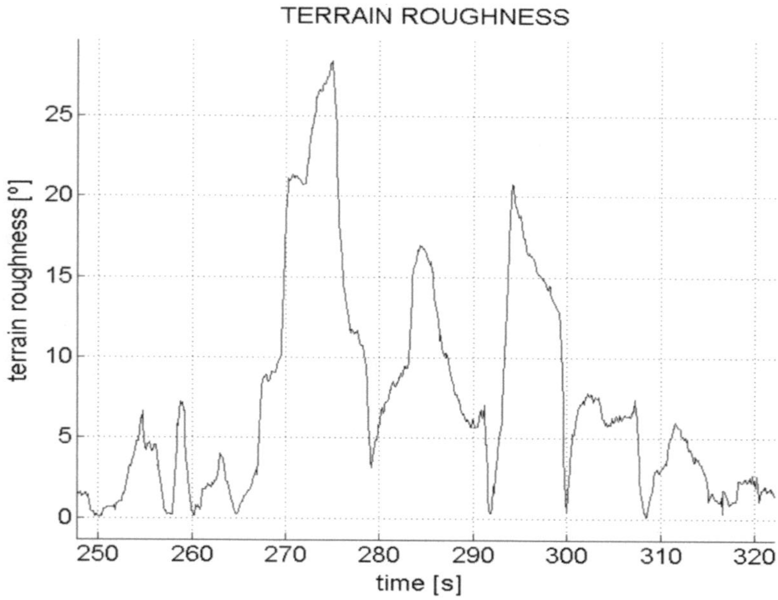

Fig. 9. WorkPartner is rolking over a wooden log, front wheels crossing the obstacle. Terrain roughness is significantly greater when a wheel is on a log. The log is not perpendicular to the driving direction and therefore here the right front wheel climbs on the log first (peak at x = 275) and comes down only at x = 290. The dip at x = ~279 is caused by the left wheel trying to climb the log but slipping back (x = 284). The left wheel comes on top of the log at x = 294.

5. Conclusion

The developed real time touch-based sensing methods for determining terrain geometry parameters such as slope and roughness, and also the characteristics of vehicle-terrain interaction through energy consumption, motion resistance and slippage are based on using the locomotion system as a sensing system. The developed terrain sensing methods can be used for mapping, path planning, terrain trafficability evaluation, obstacle detection and avoidance. These sensing methods have already been applied successfully for determining suitable modes of locomotion automatically (Leppänen, 2007). The automatic locomotion mode control uses a mode that is optimal for current terrain.

Future development should focus on combining these touch-based methods with vision and ranging. Then the robot could learn to associate image information of the terrain to the touch-based information, and then use vision to get more accurate knowledge of the terrain from afar.

6. Reference

Bekker, M. G. (1969). Introduction to terrain-vehicle systems. The University of Michigan Press. U.S.A.

Gillespie, D. G. (1992). Fundamentals of Vehicle Dynamics. Society of Automotive Engineers, Inc. U.S.A.

Halme, A.; Leppänen, I. & Salmi, S. (1999). Development of WorkPartner-robot - design of actuating and motion control system. 2nd International Conference on Climbing and Walking Robots, London, Professional Engineering Publishing Ltd, pp. 657-665.

Halme, A.; Leppänen, I.; Suomela, J.; Ylönen, S. & Kettunen, I. (2003). WorkPartner: Interactive Human-like Service Robot for Outdoor Applications. The International Journal of Robotics Research, Vol. 22, No. 7-8, pp. 627-640.

Iagnemma, K.; Kang, S.; Shibly, H. & Dubosky, S. (2004). Online terrain parameter estimation for wheeled mobile robots with application to planetary rovers. IEEE Transactions on robotics, Vol.20, No.5, pp. 921-927.

Karafiath, L. & Nowatzki, E. A. (1978). Soil mechanics for off-road vehicle engineering. Trans Tech Publications, Series on rock and soil mechanics Vol.2, No. 5.

Kemurdjian, A. L. (1990). From the Moon Rover to the Mars Rover. The Planetary Report, Vol. X, Num. 4, pp. 4-11.

Lehtinen, H. (1994). Force based motion control of a walking machine. Ph.D Dissertation, Helsinki University of Technology, 1994.

Leppänen, I.; Salmi, S. & Halme, A. (1998). WorkPartner, HUT Automation's new hybrid walking machine. CLAWAR'98 First international symposium, Brussels, Belgium, 26-28 Nov.

Leppänen, I. (2007). Automatic locomotion mode control of wheel-legged robots. PhD dissertation, 105 p, Helsinki University of Technology.

Ojeda, L.; Borenstein, J. & Witus, G. (2005). Terrain trafficability characterization with a mobile robot. Proceedings of the SPIE Defense and Security Conference, Unmanned Ground Vehicle Technology, pp. 1-9.

Selkäinaho, J. (2002). Adaptive autonomous navigation of mobile robots in unknown environments. PhD dissertation, 88 p, Helsinki University of Technology, 2002.

Virekoski, P. & Leppänen, I. (2007). Terrain-adaptive Locomotion of a Wheel-Legged Service Robot Using Actuator-based Force Measurements. CLAWAR'2007 10th International Conference on Climbing and Walking Robots and the Supporting Technologies for Mobile Machines, Singapore, 16-18 July.

Wong, J. Y. (1994). Terramechanics – Its present and future. Proceedings of 6th European Conference of the International Society of Terrain-Vehicle Systems, Vienna, Austria, pp.1-21.

Wong, J. Y. (2001). Theory of ground vehicles - 3rd ed., 528 p. New York, Wiley.

23

The Memory Game: Creating reactive, interruptible, turn-taking, human-robot interaction for ASIMO

Victor Ng-Thow-Hing , Jongwoo Lim, Joel Wormer,
Ravi Kiran Sarvadevabhatla, Carlos Rocha,
Kikuo Fujimura and Yoshiaki Sakagami
Honda Research Institute USA, Inc.
United States of America

1. Introduction

Recently, humanoid robots have begun to demonstrate an array of impressive motor skills, such as climbing stairs (Honda Motor Co., Ltd., 2000), whole body manipulations of large objects (Stilman et al., 2007), and dancing (Nakaoka et al., 2004). On the perceptual side, continued effort has been invested in object recognition and geometry reconstruction to provide robots a model of their environment. The goal for many autonomous robots is to be able to use many of these component motor and perceptual skills to accomplish complex tasks involving interaction with both objects and humans in the environment. This is especially important for humanoid robots because of their anthropomorphic body design and capability to perform many different tasks.

In addition to the motor and perceptual problems, the robot needs to be able to perform decision-making tasks that allow it to map pattern recognition modules on the perception side to the behavior-generating components on the motor side. This involves the need to recognize situational context and requires a communication subsystem for interaction with humans. For humanoid robots, speech and gesture are the main modalities available.

The ability to perform complex tasks requires the support of an established software architecture to ease integration and inter-communication of these disparate components. The successful execution of these tasks requires a system-level approach rather than focusing on individual components. Decision-making modules must make use of information or events from pattern recognition perceptual modules and be able to select appropriate behavior for the robot. Furthermore, even though components may operate error-free when tested in isolation, they can suffer unforeseen failures when operating concurrently with other modules. For example, any head-related motion may jeopardize the robustness of computer vision algorithms that assume a static input image.

For researchers, the integration process should be simple and require minimal modification to existing software. The architecture should also encourage components to be reusable, modular and exchangeable to maximize flexibility and extensibility of the robot's

capabilities. On the performance side, the architecture should be fault-tolerant so that the failure of any component module will not require restarting the entire system. Recently, we have developed a blackboard-based architecture known as the Cognitive Map (Ng-Thow-Hing et al., 2009) to meet these needs.
In this paper, we present our attempt to utilize our system architecture and component modules to create a well-defined human-robot interactive scenario that demonstrates situational context recognition, perception and reconstruction of the environment and multi-modal communication via speech and gestures, including deictic gestures that make use of spatial information obtained from the perceived environment. This scenario is realized in the form of a memory card game between Honda's humanoid robot ASIMO (Honda Motor Co., Ltd., 2000) and a human opponent.

1.1. The Memory Game Problem

The *Memory Game* features a deck of matching pairs of cards that are shuffled and placed face down on a table. Players take alternate turns picking two cards in an attempt to find and collect matching pairs. If a player successfully finds a match, he or she can take another turn. If they do not find a match, the other player starts his or her turn. When all the cards are gone, whoever has the most pairs of matches wins. A tie can occur if both players have an equal number of card pairs. In our scenario, we feature ASIMO playing against a human player, where the cards can be arranged anywhere on a table.
The choice of modeling the memory game as an interactive scenario was made due to its relatively simple rules, strategies and goals. The game requires a robust perceptual component to detect and recognize the identity of cards. ASIMO must communicate with the player using speech or gestures, including asking the player to flip cards on its behalf since ASIMO's manipulators are not dexterous enough to turn over flat cards on a table. ASIMO's perception is restricted to the two cameras located in its head. This not only makes the system more portable to different environments, but it motivates embodied behaviors such as ASIMO having to look down at the table to see the cards. The one exception to the self-contained ASIMO constraint is our use of an external projector to highlight cards that ASIMO would like to flip over (Section 4.5).

1.2. Reusability of architecture and modules

Having the concrete goal of the memory game enabled us to focus on the desired features needed for our system architecture. Modules were written in different computer languages and on different operating systems and needed to communicate with each other. The nature of interactions dictated the type of messages that modules must pass to each other. It was important to make sure the architecture and modules did not become too specialized to the memory game as we wished to reuse the system for other types of interactive applications in the future. This criteria influenced the choice of design and selection of modules to use. For example, our card detector was deliberately designed to be unaware of any game-specific rules.

1.3. Interaction characteristics

Our main research goal is to enable rich human-robot interactive experiences. The memory game is our first attempt at creating multi-modal, turn-taking based communication with an

application that requires state information to be stored and used to direct the content and style of communication. Another goal is to create robust experiences that allow people to interact with the robot in the least-constrained manner. For example in the memory game, perceptual robustness is required as cards change orientation or position on the table and undergo different illumination changes. Table detection should be robust to small changes in ASIMO's head orientation while it executes body gestures. In the memory game itself, ASIMO needed to focus on activities of interest, while ignoring unimportant events. Players were free to communicate with ASIMO using natural means such as speech and gestures.
We did not want to create a scripted experience where ASIMO only reacted as a result of speech or activities from the user. ASIMO was able to proactively perform behaviors while waiting for the player to take his or her turn. This included asking players for help if they were taking too long to pick a card. It also included expressing the same semantic message different ways to avoid monotony. For example, if ASIMO was winning the game, it could choose to say things in a more gentle manner so as not to upset the player. To solve these problems, we provided timeout mechanisms for ASIMO to proactively initiate behaviors, idle motions such as nodding acknowledgment whenever the player spoke to ASIMO, and a multi-modal communication module to generate new phrases and gestures based on requests for semantic message communication from other modules.

1.4. Previous Work

Robots interacting with humans have come in a variety of forms, ranging from baby seals like PARO (Wada and Shibata, 2007), small personal robots like PaPeRo (Osada et al., 2006), and human-like androids (Ishiguro, 2006). The form of the robot is a key factor in the design of its intended use and in people's expectations of its behaviour and capabilities. Humanoid robots by virtue of their body design being similar to our own, come with the increased expectations of human-like behaviour. In cases when the robot's behaviour did not coincide with the expectations of its visual appearance, a phenomenon of uneasiness called the *uncanny valley* (Ishiguro, 2006) has been observed in people's impressions of the robot. Consequently, humanoid robots such as the WE-4RII (Itoh et al., 2004) exhibit unique functionalities and behavior like facial expressions and arm gestures that mainly serve a communicative role rather than a utilitarian one. Combined with good perception, human-robot interaction can be seen as a novel way for people to accomplish a task with a machine using the natural communication modalities of speech, gesture and other recognized visual cues. In (Sakagami et al., 2002), a system architecture for ASIMO is introduced to connect a behavior planning system to the vision and auditory systems of the robot, enabling key capabilities important for interaction such as gesture recognition, sound localization and obstacle avoidance during navigation.
Communication between humans and embodied agents has been explored in virtual reality and in human-robot interaction. Virtual reality systems featuring synthetic actors have the advantage of more expressive computer graphics and perfect knowledge of the environment (Zhang et al., 2008). In contrast, human-robot interaction requires the additional problem of developing robust sensing and control algorithms. The Kismet system (Breazeal and Scassellati, 1999) specialized in facial expressions and featured a perceptual attention model to direct head gaze for short-term interactions.
There have been several robotic systems designed for long-term interaction with humans. The HERMES system (Bischoff and Graefe, 2002) was operated over a period of six months

in a public museum. It performed tasks such as pouring water into a glass, explaining museum exhibits and engaging in dialogs with people. Valerie the Roboceptionist (Gockley et al., 2005) is a stationary robot designed to interact with visitors over an extended period of time through social dialog. Although the robots are active for long periods of time in these systems, the duration of task interactions were relatively short. For example, the robot would respond to a query or command from the user and perform the task, ending the interaction and subsequently waiting for the next interactive event. These tasks typically require a small number of states to characterize them. In contrast, the memory game requires the modeling of many states or contexts in order for the robot to determine appropriate behavior, including exceptions where a player may ask for help or ASIMO proactively engages the player.

In the remainder of this paper, we will give an overview of our robot architecture in Section 2, followed by our design process from conceptualization of the interactive scenario to system realization in Section 3. Individual component modules are described in Section 4. Several case studies highlighting how information flows through our architecture are covered in Section 5. Discussions of several important lessons and limitations are in Section 6 with the final conclusion in Section 7.

2. System Overview

Details of our system architecture are described in (Ng-Thow-Hing et al., 2009). However, several details are reviewed here for completeness. The implementation of the memory game can be considered to be a multi-agent system, where the component modules act as independent agents that can process information concurrently with other modules. The agents can communicate with each other using messages or data streams through a shared information space based on a blackboard architecture (Hayes-Roth, 1985). Component modules can be distributed on different machines and operating systems or co-located on the same computer, subject to computational load constraints. Since communication is restricted to specific message types, different implementations of a module can be substituted as long as the same message types are posted and subscribed to. Modules can be stopped and restarted (in case of failure) without requiring other modules to restart. Fig. 1 depicts the main components of our memory game system.

Messages are typically used to denote events in the system and are accompanied by information about the properties of the related object (if applicable). For example, the card recognizer reports messages, *card-flip, card-appear, card-remove, card-touch,* to notify modules of the respective activities involving cards on the table. Each event is accompanied by a data object (called a CMobject) that describes its properties. CMobjects can also represent commands not involving any physical objects such as *Say-text* and *Do-Task* messages for text-to-speech and task commands respectively.

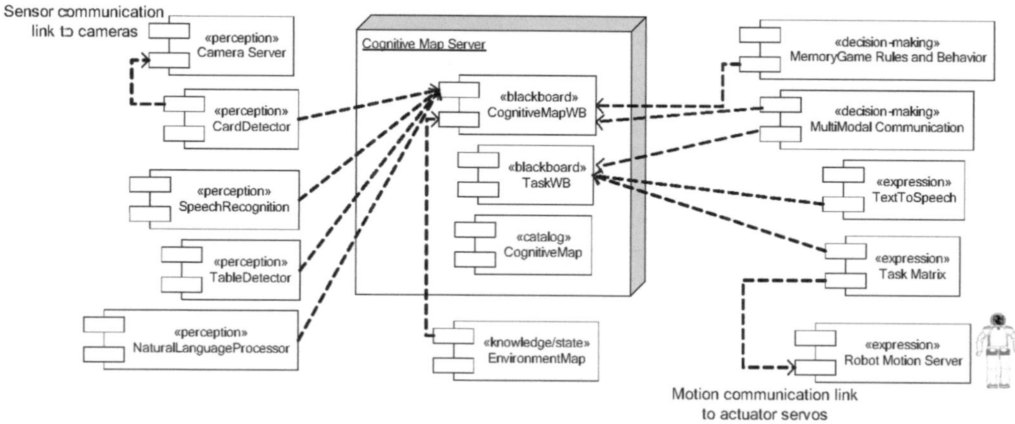

Fig. 1. Overview of the Cognitive Map robot architecture: Representative components depicted showing connectivity to the central Cognitive Map server. Component module types are labeled above name. Arrow directions depict dependencies on various interfaces (not message directions).

In general, modules should have a clearly defined role in the system. For the memory game, modules usually fall into one of three categories: pattern recognition, decision-making, and behavior-based. Pattern recognition modules typically consist of modules which take sensor input and produce perceptual features in the scene, such as card location and identity. Pattern recognition modules can also work with temporal streams of information to detect activities such as card removals or card flips. Decision-making modules accept events posted from the pattern recognition modules and consider internal state of the robot and game to generate appropriate behavior. Behavior-based or expression modules can produce perceivable robot activity such as speech utterances, gestures, or other complex motor tasks. Fig. 1 illustrates the different categories of the modules used in the memory game. Information exchange can happen in both directions between any modules in the system. In Section 5, we describe the communication flow for several situations that can occur during the game. Section 4 describes each module used in the system in more detail.

3. Design

Our design process begins with the construction of a branching storyboard (Fig. 2) that depicts the desired observed behavior of both the robot and humans involved in the scenario. Every attempt is made to predict different conditional behaviors and responses to those situations. For example, the storyboard shows how the robot would behave if it won, what happens if it failed to see a detected card or how to respond if the human player asks for help. This process allows us to identify what technical component modules are required and what information they should publish or subscribe to for accomplishing their role within the system. Relating to the Cognitive Map, the storyboard aids both in the design of individual modules and their message interfaces between each other. For example, if dialog is present in the storyboard, this implies the need for both speech recognition and speech synthesis components. The storyboard in Fig. 2 further suggests that a simple input text

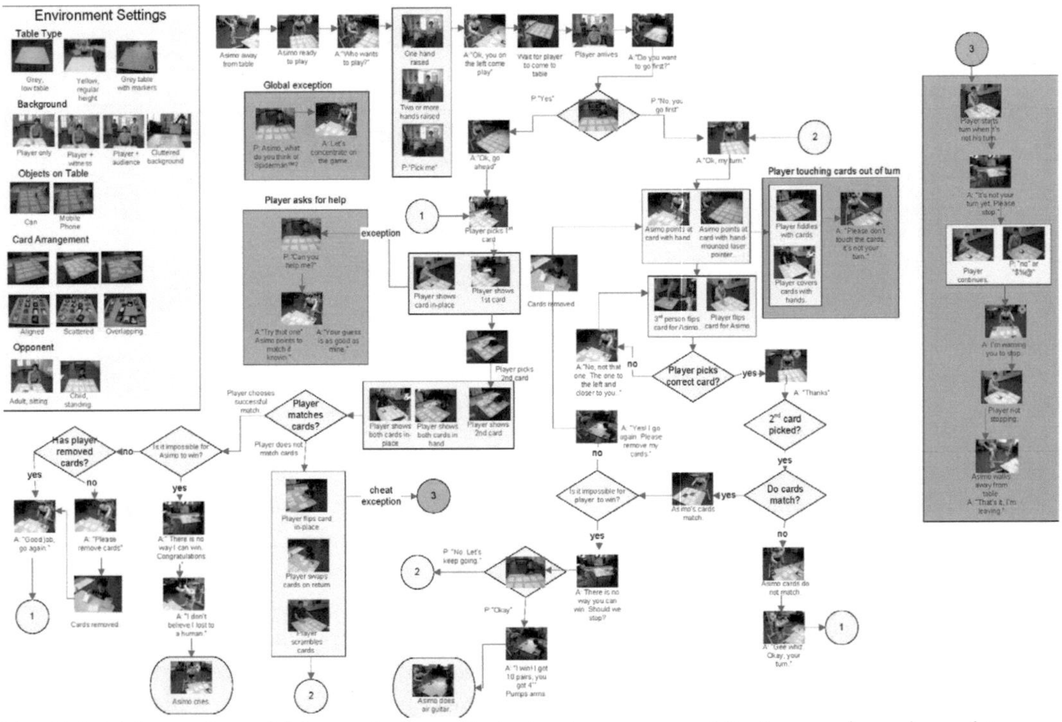

Fig. 2. Initial storyboard (shown in its entirety to give a sense of its thoroughness) used to design the memory game. The circles connect different interactive scenarios.

message interface to the speech synthesis component would suffice for the dialog requirements of the scenario, preventing the implementation of unnecessary features or an overly complex interface to the component. The types of gestures that we wish the robot to perform also has important implications for motor control and sensing. Pointing gestures require both awareness of object position in the environment and an end-effector based control scheme for positioning and orienting the robot's hand. In the design of the game behavior module described in Section 4.11, the storyboard identifies all individual states that can occur during the scenario as well as the events that can trigger transitions between these states. This suggests what would be the most appropriate decision-making scheme to adopt within the interaction component. In the memory game, a finite state machine could be used to keep track of the current contextual game state and expected transition events that are likely to occur from each state. The transition events between states in the storyboard identify what types of messages need to be generated by the components of our Cognitive Map architecture. In Fig. 2, a transition occurs when a player picks a card, implying that a perception module for detecting cards would need to detect physical card motion or flipping and publish these message events to the Cognitive Map. If the storyboard described a simpler application which does not need to keep track of history, a basic action-selection table could be implemented as the interaction component instead.

4. Modules

An interactive scenario like the memory game requires a variety of different component functionalities that when assembled together enables complex behavior of the robot with its environment.

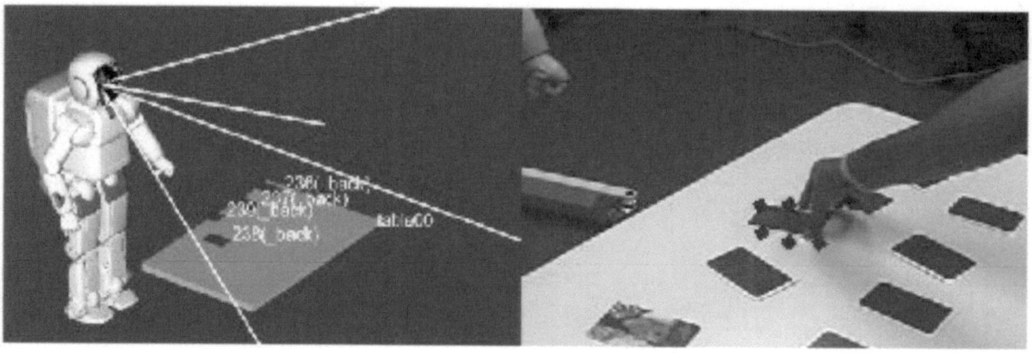

Fig. 3. Left: Environment map, Right: Projector highlighting of card (enhanced for easier visibility in image)

4.1. Environment Map

The Environment Map collects object pose information and robot configuration state to perform the necessary coordinate transformations to bring both the robot and its environment in the same frame of reference. A visual display allows monitoring of ASIMO's current estimate of its environment. The positional information is used for finding the physical locations of objects that are required to complete deictic gestures such as pointing at cards.

Incoming coordinate information can be either 2-D (such as table-top or stage coordinates) or 3-D (pose of the table). For example, the table's pose is reported in camera coordinates which is in turn transformed into the world reference system because we know the position of the camera on the robot. Cards are reported as 2-D coordinates relative to the table-top and are transformed into world coordinates since we know the table's position and orientation in the world. The left side of Fig. 3 illustrates the robot and the detected table and cards in the environment map.

4.2. Table Detection

Since the memory game is played on a table, robustly detecting and tracking the game table is very important for uninterrupted play. We use a single camera to detect and track a table. Due to large radial distortion of the wide field-of-view lens in our camera, it is quite a challenging task to extract accurate information about the table, cards and the player's actions. As shown in the top row of Fig. 4, we use color and edge information to build a response map (Fig. 4b) from a raw image frame. Once the biggest blob is detected, its border pixels are mapped to normalized image coordinates after compensating for the radial distortion of the lens (Forsyth and Ponce, 2002) (Fig. 4c). The convex hull of the border

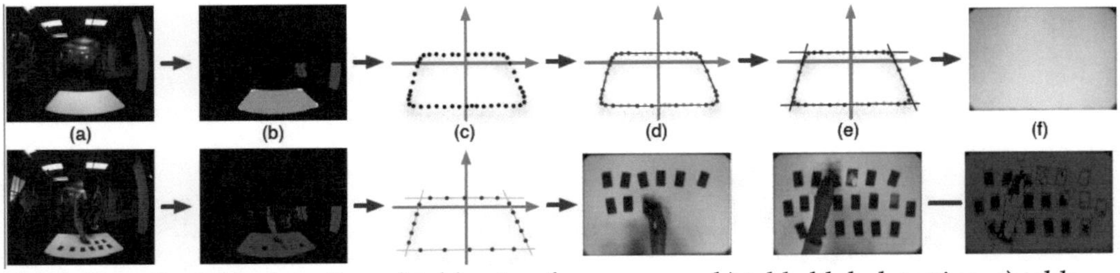

Fig. 4. Steps for table detection: a) table view from camera, b) table blob detection, c) table border points, d) table convex hull, e) table edges, f) table homography transform.
Bottom row: Same steps showing views when person and cards are present.

points are then computed (Fig. 4d) to generate table edge candidates. For the n convex hull points, we consider the $n{\cdot}k$ lines that pass through one point and its k neighbors, compute the support for each line by the table border points, and pick local-maximum lines. If four lines are detected correctly (Fig. 4f), then we compute the homography $\boldsymbol{H}$ (Hartley and Zisserman, 2004) which maps $[-w/2{:}w/2]\times[-l/2{:}l/2]$ to the area bordered by the four lines where w and l correspond to the width and length of the table. By applying this homography $\boldsymbol{H}$, we can create a virtual top-view image of the table (Fig. 4f).

Once the table is detected, we can efficiently test whether the table is in the same detected position in subsequent frames. We spread a few test points on each edge of the table, and test whether the points are lying on the boundary between the table blob and the background. If sufficiently many points are on the table edge, we use the previous homography. Otherwise, we re-detect the table. This method gives us robustness over occlusions occurring when the player's hand touches or flips cards on the table.

If we know the geometry of the table, we can reconstruct its 3-D orientation and position from the estimated homography. When the head moves and if the motion trajectory is known, it can be combined with the previously estimated homography to generate an estimated current homography without detecting the table from the image. This option is useful when some parts of the table are out of the field of view.

4.3. Card Recognition

The card recognition module was one of the most important modules for the scenario because of its responsibility for detecting the key elements of the game. Consequently, in the course of its development, we went through several iterations. The initial version was adapted from general object recognition software. The recognition was reasonably robust, especially when applied to the homography table-top view that eliminated perspective effects. However, the detection of events such as card flipping and removals was noisy and required the use of lengthy timeouts to ensure stable detection of events which decreased the responsiveness of the game.

The second version of the card recognizer used blob segmentation and direct template matching to segment and identify cards. Training images were stored as image patches. For each frame of incoming video, holes in the image blob corresponding to the table were detected as potential cards from the homography-corrected mask obtained from the table detector (Section 4.2). If the area in pixels of a hole was too large compared to the training card patches, it was split into multiple pieces using K-means clustering. For each card blob, the center and 2-D orientation were estimated by taking the mean of the pixel coordinates and the eigenvectors of the covariance matrix respectively. Using this center and rotation, the image patch from the homography-corrected image was cropped and compared to the training patches.

When the person interacted with cards, there were significant occlusions by the player. To handle this occlusion, only the holes completely surrounded by the table blob were considered as valid card blobs. The other holes were considered as occluded regions. The cards in the occluded regions were not updated until the occluding blob disappeared. If a card was removed, the region it was in before would be in the table blob, so we could detect the card had been removed during the occlusion. If a card was not touched, it would remain in the same position after the occlusion. This detector worked when there were approximately 10-20 cards with mild overlapping between cards. When there were more cards, the holes in the mask tended to become connected to each other, and this made segmentation of connected blobs harder. Another problem in the mask-based method is that the holes near the farther edge of the table tended to be larger than the nearer ones. This made the location estimate on the far cards inaccurate and sometimes resulted in incorrect detection of cards.

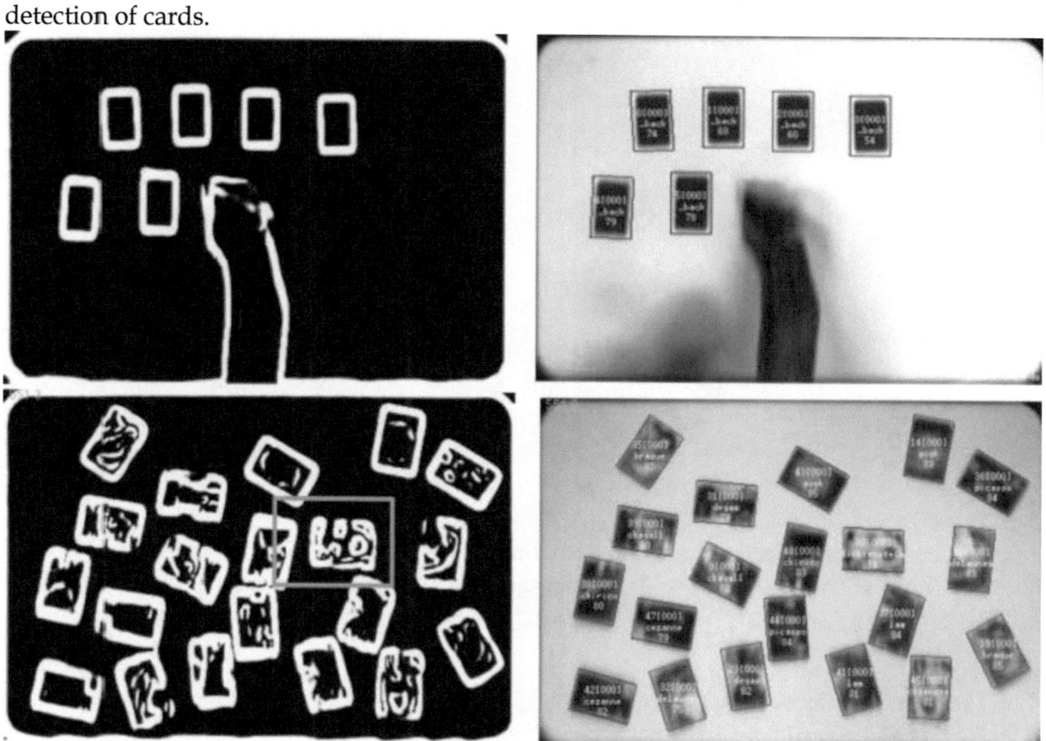

Fig. 5. In our latest card detector, the edge response maps on the left are used to detect and recognize cards (front and back) on the right. The outlined card is an example of a 'gray blob' that needs subsequent processing to determine if it is a card.

To resolve these issues, our latest card detector (Fig. 5) uses edge information extracted only from the homography-corrected image. After non-maximal suppression of sobel edge detection, we build the edge response map, whose pixel values from 0 to 1 represent the likelihood of edge existence on the pixel. Initially edge blobs higher than a certain threshold are detected as the candidates for cards. If an edge blob fits in size and shape, it is classified as a card. After this initial classification, the remaining blobs are merged and tested whether they are from one card or not. To find the proximity between edge blobs, the concept of a "gray blob" is used. Instead of the binary representation in usual blob detection methods, we find the weakest link between two blobs, and that value represents the proximity between blobs. From the strongest pairs of blobs we test if the assembled blob looks like a card, and repeat until there is no likely combination left. After detecting cards, the identification of each card is determined by comparing pixel values of the region with the trained templates.

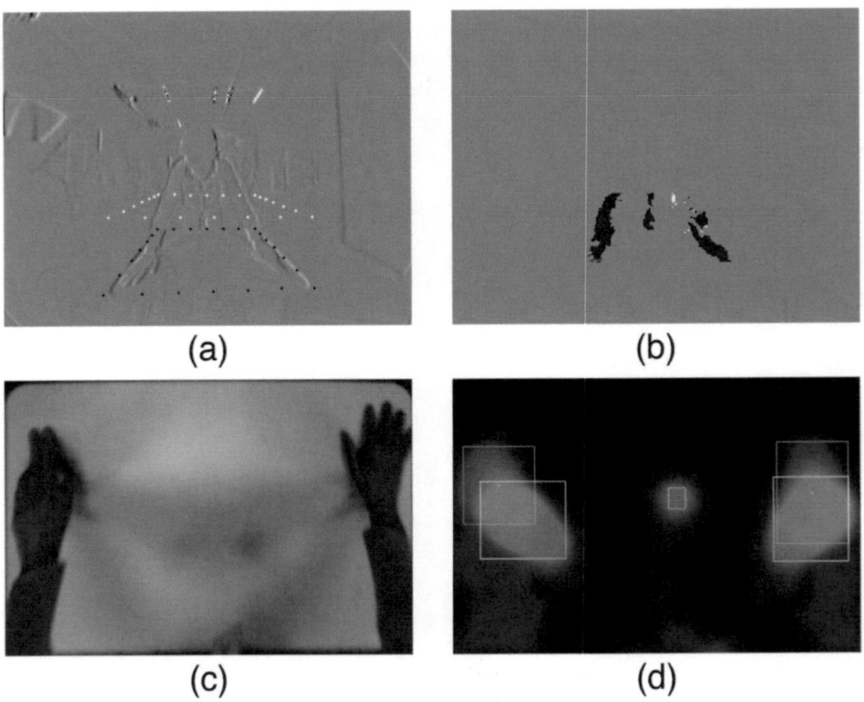

Fig. 6. Detecting hand-table touch and proximity events: (a) volume of interest (VOI) bounded by black and white dots, (b) cropped depth map in VOI, (c) hands touching table and (d) corresponding touch regions and proximity regions (lighter regions closer to table).

4.4. Table Touch Detection

ASIMO has a stereo camera pair that we use for detecting the presence of the player's hands over the table. Generally, stereo works poorly in the table region or other textureless regions. However, we have very good estimates of 3-D position and orientation of the table from the homography from Section 4.2, so we can default to the table plane when encountering ambiguous stereo matching pairs. We set a volume-of-interest (VOI) on the detected table (marked as black and white dots in Fig. 6a), and only consider 3-D blobs in the volume. The resulting depth map we get from the stereo contains only the body parts or other objects on the table. With this information we can detect whether the player is touching a card or his hand is hovering over the table by measuring the distance from the table plane to the 3-D blobs in the VOI. Fig. 6d shows the touch regions corresponding to Figure 6c.

4.5. Projector

When ASIMO points to a region of the table containing many cards, it may be difficult to tell which card it is pointing at. We contemplated having ASIMO hold a laser pointer, but the limited degrees of freedom in its arm did not provide enough precision control. Consequently to reduce ambiguity, we installed a ceiling-mounted video projector to highlight cards. The projector's display is controlled by a module that interacts with the Cognitive Map. All modules are able to interact with the projector by sending drawing commands, which will be displayed on top of the table. During the game, the projector is used to highlight selected cards with an animated pattern, complementing ASIMO's pointing gesture and verbal indications (see Fig. 3, right).

The projector system requires an initial calibration step to manually adjust the corners of a projected grid to match the corners of the table. This ensures that the projection viewport is aligned and scaled to the projection surface. This method allows the projector system to be used on a variety of rectangular tables.

4.6. Speech Recognition/Natural Language Processing

Recognizing speech utterances is important for realistic and satisfying human-robot interaction. The memory game consists of several scenarios where a player can be expected to speak. For instance, the player could seek help determining the location of a particular card. In order to recognize the words and phrases used, game-specific grammars can be defined and employed. Robustness of recognition can be achieved by re-defining the grammar to contain only relevant phrases that are expected in different contexts of the game. In order to assign semantics to the spoken utterances, messages containing the utterances from the speech-recognition module are sent and processed by the Natural Language Processing(NLP) module. This module maps all the variations of an utterance to a unique set of semantic directives. These directives are passed to the game behavior modules for processing and high-level decision making. Further examples of how this process works are described in Section 5.

4.7. Text-to-Speech

The first version of our Text-to-Speech module uses a customized commercial engine to produce ASIMO's boyish English-speaking voice. The speech engine is capable of changing intonation when posing questions and offers some speed control as well as the ability to interrupt speech in progress (barge-in capability). This module services *say* messages sent from other modules like Multi-modal Communication (Section 4.8) to convert text to an audible voice.

We recently updated our speech engine to a newer version that allows more control over intonation of any syllable in a sentence and has more emotionally expressive phrases such as "Hey!" and nonverbal expressions such as laughter. We have found that players enjoyed the more expressive speech engine, even if the functional robustness of the application suffered. We suspect that greater expression provided a stronger illusion of ASIMO having a real personality. We intend to confirm and exploit this observation in future work.

4.8. Multi-modal Communication (MC)

Human beings tend to use different modalities such as speech and body gestures in their interactions with each other. To this end, we developed a multi-modal communication (MC) module capable of communicating using body gestures and speech simultaneously to improve ASIMO's expressive ability. This is especially important as ASIMO lacks an expressive face. Since the content and style of a message are specified separately, we can avoid repetitive, identical behaviors when ASIMO wishes to communicate information with the same semantic content.

In order to allow the MC module to be reuseable in different application scenarios, care was made to avoid building in application-specific knowledge within the module. The module must act on information solely from directives it receives from other modules, as well as any shared information accessible in the Cognitive Map. The MC module takes as input high-level communication directives from other modules such as *Indicate(card10)* or *Offer(help)* and converts the directives into physical gestures and speech. Directives can be accompanied by style tags so an application module can give hints to the MC module on how to deliver the contents of the communication directive. For example, ASIMO may choose to use polite phrasing, or curt sentences to convey friendliness or impatience. For cases where an application wants complete control over how communication is delivered, there is a special *Verbatim* directive which allows a module to specify the exact words and gestures to use.

4.9. Task Matrix

The Task Matrix module, described in detail in (Drumwright and Ng-Thow-Hing, 2006) collects different motor programs that can execute motor tasks ranging from pointing to an object, playing pre-generated motion trajectories or navigation with motion planning around obstacles. Tasks are run in their own control loops and can react to changes in the environment. The main purpose of the Task Matrix is to allow other modules to launch motor tasks using a high-level command interface. For example, the MC module can send the command *Point(target=Card12)* to the Task Matrix. The corresponding pointing motor program automatically resolves the 3-D position of *Card12* and selects which arm to use based on proximity of the hand to the card. Other tasks include playing pre-generated motions like a victory celebration sequence, parametrized by duration, *Victory(duration=5 secs)* and shown in the bottom right image of Fig. 9.

4.10. Timer

The timer module produces a message echoing service that allows modules to schedule coarse-grained (minimum resolution 1 second) timeout events that can be used to trigger proactive behaviors. A module can specify when they want the message to be sent back to them (timeout interval) as well as the message type and contents. The timer module manages all timeout requests and at the appropriate time posts the messages to the Cognitive Map for the modules to receive. Since the timer module is implemented within the Cognitive Map framework, the timeout messages can be received by other modules different from the originator of the timeout request.

4.11. Game Behavior

The Game Behavior module is responsible for maintaining the game state during the entire scenario. It can be thought of as a more specific instance of an *interaction module* that encapsulates the application-specific details of an interactive scenario. It also includes functions for handling game strategy and related interaction scenarios such as asking for help, or proactively warning players of potential bad moves (see Section 5).

Finite state machines are used to handle the recognition of the game situational context as well as decide which behaviors to perform based on perceived events from other modules in the system. The finite state machine is a collection of transition sequences defined by a current state, an event message that triggers a transition to a new state, additional optional Boolean conditions or *guards* that must be true for the transition to occur, and a set of functions that are executed upon transition. This arrangement allows for a wide variety of interesting behaviors and flexibility in describing when they should be chosen. A good review of finite state machines in the context of creating games can be found in (Smed & Hakonen, 2006).

In addition to the transition sequences, local game state information is also maintained. In particular, a *tableau* is maintained of the current state of cards on a table, *card memory* keeps track of ASIMO's current knowledge of the identity of cards (those whose identification is revealed in the past), and turn-based *ply* information that monitors the current cards revealed for each player during a turn in the memory game. Various events such as card flips initiate updates to ASIMO's internal card memory.

During the construction of the memory game, we were able to build up the complexity of ASIMO's interactive behavior by simply adding new sequences to the finite state machine. The sequences are defined in the Lua scripting language (Ierusalimschy, 2006) which can be run-time interpreted to allow for quick evaluation of newly added interactions. Since all the application-specific state transitions and game-related functions of the finite state machine are implemented in the Lua scripting language, new interactive scenarios can be modeled simply by replacing this single script definition file without needing to re-compile the generic portion of the game behavior module that implements the execution of the finite state machine.

The initial version of the memory game was implemented with a single finite state machine (Fig. 7a). However, we found that this approach was not scaleable when we attempted to add more scenarios to the game since the number of new transitions that had to be added to

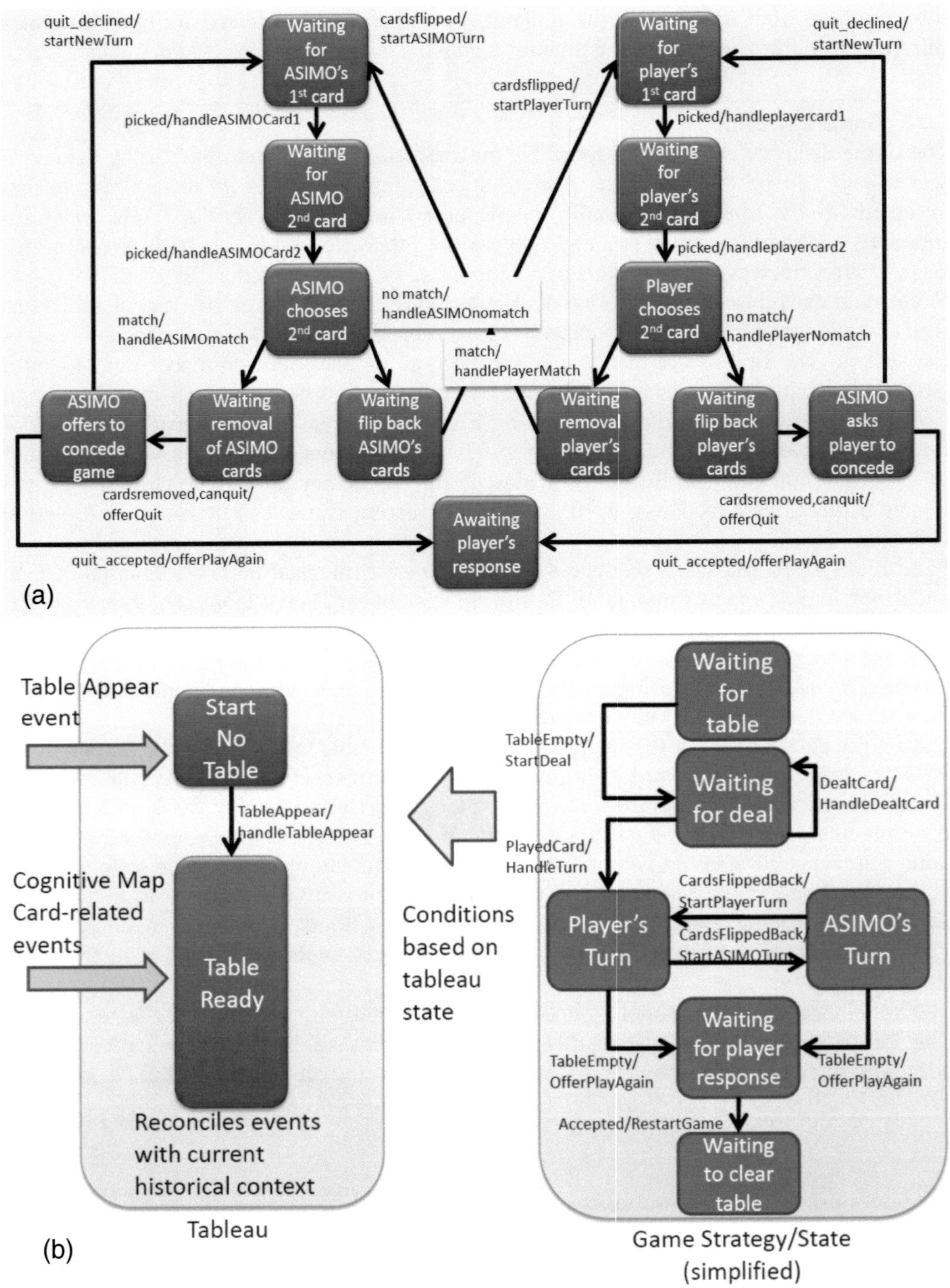

Fig. 7. Interaction component's finite state machines (simplified for clarity). Transitions denote condition/action pairs. (a) Original finite state machine. (b) Refactored finite state machine.

each newly added state also grew increasingly larger. Furthermore, any failure to perceive a condition event would result in failure of a transition to occur, often leaving the game behavior module indefinitely waiting for an event that had already occurred. We were able to refactor the game behavior module to keep multiple state machines for different entities in the game (Fig. 7b) featuring one state machine for the state of cards on the table and one for the rules of the game. This approach simplified the structure of the game, and increased robustness by allowing the card table state machine to focus on valid card layouts while the game rules state machine separately monitored whether it was currently the robot or the player's turn.

4.12. Multiple Interactive Tasks

In addition to the memory game application, we expanded the robot's ability to handle other active interaction modules simultaneously. For example, we would like the robot to answer off-topic questions from people (e.g., frequently-asked questions) while playing the memory game. There is a potential conflict if both the memory game and frequently-asked questions (FAQ) scenarios attempt to say something at the same time. Initially, we created a module called the *focus arbiter* that sorted all incoming events to one of the active interaction modules to be serviced. However, this design did not scale well as the focus arbiter would need detailed application-specific knowledge of which events each module could handle. As incoming perceived events could only be redirected to a single module by the focus arbiter, this design would not allow an event to be handled by two or more interaction modules simultaneously, creating a somewhat arbitrary restriction.

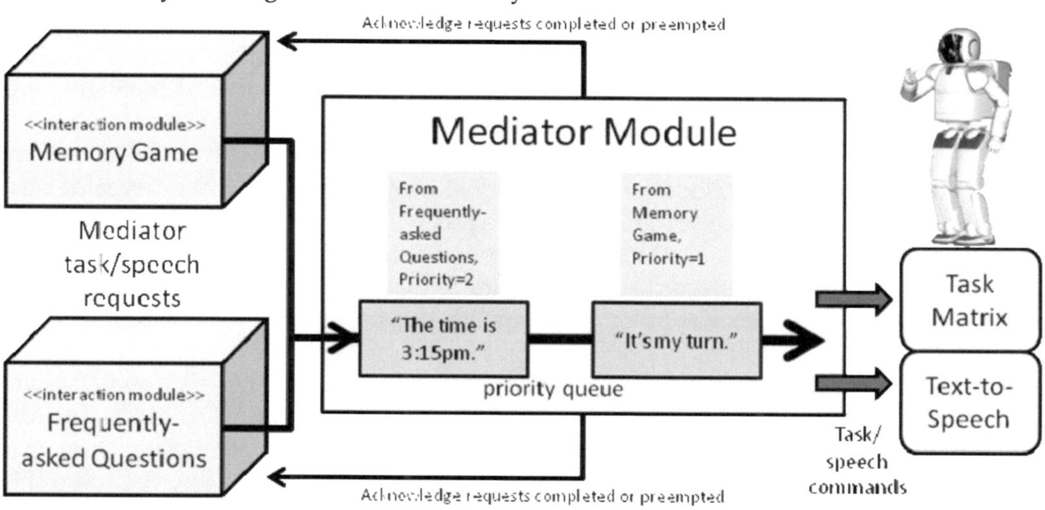

Fig. 8. The mediator module is used to arbitrate robot requests from multiple interactive applications. All messages are routed through the Cognitive Map server (not shown for simplicity).

When multiple interactive scenarios are present, the constraint should be placed on the robot's own resources for speaking and performing certain motor tasks (i.e., the robot cannot say two things at once or walk to two different locations at once), and not on the incoming perceived events. Consequently, in our revised design all interactive modules are free to receive all message events that they subscribe to, and can generate speech or task requests (see Fig. 8) that they wish the robot to service. These requests are then sent to the *mediator module* to be placed in a priority queue, with priority pre-assigned based on the relative importance of each interactive application. In Fig. 8, the memory game is assigned higher priority than the FAQ module. Requests of higher priority can pre-empt lower priority requests. The mediator also sends acknowledgement messages back to the interaction modules when their speech or task commands have been serviced or pre-empted, so that the individual modules can decide what appropriate action to take. For example, if the frequently asked questions knows it was interrupted while answering a specific question, it can choose to initiate a new speech request repeating the answer, but can also precede it with a transition phrase like "Sorry, as I was saying…".

5. Interaction Case Studies

ASIMO's style of play during the memory game focuses on offering a friendly, interactive experience for the human opponent. With this in mind, we illustrate how different modules interact with each other during the course of the memory game. Three different scenarios that can occur during the game will be described: ASIMO pointing at cards, ASIMO offering help, and ASIMO proactively warning a player if they are about to make a mistake.

5.1. Pointing at Cards

Since ASIMO does not have hands dexterous enough to pick up flat cards on a table, it must utilize its speech and gesture modalities to indicate to its human opponent which card to select on its behalf. In particular, once ASIMO decides which card to select, it verbally informs the user it wants to select the card while pointing at the card and highlighting it with the projector module.

The Game Behavior module recognizes through its state machine that it is ASIMO's turn. ASIMO then refers to its local game state memory to identify any potential unrevealed card matches on the table. In the event there are no matches, it selects an unknown card. If the card reveals a known match, ASIMO will promptly select the matching card. Otherwise, it deliberately chooses a known card so as to minimize the risk of revealing new information to its opponent. The sophistication of game strategy can be adjusted for different levels of players.

When ASIMO decides to select a card, it sends a high-level directive, *Indicate(cardID)* along with the CMobject for *cardID* containing the object's properties to the MC module. The MC module recognizes the *Indicate* directive and determines that a pointing gesture and projector highlighting are appropriate, as well as a speech utterance like: "I choose this card.". In this example, since the MC module can access the object's type through the sent CMobject, it can use this information to indicate that it is a card. Alternatively, ASIMO can choose to say something like, "I choose this one" or "This." depending on style flags as well as other state information it can access from the Cognitive Map. For pointing, a *Point(cardID)* is sent to the Task Matrix which resolves the 3-D position of *cardID* from the Environment Map and runs the motor program to generate a pointing trajectory at the 3-D location. The MC module also sends a *Project(cardID)* to the Projector module which computes the 2-D location of the card on the table to highlight the card (Fig. 9). Finally, a *Say("I choose this card")* message is sent to the Text-To-Speech module to generate speech on ASIMO's audio speakers.

5.2. Offering Help

Several game-related conversations can occur in the course of the game. For example, if ASIMO notices the player is taking a long time to perform his or her turn, it may proactively prompt the player if he or she needs help. If the player answers "No", ASIMO may apologetically say, "Sorry, just asking." . However, if the player says "Yes", ASIMO examines the state of the player's turn to determine how to proceed. If one card has already been exposed, ASIMO automatically can check its card memory to determine the location of the other card. If ASIMO does not know, ASIMO politely declines saying, "I have no idea myself." If it does know, it proceeds to indicate two cards with the MC module as described above, with one being the correct card. The rationale behind this behavior is that we do not want ASIMO to give the answer outright, but to reduce the probability the opponent will select the wrong card.

To accomplish this, when the game module enters states corresponding to the player's turn, it initiates a message to the Timer module to timeout after a predetermined number of seconds. At the end of the timeout period, the Timer module sends a message back to the Game Behavior module, prompting the state machine to enter a dialog offering help. The *Offer(help)* directive is sent to the Multi-modal communication module causing ASIMO to produce an open arm gesture while saying a phrase like: "Do you need help?". Meanwhile, the game behavior module is now awaiting a response from the user. The Speech Recognition module listens for utterances which upon receiving, it sends to the Natural Language Processing module which can convert the response to a Reject directive (if negative) or an *Acknowledge* directive (if positive). In the latter case, the NLP module can append style tags like "simple" or "lengthy" to indicate if the player responded in a curt or lengthy manner to provide hints to another module on how to appropriately respond. These directives are sent back to the Game Behavior module. Finally, if the answer is positive and ASIMO is capable of offering assistance, *Indicate* or *Verbatim* directives are sent back to the Multi-modal module to point out card choices for the user (Fig. 9).

5.3. Proactive warning of touched cards

Another proactive behavior we built into the memory game is having ASIMO warn or tease a player if they are about to flip an incorrect card. If it is the player's turn and the second card is about to be picked, ASIMO can detect the initial card touch event prior to flipping and determine if the card is the correct one to pick. If a pending mismatch is about to occur, ASIMO playfully says, "Are you sure about picking up that card?", prompting the human player to be filled with self-doubt (Fig. 9).

The touch event is created from composite information from the table detector and card recognition modules. The table detector module detects a generic table touch event, communicating position information to the card recognizer which can resolve the position information to a specific card. The card recognizer subsequently posts a message to the game behavior module indicating a card touch has occurred along with the corresponding CMobject for the card. The game behaviour module uses the submitted card information to verify if the pending card is the correct one and if not, sends a warning by issuing a *Verbatim* directive to the MC module. The game behavior state machine then returns back to the state awaiting the player's second move.

6. Discussion

Several important issues arose during development of the memory game scenario. We discuss each of these in the following subsections.

6.1. Pattern Recognition

The pattern recognition modules in our system range from the stand-alone table detector to more complex modules that combine information from several other perceptual modules. For example, identification of touched cards requires combining table detector information with card detectors. Patterns are not restricted to the spatial domain. Time-series information is an important characteristic of many activity detectors and is utilized in the memory game for identifying arrival, flipping and removal of cards.

6.2. Uncertainty

When dealing with real-world environments and sensors, handling of uncertainty becomes an issue. In our scenario, most of the uncertainty in perception is handled by individual pattern recognition modules. However, uncertainty at higher levels, related to interpretation of events is also handled by the Game Behavior module. For example, a player may simply flip a card over, or remove a card, examine it, and place it back on the table. The finite state machine can use its knowledge of the current game state to interpret the proper meaning of the sequence of events observed.

Nevertheless, our implementation of the memory game can still suffer from critical failure conditions. If a critical event is missed by any of the pattern recognition modules, the robot may be left waiting indefinitely for the missed event to occur. This problem can be handled by extending the finite state machine to include timeout events that trigger subsequent action to reassess the game state. Alternatively, using probabilistic methods to analyze time series events and sequence actions like dynamic Bayesian networks (Murphy, 2002) can be used to robustly estimate the current game state. We also experienced robustness issues with the speech recognizer while testing the memory game with children due to the higher pitch in their voices. This was due to the fact that our speech recognizer was designed with adult voice acoustic models. Creating a new voice model from children's voice data should help alleviate the problem.

6.3. Reusability

The Cognitive Map architecture facilitates reuse of many modules created for the memory game for other interactive scenarios. Since modules operate as independent agents and communicate with each other only via messages, a new interactive application can be created by changing or replacing a few key modules while keeping the rest of the system intact. Most modules should be designed to offer generic services while isolating

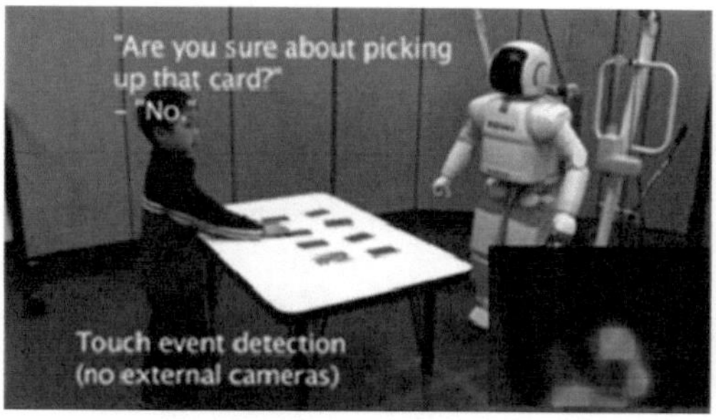

Pro-active behavior: Warning about touched cards.

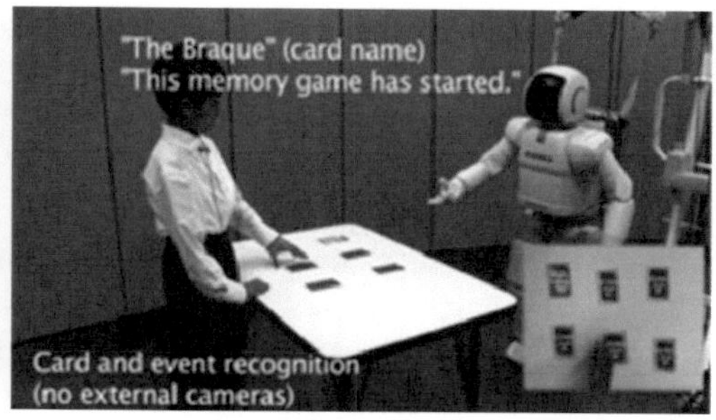

Activity detection

Speech recognition and simple conversation: Offering help to player

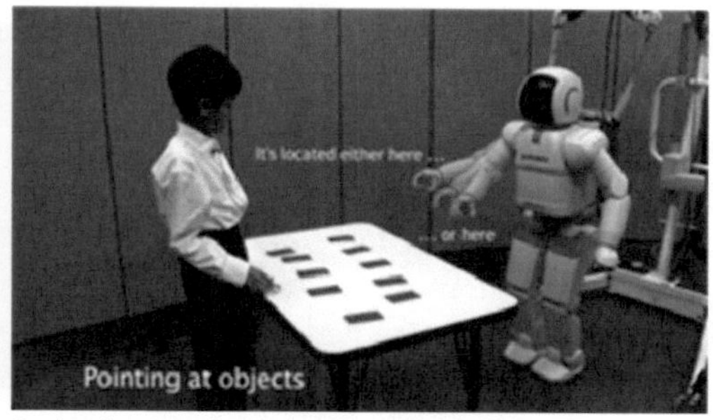

Pointing actions: uses spatial information in environment map

Multi-modal communication with speech and gestures

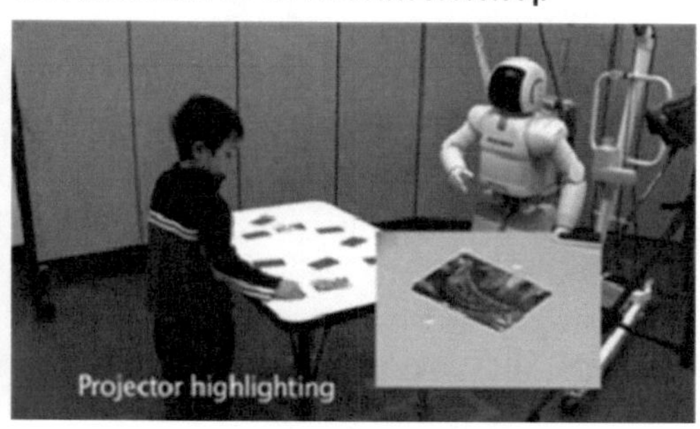

External device control: projector highlighting

Situational awareness

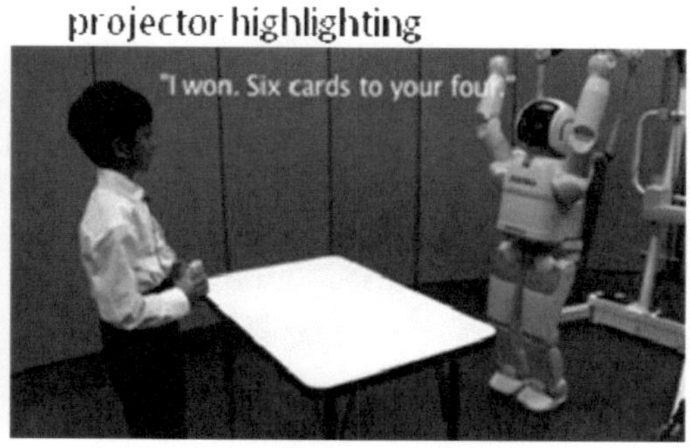

Custom motions for expressive behavior

Fig. 9. Various features of the memory game application.

application-specific details to a single module. Generic communication behavior embodied in the Task Matrix, Text-to-Speech, and Multi-Modal communication modules can be reused without modification as they are designed with no application-specific knowledge. Since the card recognizer detects generic activities such as flipping and removing, it can be reused for different card-related games. The Game Behavior module (Section 4.11) is an instance of an interaction module that encapsulates all the application-specific details of the memory game. Different card games can be developed by changing the rules and game states in this module. Messages sent from this module can contain application details used to parametrize the behavior of a generic module. For example, the Game Behavior module can send a message to the speech recognizer to configure grammar files that describe game-specific vocabulary. This incremental strategy for building new applications by reusing modules allows faster development times by focusing efforts on only new functionality.

6.4. Limitations

There can be unwanted interactions between modules that may jeopardize the performance of modules, especially when dealing with moving robots. For example, if ASIMO has excessive head motion, most pattern recognition modules may fail because they either incorrectly detect removal of objects as the field of view changes or the resulting motion blur degrades the images to be processed. We use the Cognitive Map architecture to allow the Task Matrix to send warning messages to perception-based modules of impending head motion to allow these modules to go into standby if needed during head motion so as not to post faulty information to other modules.

As with any large distributed system, the failure of critical components can still halt the application. For example, any errors in the Game Behavior module will obviously not allow the game to proceed. Although one could restart the problematic module, if the error is systematic, there is no easy way to solve the problem at run-time. These inherent problems in module functionality need to be resolved during development with thorough system-wide testing to identify failure cases early and then to test solutions to fix them.

The selection of finite state machines for decision-making features actions that are only triggered on event occurrences, creating moments of robot activity alternating with periods of inactivity. Other ways of organizing the mapping from perception to action are needed to creating complex, overlapping behaviour whose elements run over different time scales. Another problem with finite state machine decision-making is that the occurrence of events in a live system can outpace the handling of those events. In our testing, we often rapidly played out the game and ASIMO's reactions sometimes lagged with the current action of the game. For example, if a player already picked two cards, it does not make sense for ASIMO to prompt for the second card to be picked after the incident has already occurred. The solution involves ASIMO examining the incoming event queue in an efficient manner prior to processing responses so that it can decide to skip notifications if it notices the player has already performed the event.

7. Conclusion

The architectural framework we have developed based on the Cognitive Map and related modules enables us to model the memory game as a continuous, complex scenario involving two-way communication between ASIMO and its human opponent. Modules can be created for many different roles, including pattern recognition, decision-making and robot behaviour. Modules can control the behavior of other modules as seen with the mediator module (Section 4.12) prioritizing which interactive modules will have their robot commands serviced. Our design decision to make most modules application-independent allows us to reuse modules to rapidly create novel interactive applications in the future.

7.1. Future Work

The memory game has been tested informally with approximately 20 different players, but in order to systematically quantify which aspects of multi-modal communication are effective, detailed user interaction studies are planned. We are exploring new ways to infer state and organize behavior to handle uncertainty and to improve responsiveness in the robot. Many interactive sequences in the Game Behavior module can be modeled with a combination of events, state machines and behavior actions. We can potentially use the same representation as a learning template for dynamically building new state machines from observed events and responses. With the completion of this first version of the memory game, we intend to use this scenario as a research platform for ASIMO to learn new complex behaviour and to improve the richness of communication between robots and humans.

8. References

Bischoff, R. & Graefe, V. (2002). Demonstrating the humanoid robot HERMES at an exhibition: A long-term dependability test, *Proceedings of IEEE/RSJ International Conference on Intelligent Robots and Systems (IROS '02),* Workshop on Robots at Exhibitions., 2002.

Breazeal, C. & Scassellati, B. (1999). How to build robots that make friends and influence people, *Proceedings* in *IEEE/RSJ International Conference on Intelligent Robots and Systems (IROS '99),* October 1999, Kyongju, Korea.

Drumwright, E. & Ng-Thow-Hing, V. (2006). The task matrix: An extensible framework for creating versatile humanoid robots, *Proceedings of the IEEE Intl. Conf. on Robotics and Automation (ICRA '06),* Orlando, USA.

Forsyth, D. & Ponce, J. (2002). Computer Vision: A Modern Approach. Prentice Hall

Gockley, R. ; Bruce, A. ; Folizzi, J. ; Michalowski, M. ; Mundell, A. ; Rosenthal, S. ; Sellner, B. ; Simmons, R. ; Snipes, K. ; Schultz, A. & Wang, J. (2005). Designing robots for long-term social interaction, *Proceedings of IEEE/RSJ International Conference on Intelligent Robots and Systems (IROS '05),* pp. 2199–2204, Edmonton, Canada.

Hartley, R. & Zisserman, A. (2004). Multiple View Geometry in Computer Vision, 2nd ed. Cambridge University Press.

Hayes-Roth, B. (1985). A blackboard architecture for control. Artificial Intelligence, vol. 26, pp. 251–321.

Honda Motor Co., Ltd. (2000), ASIMO year 2000 model humanoid robot.

Ierusalimschy, R. (2006). Programming in Lua, 2nd Edition, Lua.org,, March 2006.

Ishiguro, H. (2006). Interactive Humanoids and Androids as Ideal Interfaces for Humans. *Proceedings of the 11th International Conference on Intelligent User Interfaces (IUI'06)*, pp. 2-9, Sydney, Australia.

Itoh, K. ; Miwa, H. ; Matsumoto, M. ; Zecca, M ; Takanobu, H. ; Roccella, S. ; Carrozza, M.C., Dario, P. & Takanishi, A. (2004), Various emotional expressions with emotion expression humanoid robot WE-4RII, *in Proceedings of the 1st IEEE Technical Exhibition Based Conference on Robotics and Automation (TExCRA '04),* November 2004, pp.35-36.

Murphy, K. (2002). Dynamic bayesian networks: Representation, inference and learning, Ph.D. dissertation, UC Berkeley.

Nakaoka, S.; Nakazawa, A.; Yokoi, K. & Ikeuchi, K. (2004). Leg Motion Primitives for a Dancing Humanoid Robot, *Proceedings of IEEE 2004 International Conference on Robotics and Automation*, pp. 610-615, April 2004, New Orleans, USA.

Ng-Thow-Hing, V.; Thórisson, K. ; Sarvadevabhatla, R. K. ; Wormer, J. & List, T. (2009) Cognitive Map Architecture: Facilitation of Human-Robot Interaction in Humanoid Robots. *IEEE Robotics and Automation Magazine*, vol. 16, No. 1, Mar. 2009, pp. 55-66.

Osada, J. ; S. Ohnaka & M. Sato. (2006) Scenario and Design Process of Childcare Robot PaPeRo. *Proceedings of the 2006 ACM SIGCHI international conference on Advances in computer entertainment technology*, vol. 266, no. 92, Hollywood, USA.

Sakagami, Y. ; Watanabe, R. ; Aoyama, C. ; Matsunaga, S. ; Higaki, N. ; & Fujimura, K. (2002) Intelligent ASIMO : System overview and integration, *Proceedings of IEEE/RSJ International Conference on Intelligent Robots and Systems (IROS '02),* pp. 2478-2483, Genova, Switzerland.

Smed, J. & Hakonen, H. (2006). Algorithms and Networking for Computer Games. John Wiley & Sons Inc.

Stilman, M.; Nishiwaki, K. & Kagami, S. (2007). Learning object models for whole body manipulation, *Proceedings of IEEE-RAS 7th International Conference on Humanoid Robots (Humanoids 2007)*, Novermber 2007.

Wada, K. & Shibata, T. (2007). Robot Therapy in a Care House - Change of Relationship among the Residents and Seal Robot during a 2-month Long Study, *Proceedings of the 16th IEEE International Symposium on Robot and Human Interactive Communication (RO-MAN '07)*, August 2007, pp. 107-112.

Zhang, L. ; Gillies, M. & Barnden, J. (2008). Emma: an automated intelligent actor in e-drama, *Proceedings of ACM International Conference on Intelligent User Interfaces 2008 (IUI 2008),* pp. 409–412.

24

Development of an Android System Integrated with Sensor Networks

Takenobu Chikaraishi, Takashi Minato and Hiroshi Ishiguro
Osaka University
Japan

1. Introduction

In recent years, research and development has been conducted on humanoid robots (Kanda et al., 2004), (Kajita, 2002) that are capable of interacting with humans in daily life, because many robot tasks that need a human-like presence, namely communication with humans in daily life. A robot equipped with a humanlike presence enables natural human-robot communication, that is, a human can naturally communicate with the robot as if he were communicating with another human. One of the approaches to learning the principles of giving a human-like presence to a robot is to develop a robot that can sense and behave like a human. Developing such a robot will reveal the principle of human-like existence. In order to develop a robot that can naturally communicate with humans through communication channels (Daibo, 1993), we have to solve the following issues.

- A robot has to be provided with a very humanlike appearance and motion to interact socially with humans (Duffy, 2003).
- A robot has to be provided with perceptional functions to communicate with humans through human communication channels. However, the degree to which a human-like nature is needed and how much perception is needed in order to realize natural human-robot communication have not yet been revealed. We have been conducting research on these issues to explore the principles for realizing natural human-robot communication using a robot with a very human-like appearance, which is called an "android" (Ishiguro, 2005). The android's human-like appearance must be sufficiently anthropomorphic to elicit a natural interpersonal reaction (Shimada et al., 2006). Thus studies on the android's human-like nature in appearance and motion have been conducted and they showed that a human-like presence is realized in short-interaction or interaction with a restricted context (Noma et al., 2006). However, the android's system is not yet sufficient to realize a human-like nature in a wider situation because the android's perceptional functions are not implemented or substituted by an experimenter in these studies. In order to clarify the perceptional functions required for natural communication, we try to realize human-like perception with an android.

The communication abilities of robots have been improved by achieving the same perception as humans using built-in sensors (e.g., (Okuno et al., 2001)). However, built-in sensors have many restrictions on range and resolution, and cause difficult issues for natural communication.

There is a technique for providing perception functions to the environment itself, that is, embedding sensors in environment (Morishita et al., 2003) (Mori et al., 2005) (Ishiguro, 1997). Such a system is quite different from a human perception system, but natural communications are possible if the robot pretends to perceive humans even though it actually does not.

In this research, to remove the restriction of built-in sensors and to achieve natural communications with humans, we propose a quite new robot sensor system. This system achieves perception and android action using as many kinds and numbers of sensors as possible including not only sensors embedded in the environment but also built-in sensors. This is called a "sensor network" in this paper. By adding various and efficient sensory system to the android with very human-like appearance, the world's first humanoid robot system can be realized, which can achieve a human-like nature integrating appearance, movement and perception.

The sensor networks give the robot a quite different perceptional modalities from a person's ones. However the modalities are not important for a human-like perception.

This means that the system needs to select the sensory information appropriate to the current situation suited for natural communication. For example, human finds where other person is by using vision or audio sensor (e.g., eyes or ear), but most important information for human-like interaction is the position (e.g., situation). Then, in order to facilitate dealing with the situation-dependent information, the system prepares various sorts of motion modules, where each module binds sensory information with an android's gesture.

This paper reports on the development of the communication system to integrate the sensors embedded in the environment and the android.

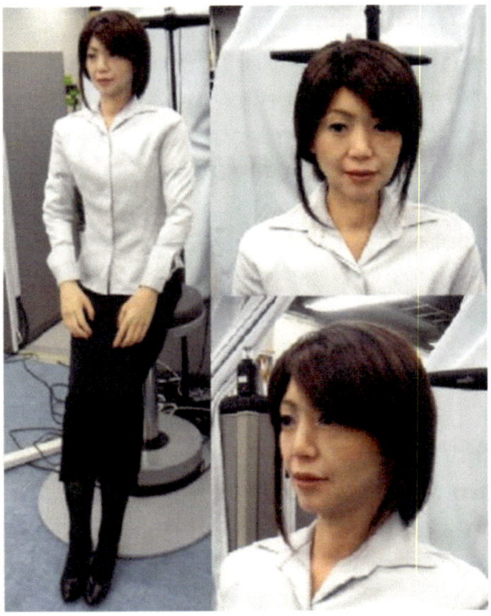

Fig. 1. Android Repliee Q2.

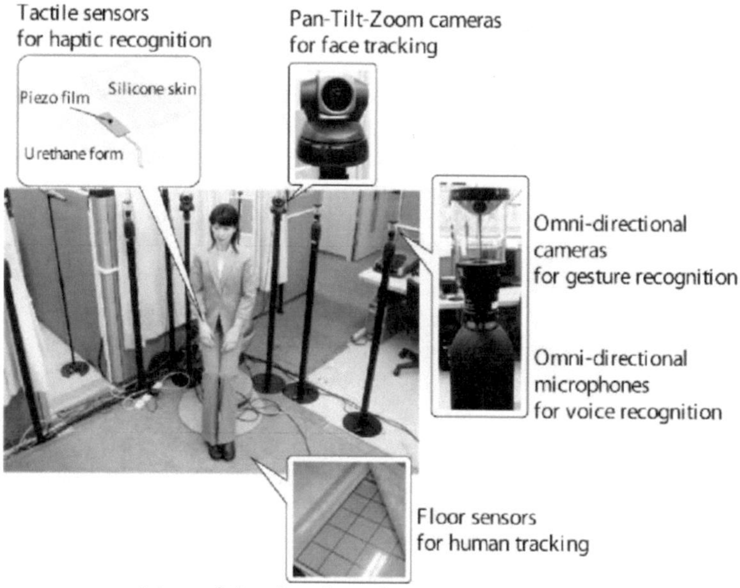

Fig. 2. Hardware composition of the developed system.

In this paper, to show that the system we have developed can achieve natural human-like behavior by an android, an experiment of evaluating impressions is conducted. We focus on a situation in which the android is sitting quietly. This situation has been implemented and evaluated in the previous studies (Shimada et al., 2006)(Noma et al., 2006). But these studies did not use sensory information. In order to make the android behavior more human-like, reactive behaviors (e.g., when a person appears, the android turns its head to the person) must be necessary. In the experiment, subjects compare two kinds of the android motion. One is a motion where the android shows reactions based on sensory information, the other is where the android does not depend on sensory information and randomly selects motions. In this paper, first we explain the hardware composition and software architecture of the system we have developed in chapter 2. Next, we report the result of the above-mentioned verification experiment in chapter 3.

2. An android system integrated with sensor networks

2.1 Hardware composition

The sensors used this system are illustrated in Figure 1. In the android's environment, omni-directional cameras, microphones, pan-tilt cameras and floor tactile sensors are arranged. Tactile skin sensors are built into the main body of the android. These sensors are connected to sensor processing computers, and sensor information is transferred between mutually connected computers through TCP/IP(Figure 3.).

One circle in Figure 3. denotes one sensor (the number of skin sensors built into the android body is 42, but only 4 circles are illustrated because all information is read by four reading circuits). In the following, first the specifications of the android are described, then the specifications of the each sensor and sensor information processing are given.

2.1.1 Android Repliee Q2

Figure 2. shows the Repliee Q2 android used in this system. To achieve a very human-like appearance, it was made by using a plaster cast of a real human. The skin material is made from silicon rubber for medical treatments, so both the appearance and the feeling of touch is close to that of a human. Movable parts all have 42 degrees of freedom, including 3 on the eyes, 1 on the eye lid, 1 on the cheek, 7 on the mouth, 3 on the neck, 9 on both arms, 2 on both hands and 4 on the waist. The android can make various shapes for the mouth and face with 13 degrees of freedom on the part of the head, thus giving it very rich expression. The android has skin sensors only in its body. Under the silicon rubber skin or clothes above the knee, 42 highly sensitive tactile skin sensors are built using PVDF film.

Each degree of freedom is driven by a pneumatic actuator. The reactions of the actuators are very natural due to air dampers against external forces without any special kind of control. This achieves a much safer interaction with humans than other actuators such as oil pressure actuators and DC servomotors. Many degrees of freedom allow the android to express the unconscious movements of humans, including movements of the shoulders and chest caused by breathing. The pneumatic actuators give it very smooth human-like motion. However, it has no ability to stand up and move. These actuators also make no large noise when driving. The pneumatic compressor, which is the source of power, generates a large noise, but it can be located far from android's main body. Because of this, the drive noise of the actuators does not disturb the communications.

The servo controller for the pneumatic actuators is located outside of the android's main body, and the android is controlled by receiving order values for each degree of freedom over a serial connection from a computer. Transmission over serial connection is executed at the speed of 20Hz. The speech is achieved by playing a recoded voice from flat speakers built into the android's chest. One computer controls the pneumatic actuators and plays the voice. The detailed specifications of the android are described in (Ishiguro, 2007)

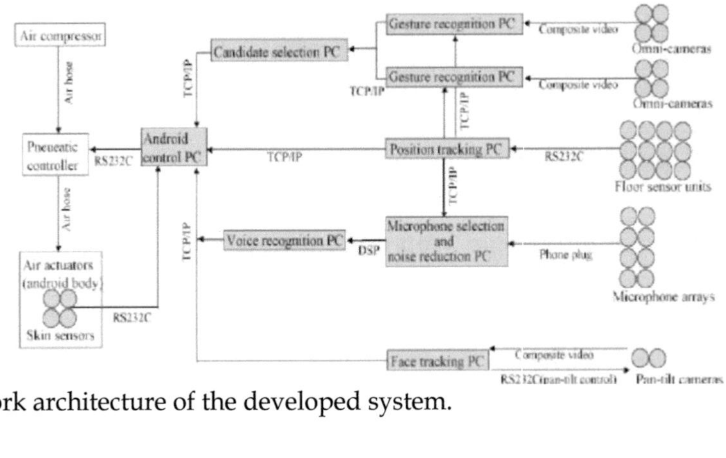

Fig. 3. Network architecture of the developed system.

Fig. 4. Floor sensor unit

2.1.2 Floor tactile sensor

Main purpose A floor tactile sensor unit is installed in front of the android in order to detect and track the position where a human stands.

Hardware We adopted floor sensor units as depicted in Figure 4. (Vstone VS-SS-SF55) for the floor sensor. The floor sensor consists of pressure-sensitive sensors that are 5 mm apart in a lattice form in order to detect and track human positions. Spreading this sensor over the floor surface allows it to detect the human position. Each floor sensor unit has a size of 500 x 500 x15 mm and a sensitivity of 200-250 g/cm^2. It has a resolution of 100 mm x100 mm, each unit can detect weight positions of 5 x 5. That is sufficiently efficient resolution for detecting a human position because adult feet generally have a size of about 200-300 x100150 mm. In this system, 16 (4x 4) parts of floor sensor units (totally 2 mx 2 m) are installed on the floor surface beneath carpet in front of the android. Each floor sensor unit is connected to a computer through a serial connection, and the transmission speed is about 37 Hz.

Data processing When a human is walking, the sensor signal is temporally and spatially discrete. The signal pattern is also greatly different when one foot is on the ground and when two feet are on the ground. Therefore that causes a many-to-one association problem when the human position is required from the floor sensor data. Murakita et al. proposed a method based on the Markov Chain Monte Carlo Method to track plural humans solving this intrinsic association problem of the floor sensor (Murakita et al., 2004). In this system, we implement this method and track several people using the floor sensor. Human detection is executed at the speed of about 17 Hz.

2.1.3 Omni-directional cameras

Main purpose We installed 8 omni-directional cameras around the front space of the android (Figure 1.), in order to recognize human gestures in addition to human tracking.

Hardware We adopted a VS-C42N-TR (Vstone Co. Ltd.) omni-directional camera. The imaging range is about 15 degrees to the upper side from the horizontal plane of all azimuths and about 60 degrees downside from the horizontal plane of all azimuths. The number of effective pixels is 768x 494. Eight these cameras are installed at a level of 1.5 m from the floor surface. Four cameras are connected to one computer through a capture device and two computers process the image data from all of the omni-directional cameras.

Data processing A human tracking and gesture recognition method is proposed with an omni-directional camera network using a triangulation of omni-directional cameras and omni-directional appearance-based model (Nishimura et al., 2001) (Ishiguro & Nishimura, 2001). This method allows it to recognize human gestures independent from the object

direction from the cameras. However, when moving object(s) exist in the image, the system may not determine a recognition region in the image, because this method employs subtraction images to detect human positions and determine the recognition region of the image. Therefore, we adjust the floor sensor information for this system. The human region of the image is estimated from the information of human position obtained from the floor sensor, and the gesture recognition is executed. The candidates for gestures recognized by the two computers and information equivalent to the reliability of the recognition are transferred to other computer. The computer selects the one candidate with the greatest reliability and this is the final result of the gesture recognition.
Therefore, three computers are used in all for gesture recognition. In this system, gesture results including "bowing" and "indicating a direction" can be recognized. The recognition is processed at a speed of 20 Hz. When several people exist in the recognition area, each person's gesture can be recognized.

2.1.4 Microphones

Main purpose In order to recognize human voices, eight omni-directional microphones are installed in the same places as the omni-directional cameras.
Hardware One omni-directional microphone consists of four built-in condenser microphones in the bottom part of the omni-directional cameras. One omni-directional microphone is achieved by synthesizing four signals from four microphones. All of the microphones are connected in parallel to one computer, and the computer extracts voice data for the voice recognition program. All of the voice data are transferred to another computer that recognizes the speech. Two computers in all are used for voice recognition.
Data processing For voice recognition, a voice signal with as little noise as possible is needed. For that, the system must select the nearest microphone to the human as the sound source based on human position information obtained from the floor sensor. Next, background noise must be filtered out of the signal obtained from the selected microphone. An existing voice recognition method (Microsoft Speech SDK) is applied to the voice signal thus obtained. Recognition grammar can be changed according to the situation.

2.1.5 Pan-tilt cameras

Main purpose The system must know the human face position and direction in order to know the direction for the android to talk to. Therefore, two cameras that can pan, tilt and zoom are put on both sides of the android.
Hardware We used EVI-D100 (Sony Co. Ltd.) pan-tilt cameras. These cameras can turn across a field of view of ± 100 degrees in the horizontal direction (maximum 300°/sec.), and ± 25 degrees in the vertical direction (maximum 125°/sec.). The two pan-tilt cameras are connected to a computer by serial connection, and the computer can control the pan, tilt and zoom functions of the camera. These cameras are connected to the same computer through a capture board, for image processing and controlling the image direction by the computer.
Data processing We implemented a face detection function based on a histogram of the edge image and extracted a skin color image (e.g. (Adachi et al., 2001)). Cameras are controlled by pan, tilt and zoom functions to keep the detected face region in the center of the sight and to keep the same size image. Therefore the face position in three-dimensional space is known from the position of the pan, tilt and zoom functions. The face direction can

be detected from three candidates (front, right and left). This process is executed at a speed of 20 Hz.

2.1.6 Skin sensors

Main purpose The skin sensors detect contact by humans. Only these sensors are installed in the body of the android, unlike other sensors, because communication by contact needs to use the body. In all, 42 skin sensors are installed beneath the skin above the knees.

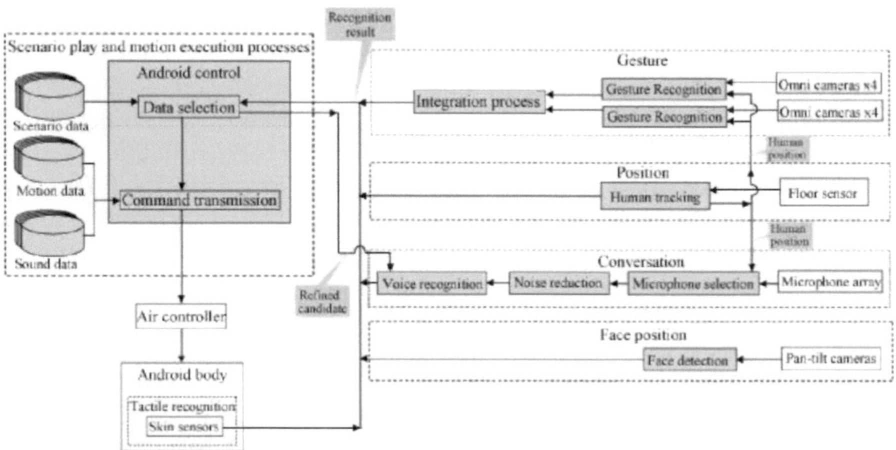

Fig. 5. Software architecture of the system.

Hardware The skin sensors are made of PVDF film. The PVDF film is sandwiched by urethane rubber and silicon rubber skin. Because each sensor element has a size of 10 x10 mm, 42 sensors are not sufficient to cover the whole surface of the android's body. However, the skin sensor can be activated even when a body part with no skin sensor is touched because the distortion of the urethane layer can easily spread on the android body when the android is touched.

Data processing A value equivalent to the contact force can be obtained by temporal integration of the output value, because the PVDF film outputs the value proportional to the transformation speed. The low distribution density of the skin sensor restricts the ability to distinguish contact behavior accompanied with movement on the skin surface (e.g., "stroking"). The android detects which part of the body is touched and how much external force is applied. Thus, contact behavior including "tapping the shoulder" "touching arms" can be recognized. The detection of contact is executed by the computer for the android's motion control. Data processing is executed at a speed of 20 Hz.

2.2 Software architecture

The flow of the recognition and determination of behavior is illustrated in Figure 5. Eight computers (Pentium IV 3GHz) are used for sensor data processing and motion control. Sensor data processing consists mainly of five parts: gesture recognition, human tracking, voice recognition, face detection and contact detection. Each data processing is explained in chapter 2.1. Each processing part is not independent, as gesture recognition and voice recognition depend on the result of human tracking. Applying the results of other recognition helps increasing recognition accuracy. Each sensor processing outputs the following information.

- Gesture recognition: Gestures of each person and human IDs.
- Human position tracking: Positions of each person and human IDs
- Voice recognition: Recognized word sequence.
- Face detection: Face position in three dimensional space and face direction.

• Contact detection: Touched body part and strength of external force. The android's motion is treated by module units gathering time-series command values for each degree of freedom and voice data. One motion module is from a few seconds to about 30 seconds (e.g. "bowing" and "waving a hand"). In this paper, we call packages of motion module and a diverging condition process "events". In an event, first the motion is executed, then the following motion is selected based on the result of sensor processing. Additionally we make a script that describes the steps of the procedure of the events. We call this a "scenario". An event includes a module to execute, sensor data processing to identify the situation after the module executes and the following event ID selected by that recognition result. If voice recognition is selected as sensor processing in the event, the recognition grammar must be specified.

When the scenario is pointed out in the android control computer, a control program is sequentially executed. The procedure is as described below.

1. Obtain the information of an event.
2. Execute the motion described and play the voice sound file.
3. Obtain the result of the recognition from the sensor processing that is pointed to. If voice recognition is pointed to, grammar ID is transferred to the voice recognition module. According to the result of the recognition, following ID of the ID will be decided according to diverging condition.
4. Back to 1.

The event limits the candidates of recognition and the proper sensor for the recognition. That increases the recognition accuracy of each sensor data processing part. The scenario is described in XML for non trained programmers. An example of a scenario is depicted in Figure 6. In the figure, a part marked by "<event>" and "</event>" means one event. For example, the "eventno="1"" means the following. The motion module subscribed by "G001.dat" is executed and the scenario goes to the "eventno="2"" with no sensor information. The "eventno="4"" means the following. The motion G004.dat is executed and the voice recognition program is executed based on "grammar_G004.xml". In the case that the result of voice recognition is "YES", the scenario goes to the "eventno="5"". In the case that the result is "NO" it goes to the "eventno="8"". In the case that the voice recognition failed, it goes to "eventno 16".

3. Natural idling motion and the evaluation

The ReplieeQ2 android realized a human-like presence in short-term and restricted situations (Shimada et al., 2006)(Noma et al., 2006). However, it is difficult to realize a human-like presence in various situations without a variety of reactions for changing surroundings. In this section, in order to prove the effectiveness of the system, experiments are conducted.

```
<?xml          versions.0"?>
<scenario>
        <event eventno="I" sensor="UNUSE" motion="G001.dat">
                <resu I tXUNUSE nextevent ="2'7></resu 11 X/event > <event
        eventno="2" sensor="UNUSE" motion="G002.dat">
                <resu I tXUNUSE nextevent = "3"/X/resu 11 X/event > <event
        eventno="3" sensor="UNUSE" motion="G003.dat">
                <resu I tXUNUSE nextevent="4VX/resultX/event>
        <event eventno-"4" sensor-"VOICE" mot ion-"G004.dat" grammar-"grammar_G004.xml "> <resultXYES nextevent="5"/>
        <N0 next event = "8"/> <ERR0R nextevent="16"/> </resu 11 X/event > <event eventno="5" sensor="UNUSE"
        motion="G005.dat">
```

Fig. 6 An example of scenario.

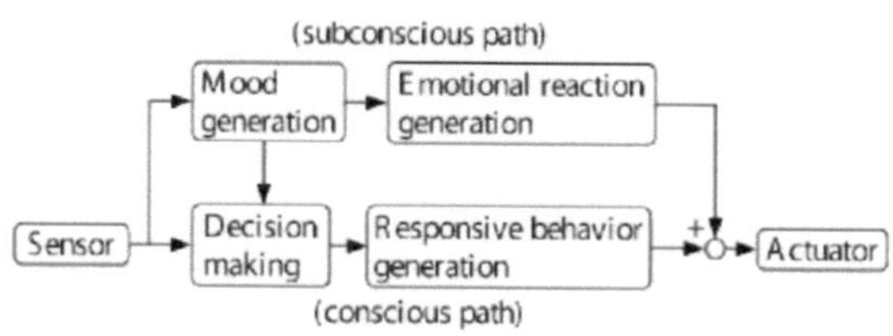

Fig. 7. A control of the android based on a mental state.

The experiments show that the human-like nature is improved by the information of a sensor network. Noma et al. evaluated the human-like nature of an android without sensor information, but sensor information could improve its human-like nature. Therefore, we prepared a situation where subjects come close to an idling android, and evaluated the impressions of the android's behavior in that situation. Two impressions of human-like presence were compared. One is the impression where the android's reactions are generated based on sensor information, and the other is the impression of idling motion without sensor information, i.e., the same as the Noma's study. The idling motion is a motion where the android stays quietly sitting and stares around or stares at something according to the surrounding situation.

This is an interaction where the active communication between a human and the robot is not established, as mentioned in chapter 1. However, it is important to achieve a natural interaction in such a situation for human-like presence for androids in our daily life. A quietly sitting person often shows a behavior in which he/she turns to a direction where some small sound is heard, but it is very difficult to achieve this behavior based on the information of a microphone. In contrast, the floor sensor allows it to detect the approach of a human and implement a motion so that the android acts as if it hears a slight sound from the human. Thus, it is necessary to observe the movement of a human's position in the distance. Before establishing active communications, it is difficult to limit the sensors and sensor data processing. A system integrated with sensor networks would improve the idling motion of the android.

3.1 Basic concept of generating idling behavior

The idling behavior to be achieved here does not contain reactions that are strongly expected, in contrast to the reaction of android in active human-robot communications. However, behaviors that are completely selected in a random manner would not give a human-like nature to the android because a quietly sitting human would show some reaction behaviors toward a change in the surroundings. Thus, a certain control model is required. For subjects to feel the human-like nature of an android, it is necessary to design the android's behavior so subjects will recognize the android's intention or mental state. For that, existing studies show that to construct mental states for the robot. For example, Miwa et al. construct a mental model integrating a conscious model (Miwa et al., 2003). This model allows a robot to clearly select a stimulus (sensor input) among several stimuli, but not randomly. Adding a reason to the selection of a stimulus among several stimuli enables a robot to show a more human-like reaction. Breazeal mentioned that the emotional expression of a robot makes interaction with humans more complex and richer (Breazeal,2003).

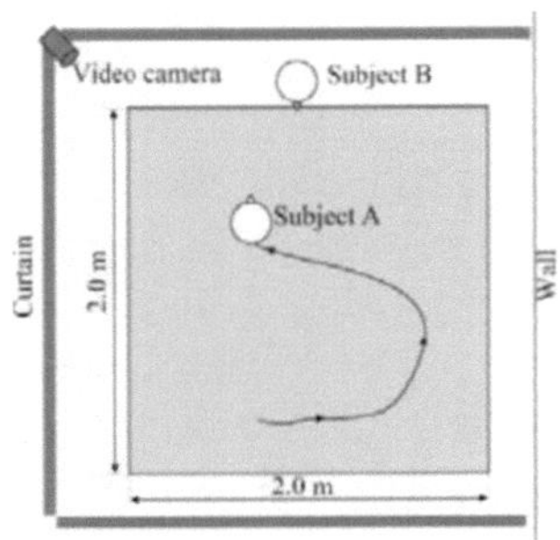

Fig. 8. Observation of a waiting person.

In this paper, we construct a motion generation model changing unconscious physical reactions according to android's mental state (Figure 7.). At first, the mental state is decided based on sensor information. The mental state is a state of emotion (e.g., fear and interest). The android choose a reaction motion based on the sensor information and the mental state. This corresponds to a conscious generation of human motion. In addition, emotional reaction is generated according to the mental state. This corresponds to unconscious emotional reaction. The emotional reactions here include frequency of blinking and change of the eye movements. Among the mental states, a neutral state is defined. The emotional reactions are expressed by difference from the neutral state (e.g., frequency of blinking, speed of the eye movement). When the mental state is neutral, one motion module is randomly selected among several motion modules prepared. As far as the mental state is neutral, this procedure will be repeated. If the mental state changes, a reaction according to the mental state is added.

The android's motion modules, mental states, emotional reactions, and idling scenario are constructed by observations of real humans and reflection reports. The motions of human without reasonable tasks should depend on individual characters. It is thinkable that extract common motions from a large number of people's motion to obtain motions independent from the characters. However, common motions are unlikely to be able to extract common motions if the motions depend on personal character. In this study, we did not research common motions of human. It is possible that results could be changed by subjects' character, but at the very least we can verify the effectiveness of the sensor network based on the models we made.

3.2 Observation of quietly sitting humans

To observe quietly sitting humans, an environment was prepared as illustrated in Figure 8. Observation was conducted in a place enclosed by curtains or walls. We took a video of the motions of a quietly sitting person and the motions of a person in his/her surrounding area by a video camera. An experimenter instructed the subject to sit in a chair and stay sitting for 20 minutes. This is subject B. Subject B knew that this experiment was a study on the motion of human-like nature in advance. The chair looks like the chair in which the ReplieeQ2 was sitting (Figure 2.). The experimenter instructed another subject to act freely for 20 minutes looking at subject B within the 2.0 m x 2.0 m area, marked by the black line Figure 8.. This was subject A. Subject A also knew that this experiment was a study on the motion of human-like nature in advance. This region of 2.0 m x 2.0 m was the same size as the floor sensor area installed in front of the android. We used two male students in age of 20's as the subject A, and two students in 20's as the subject B.

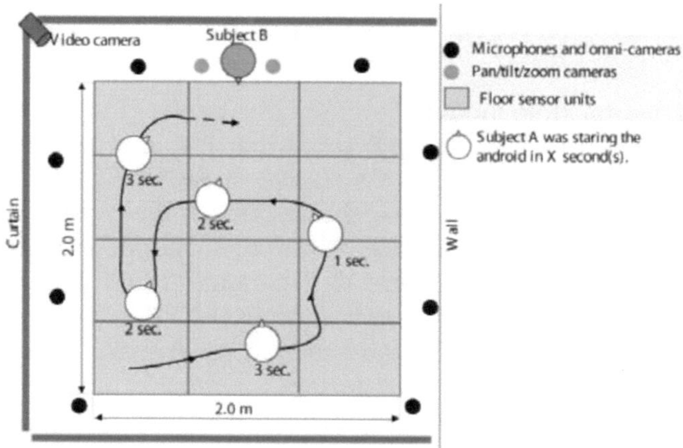

Fig. 9. Example of a behavior of the subject A.

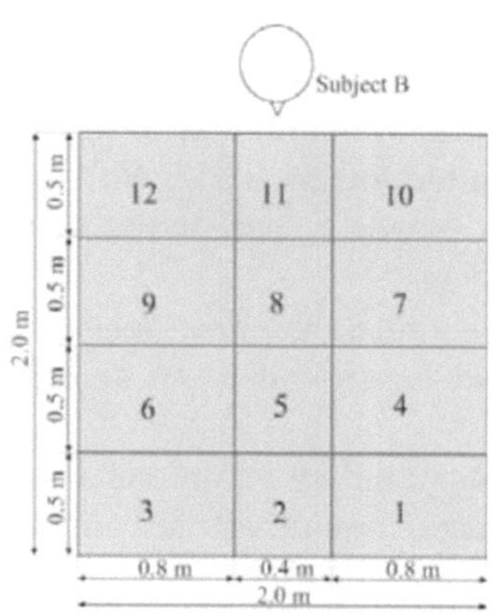

Fig. 10. ID of the position of the subjects A. Subject B was later replaced with the android in the experiment.

After the observation, subjects B gave their impressions of subjects A in order to construct a mental state model. The subjects gave as many impressions of things in these 20 minutes as they could remember.

3.3 Construction of basic motion modules

The behavior of subject A is illustrated in Figure 9. We observed the behavior in which subject B walked around subject A, sometimes stopped in a few seconds and looked at the subject A. In order to define basic motion modules, we extracted common motions frequently emerge in the motions of two subjects B. As a result of this observation, the six common motions below were obtained. These motions were defined as basic motion modules for the android, because they are frequently seen in quietly sitting humans.

- Look toward the direction subject A looks.
- Look toward subject A a few times with only eye movements and as few body movements as possible.
- Look at subject A a few times.
- Look toward the front so as not to see subject A.
- Look down to the ground.
- Keep looking at subject A (following subject A).

3.4 Construction of mental state model

The answers of the impression of subject B show that the impressions depend on the position of subject A. We thus classified the mental states based on the behavior and position of subject A. Table 1 denotes these classes. This is the mental state model. The positions of subject A are classified in Figure 10. The first column of Table 1. denotes the position IDs of subject A illustrated in Figure 10. The range "1 to 6" in this column means any number from 1 to 6. The second column denotes the behavior of subject A in the position indicated by the first column. The third column denotes the feeling of subject B for the behavior of subject A. The fourth column denotes the emotion of subject B, which is the origin of the mental state. The emotions are defined based on the emotional model of Plutchik (Plutchik, 1962). Each emotion is defined by the reasons below.

- Disgust: In the impression questionnaire, subject B reports "hate to be stared at and want it to stop" and "feeling of pressure" . We assume the disgust is caused because it is an emotion meant to remove a harmful thing.
- Acceptance: In the impression questionnaire, they report "do not care so much because subject A does not pay attention to me" . We assume the acceptance is caused because it is an emotion meant to accept the current situation without feeling comfort or discomfort.
- Fear: In the impression questionnaire, they report "feels painful to be stared at". We assume the fear is caused because it is an emotion under a situation of pain, destruction or surprise.
- Apprehension: Fear is caused because they report a feeling of discomfort to be stared at or to be in a very close place. They report that subject A has something to say. The anticipation is caused because it is an emotion under a situation of looking for the purpose. Therefore we assume the intermediate between two states of Pluichik's model, the apprehension, is caused.

- Interest: Surprise is caused because they report "caring about the direction that subject A is facing" in the impression questionnaire. It is an emotion under a situation where the other person's purpose is not known to be beneficial or not, or whether it is pleasant or not. Acceptance is caused because they tried to accept subject A's action. Therefore, we assume the intermediate between two states of Pluichik's model, interest, is caused. The mental state of a quietly sitting person is modeled based on the position of the surrounding person and his/her behavior. Employing the sensor networks we developed, this position and behaviors can be obtained more robustly than that use systems using built-in sensors on the robot body.

3.5 Construction of sub-scenarios of each mental state

The sub-scenarios for each mental state are defined based on the observation of subject B. According to the mental state, the frequency of blinking, speed of blinking and speed of eye movements changed as emotional reactions. Each value in the basic motion module is defined below.

- Frequency of blinking (BF) is once every 2-3 sec
- Speed of blinking (BF) is 0.3-0.5sec from eyes closing to opening
- Speed of eye movement is about 15 deg/sec

Sub-scenarios are explained below. The emotional reactions in each mental state are expressed by abbreviations. For example, when the case of the frequency of blinking is twice as high as the basic motion, it is described as "BFx 2".

Sub-scenario of acceptance Look at a person. BF x1, BV x1, EV x1.

Sub-scenario of fear Look at a person. BF x 3, BV x1, EV x 2.

Sub-scenario of disgust Look at a person when he/she is near (i.e., 2 and 5 in Figure 10.). Look down at the ground when a person is far (i.e., 8 and 11). BF x1, BV x 2, EV x1. **Sub-scenario of interest** Glance at a person when he/she is in a far place (i.e., 7, 9, 10 and 12). Look forward of the direction where a person looks when he/she is near (i.e., 1-6) (Two patterns: right and left). BFx 3, BVx 1, EVx 2.

Sub-scenario of apprehension Glance at a person when he/she is walking while watching the android. Look down at the ground when a person is walking in a far place (i.e., 7-12) without watching the android or when a person is looking somewhere at a far place (i.e., 8 and 11). BF x 3, BV x1, EV x 2.

3.6 Recognition of a person's behavior and decision of android's behavior

Position ID of A	Behavior of A	Feeling of B towards A	Mood of B
2,5 8 11	Stares at B Stares at B Stares at B	Unpleasant Unpleasant + Oppressive Very unpleasant + Oppressive	Disgust
1 to 6	Walks around without staring at B	Not so concerned about the action of A	Acceptance
1, 3, 4, 6	Stares at B	Harassed	Fear
7 to 12 8,11 1 to 12	Walks around without staring at B Looks at somewhere Walks around with staring at B	Anxiety Anxiety Anxiety	Apprehension
1 to7,9, 10,12	Looks at somewhere	Worry about where A is looking	Interest

Table 1. Classification of the feelings of subjects B toward subjects A. Subject B was later replaced with the android in the experiment.

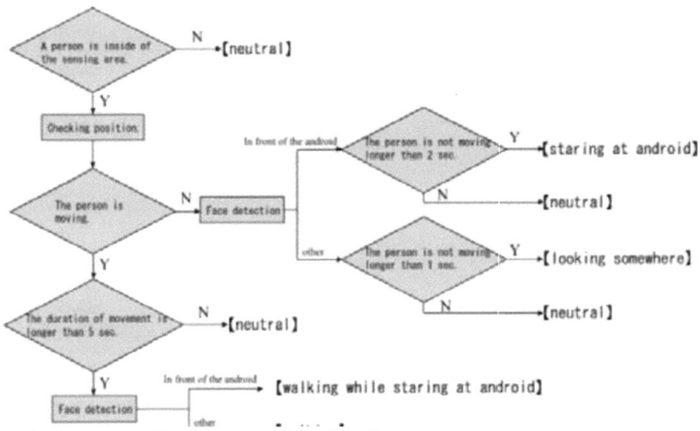

Fig. 11. Procedure for recognizing a person's behavior.

The scenario of the android's behavior in a quietly sitting state is constructed based on the mental state model and sub-scenarios we constructed. The process for recognizing the behavior of subject A in Table 1 is illustrated in Figure 11. For example, first, when a person is in the floor sensor region, his/her position is measured by the floor sensor.

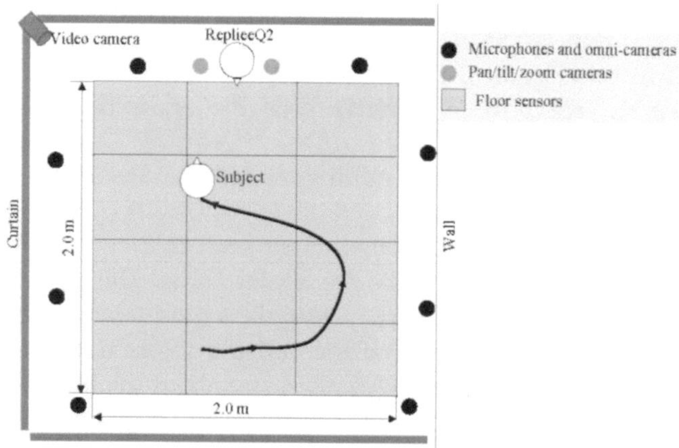

Fig. 12. Arrangement of the sensor network.

Next, whether he/she is moving or not is determined from time-series position data. When he/she is not moving, his/her face direction is measured by pan-tilt cameras. When his/her face is pointed toward the android, whether he/she is looking at the android for more than 2 sec. from the time-series data for the face direction. When he/she is looking at the android for more than 2 sec., he/she is assumed to be staring at the android. The mental state of the android is decided based on the behavior and position of a person. As mentioned in 2.2, a recognition of the situation (i.e., in this case a recognition of the person's behavior) is executed in each module. When the behavior of a person cannot be recognized, the mental state is assumed to be neutral (Figure 11). When a mental state is caused, a sub-scenario is executed accordingly. After each sub-scenario is executed, the mental state is neutral in some steps without shift to other state in order to avoid same motion is executed in sub-scenario.

Motion modules are manually made with motion-making software. Thus, the idling motion scenario is made by describing these procedures in XML.

3.7 The evaluation experiment

3.7.1 Purpose of examination

An experiment for evaluating the effectiveness of the idling scenario based on the sensor networks was conducted. The subjects evaluated the subjective human-like nature when took action around the android in an idling state. We also placed the android in idling motion without sensor information to compare. In this idling state, the android chose and repeated random motions among 15 kinds of motion modules, including "do nothing particular", "look around" and "sleepy motion". In this experiment, each subject evaluated the human-like nature of the android in both conditions. We predicted that the android that chose reactions based on the data of sensor networks would have a more human-like nature even in the situation where there was communication between the human and the android.

3.7.2 Experimental procedure

The experiment was conducted in the environment illustrated in Figure 12.. Each sensor was arranged depicted in this figure. Each subject was in the two conditions below. **Random condition:** The subject acted for 3 minutes near the android with randomly selected motions.

Sensor condition: The subject acted for 3 minutes near the android with motions selected by scenario.

The subjects were instructed before the experiment started: they were to act freely around the android and evaluate how human-like the android was. The subjects could also touch and talk to it when they acted. After participating the experiments in these two conditions, the subjects evaluate human-like nature of the android in each condition. The evaluation was on a scale from -3 to 3. Free-answer questionnaire about the impression of the android was also conducted. The order of the two conditions was changed for each subject. This was to counterbalance the order of the conditions. There was a one minute interval between the first condition and the second condition. The subjects were 16 male and female university students.

3.7.3 Result of experiment

Figure 13. shows the experimental scene. Each subject only walked around the android, and did not touch or talk to the android. The behavior exhibited by most of the subjects was to walk close to the android or to stay at a distance from it and they looked to be interested (Figure 9.). Some subjects fixed their eyes on the android almost without moving. In addition, some of them waved their hand or extended their hands toward the android's face. They appeared to be trying to confirm whether or not the android could recognize them. Next, we verified a hypothesis that the android in the sensor condition is more humanlike than the android in the random condition. In this experiment, we compared the evaluation values of human-like nature by testing only the effect the order that the viewing condition had on the subjects.

In this experiment, the subjects were divided into two groups; the first group was first presented with the sensor condition (S-R group, 8 persons) and the second group was first

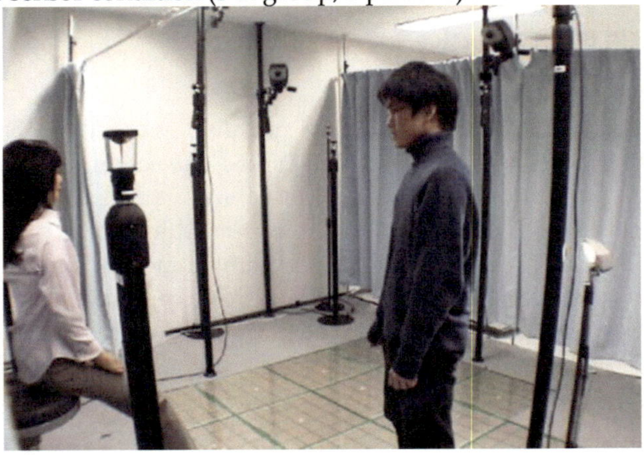

Fig. 13. Experimental scene.

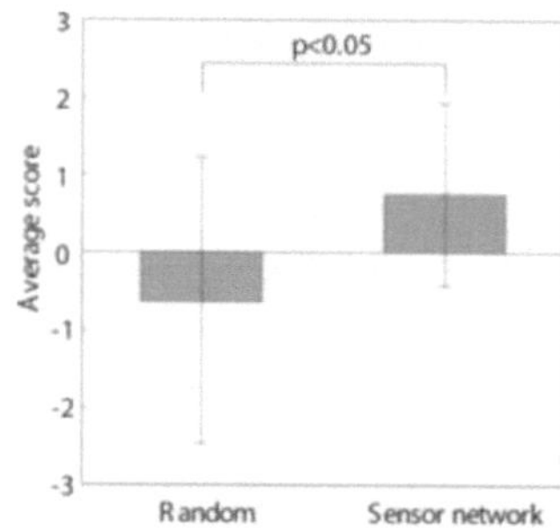

Fig. 14. Evaluation of human-like nature of the android(between-subject comparison).

presented with the random condition (R-S group, 8 persons). We compare the evaluation value of the S-R group and the R-S group. We compare the evaluation value of S-R group and R-S group. That is, we use only values of first presented condition. Figure 14. depicted the average of each condition. The error bar indicates standard deviation of each evaluation value. The independent t-test on the difference of average values shows significant difference ($t = 1.781$, degree of freedom 14, $p = 0.0483$). This result shows that the reactions based on sensor information gave the android a more human-like nature. In the free-answer questionnaire, the main reasons for the human-like nature of the android's motions were as given below.

- The motion of turning its gaze toward a person gave a human-like impression.
- When the subjects came close to the android, they felt that the android was nervous as it turned its gaze around. (This may indicate a quickening of the eye movement.)

When the subjects came close to the android, the frequency of blinking was augmented. This gave them a human-like impression.

However, some subjects mentioned these reasons as being motions which they deemed to be non human-like. This questionnaire indicates that the subjects detected changes in the emotional reaction. In addition, they used the word "nervous" and they attributed this change in emotional reaction to the change of mental state. This also indicates that the change of emotional reaction gave a human-like nature to the android.

3.8 Discussion

Only the floor sensor and pan-tilt cameras were used in this experiment, because the android's scenarios of the android were constructed based on the observation of quietly sitting real humans. As such, this experiment is not an evaluation of the entire system. However, we could verify that a quite different system from human perception can realize human-like behavior in the android. The gesture recognition with omni-cameras and the voice recognition with microphone described in section 2.3 shown made sure to function, but the impression evaluations with all of these functions are issues in the future. We built a mental state model based on the observation of two subjects in this experiment but this is only one example of a mental state model. Sensor networks allow us to realize human-like reaction in the android, but more studies are required to see: which relationships are between elements (e.g., sensor-mental state, mental state-behavior) affect the human-like nature. This mental state model should be more objective because it is constructed based on observations of the experimenter. As some methods for construct human emotion models have been proposed (e.g., (Bianchi-Berthouze & Lisetti, (2002)), it is possible to construct a more objective mental state model using these methods. In this experiment, different motion modules were used in the sensor condition and the random condition. Therefore, the differences in the evaluation of the impression for each condition were possibly not produced by reactions based on sensor information, they were possibly differences in the evaluation of the modules themselves. In other words, the motion modules in the random condition possibly obtained a low score. But we believe that the humanlike nature of these motion modules themselves were not low because they were constructed by combining motions for other studies (Shimada et al., 2006)(Noma et al., 2006). However, to clarify this point, it is necessary to conduct experiments to compare the sensor and random conditions in the future.

In the scenarios of the idling state of this experiment, human positions are detected by the floor sensor. The floor sensor is quite different from human modality and obtain the positions more precisely. The obtained data from the floor sensor were translated into the 12 level resolution and we believe that the motions of "see" of the android made the subjects believe it knew the human positions. However, suitable resolution and manner of information translation have not been revealed, studies of this issue are still necessary. In this experiment, we built the motion generation model based on the mental state aiming to make subjects attribute the motion of the android to a mental state, including the intention. Several studies aim to attribute motions of non humanoid robot to mental states to realize natural communication with human (Terada et al., 2007)(Kobayashi & Yamada, 2007). In these studies, it is possible that robots are anthropomorphized by the motions attributed to mental state. In contrast, the android would be anthropomorphized and attributed to the mental state, before observing the motions of the android because the android has a very humanlike appearance. It is possible that the appearance may promote natural communication. It is necessary to perform more study comparing the android to other robots to clarify such a supposition.

4. Conclusion

This paper proposed a system integrated an android and sensor networks in order to achieve a natural communication with human keeping human-like nature of the android. An android control system and computer networks for sensor processing of a large number of various sensors are integrated into a system. In addition, a software system developed, which achieves to obtain of sensor information and description of android's behavior in XML. Existing study already indicated that human-like appearance and motion of android causes human-like impression. In addition, reaction according to situation is required in order to cause more human-like presence in more complicated interaction. This study achieved to construct the world's first humanoid robot system which can achieve a human-like nature integrating the appearance, the movement and the perception. Configuring idling behavior of the android as an example issue indicated the developed system is effective to achieve a subjective android's human-like nature.

In order to achieve a subjective android's human-like nature in communication, it is not required to implement abilities equivalent to human including abilities to exercise and abilities to percept. The result of experiment in this paper suggests that reactions for human-like presence are efficient to achieve the subjective android's human-like nature in communication. Studies on minimum required abilities to exercise and percept for achieving human-like nature, i.e. a boundary condition of natural interaction should conduce a principle of natural communication. The system we built in this paper allow us to tackle issues focused on appearance, motion and perception simultaneously, such studies have not been started yet.

5. Acknowledgement

We developed the android ReplieeQ2 in collaboration with Kokoro Company, Ltd..

6. References

Kanda, T.; Ishiguro, H.; Ono, T.; Imai, M.; Maeda, T.; & Nakatsu, R. (2004). Development of robovie as a platform for everyday-robot research, *Electronics and Communications in Japan,* 87(4).

Kajita, S (2002). Research of biped robot and humanoid robotics project (HRP) in Japan, *The Fourth International Conference onMachine Automation (ICMA '02),* pp. 1-8.

Daibo, I. (1993). Psychology of intimacy and function of communication. *Technical Report of IEICE (The Institute of Electronics, Information and Communication Engineers),* HC-93-52:33-40, (In Japanese with English abstract).

Duffy, B. R. (2003). Anthropomorphism and the social robot. *Robotics and Autonomous Systems,* 42(3-4),177-190.

Ishiguro, H. (2005). Android science - toward a new cross - interdisciplinary framework.*Proceedings of the International Symposium of Robotics Research.*

Shimada, M.; Minato, T.; Itakura, S. & Ishiguro, H. (2006). Evaluation of android usingunconscious recognition, *Proceedings of the IEEE-RAS International Conference onHumanoid Robots,* pp. 157-162.

Noma, M.; Saiwaki, N.; Itakura, S. & Ishiguro, H. (2006). Composition and evaluation of the humanlike motions of an android, *Proceedings of the IEEE-RAS International Conference on Humanoid Robots,* pp. 163-168.

Okuno, H. G. ; Nakadai, K. ; Lourens, T. & Kitano, H. (2001). Human-robot interaction through real-time auditory and visual multiple-talker tracking, *Proceedings of IEEE/RSJ International Conference on Intelligent Robots and Systems (IROS-2001),* pp. 1402-1409.

Morishita, H.; Watanabe, K.; Kuroiwa, T.; Mori, T. & Sato, T. (2003). Development of robotic kitchen counter: A kitchen counter equipped with sensors and actuator for action-adapted and peronally-fit assistance, *Proceedings of the 2003 IEEE/RSJ International Conference on Intelligent Robots and Systems,* pp. 1839-1844.

Mori, T.; Noguchi, H. & Sato, T. (2005). Daily life experience reservoir and epitomization with sensing room, *Proceedings of Workshop on Network Robot Systems: Toward Intelligent Robotic Systems Integrated with Environments.*

Ishiguro, H. (1997). Distributed vision system: A perceptual information infrastructure for robot navigation, *International Joint Conference on Artificial Intelligence (IJCAI-97),* pp. 36-41.

Ishiguro, H. (2007). Scientific issues concerning androids, *International Journal of Robotics Research,* 26(1),105-117.

Murakita, T.; Ikeda, T. & Ishiguro, H. (2004). Human tracking using floor sensors based on the markov chain monte carlo method, *Proceedings of the 7th International Conference on Pattern Recognition,* pages 917-920.

Nishimura, T.; SoGo, T.; Ogi, S.; Oka, R. & Ishiguro, H. (2001). Recognition of human motion behaviors using view-based aspect model based on motion change, *Transactions of the IEICE*, J84-DII(10):2212-2223.

Ishiguro, H. & Nishimura, T. (2001). Vambam: View and motion based aspect models for distributed omnidirectional vision systems, *Proceedings of the International Joint Conference on Artificial Intelligence*, pages 1375- 1380.

Adachi, Y.; Imai, A.; Ozaki, M. & Ishii, N. (2001). Study on extraction of flesh-colored region and detection of face direction, *Int.J. Knowledge-Base Intelligent Eng. Syst.*, 5-2.

Miwa, H.; Itoh, K.; Takanobu, H. & Takanishi, A. (2003). Construction of mental model of humanoid robot for natural communication with human, *Proceedings of the 18th Meeting of Special Interest Group on AI Challenges*, pages 39- 44.

Breazeal, C. (2003). Emotion and sociable humanoid robots, *International Journal of Human-Computer Studies*, 59(1-2):119-155.

Plutchik, R. (1962). *The emotion: facts, theories and new model.* Random House, New York.

Bianchi-Berthouze, N. & Lisetti, C. L. (2002). Modeling multimodal expression of user's affective subjective experience, *User Modeling and User-Adapted Interaction*, 12:49-84.

Terada, K.; Shamoto, T.;Mei,H. & Ito, A. (2007). Reactive Movements of Non-humanoid Robots Cause Intention Attribution in Humans. *Proc. of IEEE/RSJ International Conference on Intelligent Robots and Systems 2007 (IROS 2007).*

Kobayashi, K. & Yamada, S. (2007) Motion Overlap for a Mobile Robot to Express its Mind, *Journal of Advanced Computational Intelligence and Intelligent Informatics*, Vol.11, No.8, pp. 964-971.

25

Human-Robot KANSEI Communication Based on Emotional Synchronization

Minoru Hashimoto and Misaki Yamano
Shinshu University
Japan

1. Introduction

In an aging society it is necessary that robots work for housekeeping and elderly care at home and hospital. Such assisting robots have to communicate with humans. Then, robots purposely designed for communicating with humans have attracted our attention. These robots are indispensable for human-robot symbiosis in the near future and need to have not only intelligence but also KANSEI to make natural communication. KANSEI is cerebral activity without logical thinking such as emotion, feeling, impression, sensitivity, intuition, and so on. Although much research has been done in artificial intelligence, there has been no consideration about KANSEI in the scientific field, since KANSEI is a subjective process. It is important to consider KANSEI in the field of communication robots.

It is well known that facial expressions play a very important role in daily communication. In order to familiarize the robot with human society, it is essential to create affinity with facial expressions. From this perspective, we have developed a head robot called KAMIN-FA1 (KAnsei MINd robot) (Hashimoto & Morooka, 2006) to enhance emotional communication with humans. The robot has a facial expression function using a curved surface display. This technique provides a facial expression easily compared with other methods of mechanical facial expression.

We may change the atmosphere of interaction by adjusting our emotion to the other accordingly. For example, the interaction field becomes lively, if a robot makes a pleased reaction when the partner is pleased. On the other hand, if we do not make any reaction, the field does not swell very much. Jonsson et al. (Jonsson et al., 2005) found that matching the car voice to the drivers' emotions had enormous consequences. Drivers who interacted with voices that matched their own emotional state had less than half as many accidents on average as drivers who interacted with mismatched voices. In communications between human beings, it was found that happy facial expressions are promoted and anger and sadness expressions are weekend by the partner's reactions with synchronized emotional expressions (Ichikawa et al., 2003). Therefore, the emotional synchronization in human-robot interaction is important to swell the interaction state.

In this chapter we propose a framework of a human-robot communication system based on emotional synchronization and examine the effectiveness of the proposed method. We develop a communication system using KAMIN-FA1 to implement the proposed framework. The robotic emotion is entrained to human emotion by using a vector field of dynamics, and then the robot makes a facial expression to express the robot emotion. In the experiments of human-robot communication with the emotional synchronization, we examine whether human feeling becomes comfortable when the robot makes the synchronized facial expression to human emotion.

The concept of the emotional synchronization is shown in Fig.1. The emotional synchronization guides human and robot to sympathy through the interaction. The ultimate purpose of this study is to find the interaction technique which makes a comfortable state through adjusting the communication field. If this study is successful, this technique will be widely utilized for human–robot communication to effect human emotional state. In this paper, we evaluate the effects of the emotional synchronization in human-robot KANSEI communications.

In the next section, we address the KANSEI communication system based on the emotional synchronization. In section III we conduct the experiment to evaluate the influence of emotional synchronization. In section IV, communication experiments using the proposed communication system with emotional synchronization are conducted to examine the effectiveness. Finally, in section V we conclude this paper.

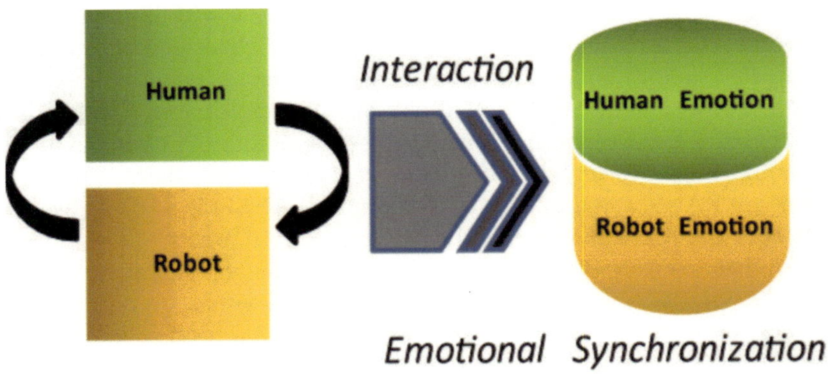

Fig. 1. Concept of Emotional Synchronization

2. Kansei Communication System

2.1 Framework of the communication system based on emotional synchronization

The framework of the proposed KANSEI communication system based on emotional synchronization is shown in Fig.2. The system consists of recognition, emotion generation and expression parts. In the recognition part, the human emotion is recognized by KANSEI analysis of human. In the emotion generation part, the robotic emotion is determined by the emotional entrainment using human emotion. The emotional entrainment is performed by a vector field of dynamics according to the strength of the synchronization. The strength of the entrainment in the emotional synchronization can be changed by the parameter "S". We use an online design method of dynamics to realize the synchronization between human and robotic emotions. The robotic emotion is mapped to an emotional symbol space, and then the vector filed of an expression space is generated according to the state of the emotional symbol space. The expression space generates the emotional expressions such as facial expression and gesture using a communication robot. These information processing systems are performed continuously by using the vector fields of dynamics, since the emotion and the expression of the robot is constantly changing.

We develop a KANSEI communication system shown in Fig.3. We use the voice analysis software, "RobEsense" made by Nemesysco Ltd. (Nemesysco), to recognize the emotions from human voice. KAMIN-FA1 is utilized as the facial expression robot. The communication system is constructed by three PC's (Usui et al., 2008) as shown in Fig.4.

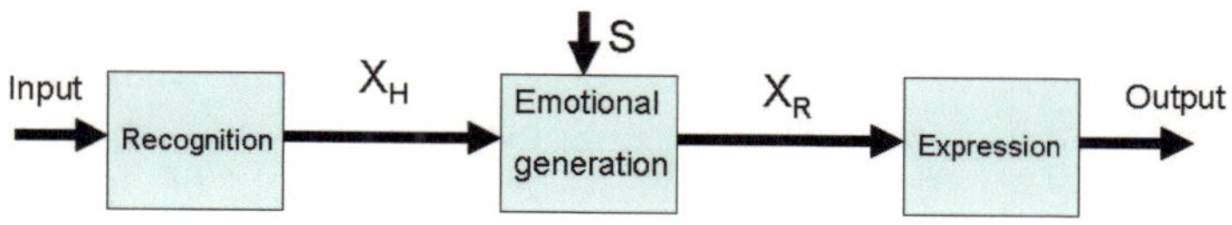

Fig. 2. Framework of the KANSEI communication system based on emotional synchronization.

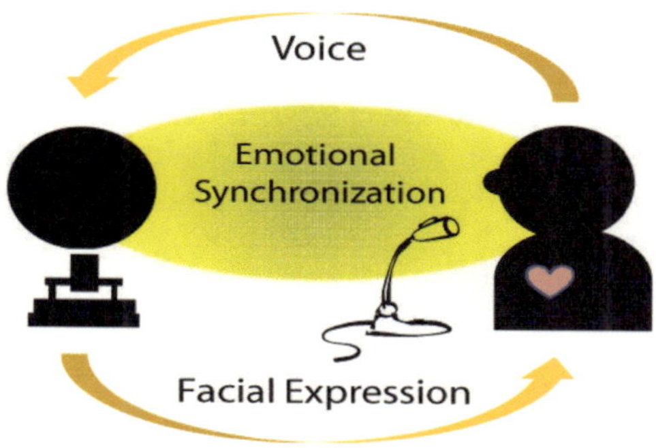

Fig. 3. Schematic drawing of human-KAMIN communication

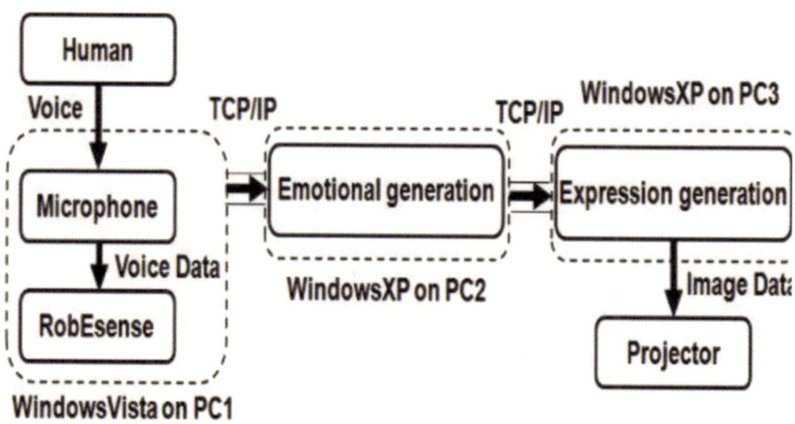

Fig. 4. Block diagram of the KANSEI communication system

2.2 Emotional recognition and generation

The voice analysis software "RobEsense" in the recognition part is a tool judging the emotional state from human voices. It is used in a call center and for a diagnosis of depression mainly. It analyses the emotion of human voice using 18 parameters which is obtained by data with high speed sampling. The result of voice analysis does not depend on language and gender.

We use two parameters, "Excited" and "Atmosphere", within 18 parameters of related emotions. Russell's circumplex model of emotion (Russell, 1980) is utilized to express human and robotic emotions in the recognition part. This model can represent the emotions based on a two dimensional space as shown in Fig.4. One of the axes in the space expresses "comfort-discomfort," and the other is "rouse-sleep". Using the numerical values of the parameters "Excited" and "Atmosphere" in "RobEsense" we express the emotional state based on Russell's model. In addition, we form a vector field of dynamics on the two dimensional space to realize the entrainment between the human and robotic emotions.

In the vector field we make a basin of attractor near the position of human emotion to entrain the robotic emotion. We call it the emotional generation space for the robot as shown in Fig.5. The human emotion obtained by "RobEsense" is plotted in the vector field, and an attractor is constructed in the vector field using the online-design method of dynamics. The attractor is updated continuously according to the result of the recognition part. In the on-line design method, the oblivion and weighted parameters explained below, as determined by the strength of the entrainment in the emotional generation space.

In the next subsection, we describe the design of vector field which is used in the emotional generation space.

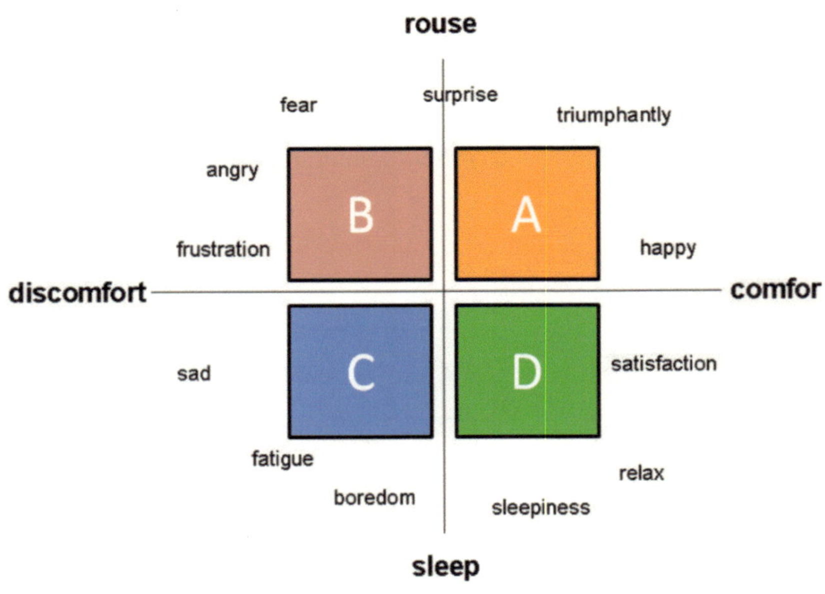

Fig. 4. Russell's circumplex model of affect

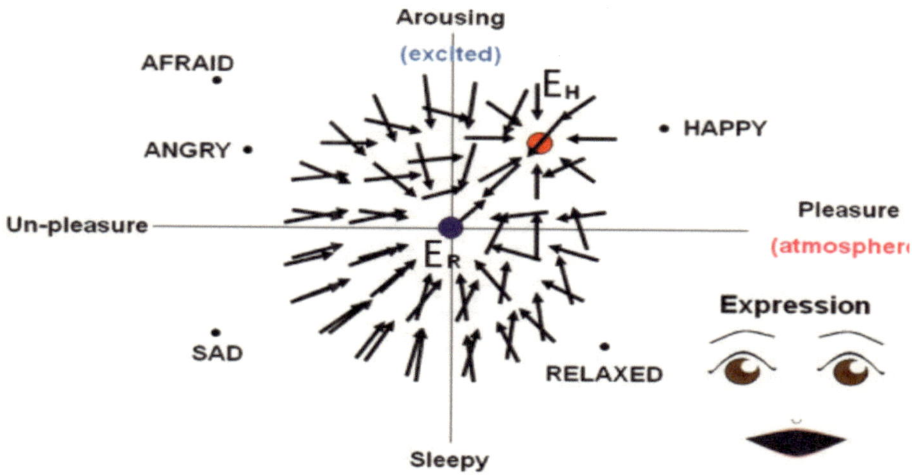

Fig. 5. Emotional generation field

2.3 Vector field design for emotion generation

We need to design dynamics to have arbitrary attractors for information processing. As one of the design methods of dynamics to have arbitrary attractors, Okada et al. (Okada & Nakamura, 2005) proposed a polynomial expression approximation of a vector field. They use dynamics to design the trajectory of the robotic motion. We apply this method to make synchronization in emotion generation. By using this method we can design a vector field around the attractor geometrically and approximate it by a polynomial expression. It is easy to design arbitrary attractors.

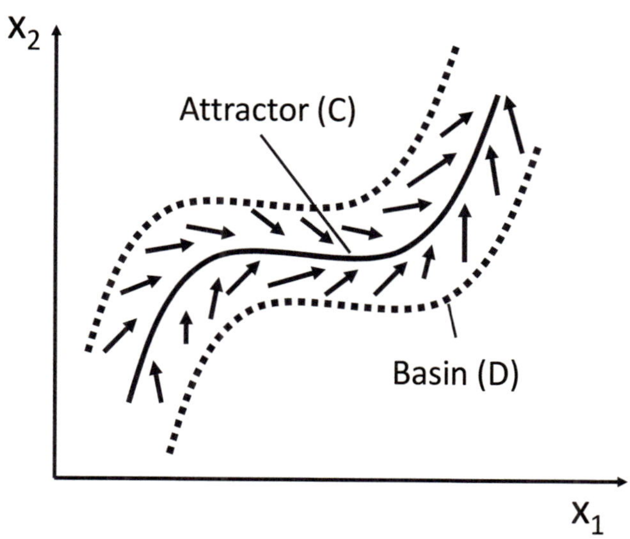

Fig. 6. Attractor and basin of a vector field

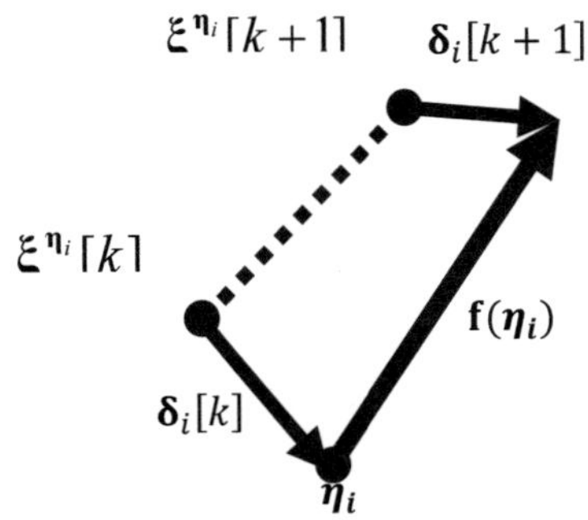

Fig. 7. Defining the vector field

An example of a vector field in the two dimensional space is shown in Fig.6. The vector filed is formed around arbitrary curve C. The curve C is assumed as an attractor, and region D is a basin of entrainment around the attractor. The curve C is a function of discrete time k and consists of $\xi[k](k=1,2,\ldots)$ as shown in Fig.7. We decided the number of sample point among the basin of entrainment around the attractor which is described as $\boldsymbol{\eta}_i(i=1,2,\ldots,m)$, where m is the number of sample points. $\xi^{\boldsymbol{\eta}_i}[k]$ are located on the attractor is the nearest point from $\boldsymbol{\eta}_i$, and $\boldsymbol{\delta}_i[k]$ is the connection vector between $\boldsymbol{\eta}_i$ and $\xi^{\boldsymbol{\eta}_i}[k]$. Then, $\boldsymbol{\delta}_i[k+1]$ and $\boldsymbol{\delta}_i[k]$ can be defined as follows:

$$\boldsymbol{\delta}_i[k+1]=(\boldsymbol{\eta}_i+\mathbf{f}(\boldsymbol{\eta}_i))-\xi^{\boldsymbol{\eta}_i}[k+1] \tag{1}$$

$$\boldsymbol{\delta}_i[k]=\boldsymbol{\eta}_i-\xi^{\boldsymbol{\eta}_i}[k] \tag{2}$$

And, the sufficient condition for convergence is:

$$\|\boldsymbol{\delta}_i[k+1]\|<\|\boldsymbol{\delta}_i[k]\| \tag{3}$$

We make vectors $\mathbf{f}(\boldsymbol{\eta}_i)$ in the basin D by using equations (1)-(3), and approximate them by a polynomial expression. When $\boldsymbol{\eta}_i\in R^2$, the polynomial expression is as follows,

$$\begin{aligned}f_x(\boldsymbol{\eta}_i)=&\,a(50)x^5+a(41)x^4y+a(32)x^3y^2+a(23)x^2y^3+a(14)xy^4+a(05)y^5\\&+a(40)x^4+a(31)x^3y+a(22)x^2y^2+a(13)xy^3+a(04)y^4\\&+a(30)x^3+a(21)x^2y+a(12)xy^2+a(03)y^3\\&+a(20)x^2+a(11)xy+a(02)y^2\\&+a(10)x+a(01)y\\&+a(00)\end{aligned} \tag{4}$$

$$\begin{aligned}f_y(\boldsymbol{\eta}_i)=&\,b(50)x^5+b(41)x^4y+b(32)x^3y^2+b(23)x^2y^3+b(14)xy^4+b(05)y^5\\&+b(40)x^4+b(31)x^3y+b(22)x^2y^2+b(13)xy^3+b(04)y^4\\&+b(30)x^3+b(21)x^2y+b(12)xy^2+b(03)y^3\\&+b(20)x^2+b(11)xy+b(02)y^2\\&+b(10)x+b(01)y\\&+b(00)\end{aligned} \tag{5}$$

Where, $a(ij), b(ij)$ are the constant parameters, and by using the vector $\boldsymbol{\eta}_i$, $\mathbf{f}(\boldsymbol{\eta}_i)$ can be described as:

$$\mathbf{f}(\boldsymbol{\eta}_i) = \boldsymbol{\Phi}\boldsymbol{\theta}(\mathbf{x}) \tag{6}$$

where,

$$\boldsymbol{\Phi} = \begin{pmatrix} a(50) & a(41) & \cdots & a(00) \\ b(50) & b(41) & \cdots & b(00) \end{pmatrix} \tag{7}$$

$$\boldsymbol{\theta}(\mathbf{x}) = [x^5\ x^4 y\ x^3 y^2 \cdots 1]^T \tag{8}$$

◆◆◆ ◆ ◆

We can determine $\boldsymbol{\Phi}$ by the least-squares method. $\mathbf{F}$ and $\boldsymbol{\Theta}$ can be defined by equations (9) and (9) :

$$\mathbf{F} = [\mathbf{f}(\boldsymbol{\eta}_1)\ \mathbf{f}(\boldsymbol{\eta}_2)\ \mathbf{f}(\boldsymbol{\eta}_3)\ \cdots\ \mathbf{f}(\boldsymbol{\eta}_m)] \tag{9}$$

$$\boldsymbol{\Theta} = [\boldsymbol{\theta}(\boldsymbol{\eta}_1)\ \boldsymbol{\theta}(\boldsymbol{\eta}_2)\ \boldsymbol{\theta}(\boldsymbol{\eta}_3)\ \cdots\ \boldsymbol{\theta}(\boldsymbol{\eta}_m)] \tag{10}$$

And,

$$\boldsymbol{\Phi} = \mathbf{F}\boldsymbol{\Theta}^{\#} \tag{11}$$

◆ ◆◆

where, $\boldsymbol{\Theta}^{\#}$ means a pseudo inverse matrix of $\boldsymbol{\Theta}$.

After designing a vector field consists of *m* sample points, by adding one sample point, the sample points become to *m+1*. Appealing to *m+1* sample point, online design can be described as equations (12)-(17) (Okada & Nakamura, 2005). If we know the values of $\mathbf{P}_m$, $\boldsymbol{\Phi}_m$, $\xi[k+1]$ and $\xi[k]$, and by determining a certain point in vector $\boldsymbol{\eta}_{m+i}(i = 1,2,\cdots,l)$ which nears to $\xi[k]$, $\boldsymbol{\eta}_{m+i}$ we can define $\mathbf{f}(\boldsymbol{\eta}_{m+i})$ based on equation (1). By this way, the input signal is given as a new sample point, and the output signal is computed by the dynamics.

$$\hat{\mathbf{F}} = [\mathbf{F}\quad \mathbf{f}(\eta_{m+1})] \tag{12}$$

$$\hat{\boldsymbol{\Theta}} = [\boldsymbol{\Theta}\quad \boldsymbol{\theta}(\eta_{m+1})] \tag{13}$$

◆◆

$$\boldsymbol{\Phi} = \hat{\mathbf{F}}\hat{\boldsymbol{\Theta}}^{\#} \tag{14}$$

$$\mathbf{P}_{m+1} = \mathbf{P}_m - \frac{\mathbf{P}_m \boldsymbol{\theta}(\eta_{m+1})\boldsymbol{\theta}^T(\eta_{m+1})\mathbf{P}_m}{1 + \boldsymbol{\theta}^T(\eta_{m+1})\mathbf{P}_m \boldsymbol{\theta}(\eta_{m+1})} \tag{15}$$

$$\boldsymbol{\Phi}_{m+1} = \boldsymbol{\Phi}_m - \{\boldsymbol{\Phi}_m \boldsymbol{\theta}(\eta_{m+1}) - \mathbf{f}(\eta_{m+1})\}\boldsymbol{\theta}^T(\eta_{m+1})\mathbf{P}_{m+1} \tag{16}$$

◆ ◆

where,

$$\mathbf{P}_m = \{\sum_{i=1}^{m} \boldsymbol{\theta}(\boldsymbol{\eta}_i)\boldsymbol{\theta}^T(\boldsymbol{\eta}_i)\}^{-1} \tag{17}$$

An oblivion parameter of online method is defined as α, and a weighted parameter for a new sample signal is defined as β. By using these two parameters equations (12)-(14) can be described as follows:

$$\hat{\mathbf{F}} = \left[a\mathbf{F} \quad \beta\mathbf{f}(\eta_{m+1})\right] \tag{18}$$ ◆◆

$$\hat{\boldsymbol{\Theta}} = \left[a\boldsymbol{\Theta} \quad \beta\boldsymbol{\theta}(\eta_{m+1})\right] \tag{19}$$ ◆◆ ◆

$$\boldsymbol{\Phi} = \hat{\mathbf{F}}\hat{\boldsymbol{\Theta}}^{\#} \tag{20}$$ ◆◆

When $0 \le \alpha \le 1$, $0 \le \beta \le 1$, we can get the following equation:

$$\boldsymbol{\Phi}_{m+1} = \left[a\mathbf{F} \quad \beta\mathbf{f}(\boldsymbol{\eta}_{m+1})\right]\left[a\boldsymbol{\Theta} \quad \beta\boldsymbol{\theta}(\boldsymbol{\eta}_{m+1})\right]^{\#} \tag{21}$$

2.4 Evaluation of voice analysis

We conducted an evaluation experiment of emotional recognition using voice analysis. 16 voice samples which are obtained in every day life are utilized to evaluate the emotion. We compared the emotional state which was declared by the speaker with the atmosphere and excited parameters of RobEsense in the Russell's model plane. Figure 8 shows the result of the evaluation experiment. The numbers are sample numbers of the voice data. The speaker's emotions and RobEsense parameters of the same sample number are located in the same area in the plane except area (B). Therefore, we can evaluate the speaker's emotion using the atmosphere and excited parameters of RobEsense.

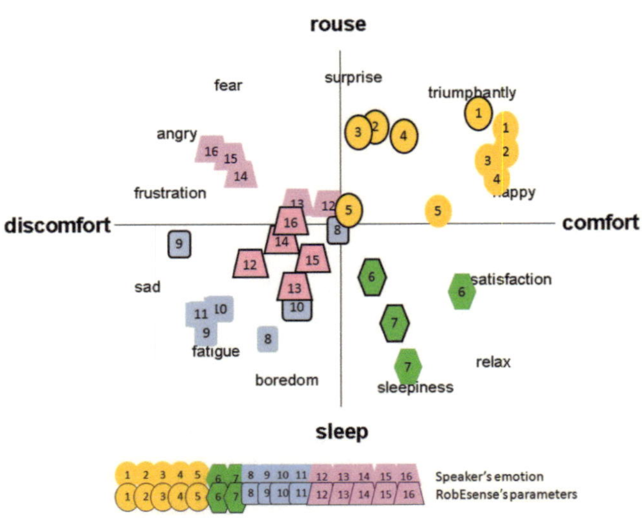

Fig. 8. Results of the evaluation experiments for emotional analysis.

2.5 Facial expression

Yamada (Yamada, 2000) found the relationship between the basic emotional category and the structural variables of facial expression based on the displacement structure of the characteristics points in the facial expression. Yamada reported that there are two kinds of the facial structural variables called "Inclination" and "Bend". "Inclination" means the displacement of the characteristic points concerning the degree of the slant of eyes and eyebrows. In the case of the mouth, it means the strength forming the shape of V and reverse V characters. "Bend" means the curvature of eyebrows, or the strength of the opening of the eyes and the mouth. In this study, we referred to this facial expression model in order to make the robotic facial expression of KAMIN-FA1. The facial image is made up of the line drawing, and it is determined numerically. The facial image consists of straight lines, bezier curves, and circles. A straight line can be created from specifying two points. Similarly a bezier curve is four points, and a circle is the center point and a radius. That is, the state of facial expression is specified with the parameter vector of the points and the radiuses. Figure 9 shows the examples of the facial expression with the characteristic points.

We make a two dimensional space whose axes indicate "Inclination" and "Bend" to express the facial image. And then we build a vector field of dynamics on the two dimensional space to make the facial expression dynamically. In this study we assume that the facial expression is not static, it is a dynamic process. The facial expression is modeled by an attractor on this vector filed. We designed the attractors of the basic six emotions based on Yamada's facial expression model. We symbolized the facial expression space and make a symbol space. In the symbol space a point expresses an emotional state and a vector filed of the facial expression space. Therefore, we can change the vector field of the facial expression by the state of the symbol space as shown in Fig.10.

An emotional state is expressed within the parameters of six basic emotions: happiness, sad, anger, fear, surprise, and disgust. The parameter vector of the facial image is computed from this emotional state. The state of the face model in six emotions is set up beforehand.

Using this method, the subtle expression between emotional states is possible by using the parameter vector. We modeled the parameter vector based on "Inclination" and "Bend". Then we can determine the parameter vector by the coordinate in the facial expression space. Figure 11 shows an example of facial expressions using the simbole space.

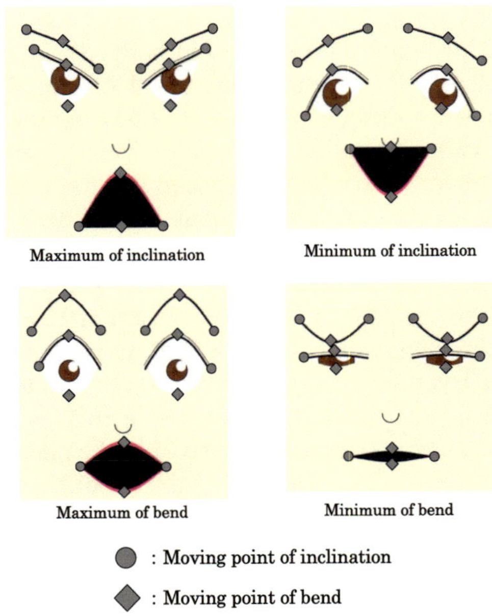

Fig. 9. Facial expressions based on Yamada's model.

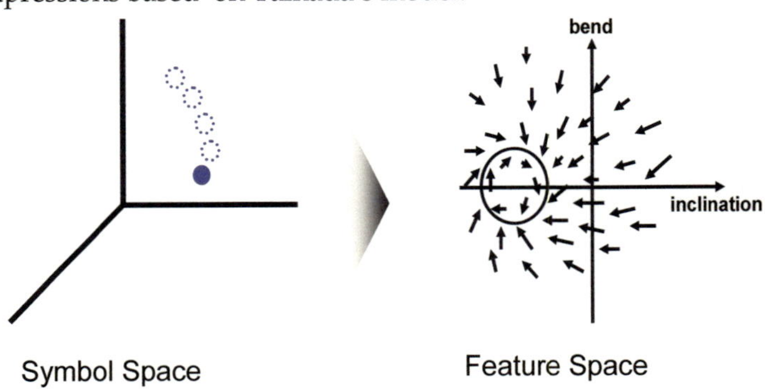

Fig. 10. Symbol space and Facial expression space.

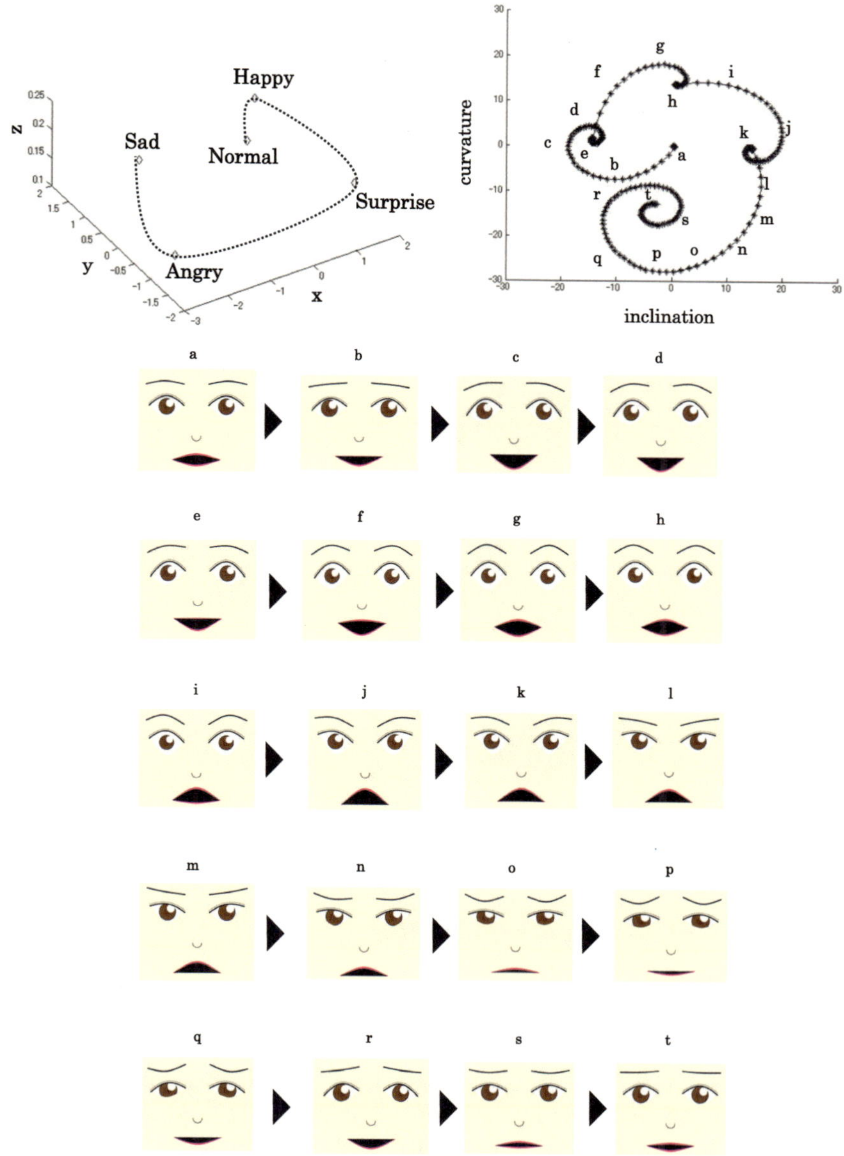

Fig. 11. Example of facail expression using the facial expression space.

2.4 KAMIN-FA1

We used the head robot KAMIN_FA1 (Hashimoto & Morooka, 2006) as shown in Fig.12. The head mechanism is a facial image display, and it consists of a dome screen, a fish-eye lens, and a projector. The face image is projected to the dome screen from the inside. The fish-eye lens is installed on the front of the projector, and it projects the picture on the dome screen.

This communication robot has several advantages. First, the ease with which the image is changed enables a wide variety of expressions. A robot having the same expressions as users may prove less advantageous than one having a standardized, predictable "robot" expression because, at the "gut" level, most human users dislike resemblance to the point of mimicry. A curved surface image has a dependable direction of observation and presence in actual space.

Additionally, this robot has the neck mechanism with 4-DOF to make head movement. In communication, head movement affects the impression of facial expression. More impressive facial expressions are possible when a facial image display and head movement are integrated.

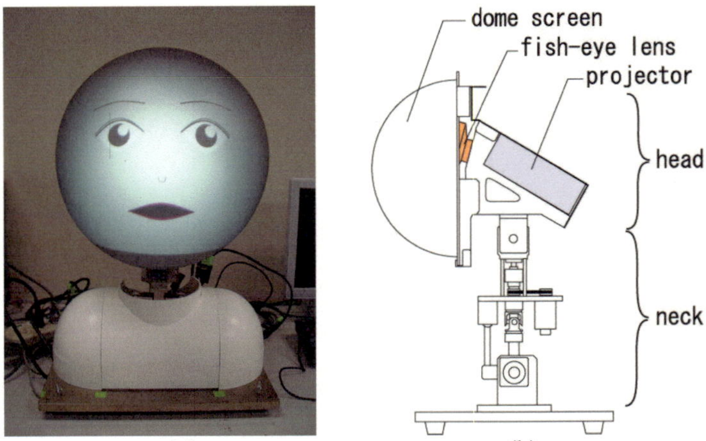

(a) (b)

Fig. 12. KAMIN_FA1. (a)Overview, (b)Structure

4. Influence of Emotional Synchronization

4.1 Experimental design

In order to examine the influence of the emotional synchronization to human emotional state, we conducted an experiment. It was necessary to have the subjects be a comfortable or uncomfortable state at the time of experiment. Therefore, either a comfortable or uncomfortable state was induced in each subject at the beginning of the experiment. This was accomplished by showing the subjects 10 six-slides of picture. Each subject observed two kinds of the pictures to induce comfortable and uncomfortable states. For comfortable sate, all of the pictures reflected happy themes; for uncomfortable state, all of the pictures reflected disturbing themes.

After the subjects were led to comfortable and uncomfortable state, the subjects communicated with KAMIN_FA1. The robot expressed the emotional state of comfort or discomfort for each subject with each emotional state. The overview of the communication experiment is shown in Fig.13. The subject sat on the position where was 100[cm] away from KAMIN-FA1 and talked to the robot. The subject talked while watching a face of KAMIN-FA1. In addition, we did not address the contents of communication to the subjects, they talked freely and free time. Each subject took the communication experiment four times; two times (emotions of the subject) and two times (emotions of the robot). We assumed that if the subject's emotion is comfortable or uncomfortable, then the robot's emotion is comfortable or uncomfortable, respectively, in the case of synchronization. In the non-synchronization case, if the subject's emotion is comfortable or uncomfortable, then the robot's emotion is uncomfortable or comfortable, respectively. The subjects filled in a questionnaire based on a semantic differential (SD) method (Inoue & Kobayashi, 1985) after the communication to examine the impression of the subject to KAMIN-FA1. Subjects answered in 7-point scoring scale. In addition, the subjects were asked which communication case you liked better to evaluate change of subject's emotion by the communication.

4.2 Results of voice analysis and subject's feeling

We compared the change of the subject's emotion with the strength of synchronization of robot using atmosphere parameter of voice analysis. The value of the atmosphere parameter indicates the strength of comfortable state. Figure14 shows the average values of the atmosphere parameter for each communication method, synchronization or non-synchronization case. In the case of synchronization, the average value of the atmosphere parameter is larger than that of the non-synchronization case. Therefore, the subjects feel better in the emotional synchronization of KAMIN-FA1.

The results of the question about the human feeling in each communication case are as follows. When the subject's emotion was comfortable, all subjects answered that the comfortable response of KAMIN-FA1 was better than uncomfortable one. On the other hand, when the subject's emotion was uncomfortable, 6 subjects answered that the uncomfortable response of KAMIN-FA1 was better than comfortable one. Consequently, the synchronized response of KAMIN-FA1 induced the human comfortable state than the unsynchronized response. That is, the human emotional state becomes more comfortable in the case of high synchronization.

4.3 Evaluation of impression to KAMIN

We examined the difference of the impression to KAMIN, when KAMIN_FA1 expresses the synchronized and non-synchronized responses. The average rated values answered in the questionnaires are shown in Fig.15. It was found that in the case of comfortable response of KAMIN the subjects had positive impressions in terms of active, happy, friendly and so on comparing with the uncomfortable case.

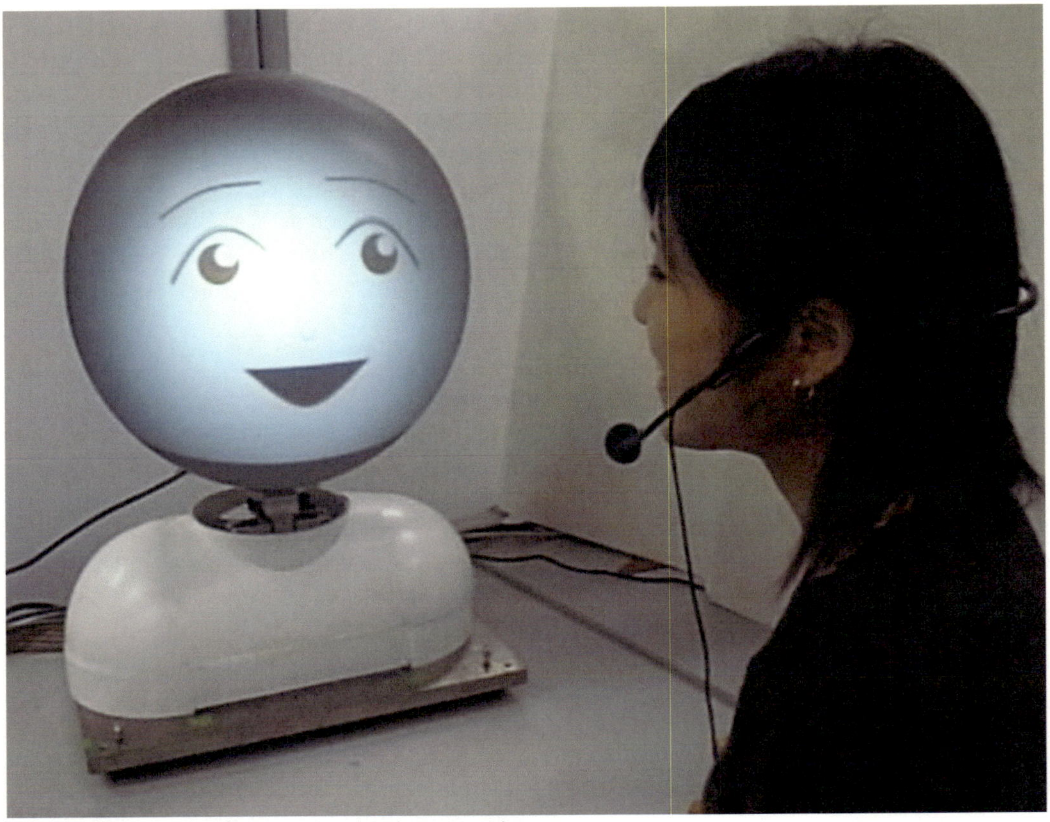

Fig. 13. Overview of a communication experiment

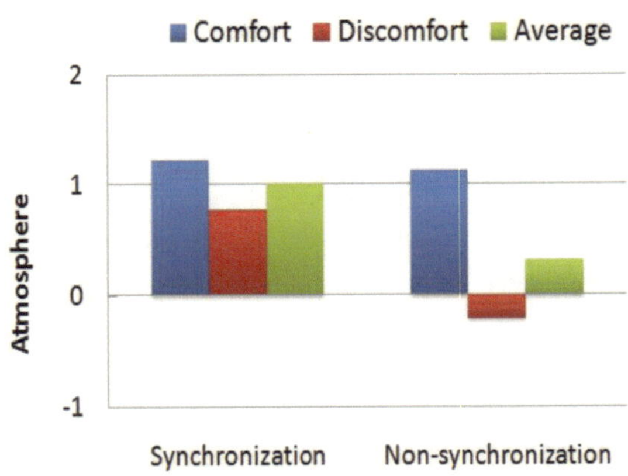

Fig. 14. Result of atmosphere parameters in voice analysis

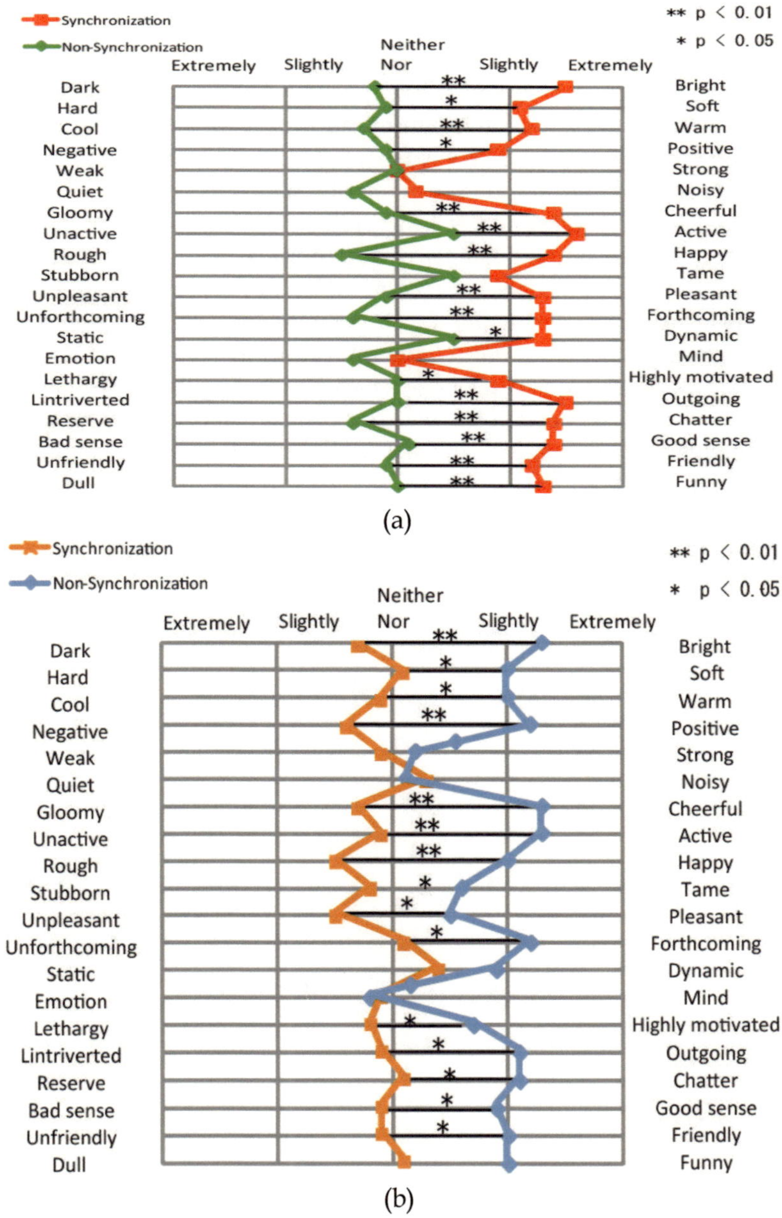

Fig. 15. Impression of KAMIN with synchronized and non-synchronized expressions. (a) Comfort (b) Discomfort

5. Communication Experiment

5.1 Experimental design

In order to examine the effectiveness of the proposed KANSEI communication system, we conducted some experiments using the communication system of KAMIN_FA1. The subject sat on the position where was 100[cm] away from KAMIN_FA1 and talked to the robot. The number of subjects was ten. The facial expression of KAMIN_FA1 changed according to the subject's emotional state. The subject talked while watching a face of KAMIN_FA1. In addition, we did not address the contents of communication to the subjects, they talked freely and free time. Each subject took the communication experiment four times. One of the experiments was conducted by using the strong emotional entrainment, the others were the non entrainment cases, normal, happy, discomfort. In the case of the high entrainment, the facial expression of KAMIN_FA1 changed much by reacting the subject's emotional state. Figures 16 and 17 show the examples of synchronized and non-synchronized normal responses, respectively. In the case of the non entrainment, the facial expression did not change so much. In addition, the subjects filled in a questionnaire based on a semantic differential (SD) method after the communication to examine the impression of the subject. Subjects answered in 7-point scoring scale.

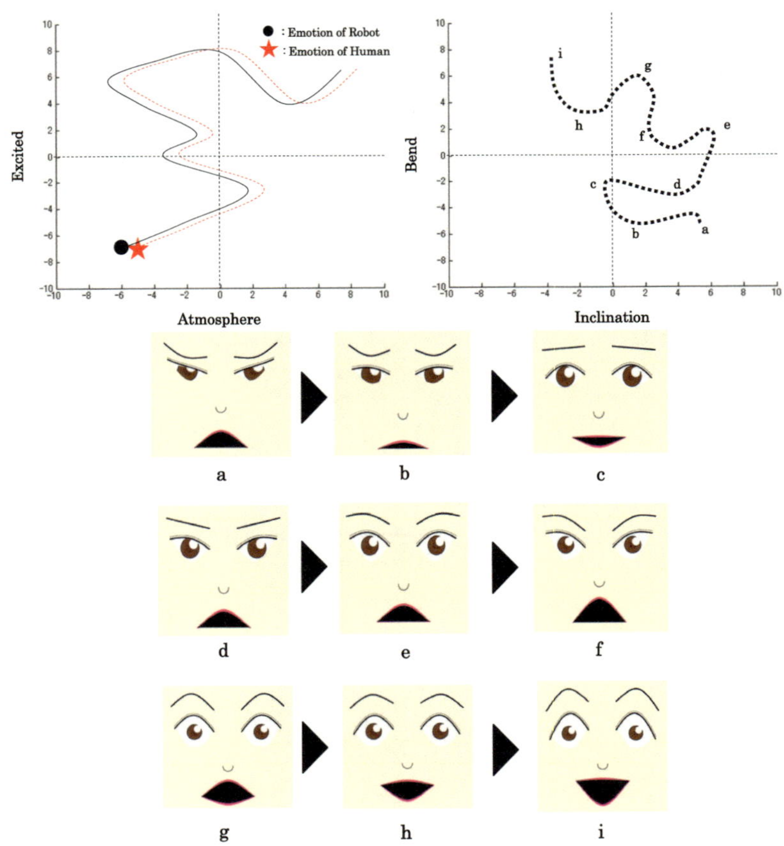

Fig. 16. Example of synchronization reaction

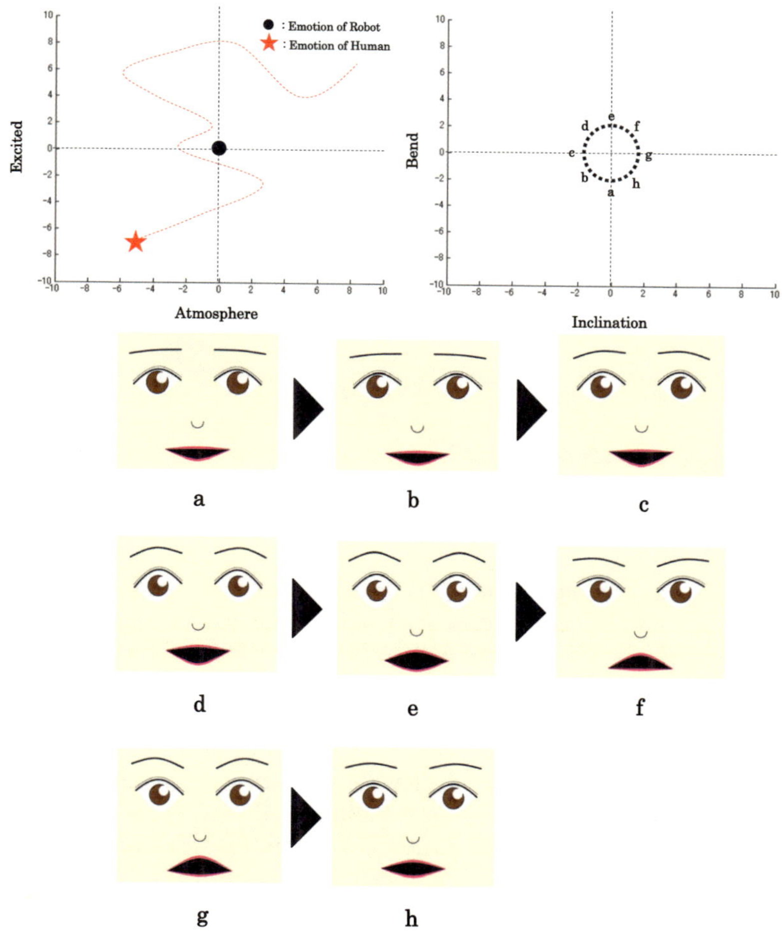

Fig. 17. Example of non-synchronization reaction

5.2 Experimental Results

The average rated values answered in the questionnaires are shown in Fig.18. Four kinds of average values are shown in the figure, a synchronization case and non-synchronization cases, normal, happy, discomfort. It was found that in the case of synchronization case the subjects felt better comparing with all the non-synchronization cases.

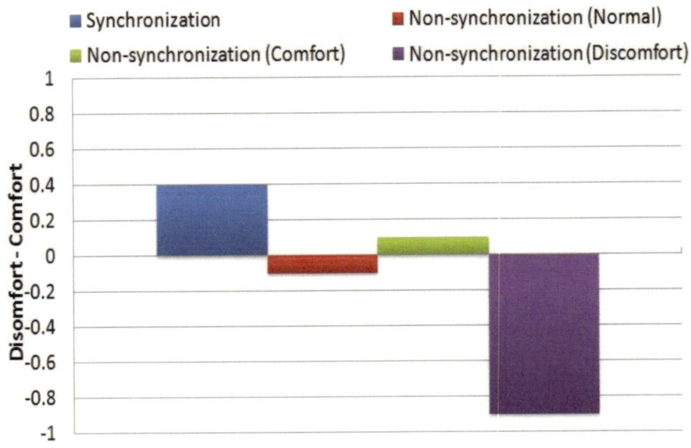

Fig. 18. Result of subject's comfortableness in communication experiments

6. Coclusions

We developed a KANSEI communication system based on emotional expressions, and its effectiveness was verified by experiments in human-robot communication. The robotic emotion was determined by an entrainment to human emotion. The entrainment was accomplished using a vector field of dynamics. The robotic facial expression using a communication robot was realized dynamically based on the emotional space.

Additionally, we investigated the influence of the emotional synchronization in human-robot KANSEI communications. We conducted experiments to evaluate the effects of the proposed system based on emotional synchronization. In the experiments of human-robot interaction using the emotional synchronization, we found that human feeling became comfortable when the robot made the synchronized facial expression to human emotion. Then it was confirmed that emotional synchronization in human-robot interaction could be effective to keep a comfortable state.

As the future work, we are planning to investigate the effect of the emotional synchronization in long-term experiments.

7. Acknowledgent

This research was supported by Toyota Tsusho Electronics Corporation. A.Prof. Honywood of Shinshu University provided critical assistance on our manuscript. We are grateful to him for helpful suggestions.

8. References

Hashimoto, M. & Morooka, D. (2006) "Robotic Facial Expression Using a Curved Surface Display," Journal of Robotics and Mechatronics, Vol.18, No.4 , pp.504-510.

http://www.nemesysco.com/

Ichikawa, H.; Chandrasiri, N.P.; Harashima,H. & Makino, J. (2003) "The effect of congruent facial expression on counterpart's facial response," Technical Report of IEICE, PRMU2002-210, pp.59-64, Tokyo,.

Inoue, M. & Kobayashi, T. (1985) "The research domain and constuction of adjective-pairs in a semantic differential method in Japan," J.of Educ.Phychol, 33, pp.253~260.

Jonsson, I.M.; Nass, C.; Harrris, H. & Takayama, L. (2005) "Matching In-Car Voice with Driver State Impact on Attitude and Driving Performance," Proceedings of the Third Informational Driving Symposium on Human Factors in Driver Assessment, Training and Vehicle Design, pp173-180, Rockport.

Okada, M. & Nakamura, Y. (2005) "Design of Continuous System Space and Whole Body Motion Generation using Dynamics-based Information Processing System," A society magazine of robot of Japan, Vol.23 No.7, pp.858~863.

Russell, J.A. (1980) "A circumplex model of affect," Journal of Personality and Society Psychology, Vol.39, PP.1161~1178.

Usui, T.; Kume, K.; Yamano, M. & Hashimoto, M. (2008) "A Robotic KANSEI Communication System Based on Emotional Synchronization," Proc. 2008 IEEE/RSJ International Conference on Intelligent Robots and Systems, Nice, France, pp.3344-3349.

Yamada, H. (2000) "Models of perceptual judgment of emotion from facial expressions," Japanese Psychological Review Vol.43, No.2, pp.245~255.